Fundamentals of ^{57}Fe Mössbauer Spectrometry

R. Justin Joseyphus · Jean-Marc Greneche

Fundamentals of ^{57}Fe Mössbauer Spectrometry

Springer

R. Justin Joseyphus
Department of Physics
National Institute of Technology
Tiruchirappalli, Tamil Nadu, India

Jean-Marc Greneche
Institute des Molécules et Matériaux du Mans
Université Le Mans-CNRS
Le Mans Cedex, France

ISBN 978-981-99-8652-1 ISBN 978-981-99-8653-8 (eBook)
https://doi.org/10.1007/978-981-99-8653-8

This Springer imprint is published by the registered company Springer Nature Singapore Pte Ltd.
The registered company address is: 152 Beach Road, #21-01/04 Gateway East, Singapore 189721, Singapore

Preface

Rudolf Ludwig Mössbauer discovered the recoilless emission and resonant absorption of nuclear gamma rays due to his research work during the years 1955–58 at the Max Planck Institute in Heidelberg, Germany. The phenomenon is known as the Mössbauer effect, the spectrum as Mössbauer spectrum and the hyperfine parameters such as isomer shift, quadrupole splitting and hyperfine field as Mössbauer parameters. The Nobel Prize in Physics for 1961 was awarded to R. L. Mössbauer "for his researches concerning the resonance absorption of gamma radiation and his discovery in this connection of the effect which bears his name". The discovery stimulated interest and opportunities for further research in cross-disciplinary areas within a short period.

Mössbauer spectrometry is used across various disciplines such as Physics, Chemistry, Biology, Metallurgy, Materials and Geology, and this book is intended for a reader to know about a few important contributions in the relevant area. Chapters 1–3 of this book introduce the nucleus, quantum mechanics and magnetism. Readers familiar with these topics can revisit the concepts related to the Mössbauer effect. Chapters 4 and 5 deal with the fundamentals of the Mössbauer effect and Mössbauer parameters, respectively. Since the focus is on ^{57}Fe nucleus, the book covers the hyperfine parameters of important Fe compounds and their application in various disciplines related to materials, Earth, planetary and biological sciences, which are covered in Chaps. 6–13. The symbols and units used in the book have their usual meaning unless explicitly mentioned in each chapter.

R. Justin Joseyphus is thankful to Prof. A. Narayanasamy, who introduced him to the Mössbauer effect and experimental intricacies while he was at the University of Madras, Chennai, India. J.-M. Greneche is very grateful to Prof. F. Varret who, as his thesis supervisor, passed on to him his passion for research and the fundamentals of Mössbauer spectrometry, with great expertise in the theoretical and instrumental aspects and the numerical analysis of spectra, in correlation with the physical properties of the materials studied. He would also like to extend his warmest thanks to all the colleagues who, in the frame of various collaborative projects, have enabled him to successfully apply ^{57}Fe Mössbauer spectrometry to different types of materials. The suggestions by Dr. J. Shebha Anandhi, Dr. R. Sankaranarayanan and Dr. R. Srinivasan are gratefully acknowledged.

The authors respect and recognize the contributions made by all the Mössbauer researchers worldwide. Nevertheless, suggestions, corrections and opinions are welcome that shall be acknowledged and incorporated in future editions.

Tiruchirappalli, India R. Justin Joseyphus
Le Mans, France Jean-Marc Greneche

Contents

Chapter 1
The Nucleus

1.1 Probing the Nucleus

Roentgen discovered X-rays in 1895, followed by the identification of other kinds of nuclear radiation in 1899 by Ernest Rutherford, known as alpha and beta rays. Paul-Villard, in 1900, found another form of radiation characterized by more penetration depth than X-rays and unaffected by electric or magnetic fields, named later on by Rutherford as γ-rays. Scattering experiments are used to probe the size and shape of the nucleus that requires energy sources over a few hundred MeV and wavelengths of less than a fm. Rutherford's analysis of the scattering of α-particles with a thin gold foil established that the atom has a positively charged nucleus, overturning the plum pudding model of J. J. Thomson. Rutherford scattering uses α particles that are difficult to produce with high energy, and therefore, electrons are used in scattering experiments to determine the nuclear size precisely. The electrons interact with the nucleus through electromagnetic interaction subject to the diffraction limits, whereas strong interaction is limited to the nucleons, and weak interaction is negligible.

The diffraction condition is satisfied if the de Broglie wavelength of the electron is equal to the size of the nucleus. Diffraction can be of two forms; far-field, known as Fraunhofer diffraction, and near-field, known as Fresnel diffraction. When the electron's kinetic energy is increased significantly and made to scatter on a nucleus, it is called relativistic scattering. The scattering cross section is then given by the Mott formula, which includes the effect of magnetic moment and recoil. The classical Rutherford scattering uses non-relativistic scattering considering only charge, neglecting recoil and magnetic moment. At very low electron energies, when the wavelength is larger than the size of the object, the interaction can be point-like, neglecting the spin. These effects are used to establish the properties of the nucleus.

R. J. Joseyphus and J.-M. Greneche, *Fundamentals of ^{57}Fe Mössbauer Spectrometry*,
https://doi.org/10.1007/978-981-99-8653-8_1

1.2 Constituents of the Nucleus

The nucleus has protons and neutrons known as nucleons, which are further made of fundamental particles called quarks. The quark structure is responsible for the positive and neutral charge of the protons and neutrons, respectively. The fundamental particles under two major classifications, known as leptons and quarks, with their charges, are shown in Table 1.1. The leptons have a charge of either -1 or 0, while the charges of quarks are either $+2/3$ or $-1/3$, in multiples of elementary charge. All ordinary matter is made up of particles comprising the leptons and quarks, whereas antiparticles are known for their opposite charges to that of particles.

The proton comprises two up quarks and a down quark where each up quark has an associated charge of $+ 2/3$, and the down quark has a charge of $- 1/3$, making a net charge of $+ 1$. Neutron has two down quarks and one up quark, resulting in a neutral charge. The quark constituents of the proton and neutron with their electric charges and another quantum number (color charge) represented by red, blue and green are shown in Fig. 1.1. The combination of the three color charges for a free particle has to be zero. Despite the neutral electric charge of the neutron, the strong nuclear force between the neutron and proton keeps them together, and the nuclear charges are uniformly distributed.

An atomic nucleus with its constituent proton (Z) and neutron (N) numbers is represented as ${}^{A}_{Z}\mathrm{X}$, where A is the mass number that is the sum of Z and N. The hydrogen nucleus has just one proton, while isotopes such as deuterium and tritium have one and two neutrons, respectively.

1.2.1 Charge Distribution

The nuclear charge distribution from scattering experiments is given by

$$\rho(r) = \frac{\rho_0}{1 + \exp\left(\frac{r-R}{t}\right)} \tag{1.1}$$

Table 1.1 The fundamental particles and their charges

Leptons			Quarks		
Particle	Name	Charge (e)	Particle	Name	Charge (e)
e^-	Electron	-1	u	up	+2/3
μ^-	Muon	-1	d	down	$-1/3$
τ^-	Tau	-1	c	charm	+2/3
ν_e	Electron neutrino	0	s	strange	$-1/3$
ν_μ	Muon neutrino	0	t	top	+2/3
ν_τ	Tau neutrino	0	b	bottom	$-1/3$

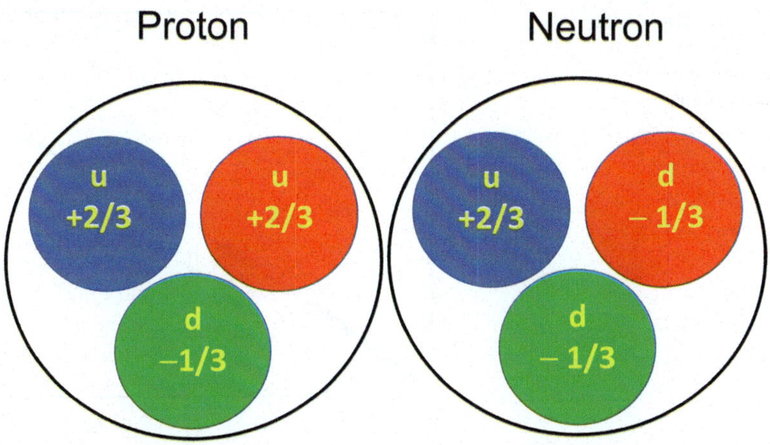

Fig. 1.1 The up (u) and down (d) quark constituents of the proton and neutron

In (1.1), ρ_o is the nuclear charge density near the nuclear core, R is the mean radius where the density decreases by half, and t is the skin thickness in which the density falls from 90 to 10%, as represented in Fig. 1.2. The density at the center of the nucleus is about $\rho_o = 0.16 \times 10^{45}$ nucleons/m^3. The radius of the nucleus is given by $R = 1.2A^{1/3} \times 10^{-15}$ m from electron scattering experiments. The nucleus can be approximated as a homogeneously charged sphere with a radius R. Another quantity often used is the mean square radius $<r^2>$ to take care of the not-so-well-defined radius and is related to R as

$$R^2 = \frac{5}{3}\langle r^2 \rangle \tag{1.2}$$

The root mean square (rms) radius of ^{56}Fe isotope is around 3.7 fm.

Fig. 1.2 The nuclear charge density variation with radial distance

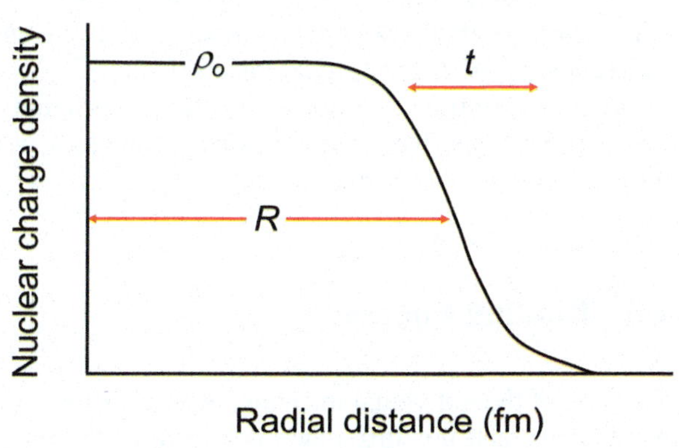

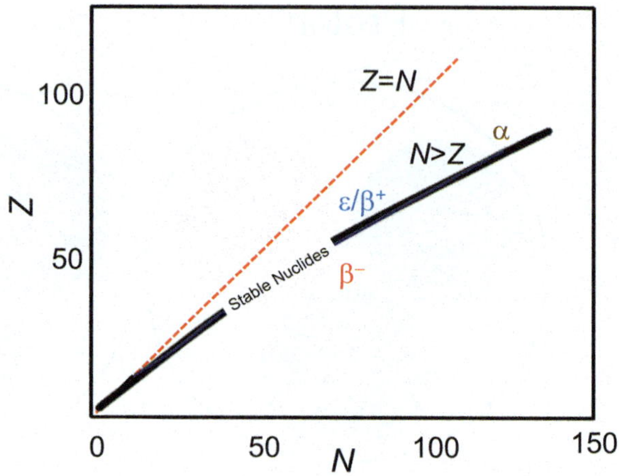

Fig. 1.3 Plot showing the deviation towards neutron-rich nuclei with higher mass numbers

As the number of protons in the nucleus increases, the number of neutrons must be added to overcome the Coulomb repulsion. The strong force thus holds the nucleons together, and as a result, most nuclei exhibit $N > Z$, as shown in Fig. 1.3. The stable nuclei are found in the valley of stability, while isotopes with excess protons and neutrons reach stability by moving towards the bottom of the valley. Those nuclei with excess protons beyond the stability line decay through β^+ or electron capture, whereas those with excess neutrons decay through β^-. The estimate of Coulomb energy for ^{56}Fe is

$$E_c = \frac{Ze^2}{4\pi\varepsilon_0 R} = \frac{Ze^2}{4\pi\varepsilon_0 R_o A^{1/3}}$$
$$= \frac{26 \times (1.602 \times 10^{-19})^2}{4\pi\varepsilon_0 R_o \times 56^{1/3}} = 1.307 \times 10^{-12}\,\text{J} = 8.2\,\text{MeV}. \quad (1.3)$$

However, this estimate of the Coulomb energy term is modified inside a nucleus due to the contribution of more proton pairs. The e and ε_0 represent the elementary charge and vacuum permittivity, respectively.

As more neutrons are added up, the mass number increases and the nucleus becomes unstable. Nevertheless, a stable nucleus in most cases is due to another important factor, the binding energy.

1.3 Binding Energy

The sum of the individual nucleon's masses does not match the actual total mass of the nucleus, and this difference is known as the mass defect. The energy is being spent for binding nucleons together to achieve stability and hence known as binding energy given as

Table 1.2 Few of the isotopes of Fe and their relative abundance

Isotope	*Z*	*N*	Stability	Relative abundance (%)
^{52}Fe	26	26	Unstable—β^+ decay	–
^{54}Fe	26	28	Stable	5.8
^{55}Fe	26	29	Unstable—electron capture	–
^{56}Fe	26	30	Stable	91.8
^{57}Fe	26	31	Stable	2.1
^{58}Fe	26	32	Stable	0.3
^{59}Fe	26	33	Unstable—β^- decay	–

$$B = \left(\{Zm_{\mathrm{H}} + Nm_{\mathrm{n}}\} - m_{Z}^{A}\mathrm{X}\right)c^2 \tag{1.4}$$

where m_{H}, m_{n} and m are the masses of the hydrogen, neutron and atom, respectively, while electrons are also part of (1.4). The binding energy could be understood easily from the case of a deuteron. The deuteron is made up of a proton and a neutron. When a deuteron is formed, it releases a gamma ray of energy 2.2246 MeV equal to the binding energy. Similarly, the energy required to break the deuteron into proton and neutron is 2.2246 MeV.

1.4 Isotopes and Their Abundance

Nuclei with the same number of protons but with different neutron numbers are known as isotopes. Table 1.2 shows a few iron isotopes, their nucleon numbers and their stability. Stable isotopes such as ^{56}Fe and ^{57}Fe are relatively abundant, where the relative fractions are 91.8 and 2.1%, respectively (Meija et al. 2016). Here, the Fe isotopes can have either even or odd neutrons, one factor affecting their stability. The ^{56}Fe, ^{57}Fe and ^{58}Fe with $N = 30$, 31 and 32, respectively, lie at the bottom of the valley of stability. The Fe isotope with $N = 29$ decays through electron capture while $N < 28$ is by β^+ and $N > 33$ by means of β^- decay.

1.5 Binding Energy Per Nucleon

The stability of the isotopes is determined by the binding energy per nucleon, which is influenced by various contributions given in the semi-empirical mass formula. Derived by von Weizsäcker, on the analogy of liquid-drop, it assumes the nucleus is made of incompressible nuclear matter. The nuclear force is identical to each nucleon and saturates, implying that nearby nucleons only interact with each other. The total binding energy of the nucleus is a sum of various positive and negative contributions that are given by,

B = volume term − surface term − Coulomb term − asymmetry term + pairing term

The total binding energy is given by

$$B = a_v A - a_s A^{2/3} - a_c \frac{Z(Z-1)}{A^{1/3}} - a_a \frac{(A-2Z)^2}{A} + \delta \tag{1.5}$$

where a_v, a_s, a_c and a_a are constants.

The first term in (1.5), the volume term, is positive due to the attractive nuclear force, which is charge-independent. Since more nucleons imply a larger volume, the binding energy contributed by the volume term is,

$$B_V \propto \frac{4}{3}\pi R^3 \propto A \tag{1.6}$$

The surface term contributes to a decrease in binding energy since

$$B_S \propto 4\pi R^2 \propto A^{2/3} \tag{1.7}$$

The proton repulsion occurs in pairs which contribute to a term $Z(Z - 1)$ and are included in the Coulomb energy contribution to the binding energy, assuming a uniformly charged sphere as

$$B_C = \frac{3}{5}\frac{e^2}{4\pi\varepsilon_0 R} Z(Z-1) = \frac{3}{5}\frac{e^2}{4\pi\varepsilon_0 R_o A^{1/3}} Z(Z-1) \tag{1.8}$$

A symmetrical number of protons and neutrons makes the nucleus more stable. To account for the binding energy contribution, the asymmetry term

$$\frac{(A-2Z)^2}{A} \tag{1.9}$$

has been introduced. An equal number of protons and neutrons eliminates the asymmetry term.

The pairing term was introduced to account for the binding energy contribution according to the odd or even nucleon contribution. If Z and N are even, then δ is positive, resulting in better stability of the nucleus. If Z and N are odd, the δ has a negative magnitude, and the term vanishes when $(N + Z)$ is odd. The various contributors to the binding energy with the mass number are shown in Fig. 1.4. The increase in the net B/A followed by the decrease at higher A is evident from the cumulative contributions.

The experimental binding energy per nucleon data for the nuclei, as shown in Fig. 1.5, could be fairly explained using the semi-empirical mass formula, although there are a few exceptions. The B/A curve results in better stability for ${}^4_2He$ nucleus, which has even protons and even neutrons and also peaks around $A \approx 60$ that happens

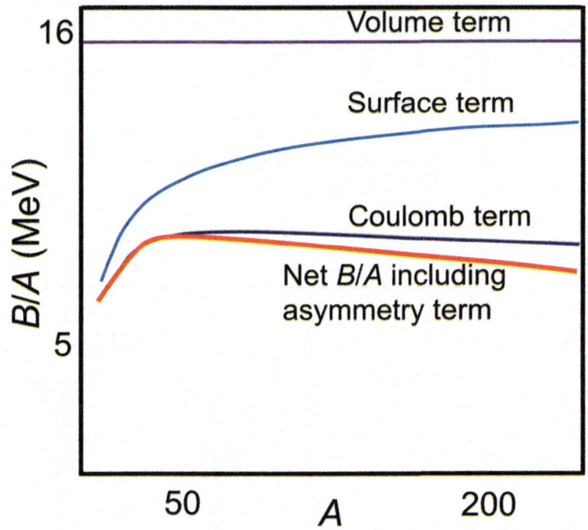

Fig. 1.4 Various cumulative contributions to the binding energy per nucleon (B/A) with mass number

to be the Fe and Ni nuclei. Thus, nuclei below $A \approx 60$ tend to fuse together and increase their B/A, whereas heavier nuclei try to shed their nucleons through the fission process.

The inset of Fig. 1.5 also shows the B/A values of a few isotopes that peak in the binding energy per nucleon curve (Fewell 1995). The largest B/A is for the ^{62}Ni isotope, while ^{56}Fe is the abundant isotope due to stellar nucleosynthesis.

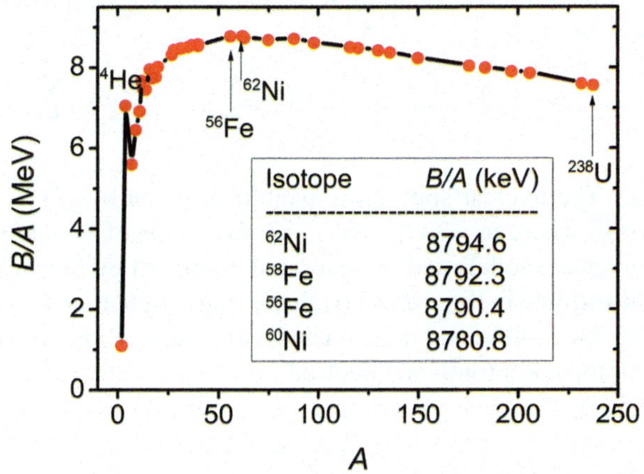

Fig. 1.5 The binding energy per nucleon (B/A) for a few isotopes with mass number. The B/A reaches a maximum at $A \approx 60$ and decreases for heavier isotopes such as ^{238}U

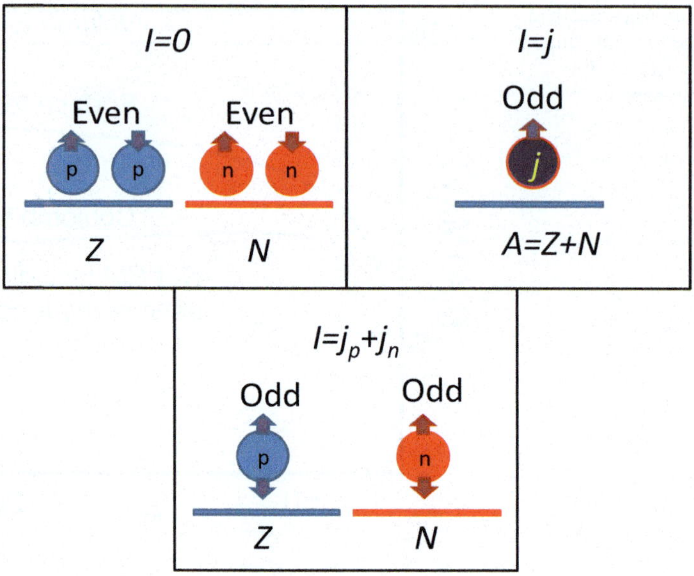

Fig. 1.6 The spin value of nuclei with even protons and neutrons is zero. The spin of the odd nucleon determines I for odd A and for nuclei with odd protons and odd neutrons

1.6 Nuclear Spin

The nuclear spin, I, is determined by the intrinsic orbital (l) and spin (s) angular momentum contribution of the nucleons (jj coupling) given as

$$I = \sum_{i=1}^{A} (l_i + s_i) = \sum_{i=1}^{A} j_i \tag{1.10}$$

The nuclear spin has a magnitude given by $\sqrt{I(I+1)}\hbar$ and its z component is $m_I\hbar$ where $m_I = +I,\ldots 0,\ldots -I$. The value of I could be obtained theoretically from nuclear models and in general, determined by the spin value of the odd nucleon as illustrated in Fig. 1.6. If: (i) Z and N are even, then $I = 0$; (ii) A is odd, I is determined by the j of the odd nucleon and; (iii) Z and N are odd, I is the sum of the spin values of the odd proton and neutron.

1.7 Nuclear Magnetic Moment

The magnetic moment of a nucleus is given by $\mu = gI\mu_N$, where g is the nuclear g-factor, I is the nuclear spin, and μ_N is the nuclear magneton. For the proton, the experimental spin g-factor is given by $g_p = 5.58569$, whereas for the neutron, it is −3.82608. The experimental magnetic moment of a proton is +2.793 μ_N, while that of a neutron is −1.913 μ_N. These values are due to the quark-gluon constituents of the nucleons requiring relativistic treatment, or else the non-zero negative magnetic moment for the neutron would not be possible. For the ^{57}Fe isotope, the experimentally measured magnetic moment is + 0.0906 μ_N corresponding to $I = 1/2$ (Fuller 1976).

1.8 The Shell Model Nucleus

Certain nuclei with Z or N values 2, 8, 20, 28, 50, 82 and 126 are exceptionally stable, and these nucleon numbers are magic numbers. In the atomic case, the stability of atoms and the ionization energy can be explained by electrons in shells with definite energy levels. The energies of shells and subshells can be calculated using the Schrödinger equation and the Coulomb potential. However, for the nuclear case, a suitable potential has to be identified since the strong nuclear force influences the nucleons, and they move in a potential of other nucleons. A realistic model, which satisfies all the magic numbers given by Woods-Saxon mean field potential, is shown in Fig. 1.7.

Since the nuclear charge distribution varies at the boundaries of the nucleus, a gradually varying nuclear potential is a good approximation. The mean radius R, skin thickness, t are incorporated in the intermediate potential of the form known as Woods-Saxon potential given by

$$V(r) = -\frac{V_o}{1 + \exp\left(\frac{r-R}{t}\right)} \tag{1.11}$$

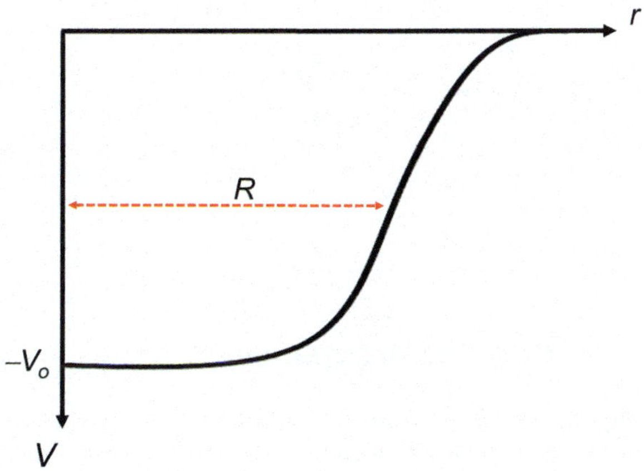

Fig. 1.7 The shell model Woods-Saxon mean field potential

where V_o is the depth of the potential well and r is the distance from the center of the nucleus. Solving the Schrödinger equation using (1.11) results in energy levels, as shown in Fig. 1.8.

The intermediate potential shown by the energy levels on the left side of Fig. 1.8 suggests magic numbers as 2, 8, 20, 40, 58 and 70, which does not satisfy the experimentally observed magic numbers. With the spin–orbit interaction term, fine

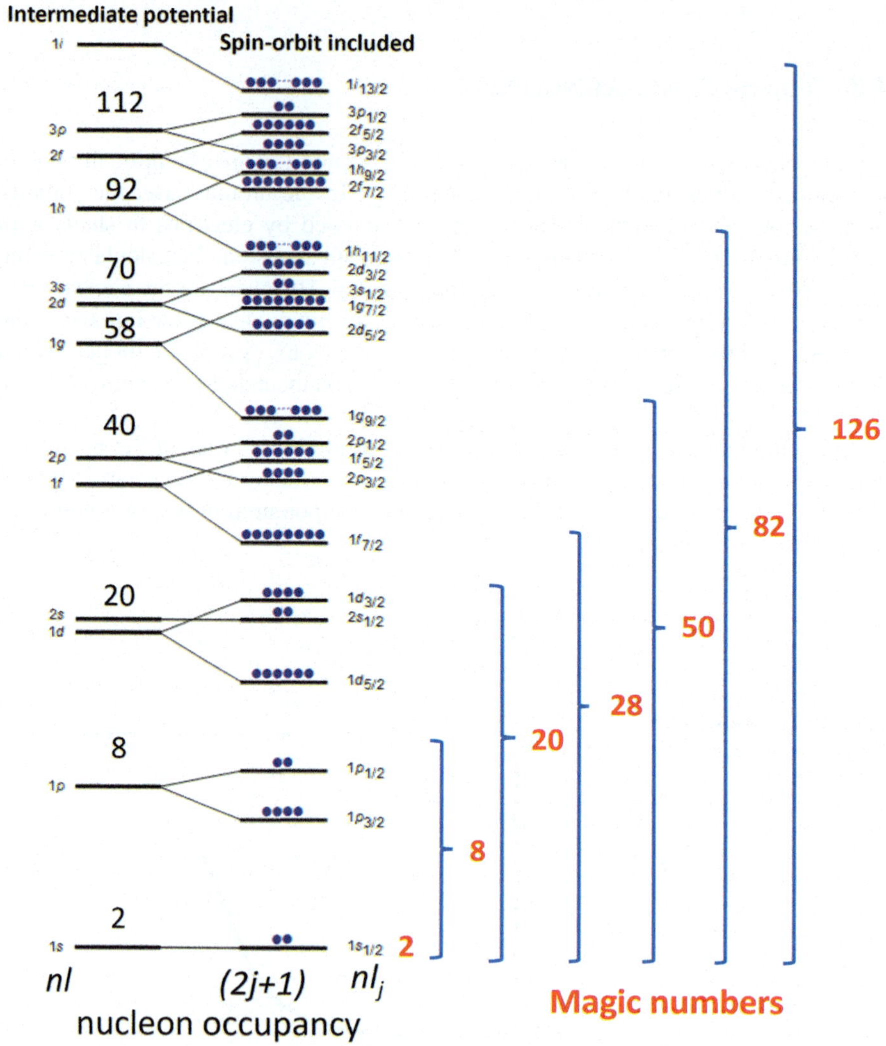

Fig. 1.8 The energy levels obtained using the intermediate potential (left) and the potential, including the spin–orbit term that satisfies all the magic numbers. The total nucleon occupancy in each level is given by $2j + 1$. Perturbations shift the energy levels. Simulated in Mathematica (Blinder 2009)

spectra were explained in atomic physics. A similar spin–orbit interaction term modifies (1.11) as $V_n(r) = V(r) + V_{ls}(r)$. Here, $V(r)$ is the Woods-Saxon and $V_{ls}(r)$ is the spin–orbit potential. The potential with spin–orbit term satisfies all the magic numbers and the energy levels as shown in Fig. 1.8. For example, the filling of energy levels corresponding to $1s_{1/2}$, $1p_{3/2}$, $1p_{1/2}$, $1d_{5/2}$, $2s_{1/2}$, $1d_{3/2}$, $1f_{7/2}$, $2p_{3/2}$, $1f_{5/2}$, $2p_{1/2}$ and $1g_{9/2}$ would satisfy the magic number 50. The occupancy of nucleons in each level is determined by the total angular momentum $j = l \pm s$. Thus, the $l = 0$ has only one $s_{1/2}$ level with an occupancy $2j + 1 = 2 \times 1/2 + 1 = 2$; $l = 1$ has levels $j = 1 + 1/2$ and $j = 1 - 1/2$ given as $p_{3/2}$ and $p_{1/2}$; $l = 2$ has levels $d_{5/2}$ and $d_{3/2}$ and so on.

Figure 1.9 shows the approximate proton and neutron potential where the nucleons are held together by the attractive strong force when $r < R$. For the proton, the Coulomb repulsion is included, which requires additional energy to overcome the barrier and get trapped in the potential well. The shell model also satisfactorily explains the spin and nuclear parity (π). All even-even nuclei have $j^\pi = 0^+$ ground states. For odd (A) nuclei, the unpaired nucleon determines the spin and parity given by j^π. Parity is given as $\pi = (-1)^l$. The spin and parity values for a few isotopes are shown in Table 1.3. The nuclei with even-even proton-neutron have zero spin, and the parity rule then gives a spin value of 0^+.

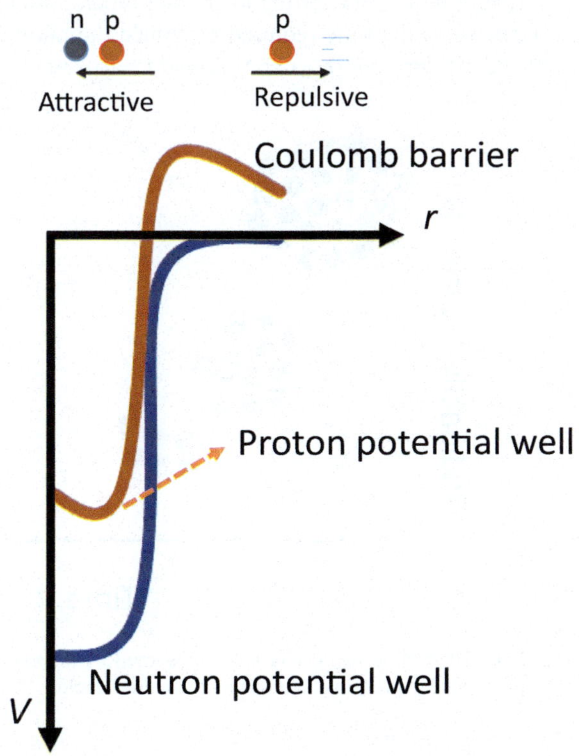

Fig. 1.9 The neutron potential well is only influenced by the strong force, whereas the strong force and Coulomb force determine the proton potential well

Table 1.3 The spin and parity values for a few Fe and Co isotopes

Isotope	N	j	Nucleons	Parity	Spin^{parity}
^{55}Fe	29	3/2	Even–odd	–	$3/2^-$
^{56}Fe	30	0	Even-even	+	0^+
^{57}Fe	31	1/2	Even–odd	–	$1/2^-$
^{58}Fe	32	0	Even-even	+	0^+
^{54}Co	27	0	Odd-odd	+	0^+
^{55}Co	28	7/2	Odd–even	–	$7/2^-$

1.9 Radioactive Decay

The number of nuclei $N(t)$ present at a given time (t) compared to the initial nuclei N_o $(t = 0)$ is given by the radioactive decay law,

$$N(t) = N_o e^{-\lambda t} \tag{1.12}$$

where λ is the disintegration constant. As illustrated in Fig. 1.10, the initial nuclei N_o transform by radioactivity, and their number decreases exponentially with time as a function of λ, converting to another type of nuclei.

Half-life is the time required to reduce the number of nuclei to half its original value. Considering $N(t_{1/2}) = N_o/2$ and $t = t_{1/2}$ in (1.12) gives

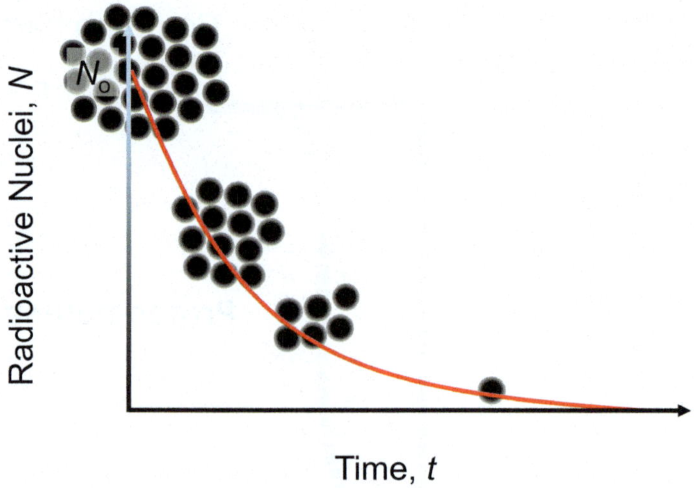

Fig. 1.10 The radioactive decay law is governed by the exponential decrease in nuclei from its initial number N_o over time, converting to another nuclei

$$t_{\frac{1}{2}} = \frac{0.693}{\lambda} \tag{1.13}$$

The mean-life is the average lifetime of nuclei given by

$$\tau = \frac{1}{\lambda} \tag{1.14}$$

The rate of radioactive decay is measured in terms of activity, which is the number of decays per unit time,

$$A = A_o e^{-\lambda t} \tag{1.15}$$

where $A_o = \lambda N_o$ is the activity at time $t = 0$. Activity is a measurable quantity in terms of cgs (centimetre–gram–second) unit curie (Ci) or in SI (Système International) unit as becquerel (Bq).

$$1\ \text{Ci} = 3.7 \times 10^{10}\ \text{decays/s} = 3.7 \times 10^{10}\ \text{Bq} = 37\ \text{GBq}.$$

Radioactive nuclei can decay using various routes giving different disintegration constants λ_1, λ_2, λ_3,... given as

$$N(\lambda_1 + \lambda_2 + \lambda_3 + \cdots) = N\lambda_t \tag{1.16}$$

The probability of various decay processes, when there is more than one, can be represented by branching ratios. ^{57}Co isotope is a category 5, low-risk source emitting gamma ray, used with a typical activity of 50 mCi.

Radioactive decay can be of three types (i) alpha (α) decay, (ii) beta (β) decay and (iii) gamma (γ) decay. The knocking out of a neutron or proton from the nucleus would also result in the transformation of the parent nucleus, accompanied by the emission of the above decay products. The α, β, and γ decays, as illustrated in Fig. 1.11, are due to quantum tunneling, weak interaction and deexcitation from higher to lower energies, respectively.

1.9.1 Alpha Decay

Alpha decay is the disintegration of the nucleus through the emission of $^4_2$He nucleus, known as α-particle. The disintegrating nucleus undergoes the following changes,

$$^{A}_{Z}\text{X}_N \rightarrow {}^{A-4}_{Z-2}\text{X}'_{N-2} + {}^4_2\text{He} \tag{1.17}$$

For the decay process (1.17), the conservation of energy implies the mass-energy and kinetic energy before disintegration should be equal to that after disintegration.

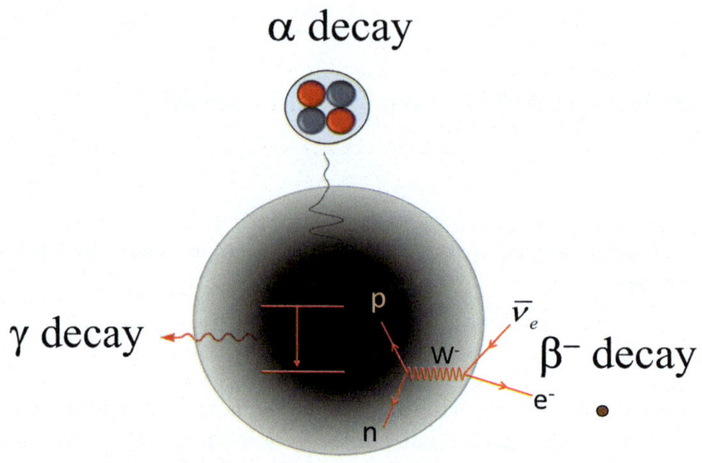

Fig. 1.11 The α-decay is due to quantum tunneling emitting positive charged ${}^4_2$He, β^- decay is a result of weak interaction, and γ-decay is caused by the transition in nuclear energy levels

The energy equation is written as

$$m_X c^2 + T_X = m_{X'} c^2 + T_{X'} + m_\alpha c^2 + T_\alpha \tag{1.18}$$

where c, m_X, T_X, $m_{X'}$, $T_{X'}$, m_α, T_α are the speed of light, mass of the initial nucleus, the kinetic energy of the initial nucleus, the mass of the final nucleus, the kinetic energy of the final nucleus, the mass of the alpha particle and the kinetic energy of the alpha particle, respectively. Assuming the initial nucleus is at rest, $T_X = 0$. Rearranging (1.18),

$$(m_X - m_{X'} - m_\alpha)c^2 = T_\alpha + T_{X'} = Q \tag{1.19}$$

The Q represented here is known as the Q-value, which is a measure of the energy released ($Q > 0$) or absorbed ($Q < 0$). A decay process will be spontaneous, provided $Q > 0$. From the conservation law, $T_\alpha + T_{X'} = Q$. Since the linear momentum of the initial nucleus is zero, the conservation of momentum gives

$$p_\alpha + p_{X'} = 0 \tag{1.20}$$

Therefore, the momenta of the α-particle and the final nucleus are equal and opposite. Note the recoil of the nucleus X′ coming into the picture here. We now obtain from (1.19),

$$T_\alpha\left(1 + \frac{T_{X'}}{T_\alpha}\right) = (m_X - m_{X'} - m_\alpha)c^2 = Q \tag{1.21}$$

Using the $T = p^2/2m$ relation, the kinetic energy of the alpha particle can be obtained as,

$$T_\alpha = \frac{Q}{\left(1 + \frac{T_{X'}}{T_\alpha}\right)} = \frac{Q}{\left(1 + \frac{m_\alpha}{m_{X'}}\right)} \tag{1.22}$$

Since $m_{X'} \gg m_\alpha$, $T_\alpha \approx Q$, implying that most of the Q value is contributed by the kinetic energy of the alpha particle. The Geiger-Nuttall law gives a typical Q value for α-particles in the range 4–10 MeV for half-life ranging from 10^{10} years to 10^{-7} s, respectively (Krane 1988). The Geiger-Nuttall law relates Q and half-life given as

$$\log t_{1/2} = a(Z) + \frac{b(Z)}{\sqrt{Q_\alpha}} \tag{1.23}$$

where a and b are functions of Z. This poses an important question. How could an alpha particle with energy as low as 4 MeV cross the nuclear potential barrier exceeding 8 MeV? This could be explained by quantum mechanics using the phenomena of barrier penetration or tunneling effect.

1.9.2 Beta Decay

The β-decay can take place through any of the following processes.

$$\begin{aligned} &\mathrm{n} \rightarrow \mathrm{p} + \mathrm{e}^- + \overline{\nu}_\mathrm{e} && \beta^- \text{ decay} \\ &\mathrm{p} \rightarrow \mathrm{n} + \mathrm{e}^+ + \nu_\mathrm{e} && \beta^+ \text{ decay} \\ &\mathrm{p} + \mathrm{e}^- \rightarrow \mathrm{n} + \nu_\mathrm{e} && \varepsilon \text{ electron capture} \end{aligned} \tag{1.24}$$

In the case of α-decay, the mass-energy of α-particle is 3.7 GeV, which is very large compared to the kinetic energy of 4 MeV and therefore, non-relativistic approach can be used there. Unlike α-particle, which has definite energy, experiments have shown that β-particle has continuous energy ranging up to a few MeV. It can be shown that the continuous distribution of kinetic energy is shared by the electron and neutrino in the β-decay process. The ^{22}Na isotope that decays to ^{22}Ne emitting e^+, with a half-life of 2.6 years, is widely used in positron annihilation spectroscopy.

The β^- decay transforms the nucleus as,

$${}^A_Z\mathrm{X}_N \rightarrow {}^{A}_{Z+1}\mathrm{X'}_{N-1} + \mathrm{e}^- + \overline{\nu}_\mathrm{e} \tag{1.25}$$

In terms of neutral atomic mass (M), nuclear mass (m) and electron mass (m_e), the mass-energy considering binding energy, is written as

$$M_X c^2 = m_X c^2 + Z m_e c^2 - \sum B \tag{1.26}$$

The antineutrino mass is negligible, electron masses and binding energies are eliminated here, and the Q value from (1.25) is

$$(M_X - M_{X'})c^2 = T_{e^-} + T_{\bar{\nu}_e} = Q \tag{1.27}$$

The β^- decay involves transforming higher to lower mass-energy due to the conversion of n → p that results in $Q > 0$. The electron and antineutrino share the kinetic energies. Let us not take this Q forward as we also define Q for nuclear quadrupole moment.

1.9.3 Gamma Decay

Any nuclear transition from a higher energy to a lower energy state involves the emission of high energy γ-ray photon. Unlike α and β decays, γ-decay does not involve any change in the nucleon number. The energies of γ-ray may overlap with the X-ray energy and the former is identified from its nuclear origin while the latter is related to the electron. Usually α and β decays produce isotopes in a higher energy state that reaches the ground state through the emission of γ-rays.

Consider a nucleus at rest emitting a γ-ray of energy E_γ after making a transition from excited state energy level, E_e to a lower level, E_g. Since the initial momentum is zero, the conservation of linear momentum should satisfy that the emitted gamma ray momentum is compensated with the recoil momentum giving

$$0 = p_R + p_\gamma \tag{1.28}$$

Using energy conservation,

$$E_e = E_g + E_\gamma + T_R \tag{1.29}$$

where T_R is the recoil energy.

Mössbauer spectroscopy deals with γ-rays, and therefore, it is important to know few of its characteristics. Unlike the α and β -rays, γ-ray has no definite absorption or penetration range because it is an electromagnetic radiation. The absorption of γ-rays of intensity I_o in a slab of thickness x can be given by the expression $I = I_o \exp(-\mu x)$ where μ is the absorption coefficient and I is the intensity after penetrating the slab. The thickness of the slab required to reduce the γ-ray intensity by half ($I/I_o = 1/2$) is known as *half-value thickness* or *half-value layer*. The absorption coefficient is proportional to the density of the material and microscopic cross section for reaction

(σ) of the gamma rays. The reaction cross section is expressed by the unit *barn* (b) which is equal to 10^{-28} m^2.

The γ-rays interact with matter in the form of the photoelectric effect, Compton effect and pair-production at low, intermediate and high energies, respectively, due to their better cross sections at the corresponding energy scales. The photoelectric effect and Compton effect are the sources of large background in the detector of a Mössbauer spectrometer. The photoelectron cross section is nearly 500 times larger than the Compton cross section for the 14.4 keV gamma ray. Apart from the inelastic, incoherent Compton scattering of the gamma rays with the electrons, coherent elastic scattering that does not involve a change in energy or phase is also possible. Two of these elastic scattering processes are; Thomson scattering and Rayleigh scattering of gamma rays with electrons, the latter being almost 10^4 times more important than the former.

1.9.3.1 Selection Rules

Oscillating electric charges would produce an electric field generating electric multipole (E*L*) type radiation, where L is the multipole angular momentum operator. Similarly, a time-varying current loop would produce magnetic multipole (M*L*) type radiation. Electromagnetic radiation carries away angular momentum subjected to selection rules. If I_i and I_f are the nuclear spin angular momenta of the initial and final states, respectively, a gamma ray photon is emitted after a nuclear transition for non-zero L. The conservation of angular momentum and parity must be respected, and the angular momentum carried by a gamma ray photon is $L\hbar$. According to the L value, the photon may be called a dipole ($L = 1$), quadrupole ($L = 2$) or multipole. The γ-ray selection rules and multipolarities are given in Table 1.4.

Electric and magnetic multipole of the same order (L) have opposite parity. L values may range from $|I_i - I_f|$ to $|I_i + I_f|$.

For electric transitions, $\Delta\pi = (-1)^L$

For magnetic transitions, $\Delta\pi = (-1)^{L-1}$

Examples of radiation types are shown in Table 1.5, corresponding to various transition states.

Table 1.4 The selection rules and multipolarities of γ-ray transitions

Multipole	$L = \Delta I$	$\Delta\pi$	Radiation type	π
Electric dipole	1	Yes	E1	Odd
Magnetic dipole	1	No	M1	Even
Electric quadrupole	2	No	E2	Even
Magnetic quadrupole	2	Yes	M2	Odd
Electric octupole	3	Yes	E3	Odd
Magnetic octupole	3	No	M3	Even

Table 1.5 The radiation types from various I^{π} transitions

Transitions	Radiation type
$2^+ \rightarrow 1^-$	E1
$3^+ \rightarrow 1^+$	E2
$2^+ \rightarrow 1^+$	M1
$3^+ \rightarrow 1^-$	M2

Generally, the most probable transition corresponds to the lowest *L*, and electric transitions are more probable than magnetic ones. Fermi's golden rule gives the rate at which the electromagnetic transition occurs (See Sect. 2.12).

Consider the ^{16}O energy levels where the ground and 6.05 MeV excited states are 0^+. Since $\Delta I(0{\rightarrow}0) = 0$, the transition is not allowed, and there shall not be gamma decay (See Sect. 2.13).

1.10 Rate of Emission

For an emitted gamma ray, from Maxwell's theory, the energy is equally distributed between the electric (*E*) and magnetic (*B*) fields where $B = E/c$. The power radiated through an electric or magnetic radiation field is given as

$$P_{\mathrm{EL}} = \frac{2(L+1)c}{\varepsilon_0 L((2L+1)!!)^2}\left(\frac{\omega}{c}\right)^{2L+2} Q_L^2 \tag{1.30}$$

where ω is the angular frequency and Q_L is the multipole moment. The multipole radiation order is given as 2^L with $L \neq 0$. The Q_L term due to oscillation of the electric charge distribution, given by the integral in the electric potential, V at a distance, R for a charge density, ρ in the z direction for a volume element, dV' is

$$V = \frac{1}{4\pi\varepsilon_0}\left(\frac{1}{R}\int \rho dV' + \frac{1}{R^2}\int \rho z dV' + \frac{1}{R^3}\int \rho\left(3z^2 - r^2\right)dV' + \cdots\right) \tag{1.31}$$

where the term outside the bracket is the Coulomb constant k. The emission probability rate is given by

$$\lambda_{\mathrm{EL}} = \frac{P_{\mathrm{EL}}}{\hbar\omega} = \frac{2(L+1)}{\hbar\varepsilon_0 L((2L+1)!!)^2}\left(\frac{\omega}{c}\right)^{2L+1} Q_L^2 \tag{1.32}$$

The lower multipoles are more probable than the higher ones. The electric multipole is more likely than the magnetic multipole by a factor of 100. The λ_{EL} for neutrons is smaller than that for protons by a factor of 2. In the quantum scenario, the multipole moments are replaced with multipole operator form over the entire volume of the nucleus given by the matrix element

$$Q_{fi} = \int \psi_f^* H_{\mathrm{EL}} \psi_i dV \tag{1.33}$$

where the initial state and final state wave functions are ψ_i and ψ_f, respectively, and H_{EL} is the operator for the electric multipole moment. The basic quantum mechanical aspects are discussed in Chap. 2.

1.11 Experimental Studies on Fe Isotopes

Consider the Fe isotopes ^{56}Fe and ^{57}Fe. Since the number of neutrons in ^{56}Fe is 30, the even-even condition gives spin 0^+. However, the shell model for ^{57}Fe gives a ground state spin value of 3/2 since the $2p_{3/2}$ is occupied with three neutrons, whereas the experimental value is 1/2. A spin value of 1/2 for ^{57}Fe was confirmed from polarized γ-rays and electron spin resonance (ESR) experiments (Trumpy 1956; Ludwig et al. 1958). Coulomb excitation experiments have shown the existence of higher excited states of ^{57}Fe, proving the energy levels 14, 137 and 365 keV and the corresponding excited state spin values as $3/2^-$, $5/2^-$ and $5/2^-$, respectively, apart from the ground state spin value of $1/2^-$ (Pieper and Keydenburg 1957).

The hyperfine structure of ^{57}Fe using polarized resonant radiation was studied on 14 keV energy level in a magnetic field (Hanna et al. 1960). The intensity ratios corresponding to the transitions from unpolarized and polarized hyperfine spectra are shown in Fig. 1.12. The six-line pattern due to the allowed transitions could be observed from the unpolarised spectrum. The application of a magnetic field under parallel and perpendicular orientation to the sample and absorber resulted in a difference in the intensity ratio of the hyperfine pattern. The experiments gave a magnetic moment of $-0.153\ \mu_{\mathrm{N}}$ for the first excited state and an effective field with an opposite sign of 33.3 T for the iron nucleus at room temperature. In subsequent chapters, details of the hyperfine interactions in Mössbauer spectroscopy are dealt with.

1.12 Internal Conversion

Internal conversion is the process of electromagnetic interaction of the nucleus with an orbital electron, and energy is transferred to the electron in the form of kinetic energy. The ejected electrons from any atomic shells, such as K, L and M, create a vacancy filled by higher shell electrons, resulting in X-ray emission. The total decay probability (λ_t) is the sum of γ-decay and internal conversion decay constants, respectively, given as

$$\lambda_t = \lambda_\gamma + \lambda_e \tag{1.34}$$

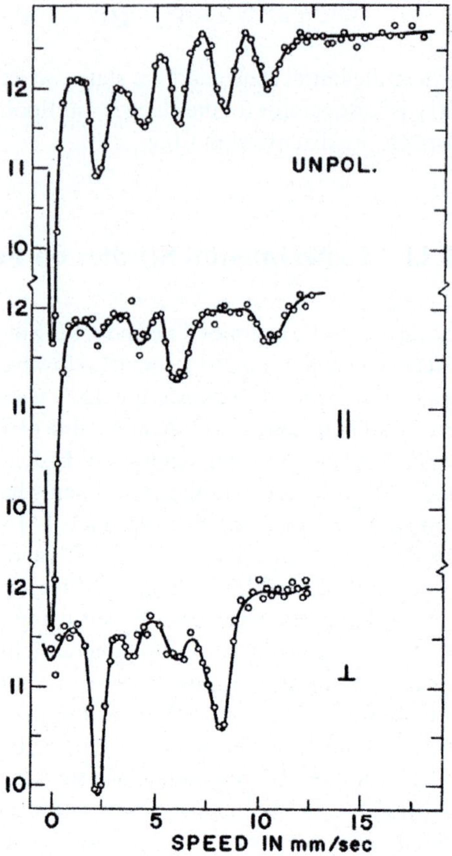

Fig. 1.12 The hyperfine spectrum of ^{57}Fe showing six resolved lines from the unpolarised γ-rays (top). The intensity difference from the polarized hyperfine spectrum could be observed under a parallel applied field (middle) and perpendicular applied field (bottom) to the source and absorber. Reprinted with permission (Hanna et al. 1960), © American Physical Society

The internal conversion coefficient is defined as $\alpha = \frac{\lambda_e}{\lambda_\gamma}$ and therefore,

$$\lambda_t = \lambda_\gamma(1 + \alpha) \tag{1.35}$$

The α is more prominent with high Z nuclei since it has Z^3 dependence, whereas it decreases with the energy. The λ_e shall have decay probability contributions from K, L, M, … shell electrons, and λ_K shall contribute the most because of the proximity of K electrons to the nucleus. The gamma decay $0^+ \to 0^+$ is forbidden, whereas such transitions can happen through internal conversion, accompanied by the emission of X-rays or Auger electrons.

1.13 Nuclear Shape

The nucleon's motion is responsible for the collective nuclear properties arising from nuclear vibration and rotation. Nuclei with magic numbers of neutrons or protons have a closed shell that acquires a spherical shape. The nuclear deformation is characterized by non-zero nuclear quadrupole moment (Q).

1.13.1 Vibrating Nucleus

A vibrating nucleus can be shaped like a liquid drop that acquires an instantaneous non-spherical shape. Nuclear vibrations can be modeled as standing waves on the surface of a liquid drop. For a charge distribution that has spherical symmetry, we can use the Laplacian in spherical polar coordinates whose solutions are harmonic functions. In terms of spherical harmonics, $Y_{\lambda m}(\theta, \varphi)$, the instantaneous nuclear radius is then given as

$$R(t) = R_{\mathrm{a}}\left\{1 + \sum_{\lambda \geq 0} \sum_{m=-\lambda}^{\lambda} \alpha_{\lambda m}(t) Y_{\lambda m}(\theta, \varphi)\right\} \tag{1.36}$$

where R_{a} is the average radius, $\alpha_{\lambda m}(t)$ is the amplitude of vibration, m is an integer, θ is the polar angle and φ is the azimuthal angle.

The various modes of vibration corresponding to λ values give the average radius with modes such as monopole ($\lambda = 0$), dipole ($\lambda = 1$), quadrupole ($\lambda = 2$) and other multipoles. The monopole mode implies only a change in the radius, the dipole mode corresponds to a shift in the center of gravity leading to translation, and the quadrupole mode implies a shape change, as represented in Fig. 1.13. The monopole excitation is also known as the breathing mode of the nucleus and requires considerable energy. Quadrupole vibration is the dominant mode in most of the nuclei. The $\lambda = 2$ vibration carries an angular momentum of magnitude 2 and even parity given by $(-1)^{\lambda}$.

1.13.2 Rotational Motion

A rotating nucleus with a non-spherical shape can be modeled using an ellipsoid of revolution. The nuclear radius is then given by

$$R(\theta, \varphi) = R_{\mathrm{average}}\{1 + \beta Y_{20}(\theta, \varphi)\} \tag{1.37}$$

where β is the eccentricity of the ellipse, which is a measure of the deformation. Note that λ and m are 2 and 0, respectively, signifying quadrupole deformation. When β

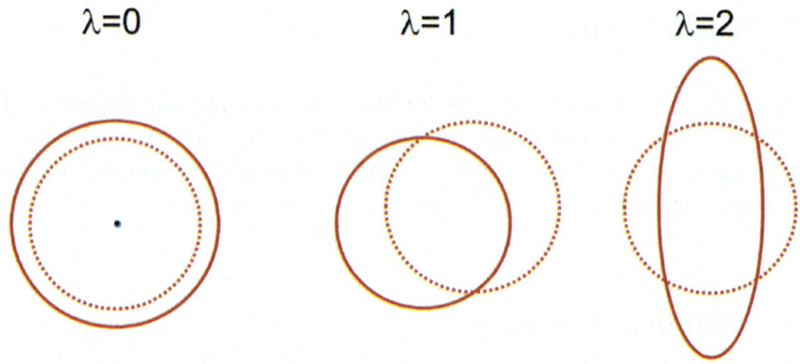

Fig. 1.13 Vibration modes of the nucleus with $\lambda = 0$ corresponding to a change in the radius, $\lambda = 1$ representing a shift in the center of gravity and $\lambda = 2$ implying a shape change

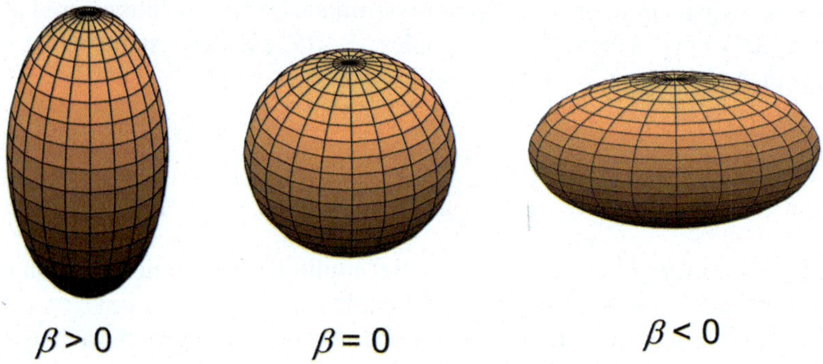

Fig. 1.14 The nuclear deformation can be prolate ($\beta > 0$) or oblate ($\beta < 0$)

> 0, the deformation is a prolate shape, whereas it becomes oblate when $\beta < 0$ as shown in Fig. 1.14.

The quadrupole moment of a point charge e can be written as eQ, where Q is the quadrupole moment measured in terms of *barn* (b). For spherical charges, $Q = 0$, whereas it can be positive and negative for charges distributed along z (prolate) and x–y (oblate), respectively. The $I = 3/2$ excited state of ^{57}Fe has a $Q \cong +0.2$ b (see Sect. 5.3 for revised values), signifying the prolate shape of the nucleus in its excited state.

The kinetic energy of a rotating object is given as $E = \frac{1}{2} I_\alpha \omega^2$, where I_α is the moment of inertia. The energy of the rotational state is given by

$$E_{\text{rot}} = \frac{I(I+1)\hbar^2}{2I_\alpha} \tag{1.38}$$

For nucleus, which has non-zero ground state spin, I_o, the rotational energy is given by

$$E_{\text{rot}} = \frac{\{I(I+1) - I_o(I_o+1)\}\hbar^2}{2I_\alpha} \tag{1.39}$$

1.14 Summary

This chapter has presented basic nuclear properties, nuclear stability, various nuclear decays, gamma decay features and the nuclear shape. The characteristics of a gamma ray emitting isotope used for Mössbauer spectrometry can be thus understood from this introductory chapter.

References

S.M. Blinder, Magic Numbers in the Nuclear Shell Model, http://demonstrations.wolfram.com/MagicNumbersInTheNuclearShellModel/, Wolfram Demonstrations Project. Published 31 July 2009.

M.P. Fewell, Am. J. Phys. **63**, 653 (1995)

G.H. Fuller, J. Phys. Chem. Ref. Data **5**, 835 (1976)

S.S. Hanna, J. Heberle, C. Littlejohn, G.J. Perlow, R.S. Preston, D.H. Vincent, Phys. Rev. Lett. **4**, 177 (1960)

K.S. Krane, *Introductory Nuclear Physics* (Wiley, Singapore, 1988)

G.W. Ludwig, H.H. Woodbury, R.O. Carison, Phys. Rev. Lett. **1**, 295 (1958)

J. Meija, T.B. Coplen, M. Berglund, W.A. Brand, P.D. Bièvre, M.G. Gröning, N.E. Holden, J. Irrgeher, R.D. Loss, T. Walczyk, T. Prohaska, Pure Appl. Chem. **88**, 293 (2016)

G.F. Pieper, N.P. Keydenburg, Phys. Rev. **107**, 1300 (1957)

G. Trumpy, Nuc. Phys. **2**, 664 (1956/57)

Chapter 2
Introduction to Quantum Mechanics

2.1 Fundamental Concepts

2.1.1 de Broglie Wavelength

The quantum concept arises due to the failure of classical physics to explain certain fundamental behaviors such as black body radiation, the photoelectric effect and the electron's stable orbit in atoms. Quantum concepts originated with the successful explanation of the black body radiation pattern by Max Planck in 1900, postulating that energy is quantized in multiples of *nhv*, and Albert Einstein's explanation of the photoelectric effect in 1905. In 1924, Louis de Broglie suggested wave-particle duality using the relation

$$\lambda = \frac{h}{p} \tag{2.1}$$

where λ is the de Broglie wavelength, *h* is the Planck constant, and *p* is the momentum. The Davisson Germer experiment in 1927 confirmed the de Broglie hypothesis using electron diffraction experiments. A free particle of mass, *m*, in motion, can be imagined as a wave packet formed by interference of individual waves, as shown in Fig. 2.1. The group velocity is the velocity of the entire wave packet, equal to that of the particle. The wave packet at a time, *t*, position, *x*, as a function of wavevector, *k* is represented by the general term,

$$\Psi(x,t) = \frac{1}{\sqrt{2\pi}} \int_{-\infty}^{+\infty} \varphi(k) \exp i(kx - \omega t) dk \tag{2.2}$$

where $\varphi(k)$ is the momentum wave function and the angular frequency, $\omega = \frac{\hbar k^2}{2m}$.

R. J. Joseyphus and J.-M. Greneche, *Fundamentals of ^{57}Fe Mössbauer Spectrometry*,
https://doi.org/10.1007/978-981-99-8653-8_2

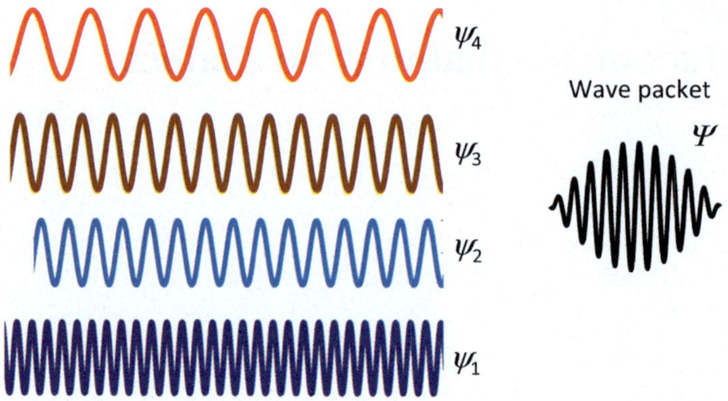

Fig. 2.1 A wave packet is obtained from the constructive interference of several waves in a small region and destructive interference elsewhere

For a trapped particle confined to a length L, standing waves give a wave packet-like scenario governed by the condition that $\lambda = 2L/n$ where $n = 1, 2, 3 \ldots$ The $n = 1$ corresponds to half the wavelength within L of the 1-dimensional (1D) box corresponding to the lowest energy state. A standing wave stationary state is possible, provided the nodes meet at the edges of the box, imagined like a standing wave pattern in a string. As we move on to the higher energies, the n value increases and the permitted wavelengths are also integral multiples of half wavelength ($n\lambda/2$) to satisfy the standing wave condition.

Using the kinetic energy relationship and inserting the de Broglie wavelength in it, we get, KE $= \frac{p^2}{2m} = \frac{h^2}{2m\lambda^2}$. The energy obtained using the standing wavelength condition for the trapped particle is

$$E_n = \frac{n^2h^2}{8mL^2} \quad \text{where } n = 1, 2, 3, \ldots \tag{2.3}$$

It is to be noted here that the lowest energy corresponding to $n = 1$ is non-zero.

Condition for Orbital Stability

The orbital stability condition from the standing wave situation and de Broglie relationship is $n\lambda = 2\pi r$ and therefore, $mvr = n\hbar$, where v is the velocity and r is the orbital radius.

2.1.2 Wave Function

The wave characteristics of a particle can be represented mathematically in terms of the wave function $\boldsymbol{\Psi}(x, t)$, where

(i) $\boldsymbol{\Psi}$ must be continuous and single-valued,
(ii) $\frac{\partial \boldsymbol{\Psi}}{\partial x}, \frac{\partial \boldsymbol{\Psi}}{\partial y}, \frac{\partial \boldsymbol{\Psi}}{\partial z}$ must also be continuous and single-valued,
(iii) normalizable such that, $\int_{-\infty}^{+\infty} |\boldsymbol{\Psi}|^2 dV = 1$
(iv) the superposition principle is obeyed such that it can be a linear combination of numerous wave functions given as $\boldsymbol{\Psi} = \sum a_n \boldsymbol{\Psi}_n$ where a_n is a linear coefficient.

The probability density of finding a real particle could be obtained using the third condition $|\boldsymbol{\Psi}|^2 = \boldsymbol{\Psi}^* \boldsymbol{\Psi}$, which will be real. For example, setting the normalization condition in 1D to $\int_0^L |\boldsymbol{\Psi}|^2 dx = 1$ indicates the probability of finding the particle within the boundaries 0 to L, whereas it is zero elsewhere. The square-integral wave function with infinite dimensions and finite probability is called *Hilbert space*. The infinite dimensions allow us to introduce numerous quantum numbers where spin is one among them. The Hilbert space, a linear, complex and inner product vector space signifies the state of a quantum object, and represented in terms of the inner product of two wave functions as

$$\langle \varphi | \boldsymbol{\Psi} \rangle = \int_{-\infty}^{+\infty} \varphi^* \boldsymbol{\Psi} dV \tag{2.4a}$$

where $\langle \varphi |$ is known as the bra, $| \boldsymbol{\Psi} \rangle$ is the ket and follows the relation $\langle \varphi | \boldsymbol{\Psi} \rangle = \langle \boldsymbol{\Psi} | \varphi \rangle^*$.

The inner product is normalizable when

$$\langle \varphi | \boldsymbol{\Psi} \rangle = \int_{-\infty}^{+\infty} \varphi^* \boldsymbol{\Psi} dV = 1 \tag{2.4b}$$

orthogonal when

$$\langle \varphi | \boldsymbol{\Psi} \rangle = \int_{-\infty}^{+\infty} \varphi^* \boldsymbol{\Psi} dV = 0 \tag{2.4c}$$

and orthonormal when it is normalizable and orthogonal, represented by the Kronecker delta function δ as

$$\langle \varphi | \boldsymbol{\Psi} \rangle = \int_{-\infty}^{+\infty} \varphi^* \boldsymbol{\Psi} dV = \delta \tag{2.5}$$

where $\delta = 1$ when $\boldsymbol{\Psi} = \varphi$ and $\delta = 0$ when $\boldsymbol{\Psi} \neq \varphi$.

2.2 Uncertainty Principle

Consider a wave packet as shown in Fig. 2.1. Although the position of a wave packet could be determined, it has a distribution of wavelengths that spread out, affecting the position. The spread in position Δx and wavenumber Δk has a minimum standard deviation, from which they are related to each other as

$$\Delta x \Delta k \geq \frac{1}{2} \tag{2.6}$$

Substituting the de Broglie wavelength in (2.6), we get the Heisenberg's uncertainty relation,

$$\Delta x \Delta p \geq \frac{\hbar}{2} \tag{2.7}$$

The uncertainty relation between momentum and position is thus given by (2.7). Similar result relating the energy and time is expressed as

$$\Delta E \Delta t \geq \frac{\hbar}{2} \tag{2.8}$$

2.3 Schrödinger Equation

Quantum concepts assign wave function characteristics even for particles, whose significance can be understood in subsequent discussions.

The classical wave propagating along the x direction with velocity v is

$$\frac{\partial^2 y}{\partial x^2} = \frac{1}{v^2}\frac{\partial^2 y}{\partial t^2} \tag{2.9}$$

The free particle wave in complex form is given by

$$y = Ae^{-i(\omega t - kx)} \tag{2.10}$$

To write the wave equation in quantum form, we start with

$$\boldsymbol{\Psi} = Ae^{-i(\omega t - kx)} \tag{2.11}$$

Replacing ω and k with $2\pi\nu$ and $2\pi/\lambda$, respectively, where ν is the frequency and using $E = h\nu$ and $\lambda = \frac{h}{p}$ we get,

$$\boldsymbol{\Psi} = Ae^{-2\pi i\left(\nu t - \frac{x}{\lambda}\right)} = Ae^{-\frac{i}{\hbar}(Et - px)} \tag{2.12}$$

Differentiating (2.12) twice w.r.t. x gives

$$p^2\boldsymbol{\Psi} = -\hbar^2\frac{\partial^2\boldsymbol{\Psi}}{\partial x^2} \tag{2.13}$$

Differentiating the wave function (2.12) w.r.t. time gives

$$E\boldsymbol{\Psi} = -\frac{\hbar}{i}\frac{\partial\boldsymbol{\Psi}}{\partial t} \tag{2.14}$$

At normal speed, the total energy (E) of a particle is the sum of kinetic ($p^2/2\ m$) and potential energy (U).

$$E = T + U = \frac{p^2}{2m} + U \tag{2.15}$$

Assigning the wave function characteristics to (2.15) gives

$$E\boldsymbol{\Psi} = \frac{p^2}{2m}\boldsymbol{\Psi} + U\boldsymbol{\Psi} \tag{2.16}$$

Substituting (2.14) and (2.13) in (2.16)

$$i\hbar\frac{\partial\boldsymbol{\Psi}}{\partial t} = -\frac{\hbar^2}{2m}\frac{\partial^2\boldsymbol{\Psi}}{\partial x^2} + U\boldsymbol{\Psi} \tag{2.17}$$

The time-dependent Schrödinger equation in 1D is given by (2.17). Using

$$\boldsymbol{\Psi} = Ae^{-\frac{i}{\hbar}(Et - px)} = \Psi(x)e^{-\frac{i}{\hbar}(Et)} \tag{2.18}$$

the time-independent 1D Schrödinger equation, from separation of variables, can be obtained as

$$E\Psi = -\frac{\hbar^2}{2m}\frac{\partial^2\Psi}{\partial x^2} + U\Psi \tag{2.19}$$

Equation (2.19) is also written as $\hat{H}\Psi = E\Psi$ where $\hat{H}$ is the Hamiltonian operator associated with physical observables, E is the energy, and ^ represents an operator. The operators arise due to the association of the wave function characteristics to the physical quantity. Although ^ may not be used extensively in this book for better readability, all the physical quantities (observables) are operators in quantum mechanics. Further, we use $\Psi(x)$ or $\psi(x)$ (italics) in this book. Table 2.1 shows a list of operators in quantum mechanics.

Table 2.1 The physical quantities and their corresponding operators in quantum mechanics

Physical quantity	Operator
Linear momentum	$\hat{p} = -i\hbar\frac{\partial}{\partial x}$
Kinetic energy	$\hat{T} = -\frac{\hbar^2}{2m}\frac{\partial^2}{\partial x^2}$
Total energy (time dependent)	$\hat{E} = i\hbar\frac{\partial}{\partial t}$
Position	$\hat{x}$
Potential energy	$\hat{U}$ or $\hat{V}$
Total angular momentum	$\hat{J}$
Orbital angular momentum	$\hat{L}_z = -i\hbar\frac{\partial}{\partial\varphi}$
Spin angular momentum	$\hat{S}$

Hermitian operators

The expectation or average value of the quantized total energy is obtained as

$$\langle H\rangle = \int \Psi^* \hat{H}\Psi dx = E \tag{2.20}$$

Here, $\hat{H}$ is an operator, and we shall represent ^ for a Hermitian operator, meaning it is a linear and self-adjoint operator such that

$$\int \Psi_1^* \hat{H}\Psi_2 dx = \int (\Psi_1 \hat{H})^* \Psi_2 dx = \int \Psi_1^* \hat{H}\Psi_2 dx \tag{2.21}$$

which holds good for any operator in quantum mechanics that obeys the superposition principle.

The expectation value for the Hermitian operator is always real. The product of two Hermitian operators is Hermitian if they commute. Two operators $\hat{A}$, $\hat{B}$ commute only if $\left[\hat{A}, \hat{B}\right] = \hat{A}\hat{B} - \hat{B}\hat{A} = 0$, which implies that A and B could be measured simultaneously as in classical mechanics. If $\left[\hat{A}, \hat{B}\right]$ is non-zero, they are quantized, follow the uncertainty relationship, and have imaginary values too. Usually, two wave functions are represented using *Dirac notation* employing the orthonormal relationship,

$$\langle\Psi_n|\Psi_m\rangle = \int_{-\infty}^{+\infty} \Psi_n^* \Psi_m dx = \delta_{nm} \tag{2.22}$$

$$\text{such that } \delta_{nm} = \begin{cases} 1 & \text{when } n = m \\ 0 & \text{when } n \neq m \end{cases}$$

where the $|\Psi_m\rangle$ and $\langle\Psi_n|$ are state vectors represented by the ket and bra notations, respectively. The Hermitian operator is now of the form,

$$\left\langle\Psi\middle|\hat{H}\middle|\Psi\right\rangle = \left\langle\hat{H}\Psi|\Psi\right\rangle \tag{2.23}$$

The solution of the Schrödinger equation is a linear combination of wave functions with coefficients, c_n is given as

$$\Psi(x,t) = \sum_{n=1}^{\infty} c_n \psi_n(x) \exp\left(\frac{-iE_n t}{\hbar}\right) \tag{2.24}$$

The energy solutions are known as energy eigenvalues and the associated wave functions are eigenfunctions. The Hamiltonian of a system can also be written in the form of a Hermitian matrix which is given for a two-level system as

$$H = \begin{pmatrix} a_1 & c - ib \\ c + ib & a_2 \end{pmatrix} \tag{2.25}$$

2.3.1 Infinite Square Well Potential

Consider a particle trapped in a box of length *L*, having a potential energy *V*, as shown in Fig. 2.2, and given by the boundary conditions,

$$V(x) = \begin{cases} 0 & \text{for } 0 \le x \le L \\ \infty & \text{elsewhere} \end{cases} \tag{2.26}$$

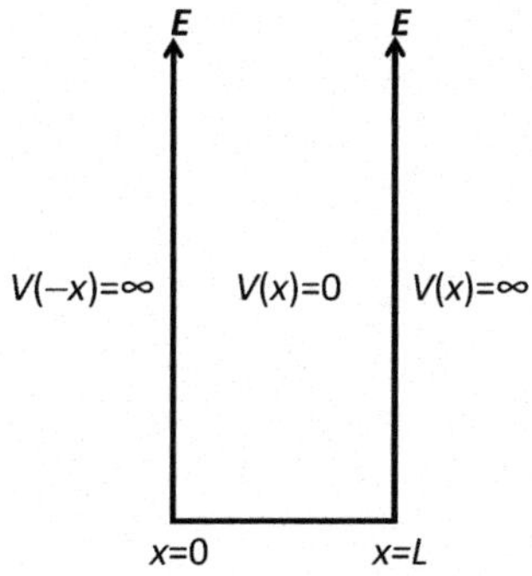

Fig. 2.2 The one-dimensional infinite potential well is represented by the boundary conditions $V(x) = 0$ for $0 \le x \le L$ and ∞ elsewhere

The time-independent Schrödinger equation in 1D becomes

$$E\Psi = -\frac{\hbar^2}{2m}\frac{d^2\Psi}{dx^2} \tag{2.27}$$

since $U = V = 0$ within the box and the particle is not allowed elsewhere.

Equation (2.27) can be rewritten as

$$\frac{d^2\Psi}{dx^2} + \frac{2mE}{\hbar^2}\Psi = 0 \tag{2.28}$$

Taking $k = \frac{\sqrt{2mE}}{\hbar}$, the second-order linear homogeneous differential wave equation has the solution given by

$$\Psi = A\sin\frac{\sqrt{2mE}}{\hbar}x + B\cos\frac{\sqrt{2mE}}{\hbar}x \tag{2.29}$$

Using the boundary condition $\Psi \to 0$ for $x = 0$ and L requires $B = 0$ and $A \neq 0$.
Hence $\Psi = A\sin\frac{\sqrt{2mE}}{\hbar}L = 0$ and $\frac{\sqrt{2mE}}{\hbar}L = n\pi$ where $n = 1, 2, 3 \ldots$
The energy eigenvalues are then given as

$$E_n = \frac{n^2h^2}{8mL^2} \quad \text{where } n = 1, 2, 3, \ldots \tag{2.30}$$

Consequently, the energy is given as n^2E_o, where $E_o = \frac{h^2}{8mL^2}$ and represented in Fig. 2.3 for various n levels. The energy is finite and non-zero for the lowest state corresponding to $n = 1$.

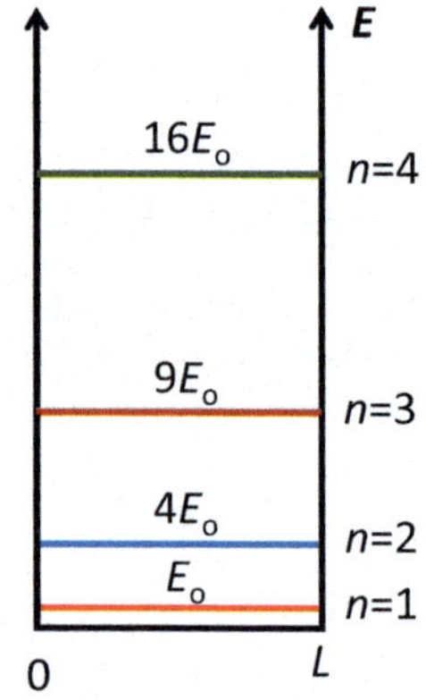

Fig. 2.3 The energy levels of infinite potential well scales on the order of n^2E_o. The lowest energy E_o is finite and non-zero, corresponding to $n = 1$

The eigenfunctions can be obtained as

$$\Psi_n = A\sin\frac{\sqrt{2mE}}{\hbar}x = A\sin\frac{n\pi}{L}x \tag{2.31}$$

The eigenfunction (2.31) has to be normalized to obtain the value of A. The normalization condition leads to

$$\int_0^L |\Psi_n|^2 dx = A^2\int_0^L \sin^2\left(\frac{n\pi}{L}x\right)dx = A^2\left(\frac{L}{2}\right) = 1 \tag{2.32}$$

Thus, $A^2\left(\frac{L}{2}\right) = 1$ or $A = \sqrt{\frac{2}{L}}$. The normalized wave function for an infinite square well now becomes

$$\Psi_n = \sqrt{\frac{2}{L}}\sin\frac{n\pi}{L}x \quad \text{where } n = 1, 2, 3, \ldots \tag{2.33}$$

The graphical representation of the wave function and the probability distribution for the first three energy states of a particle trapped in an infinite square well is shown in Fig. 2.4. The odd n wave functions ($n = 1, 3, 5 \ldots$) are symmetric from the center of the box given as $\Psi(x) = \Psi(-x)$ while the even n wave functions ($n = 2, 4, 6 \ldots$) are antisymmetric such that $\Psi(x) = -\Psi(-x)$. The parity of symmetric wave functions is even, while it is odd for the antisymmetric wave functions. The probability densities are not constant throughout the length of the box, unlike a classical particle. The probability density is a maximum at the center of the box for the $n = 1$ state, while closely spaced probability distribution occurs with higher n. The probability then approaches the classical value as $n \to \infty$.

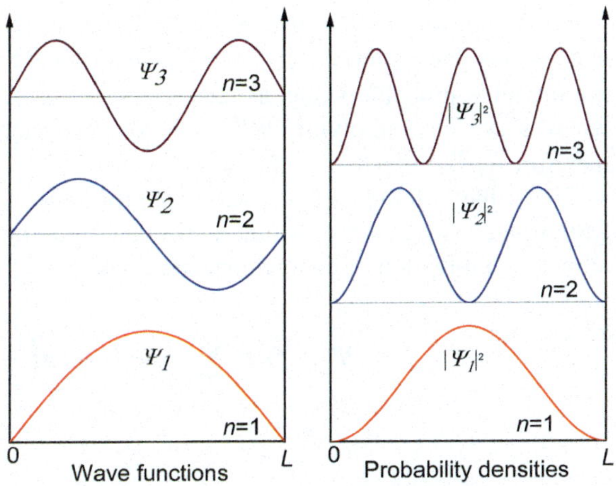

Fig. 2.4 The first three wave functions and probability densities for the particle in a one-dimensional box of infinite barrier potential, obtained from solving the Schrödinger equation

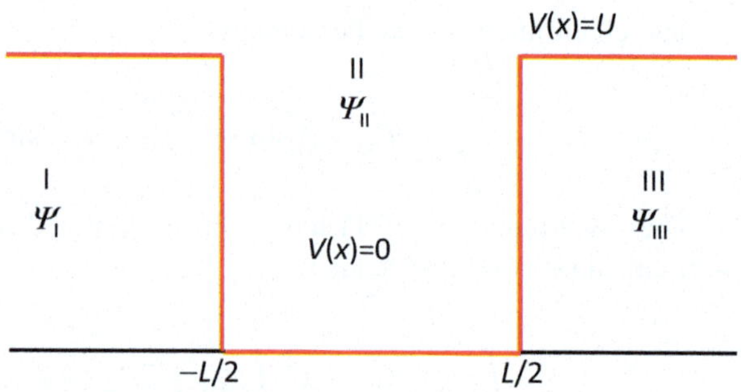

Fig. 2.5 The finite potential well considering $V(x) = 0$ in region II and $V(x) = U$ in regions I and III

The expectation values of position, momentum and square of the position and momentum for the nth state of the infinite potential well are given below.

$$\langle x\rangle = \int_0^L \Psi^* \hat{x} \Psi dx = \int_0^L \hat{x} |\Psi|^2 dx = \frac{L}{2} \tag{2.34}$$

$$\langle x^2\rangle = \int_0^L \Psi^* \hat{x}^2 \Psi dx = \int_0^L \hat{x}^2 |\Psi|^2 dx = L^2\left(\frac{1}{3} - \frac{1}{2(n\pi)^2}\right) \tag{2.35}$$

$$\langle p\rangle = \int_0^L \Psi^* \hat{p} \Psi dx = \int_0^L \hat{p} |\Psi|^2 dx = 0 \tag{2.36}$$

$$\langle p^2\rangle = \int_0^L \Psi^* \hat{p}^2 \Psi dx = \int_0^L \hat{p}^2 |\Psi|^2 dx = \frac{n^2h^2}{4L^2} = 2m\langle E\rangle = 2mE \tag{2.37}$$

It is interesting to note that while $\langle p\rangle = 0$ due to the averaging of positive and negative oscillating terms, $\langle p^2\rangle \neq 0$. The expectation value is the average value and not the most probable value, as we see for $n = 2$, $|\Psi|^2 = 0$ at *L*/2 from Fig. 2.4, whereas $\langle x\rangle = \frac{L}{2}$ implying $|\Psi|^2 \neq 0$. The expectation value of energy is *E*, and therefore, $\langle p^2\rangle = 2mE$.

Since the infinite square well was formulated using the time-independent Schrödinger equation, for the time-dependent equation, we can extend by considering a linear combination of energy eigenstates as

$$\Psi(x, t) = \sum_{n=1}^{\infty} a_n \Psi_n(x) \exp\left(-\frac{iE_n t}{\hbar}\right) \tag{2.38}$$

The expectation value of energy eigenvalue independent of time is given as

$$\langle E\rangle = \int_0^L \Psi^* \hat{E}\Psi dx = \int_0^L \hat{E}|\Psi|^2 dx = \sum_{n=1}^{\infty} E_n |a_n|^2 \quad (2.39)$$

A reasonable example of the infinite potential well can be an electron confined in a 1D nanowire of length, $L = 1$ nm. The lowest energy is then,

$$E_o = \frac{(6.626 \times 10^{-34})^2}{8 \times 9.1 \times 10^{-31} \times (1 \times 10^{-9})^2} \text{J} = 6.031 \times 10^{-20}\text{J} = 0.38 \text{ eV}.$$

This energy is larger than the thermal energy of a molecule at room temperature (~0.03 eV). As the wire length increases, $E \to 0$ approaching the classical limit, whereas the quantized energy is non-zero.

2.3.2 The Finite Potential Well

In the infinite potential barrier, we considered $V \to \infty$. Now, in the finite potential well, we shall consider the particle trapped in a barrier of height U with the following boundary conditions (Fig. 2.5).

$$V(x) = \begin{cases} 0 & \text{for } -\frac{L}{2} \le x \le \frac{L}{2} \\ U & \text{elsewhere} \end{cases} \quad (2.40)$$

Then, the steady state Schrödinger equation becomes

$$\frac{d^2\Psi}{dx^2} + \frac{2m(E-U)}{\hbar^2}\Psi = 0 \quad (2.41)$$

The differential equation can be of the form,

$$\frac{d^2\Psi}{dx^2} - k^2\Psi = 0 \quad \text{where } k = \frac{\sqrt{2m(U-E)}}{\hbar} \quad (2.42)$$

The solutions to (2.41) have real exponents in regions I and III as $U > E$

$$\Psi_{\text{I}} = Ce^{kx} + De^{-kx} \quad (2.43)$$

$$\Psi_{\text{III}} = Ge^{kx} + Fe^{-kx} \quad (2.44)$$

As $e^{\pm kx} \rightarrow \pm\infty$ when $x \rightarrow \pm\infty$, D and G must be 0 satisfying the finite wave function characteristics and, therefore,

$$\Psi_{\mathrm{I}} = Ce^{kx} \tag{2.45a}$$

$$\Psi_{\mathrm{III}} = Fe^{-kx} \tag{2.45b}$$

For region II, the solution is similar to infinite potential well giving,

$$\Psi_{\mathrm{II}} = A \sin \frac{\sqrt{2mE}}{\hbar} x + B \cos \frac{\sqrt{2mE}}{\hbar} x \tag{2.46}$$

Unlike the impermeable barrier, the wave functions penetrate regions I and III, and the continuous characteristics of the wave function demand that $\Psi_{\mathrm{II}} = C$ at $x = -L/2$ so that $\Psi_{\mathrm{I}} = \Psi_{\mathrm{II}}$, $\Psi_{\mathrm{II}} = F$ at $x = L/2$ and $\Psi_{\mathrm{II}} = \Psi_{\mathrm{III}}$ at the boundaries. Applying the boundary conditions and solving the differential equation gives (Zettili 2001)

$$k_2 \cot\left(\frac{k_2 L}{2}\right) = -k_1 \quad \text{and} \tag{2.47}$$

$$k_2 \tan\left(\frac{k_2 L}{2}\right) = k_1 \tag{2.48}$$

where $k_1 = \frac{\sqrt{2m(U-E)}}{\hbar}$ and $k_2 = \frac{\sqrt{2mE}}{\hbar}$.

$$\text{Using } \alpha_n = k_2 \frac{L}{2} \quad \text{and} \quad R = \frac{\sqrt{2mU}}{\hbar} \frac{L}{2} \tag{2.49}$$

Eq. (2.47) and (2.48) are written as

$$-\alpha_n \cot \alpha_n = \sqrt{R^2 - \alpha_n^2} \quad \text{odd parity solution} \tag{2.50}$$

$$\alpha_n \tan \alpha_n = \sqrt{R^2 - \alpha_n^2} \quad \text{even parity solution} \tag{2.51}$$

The general solutions can be obtained by numerically solving the equations. We get the energy inside the well similar to the infinite potential well case assuming perfect boundaries. The wave functions for the finite potential well are represented in Fig. 2.6. The wave functions extend into regions I and III, resulting in finite probabilities even beyond the boundaries $\pm L/2$. This behavior is known as the tunneling effect or barrier penetration. The energy eigenvalues could be obtained from the graphical solutions, where the circle given by $\eta = \sqrt{R^2 - \alpha_n^2}$ term intersects, as shown in Fig. 2.7.

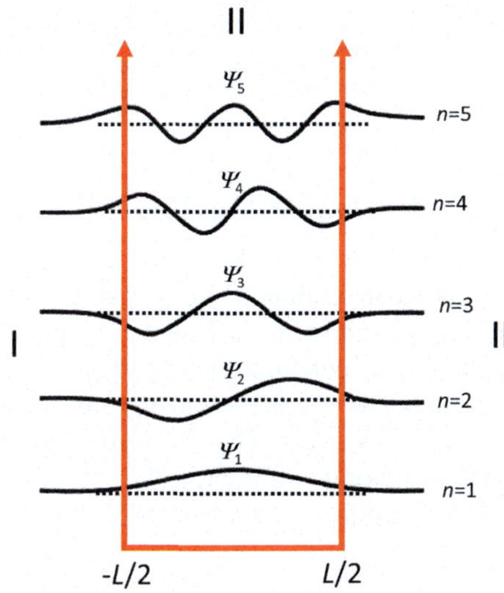

Fig. 2.6 The wave functions corresponding to the finite potential well case of the Schrödinger equation. The wave functions extend into regions I and III

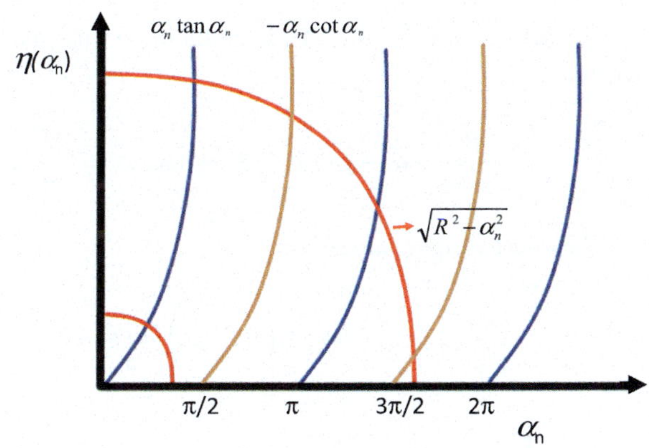

Fig. 2.7 The graphical solution for the finite potential well case of the Schrödinger equation. The intersection points of the curves determine the energy eigenvalues

From (2.49), we obtain

$$E_n = \frac{2\hbar^2\alpha_n^2}{mL^2} \tag{2.52}$$

$$R = \sqrt{\frac{mU}{2}}\frac{L}{\hbar} \tag{2.53}$$

The number of bound states depends on the radius, *R*, and hence the depth of the well, *U*, in contrast to the infinite bound states for the infinite potential well. When $\alpha_n = n\pi/2$, and the barrier height $V(x) \to \infty$, (2.52) becomes similar to (2.30) given by

$$E_n = \frac{n^2 h^2}{8mL^2}.$$

2.3.3 Harmonic Oscillator

The classical oscillator can be considered as a mass-spring system, as shown in Fig. 2.8, with a spring constant k, displaced to x from its equilibrium position that would give a potential energy of $U = \frac{1}{2}kx^2 = \frac{1}{2}m\omega^2 x^2$. It is evident that the total energy is a continuum of values according to the displacement x given by $E = \frac{1}{2}mv^2 + \frac{1}{2}kx^2$. The quantized energies obtained by solving the Schrödinger equation are quite different.

The Schrödinger equation for a 1D harmonic oscillator is given as

$$\frac{d^2\Psi}{dx^2} + \frac{2m\left(E - \frac{1}{2}kx^2\right)}{\hbar^2}\Psi = 0 \tag{2.54}$$

The differential Eq. (2.54), introducing a dimensionless variable, y or β, can be written as

$$\frac{d^2\Psi}{dy^2} + \left(\alpha - \beta^2\right)\Psi = 0, \text{ where } \alpha = \frac{2E}{h\nu} \text{ and } \beta = \sqrt{\frac{2\pi m v}{\hbar}}x.$$

The energy eigenvalues are

$$E_n = \left(n + \frac{1}{2}\right)\hbar\omega = (2n+1)\frac{\hbar\omega}{2} \tag{2.55}$$

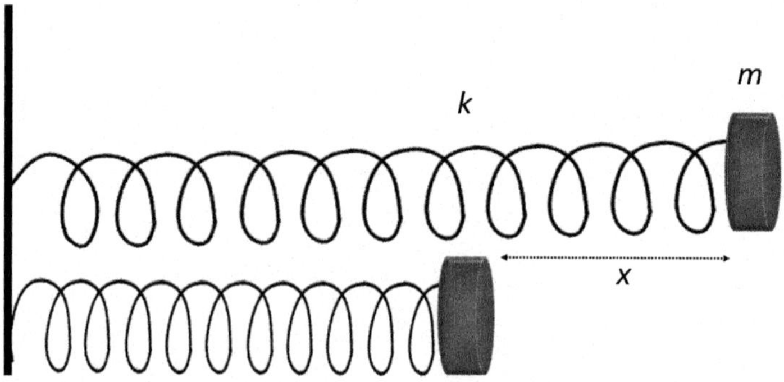

Fig. 2.8 The potential energy of a stretched or compressed spring with a spring constant k is given by $U = \frac{1}{2}kx^2$, where x is the displacement from its equilibrium position

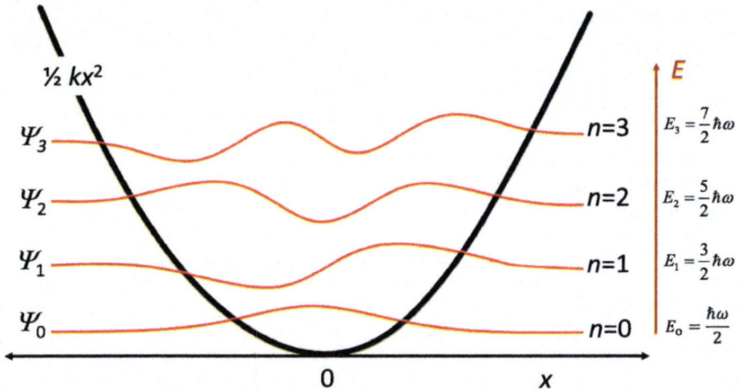

Fig. 2.9 The wave functions of a quantum harmonic oscillator. The energy levels are given by $E_n = \frac{(2n+1)\hbar\omega}{2}$ where the lowest energy is non-zero and given by $E_o = \frac{\hbar\omega}{2}$

where $n = 0, 1, 2 \ldots$

From (2.55), it could be understood that the *zero-point energy* corresponding to the ground state energy with $n = 0$ is non-zero, resulting in $E_o = \frac{1}{2}\hbar\omega$. The Schrödinger equation for the quantum harmonic oscillator can be solved using the series method to obtain a set of Hermite polynomials that produces wave functions as shown in Fig. 2.9. The energy levels are separated equally by $\hbar\omega$. The quantization of lattice vibrations leading to *phonon* energy levels is an important application of quantum harmonic oscillators.

The general formula for the normalized wave function is given as (Merzbacher 1998)

$$\Psi_n(x) = \left(\frac{m\omega}{\pi\hbar}\right)^{\frac{1}{4}} \frac{1}{\sqrt{2^n n!}} \exp\left(-\frac{m\omega x^2}{2\hbar}\right) H_n\left(\left(\frac{m\omega}{\hbar}\right)^{\frac{1}{2}} x\right) \tag{2.56}$$

where the Hermite polynomials as a general function of x are given by

$$H_n(x) = (-1)^n \exp\left(x^2\right) \frac{d^n}{dx^n} \left\{\exp\left(-x^2\right)\right\}.$$

The wave functions, as shown in Fig. 2.9, extend outside the region given by the potential 1/2 kx^2 resulting in a finite probability of finding the particle beyond the classically forbidden region. The probability densities corresponding to the ground ($n = 0$) and excited, $n = 3$ states are depicted in Fig. 2.10. The classical probability is limited to the boundaries set by the value of x, which is maximum at the boundaries due to the turning points. In contrast, the quantum probabilities extend into the classically forbidden region. As the n value increases, the mean value of the quantum probability density approaches the classical value.

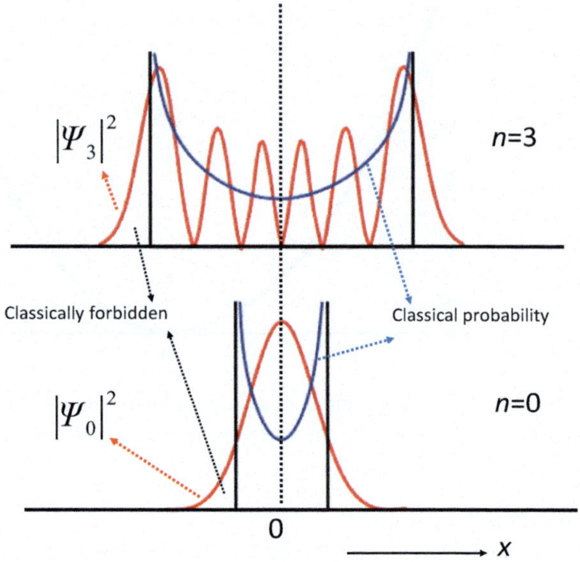

Fig. 2.10 The probability densities corresponding to $n = 0$ and $n = 3$ states of the quantum harmonic oscillator. The probability density extending beyond the classically forbidden region arises from the extension of the wave function in the region

Ladder Operator

The Hamiltonian for the harmonic oscillator can be represented in terms of the non-Hermitian ladder operators, a_+ and a_- as

$$H = \left(a_+a_- + \frac{1}{2}\right)\hbar\omega \tag{2.57a}$$

$$H = \left(a_-a_+ - \frac{1}{2}\right)\hbar\omega \tag{2.57b}$$

$$H = \left(\hat{N} + \frac{1}{2}\right)\hbar\omega \tag{2.57c}$$

The lowering or annihilation operator a_- is given by

$$a_- = \frac{1}{\sqrt{2\hbar m\omega}}(m\omega\hat{x} + i\,\hat{p}) \tag{2.58a}$$

and a_+ is the raising or creation operator (if a_- is a then a_+ is the adjoint represented as $a^\dagger$) given by

$$a_+ = \frac{1}{\sqrt{2\hbar m\omega}}(m\omega\hat{x} - i\,\hat{p}) \tag{2.58b}$$

The Hermitian number operator $\hat{N}$, also known as *phonon number operator,* is

$\hat{N} = a_+ a_-$. The eigenfunctions for the raising and lowering operators are

$$a_+ \Psi_n = \sqrt{n+1}\Psi_{n+1} \tag{2.58c}$$

$$a_- \Psi_n = \sqrt{n}\Psi_{n-1} \tag{2.58d}$$

In terms of eigenstates

$$\hat{N}|n\rangle = n|n\rangle$$

$$\hat{H}|n\rangle = E_n|n\rangle$$

$$a_+|n\rangle = \sqrt{n+1}|n+1\rangle$$

$$a_-|n\rangle = \sqrt{n}|n-1\rangle$$

The canonical commutation relations are

$$[a_-, a_+] = 1$$

$$\left[\hat{N}, a_-\right] = -a_-$$

$$\left[\hat{N}, a_+\right] = a_+$$

The lowest energy *zero-phonon state* $|0\rangle$ has the zero-point energy $\frac{\hbar\omega}{2}$ which could be increased in steps of $\hbar\omega$ by the raising operator, resulting in the creation of phonons, as represented in Fig. 2.11.

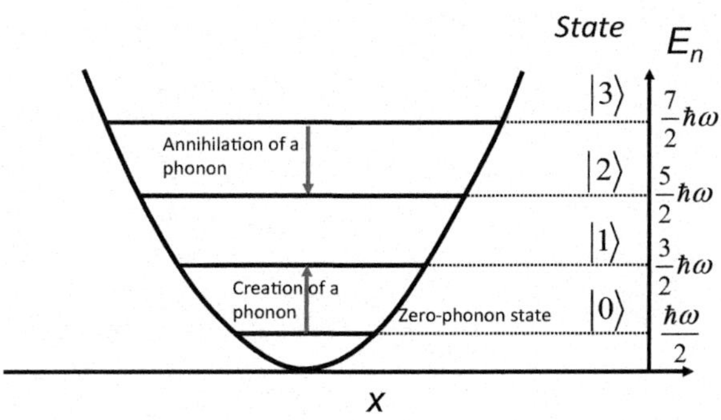

Fig. 2.11 The quantized phonon states in a harmonic oscillator. The zero-phonon state has the energy given by $\frac{\hbar\omega}{2}$. The transition to a higher energy state by one level results in the creation of a phonon, while the downward transition corresponds to the annihilation of a phonon

2.3.4 Schrödinger Equation in 3D

In spherical coordinates, the three-dimensional space is in terms of radial distance (r), polar angle (θ) and azimuthal angle (φ), as represented in Fig. 2.12.

The spherical coordinates are related to the Cartesian coordinates using the relations

$$\begin{aligned} x &= r\sin\theta\cos\varphi \\ y &= r\sin\theta\sin\varphi \\ z &= r\cos\theta \end{aligned} \tag{2.59}$$

$$\begin{aligned} r &= \sqrt{x^2+y^2+z^2} \\ \theta &= \arctan\left(\frac{\sqrt{x^2+y^2}}{z}\right) \\ \varphi &= \arctan\left(\frac{y}{x}\right) \end{aligned} \tag{2.60}$$

The (r, θ, φ) ranges are specified as

$$r = 0 \to \infty, \theta = 0 \to \pi \text{ and } \varphi = 0 \to 2\pi$$

The volume element is given by

$$dV = r^2 \sin\theta d\theta d\varphi dr$$

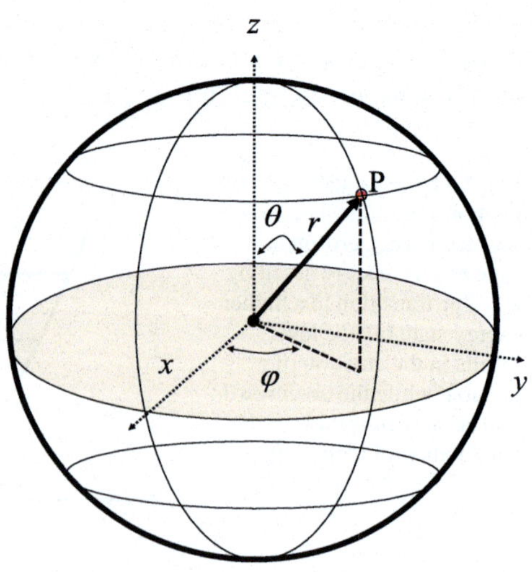

Fig. 2.12 The spherical coordinate system where a point P can be represented in terms of radial distance r, polar angle θ and azimuthal angle φ

The time-independent Schrödinger equation in 3D is written as

$$E\Psi = -\frac{\hbar^2}{2m}\nabla^2\Psi + V\Psi \tag{2.61}$$

where ∇^2 is the Laplacian given as

$$\nabla^2 = \frac{\partial^2}{\partial x^2} + \frac{\partial^2}{\partial y^2} + \frac{\partial^2}{\partial z^2} \tag{2.62}$$

In spherical coordinates

$$\nabla^2 = \frac{1}{r^2}\frac{\partial}{\partial r}\left(r^2\frac{\partial}{\partial r}\right) + \frac{1}{r^2\sin\theta}\frac{\partial}{\partial\theta}\left(\sin\theta\frac{\partial}{\partial\theta}\right) + \frac{1}{r^2\sin^2\theta}\left(\frac{\partial^2}{\partial\varphi^2}\right) \tag{2.63}$$

The time-independent Schrödinger equation, using the Laplacian becomes

$$\begin{aligned} -\frac{\hbar^2}{2m}\left\{\frac{1}{r^2}\frac{\partial}{\partial r}\left(r^2\frac{\partial}{\partial r}\right) + \frac{1}{r^2\sin\theta}\frac{\partial}{\partial\theta}\left(\sin\theta\frac{\partial}{\partial\theta}\right)\right. \\ \left. + \frac{1}{r^2\sin^2\theta}\left(\frac{\partial^2}{\partial\varphi^2}\right)\right\}\Psi + V\Psi = E\Psi \end{aligned} \tag{2.64}$$

2.3.5 *Wave Function in Spherical Coordinates*

The time-independent Schrödinger equation in spherical coordinates (2.64) can be separated into radial and angular parts given as

$$\Psi(r,\theta,\varphi) = R_{nl}(r)Y_{lm}(\theta,\varphi) \tag{2.65}$$

where $Y_{lm}(\theta,\varphi)$ is the spherical harmonics of the angular part. Using (2.65), dividing by the radial and angular parts and simplifying, we get (Griffiths 2005; Arfken and Weber 2005),

$$\frac{1}{R}\frac{d}{dr}\left(r^2\frac{dR}{dr}\right) - \frac{2mr^2}{\hbar^2}(V(r) - E) = l(l+1) \tag{2.66}$$

$$\frac{1}{Y}\left\{\frac{1}{\sin\theta}\frac{\partial}{\partial\theta}\left(\sin\theta\frac{\partial Y}{\partial\theta}\right) + \frac{1}{\sin^2\theta}\left(\frac{\partial^2 Y}{\partial\varphi^2}\right)\right\} = -l(l+1) \tag{2.67}$$

Each side of the separated radial (2.66) and angular (2.67) variables has a constant given by $l(l+1)$. Further simplification to obtain a general solution of the differential equation gives

$$\sin\theta\frac{\partial}{\partial\theta}\left(\sin\theta\frac{\partial Y}{\partial\theta}\right)+\left(\frac{\partial^2 Y}{\partial\varphi^2}\right)=-l(l+1)\sin^2\theta Y \tag{2.68}$$

Using $Y=\Theta\Phi$ we get,

$$\frac{1}{\Theta}\left\{\sin\theta\frac{\partial}{\partial\theta}\left(\sin\theta\frac{\partial\Theta}{\partial\theta}\right)\right\}+l(l+1)\sin^2\theta+\frac{1}{\Phi}\frac{d^2\Phi}{d\varphi^2}=0 \tag{2.69}$$

$$\frac{1}{\Theta}\left\{\sin\theta\frac{\partial}{\partial\theta}\left(\sin\theta\frac{\partial\Theta}{\partial\theta}\right)\right\}+l(l+1)\sin^2\theta=m^2 \tag{2.70}$$

$$\frac{1}{\Phi}\frac{d^2\Phi}{d\varphi^2}=-m^2 \tag{2.71}$$

$$\frac{d^2\Phi}{d\varphi^2}=-m^2\Phi \tag{2.72}$$

$$\Phi(\varphi)=\exp(im\varphi)=\exp(im(\varphi+2\pi))=\exp(2\pi im)=1$$
$$\text{and } m=0,\pm1,\pm2,\ldots$$

The solution for the angular component Θ is

$$\Theta(\theta)=AP_{lm}(\cos\theta) \tag{2.73}$$

where P_{lm} is the associated Legendre function given for an arbitrary x as

$$P_{lm}(x)=\left(1-x^2\right)^{\frac{|m|}{2}}\left(\frac{d}{dx}\right)^{|m|}P_l(x) \tag{2.74}$$

The associated Legendre function has the form shown in Fig. 2.13. The function is invariant for $\pm\,m$ or $+\,l$ to $-\,(l+1)$ and reaches zero if $|m|>l$. The function turns into a Legendre polynomial $P_l(x)$ when $m=0$ and l is an integer. The function becomes $P_{00}(x)=1$ for $l, m=0$ and $P_{10}(x)=x$ when $l=1$ and $m=0$.

Normalizing the wave function in the volume element $dV=r^2\sin\theta dr d\theta d\varphi$, the angular part of the wave function given by the spherical harmonics is

$$Y_{lm}(\theta,\varphi)=a\sqrt{\frac{(2l+1)(l-|m|)!}{4\pi(l+|m|)!}}\exp(im\varphi)P_{lm}(\cos\theta) \tag{2.75}$$

where $a=\begin{cases}(-1)^m & \text{for } m>0\\ 1 \quad \text{for } m\le 0\end{cases}$.

Figure 2.14 shows a few of the spherical harmonics $Y_{lm}(\theta,\varphi)$ of various degree l and order m, where the number of lobes increases with l and m. The l and m are also known as the azimuthal quantum number and magnetic quantum number,

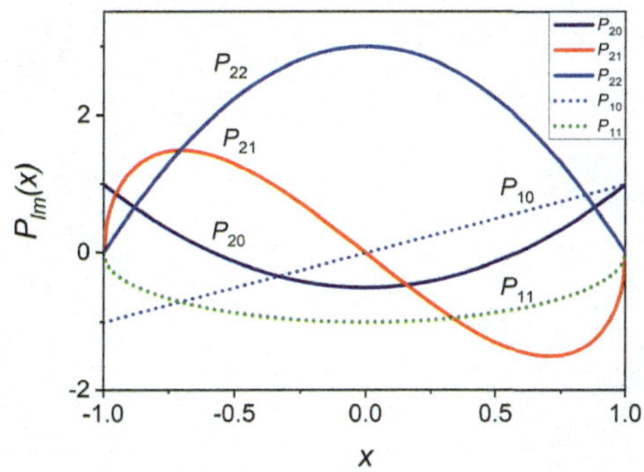

Fig. 2.13 The associated Legendre function $P_{lm}(x)$ for $l = 1, 2$ and various m

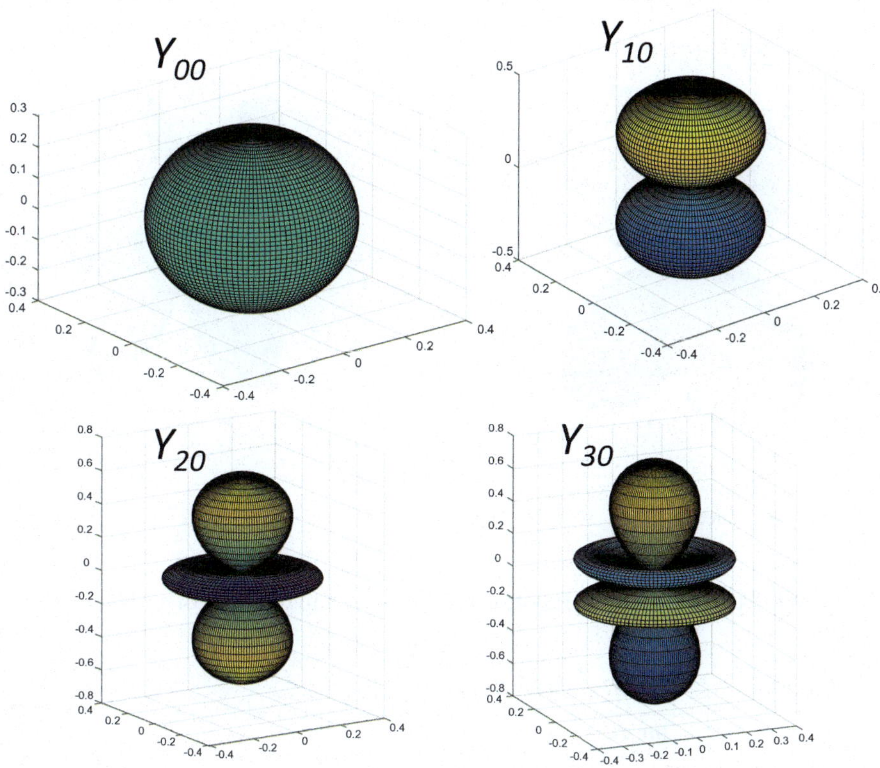

Fig. 2.14 Spherical harmonics $Y_{lm}(\theta, \varphi)$ for various l and m. Simulated in Matlab

respectively. The spherical harmonic equations for a few of the l and m are listed in Table 2.2, corresponding to various orbitals. The s orbital is spherically symmetric, whereas the shape evolution with higher orbitals could be understood from (2.75).

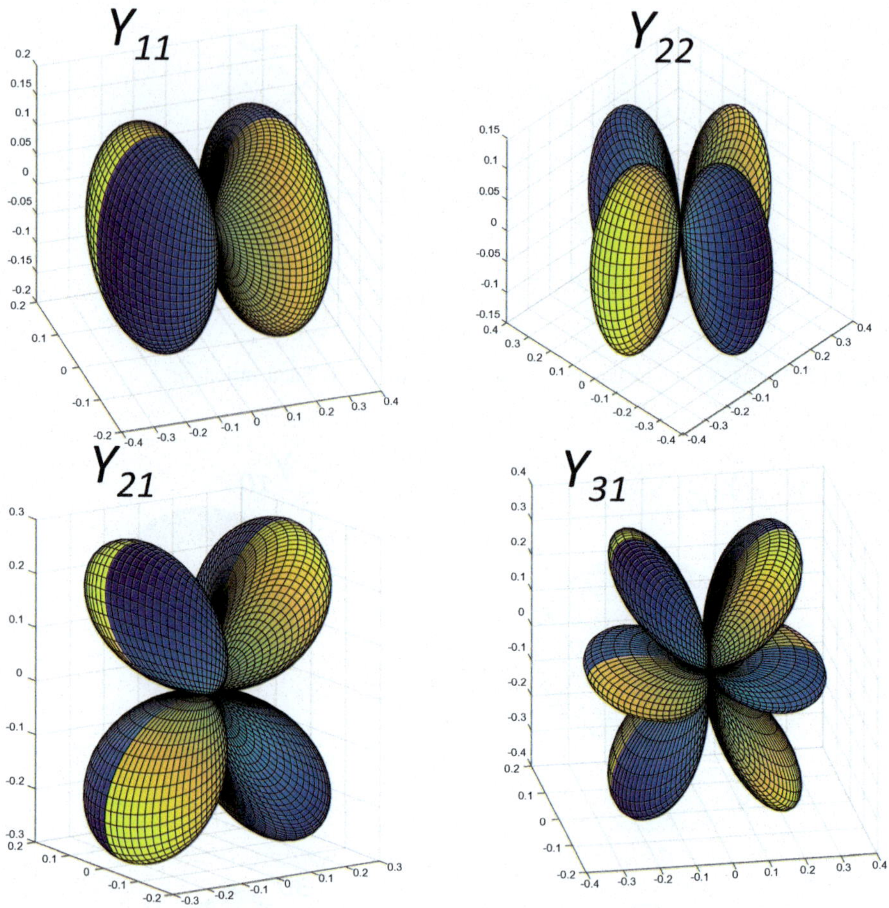

Fig. 2.14 (continued)

The radial part of the wave function can be written in the form of a *reduced radial equation* as

$$\left[-\frac{\hbar^2}{2m}\frac{d^2}{dr^2}+\left\{V+\frac{\hbar^2}{2m}\frac{l(l+1)}{r^2}\right\}\right]u(r)=Eu(r) \tag{2.76}$$

where $u(r) = rR(r)$. The effective potential is now

$$V_{\text{eff}} = V + \frac{\hbar^2}{2m}\frac{l(l+1)}{r^2} \tag{2.77}$$

Table 2.2 The spherical harmonics for various l and m

Orbital	l	m	$Y_{lm}(\theta, \varphi)$
s	0	0	$\frac{1}{2\sqrt{\pi}}$
p_z	1	0	$\frac{1}{2}\sqrt{\frac{3}{\pi}}\cos\theta$
p_x	1	1	$-\frac{1}{2}\sqrt{\frac{3}{2\pi}}\sin\theta\exp(i\varphi)$
p_y	1	−1	$\frac{1}{2}\sqrt{\frac{3}{2\pi}}\sin\theta\exp(-i\varphi)$
d_{z^2}	2	0	$\frac{1}{4}\sqrt{\frac{5}{\pi}}(3\cos^2\theta - 1)$
d_{xz}	2	1	$-\frac{1}{2}\sqrt{\frac{15}{2\pi}}\sin\theta\cos\theta\exp(i\varphi)$
$d_{(x^2-y^2)}$	2	2	$\frac{1}{4}\sqrt{\frac{15}{2\pi}}\sin^2\theta\exp(2i\varphi)$
f_{z^3}	3	0	$\frac{1}{4}\sqrt{\frac{7}{\pi}}(5\cos^3\theta - 3\cos\theta)$
f_{xz^2}	3	1	$-\frac{1}{8}\sqrt{\frac{21}{\pi}}(5\cos^2\theta - 1)\sin\theta\exp(i\varphi)$
$f_{z(x^2-y^2)}$	3	2	$\frac{1}{4}\sqrt{\frac{105}{2\pi}}\sin^2\theta\cos\theta\exp(2i\varphi)$

For $V = 0$, within a boundary L, (2.76) becomes

$$\frac{d^2u(r)}{dr^2} = \left\{\frac{l(l+1)}{r^2} - k^2\right\}u(r) \text{ where } k = \frac{\sqrt{2mE}}{\hbar} \tag{2.78}$$

When $l = 0$, the solution of the wave function becomes

$$u(r) = A\sin(kr) + B\cos(kr) \tag{2.79}$$

and we get back the energy eigenvalue similar to the infinite square well potential case as

$$E_n = \frac{n^2h^2}{8mL^2} \text{ where } n = 1, 2, 3, \ldots \tag{2.80}$$

Using the boundary conditions with $B = 0$ in (2.79), the normalized wave function corresponding to $l = 0$ in spherical coordinates becomes

$$\Psi_{nlm} = R_{n0}(r)Y_{00}(\theta, \varphi) = \frac{u(r)}{r}\frac{1}{\sqrt{4\pi}} = \frac{1}{\sqrt{2\pi L}}\frac{\sin(\frac{n\pi r}{L})}{r} \tag{2.81}$$

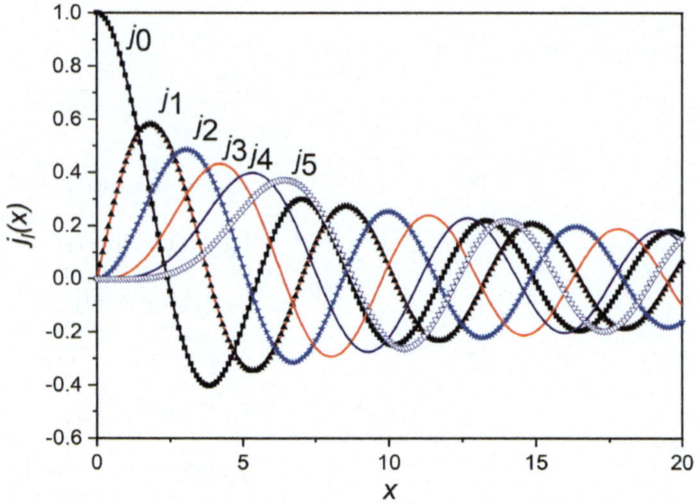

Fig. 2.15 Spherical Bessel function $j_l(x)$ of the first kind for $l \leq 5$. Simulated in Matlab

On the other hand when $l \neq 0$, the general solution of (2.78) is (Arfken and Weber 2005)

$$u(r) = Arj_l(kr) + Brn_l(kr) \tag{2.82}$$

where $j_l(kr)$ and $n_l(kr)$ are the spherical Bessel and spherical Neumann functions, respectively. The spherical Bessel function as a function of x is of the form shown in Fig. 2.15, which is oscillatory. The spherical Neumann function diverges at the origin and is not an acceptable solution. Therefore, setting $B = 0$, we get the eigenfunction for a particle in spherical coordinates as

$$\Psi(r, \theta, \varphi) = Arj_l(kr)Y_{lm}(\theta, \varphi) \tag{2.83}$$

2.4 The Hydrogen Atom

In a hydrogen atom, there is Coulomb interaction between the lone electron and proton. Using the Coulomb potential in the radial wave function gives

$$-\frac{\hbar^2}{2m}\frac{d^2u}{dr^2} + \left\{\frac{-e^2}{4\pi\varepsilon_0 r} + \frac{\hbar^2}{2m}\frac{l(l+1)}{r^2}\right\}u = Eu \tag{2.84}$$

After considerable simplification and solving the equation, we get the eigenvalues as (Griffiths 2005)

$$E_n = -\frac{1}{n^2}\frac{m}{2\hbar^2}\left(\frac{e^2}{4\pi\varepsilon_0}\right)^2 \quad \text{where } n = 1, 2, 3, \ldots \tag{2.85}$$

We also get

$$a_0 = \frac{4\pi\varepsilon_0\hbar^2}{me^2} = 0.529 \times 10^{-10}\,\text{m}, \quad \text{which is the Bohr radius.}$$

Setting $n = 1$, the lowest energy is obtained as $E_1 = -13.6$ eV, which is the ground state binding energy of the hydrogen atom.

For a transition from higher energy state n' to lower n, we get the spectral series of the hydrogen atom with the wavenumber given as

$$\frac{1}{\lambda} = \frac{k^2me^4}{4\pi c\hbar^3}\left(\frac{1}{n^2} - \frac{1}{n'^2}\right) = R_\infty\left(\frac{1}{n^2} - \frac{1}{n'^2}\right) \tag{2.86}$$

where R_∞ is the Rydberg constant, m is the mass of the electron and k is the Coulomb constant. In terms of energy, the Rydberg constant is, $\text{Ry} = R_\infty hc$.

The normalized hydrogen wave function, solving (2.84), is given as

$$\psi_{nlm}(\theta, \varphi) = \sqrt{\left(\frac{2}{na}\right)^3 \frac{(n-l-1)!}{2n((n+l)!)^3}}$$
$$\exp\left(\frac{-r}{na}\right)\left(\frac{2r}{na}\right)^l \left(L^{2l+1}_{n-l-1}\left(\frac{2r}{na}\right)\right) Y_{lm}(\theta, \varphi) \tag{2.87}$$

where $L^{2l+1}_{n-l-1}\left(\frac{2r}{na}\right)$ is the associated Laguerre polynomial.

The Laguerre polynomials are defined by the Rodrigues formula (for an arbitrary x) as

$$L_n(x) = \frac{e^x}{n!}\frac{d^n}{dx^n}(e^{-x}x^n) = \frac{1}{n!}\left(\frac{d}{dx} - 1\right)^n x^n \tag{2.88}$$

Figure 2.16 shows the Laguerre polynomials $L_n(x)$ up to $n = 5$. The associated Laguerre polynomials $L^l_n(x)$ where l is an integer, is given by

$$L^l_n(x) = \frac{e^x x^{-l}}{n!}\frac{d^n}{dx^n}(e^{-x}x^{n+l}) \tag{2.89}$$

The radial wave function part of (2.87) is

$$R_{nl}(r) = \sqrt{\left(\frac{2}{na}\right)^3 \frac{(n-l-1)!}{2n((n+l)!)^3}} \exp\left(\frac{-r}{na}\right)\left(\frac{2r}{na}\right)^l \left(L^{2l+1}_{n-l-1}\left(\frac{2r}{na}\right)\right) \tag{2.90}$$

Using the associated Laguerre polynomials (2.89), and defining the scaling factor, $a = a_0 = a_o$, the radial wave functions (2.90) take the form given in Table 2.3.

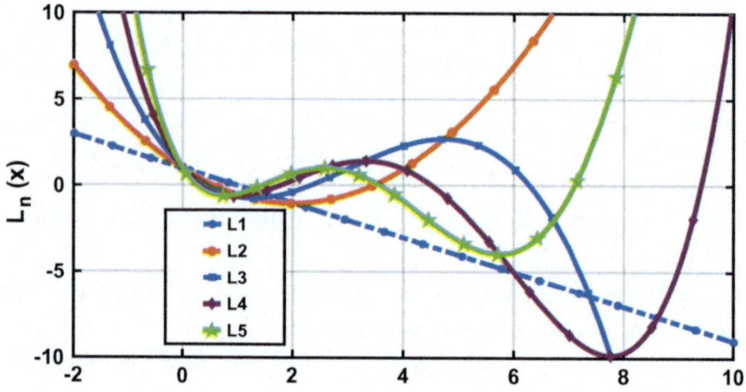

Fig. 2.16 Laguerre polynomials of order up to 5

Table 2.3 Radial wave functions for various n and l

State	n	l	$R_{nl}(r)$
1s	1	0	$2\left(\frac{1}{a_o}\right)^{\frac{3}{2}}\exp\left(-\frac{r}{a_o}\right)$
2s	2	0	$\left(\frac{1}{2a_o}\right)^{\frac{3}{2}}\left(2-\frac{r}{a_o}\right)\exp\left(-\frac{r}{2a_o}\right)$
2p	2	1	$\frac{1}{\sqrt{3}}\left(\frac{1}{2a_o}\right)^{\frac{3}{2}}\left(\frac{r}{a_o}\right)\exp\left(-\frac{r}{2a_o}\right)$
3s	3	0	$2\left(\frac{1}{3a_o}\right)^{\frac{3}{2}}\left(1-2\frac{r}{3a_o}+2\frac{r^2}{27a_o^2}\right)\exp\left(-\frac{r}{3a_o}\right)$
3p	3	1	$\frac{4\sqrt{2}}{9}\left(\frac{1}{3a_o}\right)^{\frac{3}{2}}\frac{r}{a_o}\left(1-\frac{r}{6a_o}\right)\exp\left(-\frac{r}{3a_o}\right)$
3d	3	2	$\frac{2\sqrt{10}}{135}\left(\frac{1}{3a_o}\right)^{\frac{3}{2}}\left(\frac{r}{a_o}\right)^2\exp\left(-\frac{r}{3a_o}\right)$
4s	4	0	$\frac{1}{4}\left(\frac{1}{a_o}\right)^{\frac{3}{2}}\left(1-\frac{3}{4}\frac{r}{a_o}+\frac{1}{8}\frac{r^2}{a_o^2}-\frac{1}{192}\frac{r^3}{a_o^3}\right)\exp\left(-\frac{r}{4a_o}\right)$

The radial wave functions $R_{nl}(r)$ of the hydrogen atom corresponding to a few quantum numbers are given in Fig. 2.17. The wave function has $n-1-l$ radial nodes. Thus, 3s level has 2 nodes, whereas the 1s level has no nodes. Also, the s wave function shows a maximum at $r=0$ and decreases with r, implying symmetry surrounding the origin.

The wave function, including the radial (2.90) and angular part (2.75), is thus given as in (2.65) as

$$\Psi_{nlm}(r,\theta,\varphi)=R_{nl}(r)Y_{lm}(\theta,\varphi) \tag{2.91}$$

Normalizing the wave function

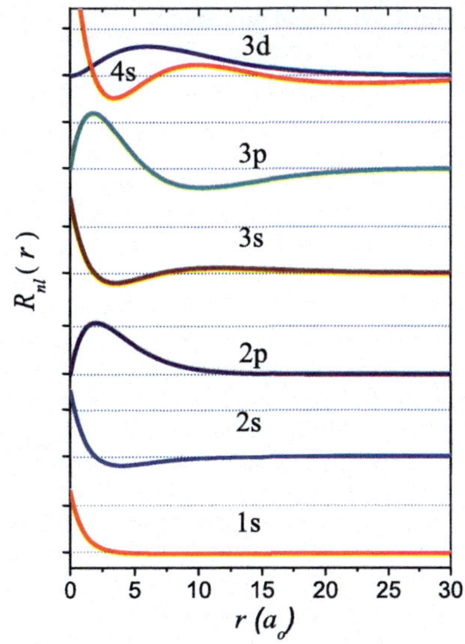

Fig. 2.17 The radial wave functions $R_{nl}(r)$ in terms of Bohr radius

$$\int \Psi^*_{nlm}\Psi_{nlm}dV = \int \Psi^*_{nlm}\Psi_{nlm}r^2 \sin\theta dr d\theta d\varphi = 1 \tag{2.92}$$

The probability of finding the electron in a hydrogen atom is given by

$$P(r)dr = r^2|R|^2 dr \int_0^{\pi} |\Theta|^2 \sin\theta d\theta \int_0^{2\pi} |\Phi|^2 d\varphi = r^2|R|^2 dr \tag{2.93}$$

The radial probability density is obtained in terms of the product of area term r^2, and the radial probability $|R_{nl}(r)|^2$. Therefore, the s wave function that shows maximum at $r = 0$ exhibits minimum probability density. The radial probability densities $r^2|R_{nl}(r)|^2$ of electrons up to 4s level are shown in Fig. 2.18. The 1s level has the peak of the radial probability density at a_o, increasing with n. Also, it is to be noted that the peak of r(2p) < r(2s) and r(3d) < r(3p) < r(3s). The average radius, from the expectation value <r> for the 1s shell, is equal to 3/2 a_o and increases to 6 a_o for 2s and 13.5 a_o for 3s.

The three-dimensional probability density of the hydrogen atom obtained by including the angular dependence of the wave functions is determined by the principal quantum number n, the azimuthal quantum number l and the magnetic quantum number m. The shape of the orbitals for various n, l and m values indicate the charge distribution of the electrons in the hydrogen atom, as represented in Fig. 2.19. The energy of the hydrogen atom depends on the n value, while the l and m values determine the shape. Unlike the hydrogen atom, the energy values for the multielectron system depend on both n and l. The probability densities are spherically symmetric for $l = 0$, dumbbell for $l = 1$ and clover leaf-like/torus for $l = 2$. The cross section of

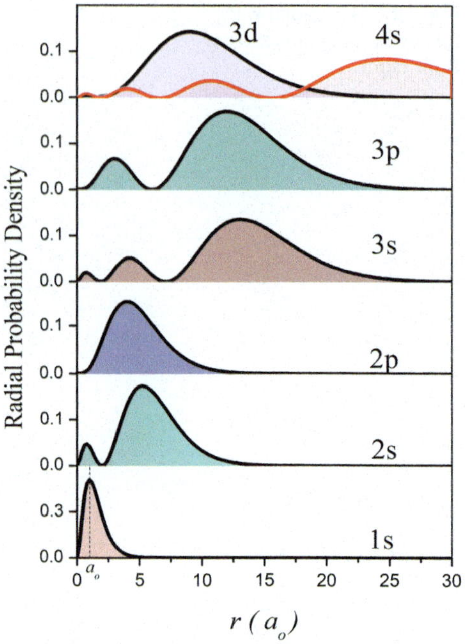

Fig. 2.18 Radial probability densities of hydrogen atom $r^2|R_{nl}(r)|^2$ for various orbitals

s orbital ($l = 0$) can be viewed as a shell-like shape with multiple shells on increasing n. The $l = 1$ takes the m_l values of $+1, 0$ and -1, giving p_x, p_z and p_y orientations due to the angular dependence for $l \neq 0$. The l values of 0, 1, 2, 3 and 4 are represented as s, p, d, f and g states, respectively.

2.5 Angular Momentum

2.5.1 Orbital Angular Momentum

The classical angular momentum is given by

$$L = r \times p = (yp_z - zp_y)\vec{i} + (zp_x - xp_z)\vec{j} + (xp_y - yp_x)\vec{k} \tag{2.94}$$

The components of L are

$$L_x = yp_z - zp_y;\ L_y = zp_x - xp_z;\ L_z = xp_y - yp_x \tag{2.95}$$

The Zeeman effect experiment in 1896, discovered by P. Zeeman, based on the splitting of spectral lines in the presence of a magnetic field, is associated with the quantized orbital angular momentum. Using the linear momentum operator $p_x = -i\hbar\frac{\partial}{\partial x}$, and other components, the angular momentum operators in quantum

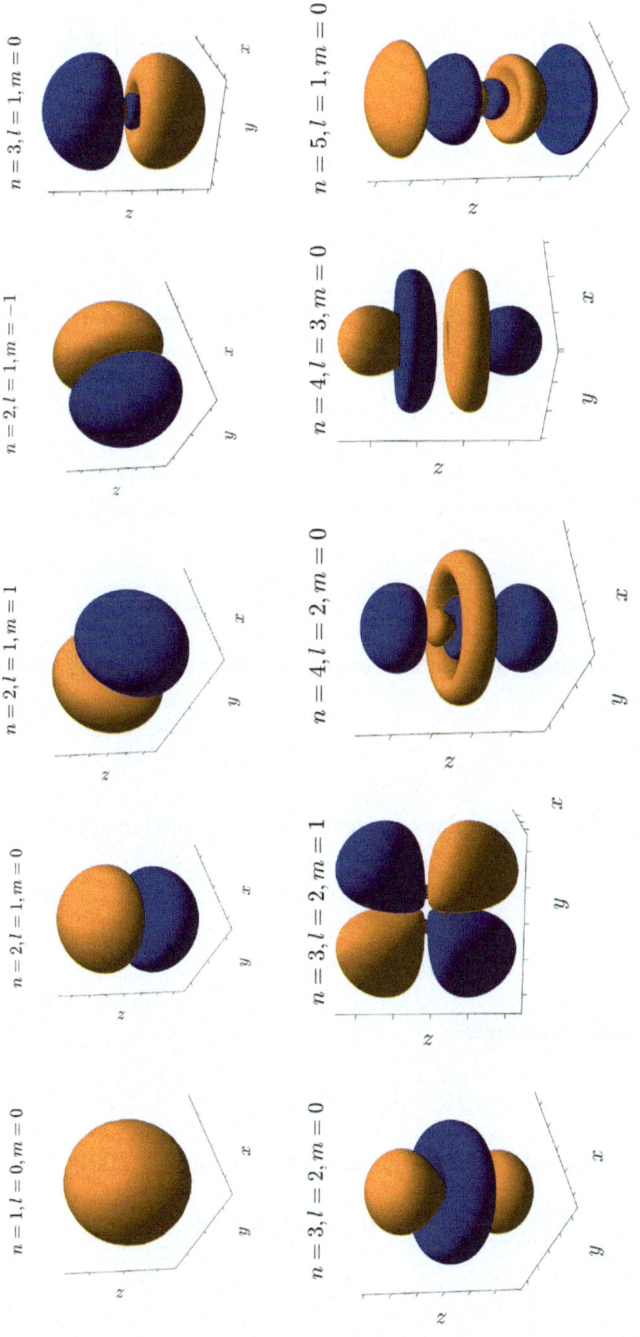

Fig. 2.19 The three-dimensional probability density plots of hydrogen atom for various *n*, *l* and *m* illustrates the shape of the charge distribution (Peter van Alem 2021)

mechanics are given as

$$L_x = -i\hbar\left(y\frac{\partial}{\partial z} - z\frac{\partial}{\partial y}\right);\ L_y = -i\hbar\left(z\frac{\partial}{\partial x} - x\frac{\partial}{\partial z}\right);\ L_z = -i\hbar\left(x\frac{\partial}{\partial y} - y\frac{\partial}{\partial x}\right) \tag{2.96}$$

Also, momentum and position operators are non-commutative such that $xp \neq px$ and we have the canonical commutation relation

$$[x, p] = i\hbar \tag{2.97}$$

Angular momentum, being an operator, follows the rules and regulations of quantum mechanics, such as

$$[L_x, L_y] = i\hbar L_z;\ [L_y, L_z] = i\hbar L_x;\ [L_z, L_x] = i\hbar L_y \tag{2.98}$$

The commutation relations for angular momentum are given by (2.98).

The total angular momentum given by $L^2 = L_x^2 + L_y^2 + L_z^2$ commute with its components such that

$$[L^2, L_x] = [L^2, L_y] = [L^2, L_z] = [L^2, L] = 0 \tag{2.99}$$

Quantization of angular momentum could be undertaken using the ladder operators defined as

$$L_+ = L_x + iL_y;\ L_- = L_x - iL_y \tag{2.100}$$

The angular momentum eigenstates or eigenkets are denoted in ket space as

$$\hat{L}^2|lm\rangle = \hbar^2 l(l+1)|lm\rangle \tag{2.101}$$

$$\hat{L}_z|lm\rangle = m_l\hbar|lm\rangle \tag{2.102}$$

and the ladder operator is given as

$$L_\pm|lm\rangle = \hbar\sqrt{l(l+1) - m(m\pm 1)}|l\ m\pm 1\rangle \tag{2.103}$$

In spherical harmonics, the eigenvalue equations are (Greiner 2001)

$$\hat{L}^2 Y_{lm}(\theta, \varphi) = \hbar^2 l(l+1) Y_{lm}(\theta, \varphi) \tag{2.104}$$

$$\hat{L}_z Y_{lm}(\theta, \varphi) = m_l\hbar Y_{lm}(\theta, \varphi) \tag{2.105}$$

where L^2 and L_z are given by the differential operators as

$$\hat{L}^2 = -\hbar^2\left\{\frac{1}{\sin\theta}\frac{\partial}{\partial\theta}\left(\sin\theta\frac{\partial}{\partial\theta}\right) + \frac{1}{\sin^2\theta}\frac{\partial^2}{\partial\varphi^2}\right\} \tag{2.106}$$

$$\hat{L}_z = -i\hbar\frac{\partial}{\partial\varphi} \tag{2.107}$$

The eigenvalues of L and its components are given as

$$\begin{aligned} &L^2 = \hbar^2 l(l+1) \quad \text{where } l = 0, 1, 2, \ldots \\ &L = \hbar\sqrt{l(l+1)} \\ &\text{and } L_z = m_l\hbar \text{ where } m_l = +l\ldots, 0, \ldots - l \end{aligned} \tag{2.108}$$

The number of m_l values is determined by l given as $2l+1$. Thus, $l = 1$ has 3 m_l components, $l = 2$ has 5, $l = 3$ has 7 and so on.

Figure 2.20 shows the orbital angular momentum components for $l = 1$ and 2, where the eigenvalues of L_z are represented in multiples of m_l along the z-axis. The magnitude of L(which precesses around z-axis) is given by the length of the arrow equal to the radius of the circle given as $|L| = \hbar\sqrt{1(1+1)} = \sqrt{2}\hbar$ for $l = 1$ state and $|L| = \hbar\sqrt{2(2+1)} = \sqrt{6}\hbar$ for $l = 2$ state. From Fig. 2.20, the z component is related to L by the angle θ as

$$\cos\theta = \frac{m_l}{\sqrt{l(l+1)}} \tag{2.109}$$

So far, the total wave function is represented in terms of the radial and angular components as

$$\Psi_{nlm}(r, \theta, \varphi) = R_{nl}(r)Y_{lm}(\theta, \varphi) \tag{2.110}$$

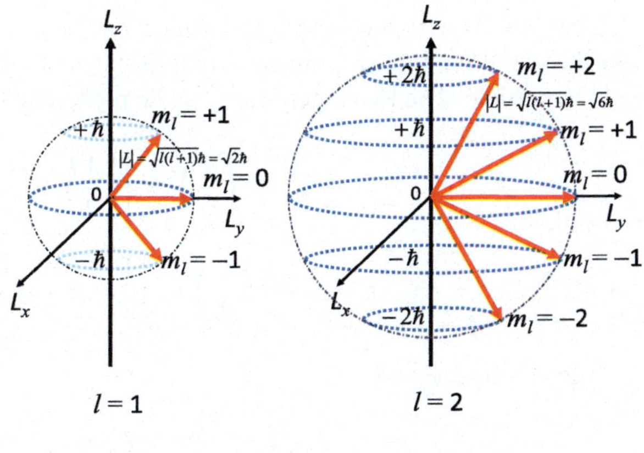

Fig. 2.20 The vector diagram for the angular momentum states $l = 1$ and 2

2.5.2 Spin Angular Momentum

The Stern–Gerlach experiment in 1922 and the proposal of the spin of the electron by Uhlenbeck-Goudsmit in 1925 established the spin angular momentum of the electron. When neutral silver atoms (configuration [Kr]$4d^{10}5s^1$) with a single outer electron were subjected to an inhomogeneous magnetic field, they were deflected equally, and the photograph of the spectrum showed two lines instead of the classically expected single line. The experiment relied on the interaction of the magnetic field with the electron's magnetic moment that deflects the atoms. Since the inner core electrons angular momenta cancel and the orbital angular momentum of the lone electron ($5s^1$) is zero, the splitting of the lines is due to the intrinsic spin of the electron. Now that the separation of the orbital angular momentum is quantized in multiples of $\hbar$, it is intuitive to assume $\pm\frac{\hbar}{2}$ for the quantized spin angular momentum so that the zero orbital angular momentum is satisfied by the addition of + 1/2 and − 1/2 spin values.

The spin angular momentum is intrinsic in nature without angular dependence eliminating $Y_{lm}(\theta, \varphi)$, whereas it has commutation relations similar to the orbital angular momentum given as

$$\left[S_x, S_y\right] = i\hbar S_z; \left[S_y, S_z\right] = i\hbar S_x; \left[S_z, S_x\right] = i\hbar S_y \tag{2.111}$$

The eigenvectors are

$$\hat{S}^2|sm\rangle = \hbar^2 s(s+1)|sm\rangle \tag{2.112}$$

$$\hat{S}_z|sm\rangle = m_s\hbar|sm\rangle \tag{2.113}$$

where $m_s = -s, -s+1, \ldots, +s$

$$S_\pm|sm\rangle = \hbar\sqrt{s(s+1) - m(m\pm 1)}|s\ m\pm 1\rangle \tag{2.114}$$

The wave function of a spin ½ fermion can be a superposition of up $|+½\rangle$ or $|\uparrow\rangle$ and down $|-½\rangle$ or $|\downarrow\rangle$ spin states. The spin states of $|s\ m\rangle$ are $|½\ ½\rangle$ and $|½\ -½\rangle$ for a spin ½ particle. The two eigenstates can be represented in terms of eigenspinor χ as

$$\chi_+ = \begin{pmatrix}1\\0\end{pmatrix} = \left|\frac{1}{2}\frac{1}{2}\right\rangle = |\uparrow\rangle \tag{2.115}$$

$$\chi_- = \begin{pmatrix}0\\1\end{pmatrix} = \left|\frac{1}{2}\ \frac{-1}{2}\right\rangle = |\downarrow\rangle \tag{2.116}$$

The eigenvalues are

$$S^2 = \hbar^2\left(\frac{1}{2}\right)\left(\frac{3}{2}\right) = \frac{3}{4}\hbar^2$$

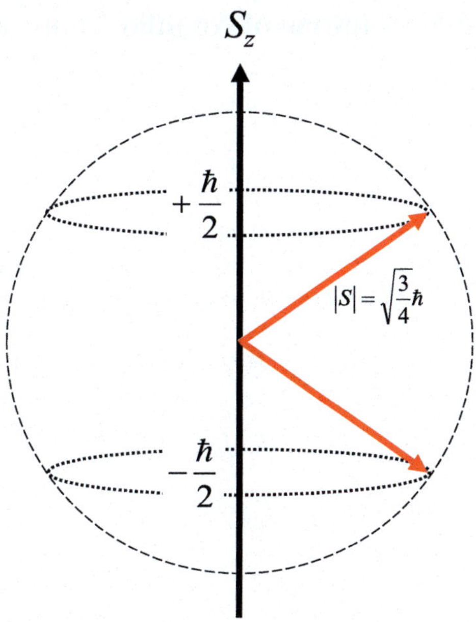

Fig. 2.21 The magnitude of spin angular momentum $|S| = \frac{\sqrt{3}}{2}\hbar$ is given by the vector equivalent to the radius of the circle. The possible S_z values are $+\frac{\hbar}{2}$ and $-\frac{\hbar}{2}$ represented along the z-axis for a spin ½ system

$$|S| = \frac{\sqrt{3}}{2}\hbar \tag{2.117}$$

and $S_z = \pm\frac{\hbar}{2}$.

The spin angular momentum is represented in the spin vector diagram shown in Fig. 2.21, where the magnitude of the vector is given by $|S| = \sqrt{\frac{3}{4}}\hbar$ and the z-component gives two values $+\frac{\hbar}{2}$ and $-\frac{\hbar}{2}$.

The spin operators can be represented by a 2×2 matrix σ as

$$\sigma_x = \begin{pmatrix} 0 & 1 \\ 1 & 0 \end{pmatrix}, \sigma_y = \begin{pmatrix} 0 & -i \\ i & 0 \end{pmatrix}, \sigma_z = \begin{pmatrix} 1 & 0 \\ 0 & -1 \end{pmatrix} \tag{2.118}$$

known as Pauli spin matrices. The total wave function now has both spatial that includes orbital angular momentum and spin parts given as

$$\Psi_{nlmm_s}(r, \theta, \varphi, s) = \Psi_{nlm}(r, \theta, \varphi)\chi_\pm \tag{2.119}$$

$$\Psi_{nlmm_s}(r, \theta, \varphi, s) = R_{nl}(r)Y_{lm}(\theta, \varphi)\chi_\pm \tag{2.120}$$

2.6 Addition of Angular Momenta

The addition of angular momenta shall be given by

$$\vec{J} = \vec{L} + \vec{S}, \vec{J} = \vec{J}_1 + \vec{J}_2 \text{ or } \vec{S} = \vec{S}_1 + \vec{S}_2 \tag{2.121}$$

The total angular momentum vector satisfies the commutation relations $[J_x, J_y] = i\hbar J_z$, $[J^2, J] = 0$ described by the $|jm\rangle$ eigenstates. The ladder operator changes the state similar to that of the orbital angular momentum given as

$$J_+|jm\rangle = \hbar\sqrt{j(j+1) - m(m+1)}|j\ m+1\rangle \tag{2.122}$$

$$J_-|jm\rangle = \hbar\sqrt{j(j+1) - m(m-1)}|j\ m-1\rangle \tag{2.123}$$

The orthonormal eigenkets are

$$\hat{J}^2|jm\rangle = \hbar^2 j(j+1)|jm\rangle \tag{2.124}$$

$$\hat{J}_z|jm\rangle = m_j\hbar|jm\rangle \tag{2.125}$$

The eigenvalues of J and its components are given as

$$\begin{aligned} &J^2 = \hbar^2 j(j+1) \quad \text{where } j = 0, 1/2, 1, 3/2, \ldots \\ &|J| = \hbar\sqrt{j(j+1)} \\ &\text{and } J_z = m_j\hbar \quad \text{where } m_j = +j\ldots 0\ldots - j \end{aligned} \tag{2.126}$$

There are ($2j$ + 1) values (taking the maximum of j), and the condition $j(j+1) \geq 0$ is applicable. The general form of the vector addition of spin and orbital angular momenta and the magnitude of the total angular momentum is represented in Fig. 2.22. Consider $S = +\ 1/2$ and $L = +\ 1$, then $J = L + S = 3/2$, and the quantized magnitude is $|J| = \sqrt{\frac{3\left(\frac{5}{2}\right)}{2}}\hbar = \sqrt{\frac{15}{4}}\hbar$. Since m_j can take + 3/2, + 1/2, 0, – 1/2, – 3/2 the z-component has values given by $J_z = \left(+\frac{3}{2}\hbar\right), \left(+\frac{\hbar}{2}\right), \left(-\frac{\hbar}{2}\right), \left(-\frac{3}{2}\hbar\right)$. Now we have ($2j$ + 1) = (2 × 3/2 + 1) = 4 values. In cases involving a set of J_z values, the j, m_j and their addition determine the intensity of spectral lines, as discussed in the next section.

2.6.1 *Clebsch-Gordan Coefficient*

If there are two states given by $|j_1 m_1\rangle$ and $|j_2 m_2\rangle$, they form a complete basis of $(2j_1 + 1) \times (2j_2 + 1)$ tensor product states

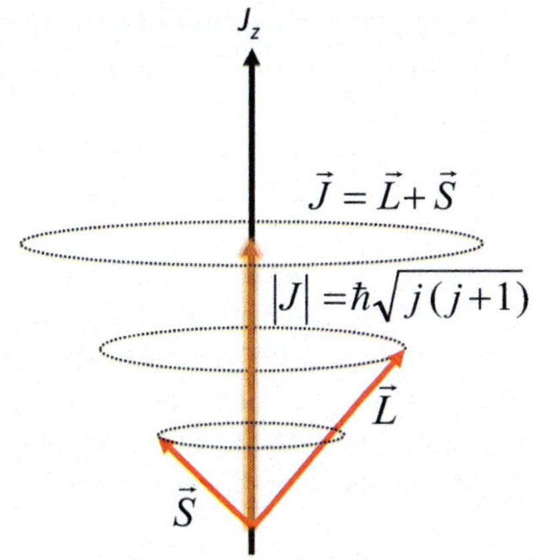

Fig. 2.22 The vector model representing the addition of angular momenta. The total angular momentum is the vector addition of orbital and spin angular momenta $\vec{J} = \vec{L} + \vec{S}$. The magnitude of J is given as $|J| = \hbar\sqrt{j(j+1)}$ and the z-component is $J_z = m_j\hbar$

$$|j_1m_1, j_2m_2\rangle = |j_1m_1\rangle \otimes |j_2m_2\rangle \tag{2.127}$$

The $|j_1m_1, j_2m_2\rangle$ bases are connected to $|jm\rangle$ using the relation (Shankar 1994)

$$|jm\rangle = \sum_{m_1=-j_1}^{j_1} \sum_{m_2=-j_2}^{j_2} |j_1m_1, j_2m_2\rangle\langle j_1m_1, j_2m_2|jm\rangle \tag{2.128}$$

$$= \sum_{m_1,m_2} |j_1m_1, j_2m_2\rangle\langle j_1m_1, j_2m_2|jm\rangle \tag{2.129}$$

$$= C^{jm}_{j_1m_1\,j_2m_2}|j_1m_1, j_2m_2\rangle \tag{2.130}$$

The coefficient $C^{jm}_{j_1m_1\,j_2m_2}$ is known as the Clebsch-Gordan (CG) coefficient. A few of their characteristics are

$$C^{jm}_{j_1m_1\,j_2m_2} \neq 0 \text{ when} \begin{cases} |j_1 - j_2| \leq j \leq j_1 + j_2 \\ m = m_1 + m_2 \end{cases}$$
$$= 1 \text{ when} \begin{cases} j = j_1 + j_2 \\ m = j \end{cases} \tag{2.131}$$

$$C^{jm}_{j_1m_1\,j_2m_2} = (-1)^{j_1-m_1}\sqrt{\frac{2j+1}{2j_2+1}}\langle jmj_1\,(-m_1)|\,j_2m_2\rangle \tag{2.132}$$

$$C^{jm}_{j_1m_1\,j_2m_2} = (-1)^{j_2+m_2}\sqrt{\frac{2j+1}{2j_1+1}}\langle j_2(-m_2)\;jm|j_1m_1\rangle \tag{2.133}$$

For the addition of angular momenta of more than two, the matrix method using Wigner 3-j or 6-j symbols is convenient.

Now, for the addition of spin angular momenta s_1, s_2, the z-components m_1 and m_2 can be written in the linear combination of the states $|s_1 m_1\rangle\,|s_2 m_2\rangle$

$$|sm\rangle = \sum_{m=m_1+m_2} C^{s_1 s_2 s}_{m_1 m_2 m}\,|s_1 m_1\rangle\,|s_2 m_2\rangle \tag{2.134}$$

where $C^{s_1 s_2 s}_{m_1 m_2 m}$ is the CG coefficient.

The Clebsch-Gordan coefficient can be conveniently found using the CG table, as shown in Fig. 2.23. The CG coefficient can be used to find the transition probability and spectral line's intensity. The forbidden transitions have zero coefficients or are not represented in the CG matrix table. The CG coefficient determines the intensity pattern of the Mössbauer spectrum.

2.7 Multielectron Atom

A hydrogen-like atom can be thought of as an atom or ion having only one valence electron. The Hamiltonian for a neutral atom with atomic number Z is given as

$$H = \sum_{j=1}^{Z}\left(-\frac{\hbar^2}{2m}\nabla_j^2 - \frac{Ze^2}{4\pi\varepsilon_0 r_j}\right) + \frac{1}{2}\frac{1}{4\pi\varepsilon_0}\sum_{j\neq k}^{Z}\frac{e^2}{|r_j - r_k|} \tag{2.135}$$

The Hamiltonian for the helium atom ($Z = 2$), including the electron–electron interaction, is given as

$$H = \left(-\frac{\hbar^2}{2m}\nabla_1^2 - \frac{2e^2}{4\pi\varepsilon_0 r_1}\right) + \left(-\frac{\hbar^2}{2m}\nabla_2^2 - \frac{2e^2}{4\pi\varepsilon_0 r_2}\right) + \frac{1}{4\pi\varepsilon_0}\frac{e^2}{|r_1 - r_2|} \tag{2.136}$$

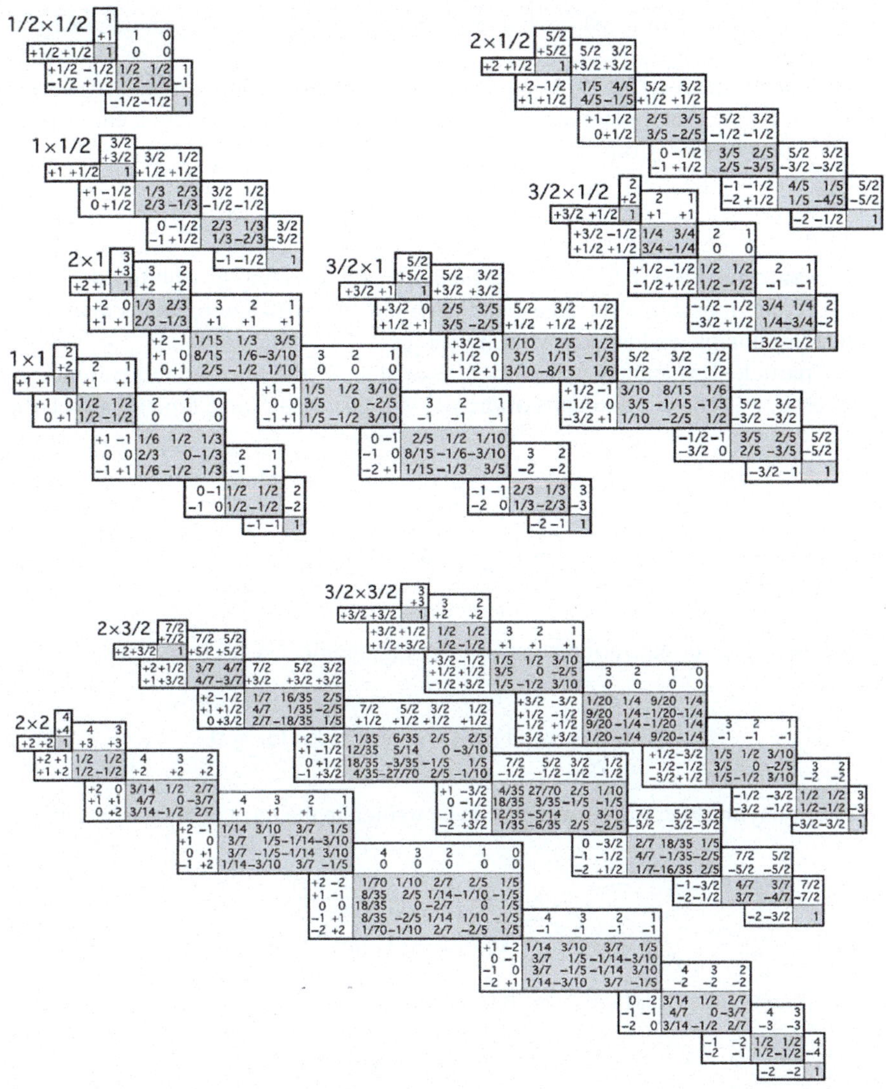

Fig. 2.23 Table of Clebsch-Gordan coefficients. The appropriate table could be identified from the $j_1 \times j_2$ value followed by m_1, m_2 row; j, m column and the intersecting cell gives the CG coefficient after taking its $\surd$. For a negative sign, it is $-\surd$(value). (Zyla et al. 2020)

The complete wave function of a system of non-interacting particles can be represented as the product of wave functions

$$\Psi(1, 2, 3, \ldots n) = \psi(1)\psi(2)\ldots\psi(n) \tag{2.137}$$

For a two-particle system, there can be two possibilities

$$\Psi = \psi(1)\psi(2) = \psi(2)\psi(1) \tag{2.138}$$

The wave function (2.138) is *symmetric,* representing equal probability density in the same state. Such indistinguishable particles occupying the same energy state are known as bosons. Consider another case where

$$\Psi = \psi(1)\psi(2) = -\psi(2)\psi(1) \tag{2.139}$$

The wave function (2.139) is *antisymmetric,* representing probability density in two different antisymmetric states, and such indistinguishable particles obey Pauli exclusion principle and are known as fermions.

For particle 1 in state *a* and particle 2 in state *b,* the linear combinations give symmetric, Ψ_S, bosonic and antisymmetric, Ψ_A, fermionic wave functions such that

$$\Psi_S = \frac{1}{\sqrt{2}}|\psi_a(1)\psi_b(2) + \psi_a(2)\psi_b(1)| \tag{2.140a}$$

$$\Psi_A = \frac{1}{\sqrt{2}}|\psi_a(1)\psi_b(2) - \psi_a(2)\psi_b(1)| \tag{2.140b}$$

where $\frac{1}{\sqrt{2}}$ is the normalization constant. Setting identical states $a = b$, we get

$$\Psi_A = \frac{1}{\sqrt{2}}|\psi_a(1)\psi_a(2) - \psi_a(2)\psi_a(1)| = 0 \tag{2.141}$$

The antisymmetric wave function can be written in the form of *Slater determinant* as

$$\Psi_A = \frac{1}{\sqrt{2!}}\begin{vmatrix} \psi_a(1) & \psi_a(2) \\ \psi_b(1) & \psi_b(2) \end{vmatrix} \tag{2.142}$$

For a system with *N* particles, this becomes

$$\Psi_A = \frac{1}{\sqrt{N!}}\begin{vmatrix} \psi_a(1) & \psi_a(2) & \ldots\psi_a(N) \\ \psi_b(1) & \psi_b(2) & \ldots\psi_b(N) \\ . & . & . \\ \psi_N(1) & \psi_N(2) & \ldots.\psi_N(N) \end{vmatrix} \tag{2.143}$$

If two particles occupy the same energy state, then the fermionic wave function vanishes, which is the *Pauli exclusion principle*, implying that two fermions cannot have all the same quantum numbers. If the state of a system has degenerate energy levels specified by the quantum numbers n, l, m_l and m_s are of the same energy with the spin state $|\uparrow\rangle$, then another fermionic particle will occupy the spin $|\downarrow\rangle$ state. Thus, if Ψ_S is symmetric, the spin wave function χ has to be antisymmetric (χ_A) and vice-versa given as

$$\Psi = \Psi_S \chi_A = \frac{1}{\sqrt{2}} |\psi_a(1)\psi_b(2) + \psi_a(2)\psi_b(1)| \chi_A \tag{2.144}$$

$$\Psi = \Psi_A \chi_S = \frac{1}{\sqrt{2}} |\psi_a(1)\psi_b(2) - \psi_a(2)\psi_b(1)| \chi_S \tag{2.145}$$

For two electrons, there can be four basis spin states with one antisymmetric $S = 0$ singlet and symmetric spin $S = 1$ triplet with the spin configurations

$$\chi_S = \begin{cases} \uparrow\uparrow \\ \frac{1}{\sqrt{2}}(\uparrow\downarrow + \downarrow\uparrow) \\ \downarrow\downarrow \end{cases} \tag{2.146}$$

$$\chi_A = \left\{ \frac{1}{\sqrt{2}}(\uparrow\downarrow - \downarrow\uparrow) \right. \tag{2.147}$$

The total wave function thus includes the spatial and spin part where the spin term allows two electrons in He atom to occupy the same ground state energy but with opposite spins, leading to the singlet configuration as per the experiments.

The energy estimate of He atom ($Z = 2$) with both electrons in the same level ($n = 1$) gives $E_1 = -2\frac{z^2}{n^2} E_{\mathrm{H}} \approx -109$ eV, where E_{H} is the ground state energy of the hydrogen atom, which is − 13.6 eV. This value is far from the experimental binding energy of − 79 eV for the He atom. The deviation arises due to the shielding by the electrons and the effective nuclear charge felt by the electrons, which is not 2, but lower.

The variational principle, an approximation method employing a trial wave function estimates $Z_{\mathrm{eff}} = Z - \sigma = Z - \frac{5}{16} = 1.69$, where σ is the shielding constant that gives a close estimate of − 78 eV for the ground state energy of He. The shielding constant and Z_{eff} can be estimated using Slater's rules and presented in Table 2.4 for Fe (Slater 1930). It could be understood that 4s electrons are significantly shielded such that the Z_{eff} of the valence electron is only 3.75.

Table 2.4 Shielding constants from various orbitals of Fe ($Z = 26$) and the effective charges (Z_{eff})

Shielding orbital	Shielding constant σ	Effective charge Z_{eff}
1s	0.30	25.70
2s, p	4.15	21.85
3s, p	11.25	14.75
3d	19.75	6.25
4s	22.25	3.75

2.8 The Periodic Table

Elements in the periodic table are arranged in the order of increasing atomic number, and the filling of the quantum levels *nlm* gives the electronic configuration of an element. As shown in Fig. 2.24, the periodic table is arranged as s-block, p-block, d-block and f-block elements where the block specifies the valence electrons *l* level. The horizontal rows are known as periods, and the vertical columns are known as groups. The groups with the full orbitals are in the rightmost, known as the noble gases, while the first group starting with Li are known as alkali metals. The f-block elements are placed separately in the periods as lanthanides and actinides, except that La/Lu and Ac/Lr are often positioned in the d-block.

The electronic configuration of the elements is determined by (i) Pauli exclusion principle, (ii) Aufbau principle and (iii) Hund's rule, apart from the allowed quantum numbers.

(a) The energy level determined by the principal quantum number n is known as a *shell*. The total number of electrons in an atom's n^{th} shell is $2n^2$. In X-ray notation, in the increasing order of n, the shells are represented as K, L, M, N, …

(b) The electrons occupying the energy level, determined by the azimuthal quantum number l, is the *subshell*. The maximum number of electrons allowed in a subshell is $2(2\ l + 1)$. The l values are from 0 to $(n - 1)$ mentioned as s, p, d, f, g, h, … subshells in the order of increasing l. For example, the d subshell corresponds to $l = 2$ and can hold $2(2 \times 2 + 1) = 10$ electrons.

s block | d block | p block

Groups ↓ Periods →

s block		d block										p block					
1 H																	2 He (⇦ s)
3 Li	4 Be											5 B	6 C	7 N	8 O	9 F	10 Ne
11 Na	12 Mg											13 Al	14 Si	15 P	16 S	17 Cl	18 Ar
19 K	20 Ca	21 Sc	22 Ti	23 V	24 Cr	25 Mn	26 Fe	27 Co	28 Ni	29 Cu	30 Zn	31 Ga	32 Ge	33 As	34 Se	35 Br	36 Kr
37 Rb	38 Sr	39 Y	40 Zr	41 Nb	42 Mo	43 Tc	44 Ru	45 Rh	46 Pd	47 Ag	48 Cd	49 In	50 Sn	51 Sb	52 Te	53 I	54 Xe
55 Cs	56 Ba		72 Hf	73 Ta	74 W	75 Re	76 Os	77 Ir	78 Pt	79 Au	80 Hg	81 Tl	82 Pb	83 Bi	84 Po	85 At	86 Rn
87 Fr	88 Ra		104 Rf	105 Db	106 Sg	107 Bh	108 Hs	109	....								

f block														d
57 La	58 Ce	59 Pr	60 Nd	61 Pm	62 Sm	63 Eu	64 Gd	65 Tb	66 Dy	67 Ho	68 Er	69 Tm	70 Yb	71 Lu
89 Ac	90 Th	91 Pa	92 U	93 Np	94 Pu	95 Am	96 Cm	97 Bk	98 Cf	99 Es	100 Fm	101 Md	102 No	103 Lr

Fig. 2.24 The periodic table of elements, arranged in s, p, d and f -block groups

(c) Each subshell has different quantum levels oriented in space, determined by the quantum number m, which can take values from $+ l, \ldots,0,\ldots, - l$.
(d) The electron's spin shall be determined by Pauli exclusion principle and Hund's rule.
(e) The filling of the orbital energy levels is of the order given by the Aufbau principle. The relative energies of orbitals can be predicted using the $(n + l)$ rule. The filling will be in the order of increasing energy. If the subshell $>n + l$, then the filling will be the next n^{th} level. For example, the 3d level has $n + l$ value as $(3 + 2) = 5$ and therefore, the electrons occupy the 4s level $(4 + 0)$ first as $E(4\text{s} < 3\text{d})$. In the case of 5p and 6s levels, $E(5\text{p} < 6\text{s})$ and n take precedence, filling the 5p level first.
(f) The energy levels are further shifted and split due to the internal and external magnetic fields, relativistic effects and nuclear interaction. The electronic state, assuming *LS* coupling except for the heavy elements, is represented by the term symbol ${}^{2S+1}L_J$. The orbital quantum number (L) is given by the spectroscopic notation as *S, P, D, F,* ... corresponding to $L = 0, 1, 2, 3, \ldots$, respectively. For full shell and sub-shells, $S = 0$, $L = 0$ and $J = 0$ and therefore denoted as ${}^1S_0$ for elements such as noble gases, Be and Zn groups.

2.8.1 Hund's Rule

Hund's rule is used to obtain the ground state configuration of an atom or molecule. It is also helpful to find the magnetic state of localized electron systems.

Rule 1: The highest spin (S) will have the lowest energy.
Rule 2: The highest orbital angular momentum (L) will have the lowest energy.
Rule 3: For less than half-filled subshells, $J = L - S$ and for more than half, it is $J = L + S$.

From Hund's first and second rule, it is clear that all the m_l levels in the subshells will fill up with a single electron to maximize their spin and orbital angular momentum. After allotting an electron to each m_l state, the subshell becomes half full, and the excess electron occupies the higher m_l state with the opposite spin due to Pauli exclusion principle. The net J value could be obtained from Hund's third rule.

As the number of electrons increases, the energy of the orbitals is increased due to the l value by a factor of $l(l + 1)$. The farther outer electron causes the s electron to be closer to the nucleus, producing *screening effect*, while the d electron experiences weak nuclear attraction. Therefore, the filling of the energy levels in the order given by the Aufbau principle is

$$1s^2 2s^2 2p^6 3s^2 3p^6 4s^2 3d^{10} 4p^6 5s^2 4d^{10} 5p^6 6s^2 4f^{14} 5d^{10} 6p^6 7s^2 5f^{14} \ldots$$

Table 2.5 The electronic configurations of a few elements and their ions

Element/Ion	Electrons	Electronic configuration
Fe^0	26	$1s^2 2s^2 2p^6 3s^2 3p^6 4s^2 3d^6$
Fe^{2+}	24	$1s^2 2s^2 2p^6 3s^2 3p^6 3d^6$
Fe^{3+}	23	$1s^2 2s^2 2p^6 3s^2 3p^6 3d^5$
Cu^0	29	$1s^2 2s^2 2p^6 3s^2 3p^6 4s^1 3d^{10}$
Cu^{1+}	28	$1s^2 2s^2 2p^6 3s^2 3p^6 3d^{10}$
Zn^0	30	$1s^2 2s^2 2p^6 3s^2 3p^6 4s^2 3d^{10}$
Zn^{2+}	28	$1s^2 2s^2 2p^6 3s^2 3p^6 3d^{10}$

The 4s subshell shows lower energy than 3d, whereas when the electrons fill up, the 4s level become higher in energy and become valence level. Thus, the 4s electrons are the first to fill up and come out, which determines the ionic state and the magnetic properties. The electronic configurations of a few elements and their ions are listed in Table 2.5. It has to be noted that 3d subshells are fully occupied in Cu before 4s to acquire stability.

As the electrons are filled as per the sequence, they attain better stability when all the allowed orbitals in a given n level are full. As a consequence of the stability of atoms with filled orbitals, as in noble gases, their first ionization energy is also larger, as shown in Fig. 2.25. The next element, following noble gases, such as Li, Na, K, Rb and Cs, has a larger atomic radius because the electron starts to fill the next n level. The atomic radii decrease as we move towards the right across the period and increase down within each group. The ionization energy increases from left to right within a period and drops downward in a group. These general trends are due to the difference in the Coulombic attraction felt by the electrons due to the Z_{eff} or increasing distance. The increase in the Z_{eff} within a period towards the right causes strong electron binding. The increase in distance of the electron within a group downwards causes easy removal of the electron. Indeed, there are exceptions due to the electronic configuration of the element.

2.9 Fine Structure Splitting

The spin–orbit interaction, kinetic energy relativistic correction and the Darwin term determine the fine structure splitting of the spectral lines. The non-relativistic expansion of the relativistic Dirac wave equation of fermions is known as the Darwin term.

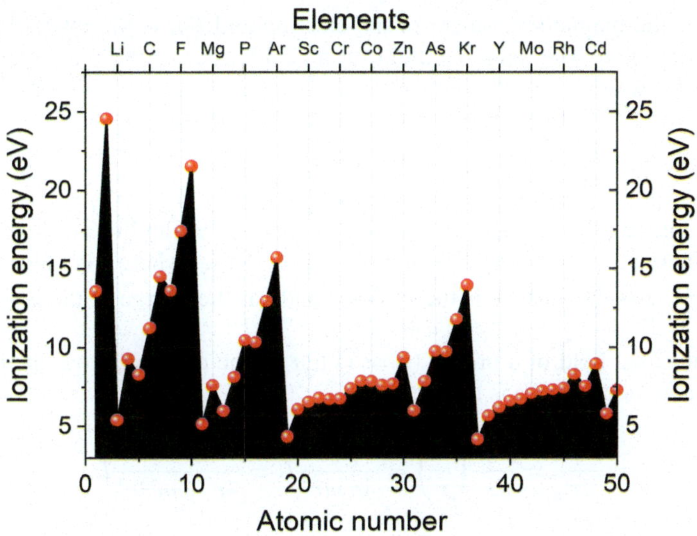

Fig. 2.25 The first ionization energy of neutral elements up to $Z = 50$. Noble gases with closed shells show higher ionization energies

2.9.1 *Spin–Orbit Coupling*

The electron experiences a magnetic field by the protons since it perceives that the nucleus is gyrating from its rest frame of reference, as shown in Fig. 2.26.

The nucleus exerts a torque on the electron according to the magnetic field. The Hamiltonian for the magnetic field experienced by the electron is

$$H = -\mu \cdot B_{\text{int}} \tag{2.148}$$

Fig. 2.26 The electron from its rest frame of reference experiences an internal magnetic field due to the positive nucleus. The interaction of the spin magnetic moment of the electron with the orbital magnetic field is the basis for spin–orbit coupling

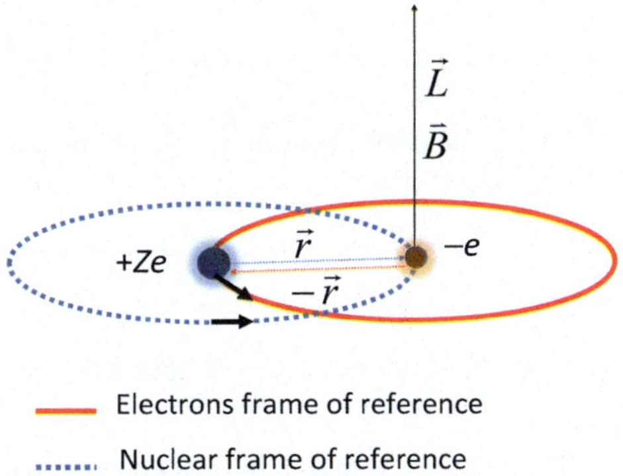

where μ is the magnetic moment of the electron and B_{int} is the internal magnetic field of the proton. The magnetic field for a circular current loop is obtained from Biot-Savart law as

$$B = \frac{\mu_0 I}{2r} \tag{2.149}$$

where I is the current, μ_0 is the vacuum magnetic permeability and r is the radius. We use the relations $L = mvr = mr\left(\frac{2\pi r}{t}\right)$; $c = \frac{1}{\sqrt{\epsilon_0 \mu_0}}$, where t is the time period of the orbit, m is the mass, v is the velocity, and e is the charge of the proton that is positive.

The magnetic field of a proton ($Z = 1$) felt by the electron, from the electron's inertial frame of reference is then,

$$B_{\text{int}} = \frac{\mu_0 I}{2r} = \frac{\mu_0 e}{2r}\left(\frac{1}{t}\right) = \frac{\mu_0 e}{4\pi m r^3} L = \frac{e}{4\pi \varepsilon_0 m c^2 r^3} L \tag{2.150}$$

The spin–orbit interaction Hamiltonian takes the form

$$H_{\text{so}} = -\mu_S \cdot B_{\text{int}} = -\left(\frac{-e}{m}\right) S \cdot B_{\text{int}} = \frac{e^2}{4\pi \varepsilon_0 m^2 c^2 r^3} S \cdot L \tag{2.151}$$

An accelerated particle with an angular velocity undergoes Thomas precession given by

$$\omega_{\text{T}} = \frac{\sqrt{1 - v^2/c^2}}{2c^2\left(1 - v^2/c^2\right)} a \times v \tag{2.152}$$

where a is the acceleration and v is the velocity, considered lower than c here.

In the non-relativistic limit as $v \ll c$, we get

$$\omega_{\text{T}} \approx \left(\frac{1}{2}\right) \frac{1}{c^2} a \times v \tag{2.153}$$

Using the correction factor of 1/2 in H_{so},

$$H_{\text{so}} = \frac{e^2}{8\pi \varepsilon_0 m^2 c^2 r^3} S \cdot L = \xi(r) S \cdot L \tag{2.154}$$

$$\text{where } \xi(r) = \frac{e^2}{8\pi \varepsilon_0 m^2 c^2 r^3} \tag{2.155}$$

The Hamiltonian in a hydrogen atom with one proton and one electron, including the spin–orbit interaction term, takes the form

$$\hat{H} = \hat{H}_o + \hat{H}_{\mathrm{so}} \tag{2.156}$$

where H_o is the unperturbed Hamiltonian and H_{so} is the perturbation due to spin–orbit interaction. Since H_o, L^2, S^2, J_z could only commute with each other in perturbation theory,

$$\begin{aligned} J &= L + S \\ J^2 &= (L + S) \cdot (L + S) = L^2 + S^2 + 2L \cdot S \end{aligned} \tag{2.157}$$

$$L \cdot S = \frac{1}{2}\left(J^2 - L^2 - S^2\right) \tag{2.158}$$

The eigenvalue of $L{\cdot}S$ in the basis $|nljm\rangle$ is then

$$\langle nljm|L \cdot S|nljm\rangle = \frac{\hbar^2}{2}[j(j+1) - l(l+1) - s(s+1)] \tag{2.159}$$

The energy eigenvalue is obtained using

$$\hat{H}_{\mathrm{so}}|nljm\rangle = E_{\mathrm{so}}|nljm\rangle \tag{2.160}$$

$$E_{\mathrm{so}} = \frac{e^2}{8\pi\varepsilon_0 m^2c^2}\frac{\hbar^2}{2}[j(j+1) - l(l+1) - s(s+1)]\langle\frac{1}{r^3}\rangle \tag{2.161}$$

The expectation value $\langle\frac{1}{r^3}\rangle$ is

$$\langle\frac{1}{r^3}\rangle = \frac{1}{a_o^3}\frac{1}{n^3 l(l+1)\left(l+\frac{1}{2}\right)} \tag{2.162}$$

and the energy eigenvalue is

$$E_{\mathrm{so}} = \frac{e^2}{8\pi\varepsilon_0 m^2c^2}\frac{\hbar^2}{2}[j(j+1) - l(l+1) - s(s+1)]\left(\frac{2}{n^3 l(l+1)(2l+1)a_o^3}\right) \tag{2.163}$$

The Bohr radius is given as $a_0 = \frac{4\pi\varepsilon_0\hbar^2}{me^2}$ and (2.163) can be modified as

$$E_{\mathrm{so}} = \frac{e^2}{8\pi\varepsilon_0}\frac{\hbar^2}{m^2c^2}\left(\frac{me^2}{4\pi\varepsilon_0\hbar^2}\right)^3\left(\frac{1}{n^4}\right)n\left(\frac{j(j+1) - l(l+1) - s(s+1)}{l(l+1)(2l+1)}\right) \tag{2.164}$$

Using the hydrogen energy levels given by $|E_n| = \frac{1}{n^2}\frac{m}{2\hbar^2}\left(\frac{e^2}{4\pi\varepsilon_0}\right)^2$,

$$E_{so} = \frac{E_n^2}{mc^2}\left(\frac{n\left[j(j+1) - l(l+1) - \frac{3}{4}\right]}{l\left(l+\frac{1}{2}\right)(l+1)}\right) \tag{2.165}$$

Rearranging (2.164),

$$E_{so} = \frac{1}{n^2}\frac{e^2}{8\pi\varepsilon_0 a_o}\left(\frac{\hbar}{mca_o}\right)^2\left(\frac{1}{n}\right)\left(\frac{j(j+1) - l(l+1) - s(s+1)}{l(l+1)(2l+1)}\right) \tag{2.166}$$

$$E_{so} = |E_n|\left(\frac{\alpha^2}{n}\right)\left(\frac{j(j+1) - l(l+1) - s(s+1)}{l(l+1)(2l+1)}\right) \tag{2.167}$$

where $\alpha = \frac{\hbar}{mca_o} = \frac{ke^2}{\hbar c} = \frac{1}{137}$ is a dimensionless constant known as the fine structure constant ($k = 1$ in cgs). The spin–orbit interaction energy vanishes for $l = 0$ when the nucleus is considered stationary or when $s = 0$ for the electron.

In a multielectron atom, the total angular momentum can be given by the vector addition of orbital and spin angular momenta determined by the electron–electron interaction, and spin–orbit interaction, resulting in two major configurations known as *LS* coupling and *jj* coupling.

LS coupling or Russell–Saunders coupling is applicable when the spin–orbit interaction is weaker than the electron–electron interaction and is given by the summation of the orbital angular momenta and spin angular momenta to obtain the total angular momentum

$$\vec{J} = \vec{L} + \vec{S} = \sum_i \vec{l}_i + \sum_i \vec{s}_i \tag{2.168}$$

In the *jj* coupling, the individual spin and orbital angular momenta are first added together to obtain the angular momentum of the single electron, and all these *j* values are then added together. The *jj* coupling occurs only for heavy atoms such as Pb and Bi where the spin–orbit interaction is stronger than the *l*-*l* and *s*-*s* interactions.

$$\vec{J} = \sum_i \vec{j}_i = \sum_i \left(\vec{l}_i + \vec{s}_i\right) \tag{2.169}$$

The vector model representing the *LS* and *jj* coupling schemes for two electrons is shown in Fig. 2.27.

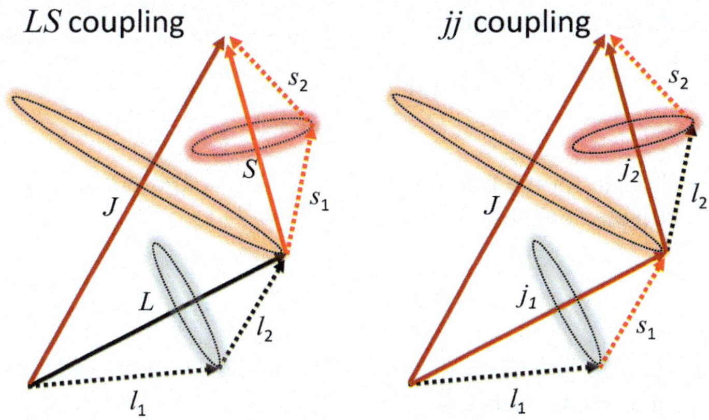

Fig. 2.27 The *LS* coupling scheme is given by the vector addition $\vec{J} = \vec{L} + \vec{S} = \left(\vec{l}_1 + \vec{l}_2\right) + (\vec{s}_1 + \vec{s}_2)$ while the *jj* coupling is $\vec{J} = \vec{j}_1 + \vec{j}_2 = \left(\vec{l}_1 + \vec{s}_1\right) + \left(\vec{l}_2 + \vec{s}_2\right)$, represented for a two-electron system

2.9.2 Relativistic Kinetic Energy Correction

The Hamiltonian for the hydrogen atom can be written as

$$H_o = -\frac{\hbar^2}{2m}\nabla^2 - \frac{e^2}{4\pi\varepsilon_0 r} \tag{2.170}$$

Consider relativistic energy and momentum relation,

$$E = \sqrt{p^2c^2 + m^2c^4} = mc^2 + \frac{p^2}{2m} - \frac{1}{8}\frac{(p^2)^2}{m^3c^2} + \cdots \tag{2.171}$$

The relativistic correction to the kinetic energy modifies the Hamiltonian as

$$H = H_o + H_{\text{re}} = -\frac{\hbar^2}{2m}\nabla^2 - \frac{e^2}{4\pi\varepsilon_0 r} - \frac{p^4}{8m^3c^2} \tag{2.172}$$

$$H_{\text{re}} = -\frac{p^4}{8m^3c^2} \tag{2.173}$$

Using the H_{re} as a perturbation, the first-order correction to the Hamiltonian results in eigenvalue for the relativistic energy part as

$$E_{\text{re}} = -\frac{E_n^2}{2mc^2}\left(\frac{4n}{l + \frac{1}{2}} - 3\right) \tag{2.174}$$

In terms of α, (2.174) becomes

$$E_{\mathrm{re}} = -\frac{\alpha^2 |E_n|}{4n^2}\left(\frac{8n}{2l+1} - 3\right) \tag{2.175}$$

The Hamiltonian for the Darwin term due to the rapid oscillatory motion of the electrons contributed only by s orbital is given as

$$H_{\mathrm{Darwin}} = \frac{2n}{mc^2} E_n^2 \tag{2.176}$$

The Hamiltonian, including the spin–orbit interaction term and the relativistic correction to the kinetic energy is

$$H = H_o + H_{\mathrm{so}} + H_{\mathrm{re}} + H_{\mathrm{Darwin}} = H_o + H_{\mathrm{FS}} \tag{2.177}$$

where the Hamiltonian H_{FS} corresponds to the fine structure.

The spin–orbit interaction energy, the relativistic kinetic energy and Darwin terms lead to the fine splitting of hydrogen spectral lines, known as *fine structure*. Since the magnitude of the total contributions is proportional to $\alpha^2 E_n$, the fine structure splitting energy is of the order $10^{-4} - 10^{-3}$ eV. Note that the spin–orbit interaction field is contributed by the internal magnetic field (B_{int}) of the charged particle in motion. The shell model nucleus describes a similar effect on nuclear energy levels.

2.10 Zeeman Effect

The Zeeman effect is the splitting of the spectral lines in an applied magnetic field (B). It arises due to the interaction of the external magnetic field with the spin and orbital magnetic moment of the electron. It can be classified as (i) weak field Zeeman effect ($B \ll B_{\mathrm{int}}$), (ii) strong field Zeeman effect ($B \gg B_{\mathrm{int}}$) or Paschen-Back effect and (iii) intermediate field Zeeman effect. In the Zeeman effect, if the spin contribution is absent, it is known as the normal Zeeman effect, while a non-zero spin gives anomalous Zeeman effect.

In the weak field Zeeman effect, since the applied field is smaller than the internal field, the spin–orbit interaction term dominates. The L and S are not separately conserved, whereas the total angular momentum J is conserved. The Hamiltonian is given by

$$H = H_o + H_{\mathrm{FS}} + H_{\mathrm{Z}} \tag{2.178}$$

where the Zeeman Hamiltonian H_{Z} is treated as a perturbation. The orbital magnetic moment of an electron is

$$\mu_L = \frac{-e}{2m} L \tag{2.179}$$

The spin magnetic moment is

$$\mu_S = \frac{-e}{m} S \tag{2.180}$$

The total magnetic moment is then

$$\mu = \mu_L + \mu_S = \frac{-e}{2m}(L + 2S) \tag{2.181}$$

The Hamiltonian, due to the interaction of the magnetic moment with an external field B, is then given as

$$H_Z = -\mu \cdot B = \frac{e}{2m}(L + 2S) \cdot B \tag{2.182}$$

From the vector addition, $L + 2S = (L + S) + S = J + S$. The interaction now depends on $J + S$ rather than J alone. The expectation value of S could be obtained from its projection along J. Since $\langle S \rangle \propto J$, we can write $\langle S \rangle = \beta J$ where β is a constant. Using $\langle S \rangle \cdot J = \langle S \rangle \cdot J = \beta_J J \cdot J$, the constant β_J can be evaluated as $\beta_J = \frac{\langle S \cdot J \rangle}{J \cdot J} = \frac{\langle S \cdot J \rangle}{J^2}$. Similar to (2.158), we can evaluate $S{\cdot}J$ as $S \cdot J = \frac{1}{2}\left[J^2 + S^2 - L^2\right]$ and the eigenvalues are obtained as

$$\begin{aligned} \langle S \cdot J \rangle &= \frac{\hbar^2}{2}[j(j+1) + s(s+1) - l(l+1)] \\ J^2 &= \hbar^2 j(j+1) \end{aligned} \tag{2.183}$$

hence

$$\beta_J = \frac{1}{2}\left[\frac{j(j+1) + s(s+1) - l(l+1)}{j(j+1)}\right] \tag{2.184}$$

$$\begin{aligned} \langle L + 2S \rangle = \langle J + S \rangle &= \left\langle \left(1 + \frac{S}{J}\right) J \right\rangle = \left\langle \left(1 + \frac{S \cdot J}{J \cdot J}\right) J \right\rangle \\ &= (1 + \beta_J)\langle J \rangle = g_J \langle J \rangle \end{aligned} \tag{2.185}$$

where

$$g_J = 1 + \frac{1}{2}\left[\frac{j(j+1) + s(s+1) - l(l+1)}{j(j+1)}\right] \tag{2.186}$$

is known as the *Landé g-factor*. For pure orbital contribution, the g_J value is 1, while it is 2 for spin. The energy eigenvalue is given as

$$E_Z = \langle nljm|H_Z|nljm\rangle \tag{2.187}$$

where the quantum numbers represented here are n, l, j and $m = m_j$. If B is along the z component, then

$$E_Z = g_J\left(\frac{e\hbar}{2m}\right)(m_j)B = g_J m_j \mu_B B \tag{2.188}$$

where

$$\mu_B = \frac{e\hbar}{2m} \tag{2.189}$$

is known as the Bohr magneton.

For the weak field Zeeman effect, as described above, the energy eigenvalues are determined by the g-factor and the quantum numbers j and m_j. In the strong field Zeeman effect, J is not a conserved quantity; rather, J_z and S_z are. L^2, S^2 and J_z are always good for weak and strong field cases. Since the external field dominates ($B \gg B_{int}$), the spin–orbit interaction term can be neglected in the strong field Zeeman effect. Figure 2.28 shows the normal Zeeman splitting of the hydrogen spectrum where the 2p degeneracy is lifted, giving three emission lines in an applied magnetic field whose frequencies are determined by (2.188). The splitting energy is of the order 10^{-4} eV, using the field produced by a typical electromagnet. Equally separated $2l+1$ components arise due to the interaction of the applied magnetic field with the orbital magnetic moment.

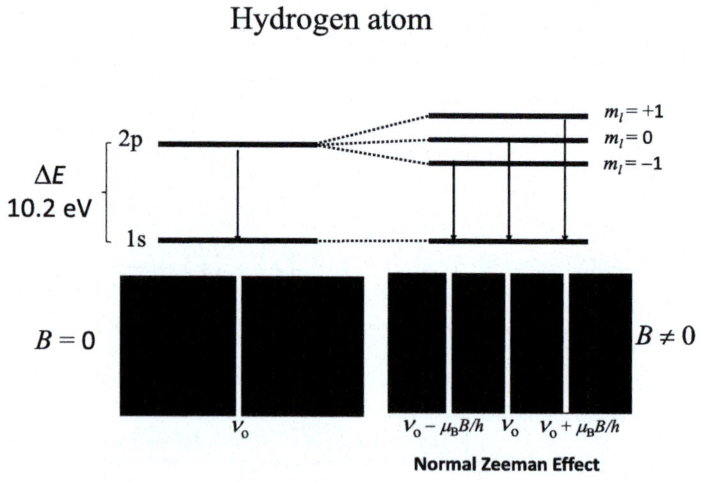

Fig. 2.28 The hydrogen atom 2p → 1s emission wavelength line in the uv region (≈ 122 nm) is split into three lines in the normal Zeeman effect. The 2p degeneracy is lifted and gives $2l+1$ components in an applied magnetic field

2.11 Hyperfine Splitting

Hyperfine splitting arises due to the interaction between the magnetic moments of the nucleus and electrons in the atom. Also known as spin–spin or magnetic dipole interaction, it may shift and split the energy levels of the order 10^{-5}–10^{-6} eV. The magnetic field of an ideal magnetic dipole at the origin of a uniformly magnetized spherical volume is given as

$$B_i = \frac{\mu_0}{4\pi}\frac{1}{r^3}\left[3(\mu_i \cdot \hat{r})\hat{r} - \mu_i\right] + \frac{2}{3}\mu_0\mu_i\delta^3(r) \tag{2.190}$$

where μ_i is the magnetic dipole moment, r is the distance and the unit vector $\hat{r} = \frac{\vec{r}}{|r|}$. The first term of (2.190) originates from the magnetic field outside the uniformly magnetized sphere and the second term corresponds to the dipole as $r \to 0$ where the sphere becomes a 3D $\delta^3(r)$ function (Griffiths 1982).

The Hamiltonian for the magnetic field experienced by the electron due to the protons' magnetic moment is

$$\begin{aligned} H_{\text{hf}} &= -\mu_j \cdot B_i \\ &= -\frac{\mu_0}{4\pi}\frac{1}{r^3}\left[3\frac{(\mu_{\text{p}} \cdot \vec{r})(\mu_{\text{e}} \cdot \vec{r})}{r^2} - \mu_{\text{p}} \cdot \mu_{\text{e}}\right] \\ &\quad - \frac{2}{3}\mu_0\mu_{\text{p}} \cdot \mu_{\text{e}}\delta^3(r) \end{aligned} \tag{2.191}$$

In (2.191), the $\delta^3(r)$ is relevant only to the nucleus known as the *Fermi contact term.* The spin magnetic moment of a proton of mass m_{p}, spin S_p and g-factor g_{p} is

$$\mu_{\text{p}} = g_{\text{p}}\frac{e}{2m_{\text{p}}}S_p \tag{2.192a}$$

while that of an electron is

$$\mu_{\text{e}} = \frac{-e}{m_{\text{e}}}S_e \tag{2.192b}$$

where m_{e} and S_{e} are the mass and spin of the electron, respectively. Using (2.192a, b) in (2.191), the Hamiltonian is

$$H_{\text{hf}} = \left(\frac{\mu_0 g_{\text{p}} e^2}{m_{\text{p}} m_{\text{e}}}\right)\left\{\frac{1}{8\pi r^3}\left[3\frac{(S_p \cdot \vec{r})(S_e \cdot \vec{r})}{r^2} - S_p \cdot S_e\right] + \frac{1}{3}S_p \cdot S_e\delta^3(r)\right\} \tag{2.193}$$

Neglecting the orbital contribution to the magnetic field in a neutral hydrogen atom shall lead to the consideration of only the spherically symmetrical s wave function. Accordingly, from the first-order perturbation theory, the expectation value of the second part of (2.193) alone contributes as $\delta^3(r) = |\Psi_s|^2$. The hyperfine interaction energy is now given by

$$E_{\text{hf}} = \left(\frac{\mu_0 g_p e^2}{3m_p m_e}\right)\langle S_p \cdot S_e\rangle|\Psi_s|^2 = \left(\frac{\mu_0 g_p e^2}{3m_p m_e}\right)\left(\frac{1}{\pi a_o^3}\right)\langle S_p \cdot S_e\rangle \tag{2.194}$$

where the ground state s wave function is (see Sect. 2.3.5 and Sect. 2.4)

$$\Psi_s = \Psi_{nlm} = \Psi_{100} = \frac{1}{\sqrt{\pi a_o^3}} \exp(-r/a_o) \tag{2.195}$$

The energy eigenvalue (2.194) can be written as

$$E_{\text{hf}} = A S_p \cdot S_e \tag{2.196}$$

where A is the coupling constant.

The total spin $S = S_p + S_e$ and can take spin values such that $S = 0$ (singlet) and $S = 1$ (triplet). Evaluation of $\langle S_p \cdot S_e\rangle$ gives $-\frac{3}{4}\hbar^2$ for the singlet and $\frac{1}{4}\hbar^2$ for the triplet. The difference in energy between the singlet and the triplet state could be given in terms of the fine structure constant (α) as

$$\Delta E_{\text{hf}} = A = \frac{4 g_p \hbar^4}{3m_p m_e^2 c^2 a_o^4} = \frac{8}{3k^2} g_p \frac{m_e}{m_p} \alpha^2 E_1 \approx 6 \times 10^{-6} \text{ eV}$$

The hyperfine splitting energy difference corresponds to a frequency of $\nu = \frac{\Delta E}{h} = 1420$ MHz and wavelength $\lambda = 21$ cm. The singlet ($F = 0$) and triplet ($F = 1$) energy states and their magnitude of energy splitting for neutral hydrogen are shown in Fig. 2.29.

The ground state singlet with antiparallel spin arrangement between the proton and the electron has low energy. The spin magnetic moments of the proton and electron in the singlet state are opposite due to the charges. Since the spin-flip transition is forbidden as per the selection rules, the spontaneous emission rate is very low, and the atom de-excites through atomic collision. However, the interstellar medium has a vast amount of neutral hydrogen, sufficient time for random slip flip and fewer chances for collision and therefore, the $\lambda = 21$ cm radiation could be observed using radio astronomy. The triplet-singlet hyperfine transition of neutral hydrogen is a well-known example of magnetic dipole transition.

For the nucleus with angular momentum I and the electron with angular momentum J, the total angular momentum is given as $F = I + J$. We now consider the single proton and single electron situation for the neutral hydrogen. The hyperfine energy for the neutral hydrogen atom for $l = 0$ becomes

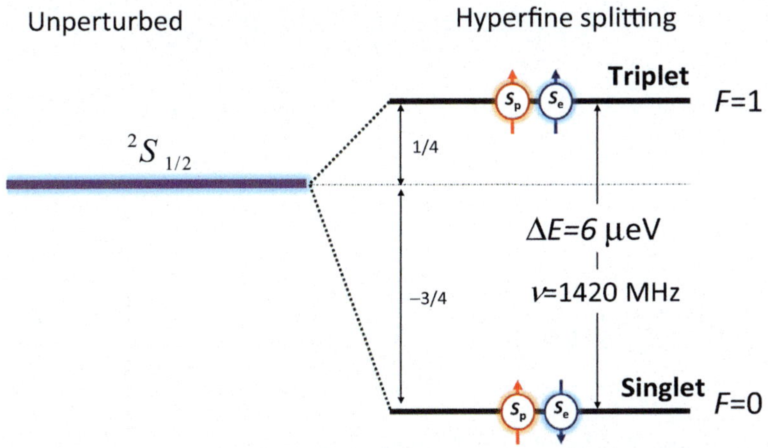

Fig. 2.29 The hyperfine structure of neutral hydrogen has high energy triplet and low energy singlet states separated by an energy of 6 μeV. The transition from high to low energy state results in an emission frequency of 1420 MHz

$$E_{\mathrm{hf}} = \left(\frac{\mu_0 g_{\mathrm{p}} e^2}{3m_{\mathrm{p}} m_{\mathrm{e}}}\right)\langle I \cdot J\rangle|\Psi_{\mathrm{s}}|^2 = \left(\frac{\mu_0 g_{\mathrm{p}} e^2}{3m_{\mathrm{p}} m_{\mathrm{e}}}\right)\left(\frac{1}{\pi a_o^3}\right)\langle I \cdot J\rangle \tag{2.197}$$

For a one-electron system with a nucleus, we can write (2.197) as

$$E_{\mathrm{hf}} = \left(\frac{2\mu_0 g_I g_{\mathrm{e}} \mu_{\mathrm{B}} \mu_{\mathrm{N}}}{3}\right)\frac{\langle I \cdot S_e\rangle}{\hbar^2}|\Psi_{\mathrm{s}}|^2 \tag{2.198}$$

where g_I, g_{e}, μ_{B}, μ_{N} are the nuclear g-factor, electron g-factor, Bohr magneton and nuclear magneton, respectively. Since $L = S$, using

$$\langle I \cdot S\rangle = \frac{\hbar^2}{2}(F(F+1) - I(I+1) - S(S+1)) \tag{2.199}$$

the hyperfine energy is

$$E_{\mathrm{hf}} = \left(\frac{\mu_0 g_I g_{\mathrm{e}} \mu_{\mathrm{B}} \mu_{\mathrm{N}}}{3}\right)(F(F+1) - I(I+1) - S(S+1))|\Psi_{\mathrm{s}}|^2 \tag{2.200}$$

and the coupling constant for the split 1s levels of hydrogen atom is given as

$$A = \left(\frac{2\mu_0 g_I \mu_{\mathrm{B}} \mu_{\mathrm{N}}}{3}\right)|\Psi_{\mathrm{s}}|^2 \tag{2.201}$$

For a one-electron system with $l > 0$, the coupling constant is

$$A = \left(\frac{\mu_0 g_I \mu_B \mu_N}{4\pi} \right) \frac{2l(l+1)}{J(J+1)} \langle nl | r^{-3} | nl \rangle \tag{2.202}$$

The first term in (2.193) is the dipolar Hamiltonian given in terms of tensor product as

$$H_{\text{dip}} = S^T \cdot \tilde{A} \cdot I \tag{2.203}$$

$$H_{\text{dip}} = \frac{\mu_0 g_I g_e \mu_B \mu_N}{4\pi} [\, S_x \; S_y \; S_z \,] \cdot \begin{bmatrix} \left(\frac{3x^2 - r^2}{r^5} \right) & \left(\frac{3xy}{r^5} \right) & \left(\frac{3xz}{r^5} \right) \\ \left(\frac{3yx}{r^5} \right) & \left(\frac{3y^2 - r^2}{r^5} \right) & \left(\frac{3yz}{r^5} \right) \\ \left(\frac{3zx}{r^5} \right) & \left(\frac{3zy}{r^5} \right) & \left(\frac{3z^2 - r^2}{r^5} \right) \end{bmatrix} \cdot \begin{bmatrix} I_x \\ I_y \\ I_z \end{bmatrix} \tag{2.204}$$

where the matrix A is traceless and symmetric across the diagonal.

The average dipolar interaction hyperfine tensor is given as (Hertel and Schulz 2015)

$$\langle \tilde{A} \rangle = \frac{\mu_0 g_I g_e \mu_B \mu_N}{4\pi \hbar^2} \begin{pmatrix} \left(\frac{3x^2 - r^2}{r^5} \right) & \left(\frac{3xy}{r^5} \right) & \left(\frac{3xz}{r^5} \right) \\ \left(\frac{3yx}{r^5} \right) & \left(\frac{3y^2 - r^2}{r^5} \right) & \left(\frac{3yz}{r^5} \right) \\ \left(\frac{3zx}{r^5} \right) & \left(\frac{3zy}{r^5} \right) & \left(\frac{3z^2 - r^2}{r^5} \right) \end{pmatrix} \tag{2.205}$$

Although we have discussed the atomic hyperfine splitting so far that is utilized in electron paramagnetic resonance (EPR) spectroscopy, its extension to the nuclear energy levels is the basis for Mössbauer spectrometry (see Sect. 5.4).

2.12 Transition Probability

The transition probability or decay constant for a transition from one state to another could be obtained from quantum mechanics. Consider a transition from an initial unperturbed state to a final state. The transition probability corresponding to these states can be found in the time-dependent perturbation theory.

The Hamiltonian has a time-independent part H_o and a time-dependent perturbation part $V'(t)$, which is represented in the interaction form as

$$H = H_o + V'(t) \tag{2.206}$$

The state vector in the stationary state is given by

$$|\psi_n(t)\rangle = \exp\left(-\frac{i H_o t}{\hbar} \right) |\psi_n\rangle \tag{2.207}$$

The Schrödinger equation in the perturbed state is

$$H|\psi(t)\rangle = H_o + V'(t)|\psi(t)\rangle = i\hbar\frac{\partial|\psi(t)\rangle}{\partial t} \tag{2.208}$$

The interaction representation is

$$\begin{aligned} i\hbar\frac{\partial}{\partial t}|\psi(t)\rangle &= e^{i\frac{H_o t}{\hbar}}(H - H_o)|\psi(t)\rangle \\ &= e^{i\frac{H_o t}{\hbar}}V'(t)e^{-i\frac{H_o t}{\hbar}}|\psi(t)\rangle = V_I'(t)|\psi(t)\rangle \end{aligned} \tag{2.209}$$

Using linear combinations

$$|\psi(t)\rangle = \sum c_n(t)|\psi_n\rangle \exp\left(-\frac{iE_n t}{\hbar}\right) \tag{2.210}$$

The time derivative of $c_n(t)$ is then

$$\frac{\partial c_n(t)}{\partial t} = \frac{1}{i\hbar}\sum_k c_k(t)V_{nk}'(t)\exp(i\omega_{nk}t) \tag{2.211}$$

where $V_{nk}'(t) = \langle\psi_n|V'(t)|\psi_k\rangle$ is the perturbation matrix element.

Using $i\hbar\frac{\partial|\psi(t)\rangle}{\partial t} = i\hbar\frac{\partial}{\partial t}\left(\sum c_n(t)|\psi_n\rangle\exp\left(-\frac{iE_n t}{\hbar}\right)\right)$, we get the final transition probability corresponding to initial state i and final state f

$$P_{\text{if}}(t) = \left|-\frac{i}{\hbar}\int_0^t \langle\psi_f|V'(t')|\psi_i\rangle\exp(i\omega_f t')dt'\right|^2 \tag{2.212}$$

When V' is time independent and $t \to \infty$, then

$$P_{\text{if}}(t) = \frac{2\pi t}{\hbar}|\langle\psi_f|V'|\psi_i\rangle|^2\delta(E_f - E_i) \tag{2.213}$$

The transition rate is given by

$$\Gamma_{\text{if}} = \frac{P_{\text{if}}(t)}{t} = \frac{2\pi}{\hbar}|\langle\psi_f|V'|\psi_i\rangle|^2\delta(E_f - E_i) \tag{2.214}$$

The total transition rate for a transition from an initial state to a continuum of final states is given by

$$W_{\text{if}} = \int\frac{P_{\text{if}}(t)}{t}\rho(E_f)dE_f$$

$$= \frac{2\pi}{\hbar}|\langle\psi_f|V'|\psi_i\rangle|^2 \int \rho(E_f)\delta(E_f - E_i)dE_f \tag{2.215}$$

where ρ is the density of the particular state. The total time-independent transition rate is given as

$$W_{\text{if}} = \frac{2\pi}{\hbar}|\langle\psi_f|V'|\psi_i\rangle|^2 \rho(E_i) \tag{2.216}$$

known as *Fermi's golden rule*.

The absorption transition rate is

$$W_{\text{abs}} = \frac{2\pi}{\hbar}|\langle\psi_f|V'|\psi_i\rangle|^2 \rho(E_g - E_e = \hbar\omega) \tag{2.217}$$

The emission transition rate is

$$W_{\text{emiss}} = \frac{2\pi}{\hbar}|\langle\psi_f|V'|\psi_i\rangle|^2 \rho(E_e - E_g = \hbar\omega) \tag{2.218}$$

where the excited and ground state energies are represented as E_e and E_g, respectively.

Oscillating charges produce oscillating electric and magnetic field lines generating electromagnetic waves that travel at the speed of light in vacuum. A dipole antenna with oscillating electric charges radiates electromagnetic waves and receives them as the waves oscillate the charges. The electromagnetic waves propagate in a direction orthogonal to the conductor of oscillating charges, as shown in Fig. 2.30 (left). The electric and magnetic field components and direction of propagation are mutually orthogonal (Fig. 2.30 right). The oscillating electric field component interacts with a charged particle, such as an electron bound to the positively charged nucleus, resulting in a change of energy states known as electric dipole transition. The interaction of magnetic field components, in turn, results in magnetic dipole transition. The nucleus and electron pair constitute a dipole, and the feasible transitions to various available quantized energy states can be approximated. The potential energy for an electric dipole is given as $U = -\mu_E \cdot E$ where μ_E is the electric dipole moment and E is the electric field.

The electric and magnetic field components of electromagnetic waves are given as

$$E = E_o \exp i(k \cdot r - \omega t) \text{ and } B = B_o \exp i(k \cdot r - \omega t) \tag{2.219}$$

Consider the electric field component, which is given as

$$E = E_o \exp i(k \cdot r) \exp(-i\omega t) \tag{2.220}$$

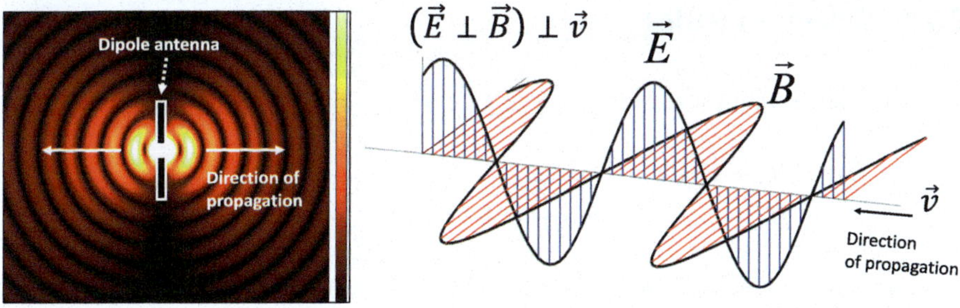

Fig. 2.30 The propagation of electromagnetic waves from a dipole antenna (left). Electromagnetic waves are made of oscillating electric and magnetic field components, propagating in a direction orthogonal to each other

The Taylor expansion of the first exponential term of (2.220) is

$$\exp i(k \cdot r) = 1 + i(k \cdot r) + \frac{[i(k \cdot r)]^2}{2!} + \frac{[i(k \cdot r)]^3}{3!} + \cdots \tag{2.221}$$

For radio waves, $k = 2\pi/\lambda$ is less than the atomic size determined by r (r ~ few Å; λ > hundreds of Å), $k \cdot r \ll 1$ and therefore the higher order terms could be neglected to obtain $\exp i(k \cdot r) = 1$. This approximation is also good for nuclear gamma radiation as $\lambda \gg$ nuclear size. The electric dipole approximation holds well when expanded in terms of multipole order (l) with $l = 1$, whereas if $i(k \cdot r)$ and higher order terms are considered, they give quadrupole ($l = 2$), octupole ($l = 3$) and other multipoles. The most probable transition corresponds to the electric dipole term, and the next higher-order term becomes prominent when the lower order is absent. The transition probability is reduced by a factor of $2\pi r/\lambda$ for each higher-order expansion. The Hamiltonian considering only the electric part of the dipole approximation is called the electric dipole Hamiltonian.

The expectation value of the energy due to the electric field from the initial state to the final state wave function is given as $\langle U \rangle = \int \psi_i^*(-\mu_E \cdot E)\psi_f dr$. For a constant electric field, we can write $\langle U \rangle = -E \int \psi_i^*(\hat{\mu}_E)\psi_f dr$ and hence $\langle \mu_E \rangle = \int \psi_i^* \hat{\mu}_E \psi_f dr$. The expectation value $\langle \mu_E \rangle$ for all possible transitions can be arranged in an $n \times n$ matrix. The dipole moment operator is then given as an off-diagonal matrix element $M_{\text{if}} = -e\langle \psi_i | r | \psi_f \rangle$ using $\mu_E = q \cdot r$, where r is the distance from the nucleus and charge $q = -e$. Selection rules emanate from the evaluation of matrix element, and a transition is allowed if $|M_{\text{if}}|^2 \neq 0$ and forbidden if $|M_{\text{if}}|^2 = 0$. The magnetic part of the electromagnetic wave interacts with the intrinsic magnetic moment of the electron that contributes to the magnetic dipole transition. The magnetic moment operator of an atom is assumed to be of the form $\hat{\mu} = -\frac{e}{2m}(L + 2S)$.

2.13 Selection Rules

The interaction Hamiltonian is given as

$$\begin{aligned} H &= H_o + V_{EI}(t) + V_{MI}(t) + \cdots \\ &= H_o + [V_{E1}(t) + V_{M1}(t)] + [V_{E2}(t) + V_{M2}(t)] + \cdots \end{aligned} \quad (2.222)$$

$V_{E1}(t)$ is the electric dipole transition (E1), $V_{M1}(t)$ is the magnetic (M1) dipole transition, $V_{E2}(t)$ is the electric quadrupole and $V_{M2}(t)$ is the magnetic quadrupole interaction terms. The selection rules allow transitions involving changes from one quantum state to another, while certain other transitions may be less probable and termed forbidden transitions. The summary of selection rules for the prominent electric (E1, E2) and magnetic dipole (M1) transitions (Foot 2005) is shown in Table 2.6.

Table 2.6 The selection rules for the electric and magnetic dipole transitions

Electric dipole transitions (E1)		**Magnetic dipole transitions (M1)**	
	$\Delta J = 0, \pm 1$		$\Delta J = 0, \pm 1$
$J = 0 \leftrightarrow J = 0$	Not allowed	$J = 0 \leftrightarrow J = 0$	Not allowed
	$\Delta m_J = 0, \pm 1$		$\Delta m_J = 0, \pm 1$
$\pi_i \neq \pi_f$	Parity changes	$\pi_i = \pi_f$	No parity change
LS coupling		*LS* coupling	
$\Delta S = 0$		$\Delta S = 0$	
$\Delta L = 0, \pm 1$		$\Delta L = 0$	
$L = 0 \leftrightarrow L = 0$	Not allowed		
$\Delta l = \pm 1$	One electron jump	$\Delta l = 0$	No electron jump
Δn	Any value	Δn	0
Electric dipole matrix element		Magnetic dipole matrix element	
$D_{\text{if}} = \langle \psi_i \vert x \vert \psi_f \rangle$		$M_{\text{if}} = \langle \psi_i \vert L + 2S \vert \psi_f \rangle$	

Electric quadrupole transitions (E2)	
$\Delta J = 0, \pm 1, \pm 2$	
$J = 0 \leftrightarrow J = 0$, $J = 0 \leftrightarrow J = 1$, $J = 1/2 \leftrightarrow J = 1/2$	Not allowed
$\Delta m_J = 0, \pm 1, \pm 2$	
$\pi_i = \pi_f$	No parity change
$\Delta l = 0, \pm 2$	No (or) one electron jump
$\Delta L = 0, \pm 1, \pm 2$	For heavy atoms
$\Delta S = 0$	For *LS* coupling
$L = 0 \leftrightarrow L = 0$, $L = 0 \leftrightarrow L = 1$	Not allowed
Electric quadrupole matrix element	
$Q_{\text{if}} = \langle \psi_i \vert x_j x_k - r^2 \delta_{jk}/3 \vert \psi_f \rangle$	

2.14 Spectral Linewidths

The energy difference between the transition from an initial state with lower energy (E_i) to the final state with higher energy (E_f) is $\Delta E = E_f - E_i$, which is expected to produce a discrete value. However, we obtain a distribution of energies producing a linewidth instead of the ideal single line. The shape of such linewidths is determined by the intensities I_i and I_f corresponding to the respective energy levels E_i and E_f. An atom emits radiation spontaneously within a typical lifetime of 1–10 ns. The lifetime of states will result in a natural linewidth following the uncertainty relation.

The radiation emission characteristics due to atomic electrons can be modeled using a damped harmonic oscillator, as shown in Fig. 2.31. Here, the number of oscillators can be considered for a system of N atoms oscillating with a central frequency ω_o corresponding to the underdamped case. The mass, spring constant, angular frequency and damping constant are m, k, ω and γ, respectively. The corresponding differential equation of motion for a single atom is given as

$$\begin{aligned} m\frac{d^2x}{dt^2} + b\frac{dx}{dt} + kx = 0 \\ \frac{d^2x}{dt^2} + \gamma\frac{dx}{dt} + \omega_o^2 x = 0 \end{aligned} \tag{2.223}$$

where $\omega_o^2 = \frac{k}{m}$ and for a damping coefficient b, $\gamma = b/m$.

Using the initial condition at $t = 0$, $x(t) = x_o$ and $\gamma < \omega_o$, the solution is

$$x(t) = x_o \exp(-\gamma t/2)(\cos \omega t + \varphi) \tag{2.224}$$

where φ is the phase term that is a sine function. The above decay has an exponential decrease in magnitude given by $x_o\exp(-\gamma t/2)$ as shown in Fig. 2.32.

For very low damping and $\omega \cong \omega_o$,

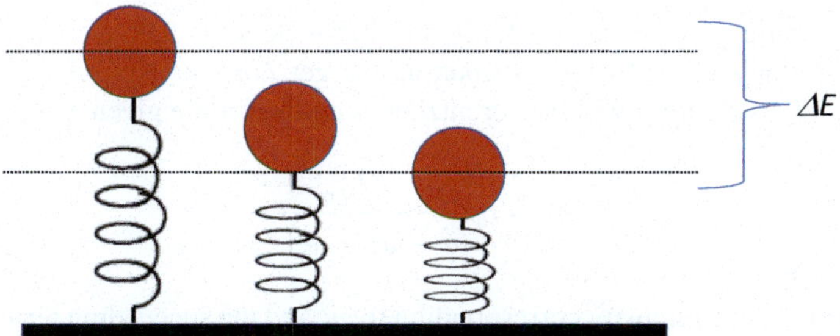

Fig. 2.31 The damped harmonic oscillator model of atom emitting radiation where the energy width (ΔE) is determined by the magnitude of the oscillations

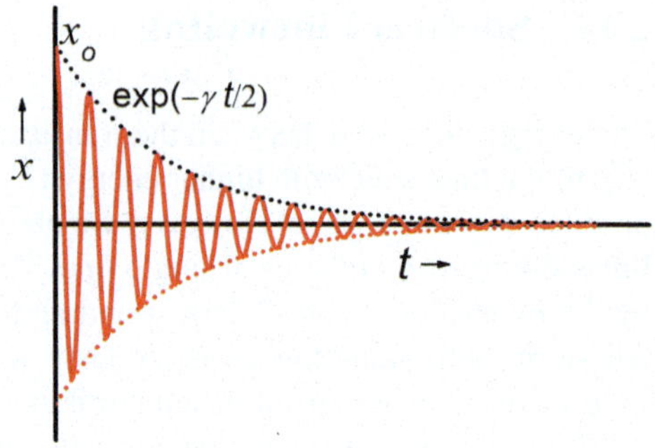

Fig. 2.32 The underdamped harmonic oscillator model with a damping coefficient γ shows exponential decay of the periodic oscillations with time t from a maximum amplitude of x_o

$$x(t) = x_o \exp(-\gamma t/2)(\cos\omega_o t) \tag{2.225}$$

where the central frequency $\omega_o = 2\pi \nu_o$ or $\omega_o = \frac{(E_f - E_i)}{\hbar}$.

Now, for a system of N oscillators of different frequencies, the solution can be written as

$$x(t) = \frac{1}{2\sqrt{2\pi}} \int_0^{\infty} A(\omega) \exp(i\omega t) d\omega \tag{2.226}$$

where we have used the integration for ω from 0 to ∞. The amplitude $A(\omega)$ is obtained from the Fourier transform as

$$A(\omega) = \frac{1}{\sqrt{2\pi}} \int_0^{\infty} x_o \exp(-\frac{\gamma t}{2}) \cos(\omega_o t) \exp(-i\omega t) dt \tag{2.227}$$

$$A(\omega) = \frac{x_o}{\sqrt{8\pi}} \left(\frac{1}{i(\omega - \omega_o) + \frac{\gamma}{2}} + \frac{1}{i(\omega + \omega_o) + \frac{\gamma}{2}} \right) \tag{2.228}$$

Neglecting the $\omega + \omega_o$ second term in (2.228) due to its feeble contribution and using the normalized intensity distribution, we get $I(\omega) = \int_{-\infty}^{+\infty} |A(\omega)|^2 d\omega = 1$. The intensity distribution will be Lorentzian, with the profile given as

$$I(\omega) = I_o \frac{\gamma/2\pi}{(\omega - \omega_o)^2 + \left(\frac{\gamma}{2}\right)^2} \tag{2.229}$$

where I_o is the peak intensity corresponding to ω_o and the succeeding term is known as the normalized Lorentzian profile. The full width at half maximum (FWHM) is given as $\Gamma = \Delta\omega_n$ where the intensity of the profile is reduced to a value of *1/e*. Figure 2.33 (left) shows the time-dependent oscillations where the damping term γ determines the time to decay as per (2.225). The Fourier transform of the oscillations given in the frequency domain (Fig. 2.33 right) shows a Lorentzian profile with a sharp peak at ω_o where the width of the profile is affected by γ according to (2.229).

Generally, the Lorentzian profile is a probability density function of the form

$$L(x) = \frac{1}{\pi} \frac{\Gamma/2}{(x - x_o)^2 + \left(\frac{\Gamma}{2}\right)^2} \tag{2.230}$$

where Γ is the FWHM, x_o is the location parameter or median and the profile is also known as Cauchy distribution or Breit-Wigner distribution. The intensity of the profile comes down as the FWHM increases, as shown in Fig. 2.34. Apart from the natural broadening that determines the minimum FWHM, various other contributions, such as thermal Doppler broadening, pressure broadening and inhomogeneous broadening, may increase the FWHM, resulting in a Voigt line profile.

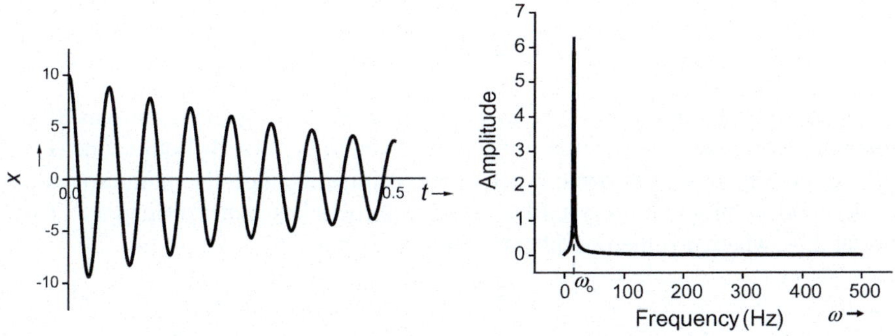

Fig. 2.33 The time-dependent oscillations of a harmonic oscillator (left) will take a longer time to decay if the damping constant γ is small. The Fourier transform of the oscillations shows a Lorentzian profile with sharp amplitude centered at a frequency ω_o (right)

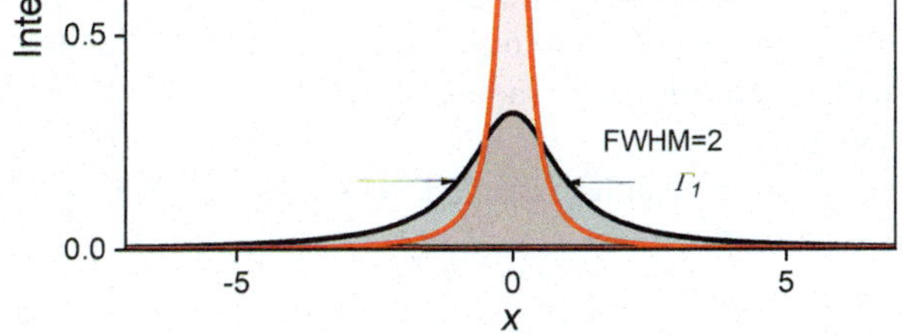

Fig. 2.34 The Lorentzian profiles with $\Gamma_1 = 2$ and $\Gamma_2 = 0.5$. The line profile is narrow when the FWHM is smaller, whereas the intensity decreases, and the Lorentzian becomes broader with a large FWHM

2.15 The Solid State

The lattice vibrations influence the recoil-free emission of gamma rays to observe the Mössbauer effect in solids. The quantization of lattice vibrations and their impact on the heat capacity are described in this section.

For a Bravais lattice vector *R*, the electron is in a periodic potential given by

$$U(r+R) = U(r) \tag{2.231}$$

The Hamiltonian for such a system is of the form

$$H = -\frac{\hbar^2}{2m}\nabla^2 + U(r).$$

The eigenstate is

$$\Psi_{nk}(r) = \exp(ik \cdot r)u_{nk}(r) \tag{2.232}$$

where it can be written as $\Psi_{nk}(r+R) = \exp(ik \cdot r)u_{nk}(r)$, which is known as Bloch's theorem. Here, the wavevector $k = \frac{2\pi}{\lambda}$ is confined to the first Brillouin zone (FBZ). FBZ is the Wigner–Seitz primitive cell of the reciprocal lattice, the minimum volume basic building block of a crystal lattice in reciprocal space. The boundary of FBZ lies at $\pm\frac{\pi}{a}$, where a is the interplanar spacing.

2.15.1 Phonons

Phonons are quantized lattice vibrational modes in a lattice. In three dimensions, the vibrations can occur along the direction of propagation in the form of longitudinal waves and perpendicular to the direction of propagation in the form of transverse waves. There can be two types of phonons based on the displacement of the neighboring lattice, acoustic and optic phonons, as shown in Fig. 2.35. If there is a coherent movement of the atoms along the direction of propagation, it is known as a longitudinal acoustic (LA) phonon, whereas if it happens perpendicular to the direction of propagation, it is a transverse acoustic (TA) phonon. If the displacement of the atoms is in the opposite direction, then similar to previous cases, we have longitudinal optical (LO) and transverse optical (TO) modes. Optical modes can be excited with electromagnetic radiation, whereas acoustic modes can be excited by heat and sound.

As atoms can be assumed to be a mass-spring system whose displacement along x direction is given as u_s, the displacement between successive planes is given by

$$\frac{u_{s+1}}{u_s} = \exp(ika) \tag{2.233}$$

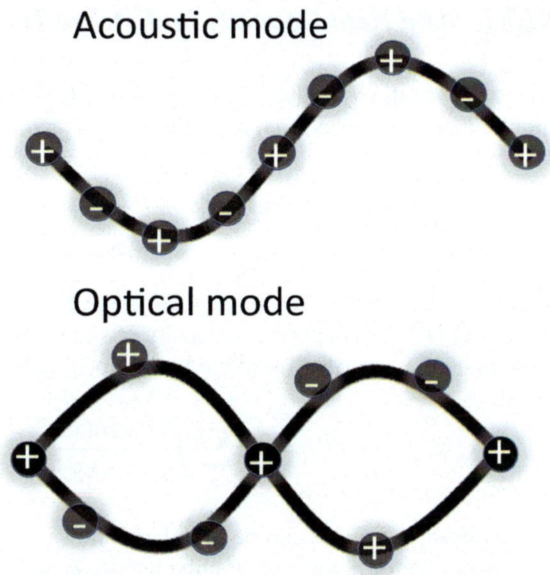

Fig. 2.35 The acoustic and optic phonons are based on the displacement of the neighboring lattice. The optical mode shows the opposite displacement of neighboring atoms in the lattice

The frequency versus k relationship, known as the dispersion relation, is given by

$$\omega = \sqrt{\frac{4K}{M}} \sin\left(\frac{ka}{2}\right) \tag{2.234}$$

for a monoatomic lattice of force (spring) constant, K and mass of the atom, M.

There are 3 acoustical and $3N - 3$ optical branches for N atoms in the primitive cell. Thus, the acoustic branch requires a minimum of two atoms per primitive cell. The frequencies at FBZ and $ka \ll 1$ for masses M_1 and M_2 corresponding to the optical branch is

$$\omega_{\mathrm{op}} = \sqrt{2K\left(\frac{1}{M_1} + \frac{1}{M_2}\right)} \tag{2.235}$$

The acoustic branch gives a frequency

$$\omega_{\mathrm{ac}} = \sqrt{\frac{1}{2}\left(\frac{K}{M_1 + M_2}\right)}ka \tag{2.236}$$

2.15.2 The Heat Capacity and Debye Temperature

The energy of the oscillators of frequency $\omega(k, p)$ using Planck distribution and the phonon modes is given as (Kittel 2005)

$$U = \sum_p \int \frac{\hbar\omega(k)}{\exp\left(\frac{\hbar\omega(k)}{k_B T}\right) - 1} D_p(\omega) d\omega \tag{2.237}$$

where p is the polarization modes, k_B is the Boltzmann constant (ISO symbol is k or k_B) and $D_p(\omega)$ is the phonon density of states. Putting $x = \frac{\hbar\omega}{k_B T}$ and using $\frac{\partial U}{\partial T}$, we get the phonon contribution to the lattice specific heat as

$$C_L = k_B \sum_p \int \frac{\exp(x)x^2}{(\exp(x) - 1)^2} D_p(\omega) d\omega \tag{2.238}$$

The density of states can be found as follows:
The number of modes, N, of cube length, L is

$$N = \left(\frac{L}{2\pi}\right)^3 \left(\frac{4}{3}\pi k^3\right) \tag{2.239}$$

Using the Debye model, where the particle velocity (v) is taken as constant and assuming acoustic mode contribution, we get $\omega = vk$. The density of states is

$$D(\omega) = \frac{dN}{d\omega} = \left(\frac{L^3}{2\pi^2}\right)\left(k^2 \frac{dk}{d\omega}\right) = \left(\frac{L^3}{2\pi^2}\right)\left(\frac{\omega^2}{v^3}\right) \tag{2.240}$$

Using the integration limits of 0 at a cutoff frequency ω_D, the thermal energy then becomes

$$U = \int_0^{\omega_D} \left(\frac{L^3}{2\pi^2}\right)\left(\frac{\omega^2}{v^3}\right)\left(\frac{\hbar\omega}{\exp\left(\frac{\hbar\omega}{k_B T}\right) - 1}\right) d\omega \tag{2.241}$$

Considering the three phonon polarization modes (1 longitudinal and 2 transverse),

$$U = \frac{3V\hbar}{2\pi^2 v^3} \int_0^{\omega_D} \left(\frac{\omega^3}{\exp\left(\frac{\hbar\omega}{k_B T}\right) - 1}\right) d\omega \tag{2.242}$$

where we have taken $L^3 = V$. Further,

$$U = \frac{3Vk_B^4T^4}{2\pi^2 v^3\hbar^3}\int_0^{x_D}\left(\frac{x^3}{e^x-1}\right)dx \tag{2.243}$$

where $x_D = \frac{\hbar\omega_D}{k_B T} = \frac{\Theta_D}{T}$. The Debye temperature Θ_D is limited by a cutoff frequency ω_D (ISO symbol) given by

$$\omega_D = k_D v = \left(\frac{6\pi^2 v^3 N}{V}\right)^{1/3} \tag{2.244}$$

where N is the number of acoustic phonon modes determined by the number of primitive cells. The thermal energy from (2.243) with Θ_D is

$$U = 9Nk_B\frac{T^4}{\Theta_D^3}\int_0^{x_D}\left(\frac{x^3}{e^x-1}\right)dx \tag{2.245}$$

The heat capacity then becomes

$$C_V = \frac{dU}{dT} = 9Nk_B\left(\frac{T^3}{\Theta_D^3}\right)\int_0^{x_D}\frac{x^4e^x}{(e^x-1)^2}dx \tag{2.246}$$

At $T \ll \Theta_D$

$$C_V = 36Nk_B\left(\frac{T^3}{\Theta_D^3}\right)\frac{\pi^4}{15} \tag{2.247}$$

At $T \gg \Theta_D$

$$C_V = 3Nk_B \tag{2.248}$$

which is the Dulong-Petit law. The specific heat capacity contribution by free electron Fermi gas from the Fermi–Dirac distribution function is

$$C_{ele} = k_B^2 T D(E_F)\int_{-E_F/k_BT}^{\infty}\frac{x^2e^x}{(e^x+1)^2}dx \tag{2.249}$$

where
$x = \frac{(E-E_F)}{k_BT}$ and E, E_F and $D(E_F)$ are the orbital energy, Fermi energy and density of states at the Fermi energy, respectively.

At low temperatures

$$C_{\text{ele}} = k_{\text{B}}^2 T D(E_{\text{F}}) \frac{\pi^2}{3} = k_{\text{B}}^2 T \left(\frac{3N}{2k_{\text{B}} T_{\text{F}}} \right) \frac{\pi^2}{3} = \frac{\pi^2}{2} N k_{\text{B}} \left(\frac{T}{T_{\text{F}}} \right) \tag{2.250}$$

where we have used $D(E_{\text{F}}) = \frac{3N}{2k_{\text{B}} T_{\text{F}}}$ and the Fermi temperature $T_{\text{F}} = E_{\text{F}}/k_{\text{B}}$. The heat capacity of metals is thus a sum of electron and phonon contribution given by

$$C = \gamma T + A T^3 \tag{2.251}$$

where γ and A are constants.

2.16 Summary

Fundamentals of quantum mechanics that include the Schrödinger equation, its applications in an infinite and finite potential well, and the hydrogen atom are introduced. The role of angular momentum quantum numbers in determining the characteristics of electromagnetic transitions, periodic table and fine structures is elaborated. This chapter has given an overview of the quantum aspects of photons and phonons essential for understanding Mössbauer spectroscopy.

References

G.B. Arfken, H.J. Weber, *Mathematical Methods for Physicists*, 6th edn. (Elsevier, MA, 2005)
C.J. Foot, *Atomic Physics* (Oxford University Press, NY, 2005)
W. Greiner, *Quantum Mechanics: An Introduction*, 4th edn. (Springer, Heldelberg, 2001)
D.J. Griffiths, Am. J. Phys. **50**, 698 (1982)
D.J. Griffiths, *Introduction to Quantum Mechanics*, 2nd edn. (Pearson Prentice Hall, NJ, 2005)
I.V. Hertel, C.P. Schulz, *Atoms, Molecules and Optical Physics 1: Atoms and Spectroscopy* (Springer, Heidelberg, 2015)
C. Kittel, *Introduction to Solid State Physics*, 8th edn. (John Wiley, NJ, 2005)
E. Merzbacher, *Quantum Mechanics*, 3rd edn. (John Wiley, NY, 1998)
R. Shankar, *Principles of Quantum Mechanics*, 2nd edn. (Plenum Press, NY, 1994)
J.C. Slater, Phys. Rev. **36**, 57 (1930)
P. van Alem, *Hydrogen Orbitals*, https://www.mathworks.com/matlabcentral/fileexchange/64274-hydrogen-orbitals. MATLAB Central File Exchange. Retrieved 16 Feb 2021
N. Zettili, *Quantum Mechanics: Concepts and Applications* (John Wiley, NY, 2001)
P.A. Zyla et al. (Particle Data Group), Prog. Theor. Exp. Phys. **2020**, 083C01 (2020)

Chapter 3
Introduction to Magnetism

3.1 Magnetic Moment

Circulating charges produce magnetic moment (μ) given by $\mu = IA$ where I is the current and A is the cross sectional area. An electric dipole is a pair of positive and negative charges, while a magnetic dipole is made of north and south poles, which cannot be separated into a monopole. A magnetic dipole placed in a uniform magnetic field will experience a torque acting on the direction of the field. The current loop producing a magnetic moment and a magnetic dipole in an external magnetic field B are shown in Fig. 3.1a and b, respectively.

The magnetic moment is a physical quantity determined by the work done by a dipole in a magnetic field. In an applied magnetic field, the dipole parallel to B is taken as the minimum energy position, and therefore, the magnetic potential energy (E_p) is related to μ and B as

$$E_p = -\mu \cdot B \tag{3.1}$$

The SI units of E_p, μ and B are J, A·m^2 and tesla (T), respectively. The unit for magnetic moment is given as J/T or A·m^2 in SI and erg/Oe or emu in cgs unit. The external magnetic fields B and H are related by

$$B = \mu_0 H \tag{3.2}$$

where $\mu_0 = 4\pi \times 10^{-7}$ H/m (henry/m) and H has the SI unit of A/m. Note that henry has further derived units J/A^2 = T·m^2/A = Wb/A = V·s/A. Mass, volume and molar magnetizations are defined with their SI units, as given below:

Mass magnetization,

$$M_m = \frac{\sum \mu}{m} \tag{3.3}$$

R. J. Joseyphus and J.-M. Greneche, *Fundamentals of ^{57}Fe Mössbauer Spectrometry*,
https://doi.org/10.1007/978-981-99-8653-8_3

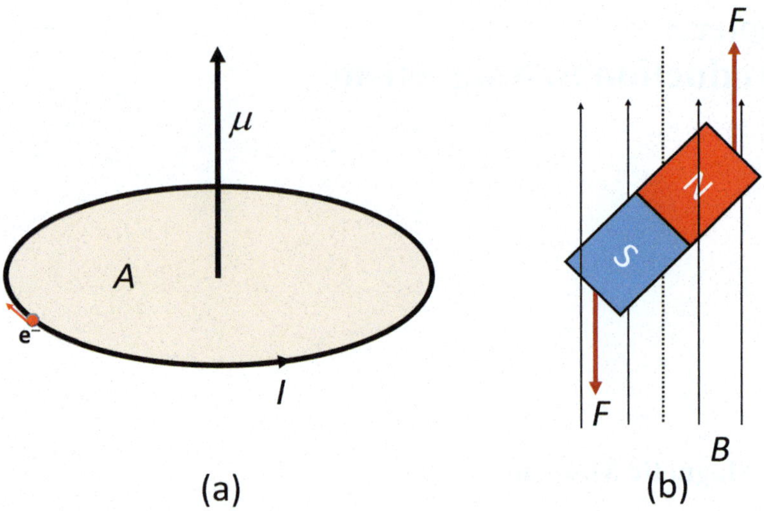

Fig. 3.1 **a** Current loop producing a magnetic moment, **b** bar magnet experiencing a torque along the applied magnetic field

which is the total magnetic moment per unit mass of the substance with the unit A·m^2/kg. Volume magnetization,

$$M_V = \frac{\sum \mu}{V} \tag{3.4}$$

which is the sum of magnetic moment per unit volume of the substance with the unit A/m. The volume magnetization has the unit of T from the relation

$$M = \mu_0 M_V \tag{3.5}$$

Molar magnetization,

$$M_{mol} = \frac{\sum \mu}{mol} \tag{3.6}$$

which is the sum of magnetic moment per unit molar concentration with the unit A·m^2/mol. In case magnetization is represented as M, the volume or mass magnetization can be inferred from their units. A solenoid of length L, having N number of turns, on passing a current, I through it generates a magnetic field (H) given in terms of the unit A/m by

$$H = \frac{NI}{L} \tag{3.7}$$

Table 3.1 The cgs and SI units of various physical quantities used in magnetism

Physical quantity	cgs	SI
magnetic moment (m, μ)	emu	A·m^2
Mass magnetization (M)	emu/g	A·m^2/kg
Volume magnetization (M)	emu/cc	A/m
Magnetic field (H)	Oe	A/m
Magnetic field (B)	G	T
Permeability ($\mu = B/H$)	No unit	H/m
Energy (E)	erg	J
Magnetic flux ($\varphi = BA$)	Mx	Wb
Magnetic susceptibility ($\chi = M/H$)	emu/(Oe cc)	No unit

The H field is related to the B field in tesla (T) using the relation

$$B = \mu_0 H = \mu_0 \frac{NI}{L} \tag{3.8}$$

The volume magnetization given by (3.5) and B field has the same unit of T. On inserting a magnetic material of volume magnetization, M, into a solenoid, the B field generated is

$$B = \mu_0 H + \mu_0 M = \mu_0 (H + M) \tag{3.9}$$

The B field in the cgs unit is gauss (G) with H in oersted (Oe) given as

$$B = H + 4\pi M \tag{3.10}$$

When $M = 0$, we have $B = H$ giving G = Oe. Often, cgs units are used for easy comparison (G and Oe have the same Gaussian unit). The physical quantities and their cgs and SI units are shown in Table 3.1. Although the definition of M or μ may vary with each textbook as magnetization may represent mass or volume magnetization and μ maybe permeability or magnetic moment, they could be identified from the units.

The magnetic moment is represented as μ if m is defined as mass. χ is the volume susceptibility here.

3.2 Magnetic Field

3.2.1 Biot-Savart Law

The flow of electrons in a wire produces a magnetic field, which forms the basis of the Biot-Savart law.

Consider current (I) flowing through a wire, and we want to measure the magnetic field element dB at point P. Each infinitesimal current element ($I\,dl$) contributes to the magnetic field at a distance perpendicular to the radius vector (r), as illustrated in Fig. 3.2. Based on the experimental observation, an empirical relationship is obtained as

$$dB \propto Idl \quad \propto 1/r^2$$

$$dB = \frac{\mu_0}{4\pi}\frac{\overrightarrow{dl} \times \vec{n}}{r^2} \tag{3.11}$$

where dl is the infinitesimal length of the conductor carrying current, I. The unit vector, $\vec{n}$, specifies the direction of r from the current element. A straight wire carrying current produces a circular magnetic field and vice-versa. The Biot-Savart law can be used to obtain the magnetic field generated by various current-carrying conductors, as given below.

(i) *Field due to a wire*

$$B = \frac{\mu_0 I}{2\pi R} \tag{3.12a}$$

where R is the distance from the wire carrying current I. Inside the wire, the field is given as

$$B = \frac{\mu_0 I r}{2\pi R^2} \tag{3.12b}$$

where r is the distance from the center of the wire and $r < R$.

(ii) *Field on the axis of a current loop*

$$B = \frac{\mu_0}{4\pi}\frac{2\pi R^2 I}{(z^2 + R^2)^{3/2}} \tag{3.13a}$$

where R is the radius of the wire and z is the axial distance from the center of the loop. At the center of the loop, $z = 0$ giving, $B = \frac{\mu_0 I}{2R}$.

(iii) *The field at the center of a square wire*

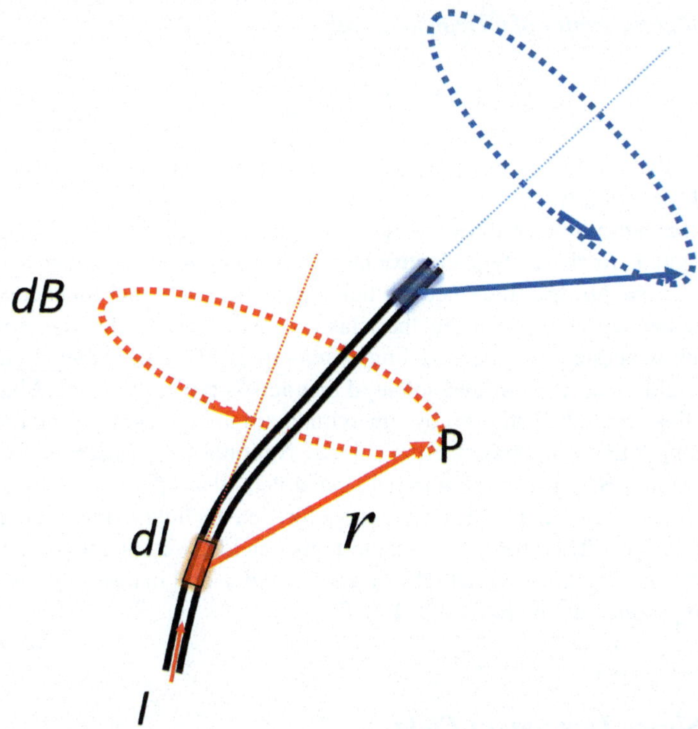

Fig. 3.2 A current-carrying conductor produces a magnetic field that could be determined using the Biot-Savart law

$$B = \frac{\sqrt{2}\mu_0 I}{\pi a} \tag{3.13b}$$

where a is the side of the square loop.

(iv) *Field due to a hollow cylinder*

$$B = \frac{\mu_0}{2\pi R} \frac{I\left(R^2 - a^2\right)}{\left(b^2 - a^2\right)} \tag{3.14}$$

in the region $a < R < b$ where R is the radius, a and b are inner and outer radii, respectively. R lies at a distance of $(a + b)/2$.

(v) *Field at the center of a solenoid*

$$B = \frac{\mu_0 N I}{L} \tag{3.15}$$

where N is the number of turns and L is the length of the solenoid. $L >> D$, where D is the diameter of the solenoid.

(vi) *Field at the center of a Helmholtz coil*

$$B = 0.716\frac{\mu_0 N I}{R} \tag{3.16}$$

where N is the number of turns in each coil and R is the distance between the coil pair equal to its radius.

Other magnetic field sources are electromagnets, superconducting magnets and pulsed magnets. Resistive-type electromagnets produce fields of a few T, superconducting coils can produce intermediate fields, and a much higher magnetic field can be produced by pulsed magnets, but these last only a few μs. As electromagnets using copper wire generate I^2R loss, a cooling water supply is required and the maximum magnetic field using wire wound solenoid magnets is typically 3–4 T. Much higher magnetic field could be achieved by superconducting magnets, with the record field of 32 T using YBCO superconducting coils by National High Magnetic Field Laboratory, Florida, USA. The maximum field generated by a superconducting magnet is limited by the critical field of the superconductor, i.e., the field required to destroy the superconductivity. Therefore, to generate higher fields, we opt for pulsed magnets in which capacitors discharge a high current through a coil in a short period of time, which can produce up to around 95–100 T.

3.2.2 Shape Dependent Fields

The hysteresis loop of a toroid sample saturates at a lower applied field than an open toroid with an air gap. The normal component of flux lines, which emerge through the surface as free poles, reenter the sample and oppose the magnetization. The opposing field is known as the *demagnetizing field* (H_d), which depends on the sample's shape and is given as $H_d = -N_d M$, where N_d is the dimension-dependent tensor known as the *demagnetization factor*. The demagnetization factor in three orthogonal directions is given as $N_x + N_y + N_z = 1$. Consequently, the spherical shape has $N_d = 1/3$ while it is 0 along the axis of a long cylinder. The magnetization is non-uniform according to the shape of the magnetic material, and N_d is not viable analytically for an arbitrary shape. The hysteresis loop of such samples shall be sheared towards a higher field since the effective field required for saturation is the sum of the external magnetic field and the negative demagnetizing field.

If there is a cavity inside the magnetic material, it will create opposite poles with N_d corresponding to the shape of the cavity. In the case of a spherical cavity, the internal magnetic field encountered is $M/3$ along the direction of M, called the *Lorentz field*. Consider a dipole of magnetic moment μ whose size is significantly small compared to a distance r where the field is measured. The dipolar field strength is given as

$$B = \frac{\mu_0}{4\pi}\left[\frac{3r(\mu \cdot r)}{r^5} - \frac{\mu}{r^3}\right] \tag{3.17}$$

The potential energy of two dipoles of magnetic moments μ_1 and μ_2 separated by a distance r is obtained as

$$E_\mathrm{d} = \frac{\mu_0}{4\pi r^3}\left[\mu_1 \cdot \mu_2 - \frac{3(\mu_1 \cdot r)(\mu_2 \cdot r)}{r^2}\right] \tag{3.18}$$

3.3 Origin of Magnetism

Atoms with closed shells where the electrons are entirely paired, such as noble gases, exhibit diamagnetism. Therefore, we require unpaired electrons to obtain a net magnetic moment. Diamagnetic materials possess negative susceptibility, implying a negative magnetization with a positive applied field. The negative magnetization arises from the induced magnetic moment due to the Lenz law effect in an applied magnetic field. Lenz's law indicates the magnitude and direction of the induced emf by a magnetic field in a current-carrying loop. Since atoms have circulating electrons, it causes an emf opposing the applied field both by the up and and down-spin electrons.

The diamagnetic susceptibility is

$$\chi = -\frac{N}{V}\frac{e^2\mu_0}{6m}\sum_i \langle r_i^2 \rangle \tag{3.19}$$

where N is the number of atoms in volume V, m is the mass of the electron and r_i is the orbital radius. The susceptibility is temperature-independent and depends on the orbital radius r_i. The noble gases that show diamagnetism exhibit negative susceptibility that increases down the period as the radius increases with atomic number.

3.3.1 Orbital and Spin Magnetic Moment

The electron in orbit can be modeled as equivalent to a current loop, giving magnetic moment $\mu = IA$, where I is the current and A is the area enclosed by the loop. The electron makes an angular frequency of ω. The charge of the electron, *e,* is inferred as negative. The magnetic moment is given as

$$\mu = I\pi r^2 = \frac{e\omega r^2}{2} \tag{3.20}$$

The quantization of angular momentum gives $mr^2\omega = L$ and $r^2\omega = \frac{L}{m}$. Using (3.20), we get, (see Sect. 2.10),

$$\mu_L = \frac{eL}{2m} \tag{3.21}$$

$L = n\hbar$ is the orbital angular momentum that gives

$$\mu_L = \frac{ne\hbar}{2m} \tag{3.22}$$

For $n = 1$ we get the smallest possible orbital magnetic moment contribution given in terms of Bohr magneton μ_{B} as

$$\mu_{\text{B}} = \frac{e\hbar}{2m} \tag{3.23}$$

The value of $\mu_{\text{B}} = 9.27 \times 10^{-24}$ A·m^2 $= 9.27 \times 10^{-21}$ emu.

The spin, another quantum behavior of an electron established through experiments and theory, gives the smallest possible magnetic moment equal to that of the orbital magnetic moment. With the spin angular momentum component, $S = \hbar/2$,

$$\mu_S = \frac{eS}{m} = \frac{e\hbar}{2m} \tag{3.24}$$

The spin contribution has a magnitude twice that of the orbital contribution, which results in 1 μ_{B}. The ratio of the magnetic moment to the orbital or spin magnetic moment is thus defined by the *g*-factor, which determines the orbital or spin contribution. The relationship between magnetic moment and angular momenta are

$$\mu_L = g_l \frac{e}{2m} L \tag{3.25}$$

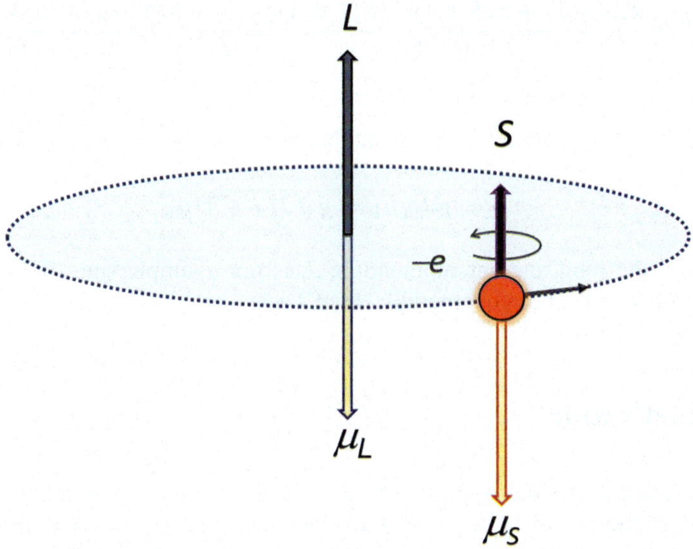

Fig. 3.3 The orbital and spin angular momenta of an electron that are pointed up have their magnetic moments pointed downwards. The magnitude of the magnetic moment of orbital and magnetic moments is equal for an electron with the quantum numbers $L = 1$ and $S = 1/2$. The spin contribution is proportional to the g-factor of 2

and

$$\mu_S = g_s \frac{e}{2m} S \tag{3.26}$$

where g_l and g_s are 1 and 2, respectively. The orbital and spin magnetic momenta contributed by an electron are schematically shown in Fig. 3.3.

The overall magnetic moment is thus the sum of spin and orbital contribution given as

$$\mu = \frac{e}{2m}(g_l L + g_s S) = \frac{\mu_B}{\hbar}(L + 2S) \tag{3.27}$$

Due to the precession of orbital and spin angular momentum, the net magnetic moment has the magnitude component of g given as

$$\mu_{avg} = \frac{e}{2m} g(L + S) = \frac{gJ\mu_B}{\hbar} \tag{3.28}$$

where the total angular momentum $J = L + S$ and

$$g = 1 + \frac{J(J+1) + S(S+1) - L(L+1)}{2J(J+1)} = \frac{3}{2} + \frac{S(S+1) - L(L+1)}{2J(J+1)} \tag{3.29}$$

The effective magnetic moment using the value of J (see Sect. 2.6) is now

$$\mu_{\text{eff}} = gJ_{\text{eff}}\mu_{\text{B}} = g\sqrt{J(J+1)}\mu_{\text{B}} \tag{3.30}$$

However, the total angular momentum, J, is not a simple sum of $L + S$ for a multielectron system, but governed by Hund's rule.

3.4 Hund's Rule

Hund's rule state that the highest spin and orbital angular momentum will have the lowest energy. Also, $J = L - S$ for less than half-filled subshells, and it is $J = L + S$ for more than half-filled subshells (see Sect. 2.8.1). The net magnetic moment is proportional to the effective quantum number obtained using Hund's rule. Hund's rule implies that atoms acquire as many unpaired spins as possible. However, crystal field energy could turn the high spin state to a low spin state, altering the ground state magnetic moment obtained from Hund's rule. The L, S, J, g, and μ_{eff} for rare-earth ions are given in Table 3.2 along with the experimental values. The maximum value for the effective magnetic moment is obtained for Dy^{3+} whereas the μ_{avg} is equal for both Dy^{3+} and Ho^{3+}. Experiments have shown that Dy^{3+} has the highest magnetic moment, suggesting μ_{eff} is a good approximation, obtained through quantum principles, that predict the magnetic moment in a better way. Although Eu^{3+} has 6 unpaired electrons, the net magnetic moment is zero due to Hund's 3rd rule. However, there are discrepancies for Sm^{3+} and Eu^{3+} with large magnetic moments from experiments. The high magnetic moment is due to the van Vleck contribution term that alters Hund's ground state.

Figure 3.4 shows the trend of the parameters given in Table 3.2 for the rare-earth ions. Following Hund's rule, the spins do not pair up but fill up to seven 4f electrons where $J = S$ and $L = 0$. Above 7 electrons, we have more than half-filled subshells having a higher magnetic moment as $J = L + S$, giving larger J values. In the case of 3d ions, the orbital angular momentum is quenched due to the crystal field effects, and the total magnetic moment contribution comes only from spin angular momentum. Therefore, the experimental magnetic moment of transition metal ions with 3d electrons shows a good correlation considering spin contribution alone (Coey 2009), as shown in Fig. 3.5.

The net magnetic moment of atoms or ions obtained from Hund's rule could combine in different ways to obtain paramagnetism or ferro/ferri/antiferromagnetism. If the atoms or ions in a lattice are non-interacting, it results in paramagnetism.

Table 3.2 The magnetic properties of the rare-earth ions and their correlation to the quantum numbers

Ion	4f electrons	L	S	J	g	$g\sqrt{J(J+1)}(\mu_B)$	$m_{exp}(\mu_B)$
La^{3+}	0	0	0	0	0	0	0
Ce^{3+}	1	3	1/2	5/2	6/7	2.54	2.5
Pr^{3+}	2	5	1	4	4/5	3.58	3.5
Nd^{3+}	3	6	3/2	9/2	8/11	3.62	3.5
Pm^{3+}	4	6	2	4	3/5	2.68	Unstable
Sm^{3+}	5	5	5/2	5/2	2/7	0.84	1.7
Eu^{3+}	6	3	3	0	0	0	3.4
Gd^{3+}	7	0	7/2	7/2	2	7.94	8.0
Tb^{3+}	8	3	3	6	3/2	9.72	9.7
Dy^{3+}	9	5	5/2	15/2	4/3	10.63	10.6
Ho^{3+}	10	6	2	8	5/4	10.6	10.4
Er^{3+}	11	6	3/2	15/2	6/5	9.59	9.5
Tm^{3+}	12	5	1	6	7/6	7.57	7.6
Yb^{3+}	13	3	1/2	7/2	8/7	4.54	4.5
Lu^{3+}	14	0	0	0	0	0	0

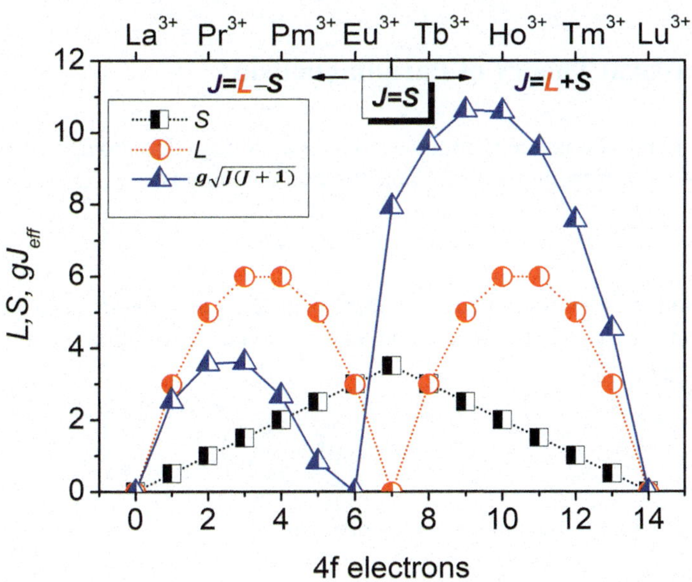

Fig. 3.4 The L, S, and gJ_{eff} varies with the number of 4f electrons. The magnetic moment of rare-earth ions is proportional to $g\sqrt{J(J+1)}$, obtained using Hund's rule

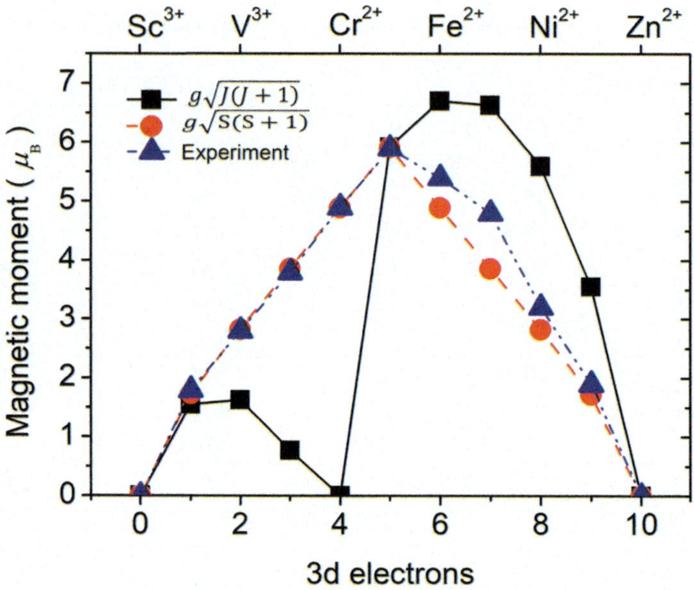

Fig. 3.5 The magnetic moment contributed by transition metal ions with 3d valence electrons is in close agreement with the experiment, considering only the spin contribution

3.5 Classical Theory of Paramagnetism

Consider a unit volume containing n atoms with a magnetic moment μ distributed in a sphere of unit radius. The magnetic potential energy is

$$E_p = -\mu \cdot B = -\mu B \cos\theta \tag{3.31}$$

The number of spins aligned at an angle between θ and $\theta + d\theta$ is proportional to the elemental area dA in the sphere and the Boltzmann distribution function given as $\exp(-E_p/k_B T)$. Then

$$dn \propto dA \exp\left(-\frac{E_p}{k_B T}\right) \tag{3.32}$$

Magnetization in the unit volume is given as

$$M = \int_0^n \mu \cos\theta dn \tag{3.33}$$

Substituting the values taking $a = \frac{\mu B}{k_B T}$ and evaluating the expression, we get

$$M = n\mu\left(\coth a - \frac{1}{a}\right) \tag{3.34}$$

$$M_o = n\mu \tag{3.35}$$

The saturation magnetization is

$$\frac{M}{M_o} = \left(\coth a - \frac{1}{a}\right) = L(a) \tag{3.36}$$

known as the Langevin function. The Taylor series expansion gives

$$\left(\coth a - \frac{1}{a}\right) = L(a) = \left(\frac{1}{a} + \frac{a}{3} - \frac{a^3}{45} + \frac{2a^5}{945} - \cdots\right) - \frac{1}{a}$$

$$L(a) = \frac{a}{3} - \frac{a^3}{45} + \frac{2a^5}{945} - \cdots \quad \text{where } a \leq 1 \tag{3.37}$$

When a is very small, the higher-order terms can be neglected, and we get

$$L(a) = \frac{a}{3} = \frac{\mu B}{3k_B T} \tag{3.38}$$

Figure 3.6 shows the plot of the Langevin function in terms of a where the function $L(a)$ approaches 1 for larger a. A large value of a implies a higher B and a smaller T.

The susceptibility can be obtained as

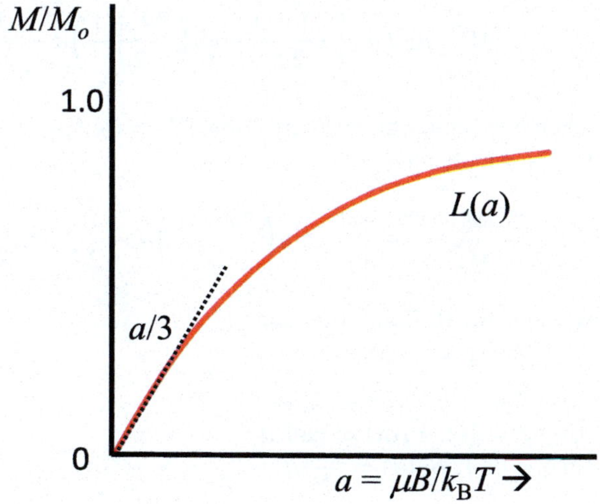

Fig. 3.6 The Langevin function $L(a) = a/3$ for small a and $L(a) \longrightarrow 1$ when $a \longrightarrow \infty$

$$\chi = \frac{M}{H} = M_o L(a)\frac{1}{H} = \frac{n\mu}{3H}\left(\frac{\mu B}{k_B T}\right) = \frac{n\mu_0\mu^2}{3k_B T} = \frac{C}{T} \tag{3.39}$$

where

$$C = \frac{n\mu_0\mu^2}{3k_B} \tag{3.40}$$

Note that χ is in SI units here. In cgs, it is

$$\chi = \frac{M}{H} = \frac{n\mu^2}{3k_B T} \tag{3.41}$$

The susceptibility versus temperature relation (3.41) is known as the Curie law of paramagnetism. Here, n is the number of spins per unit volume.

3.6 Quantum Theory of Paramagnetism

In the quantum theory of paramagnetism, instead of continuous magnetic potential energy, only specific energies are allowed that are

$$E_p = -gM_J\mu_B B \tag{3.42}$$

where $M_J = +J, (J-1)\ldots-(J-1), \ -J$. The magnetization now becomes

$$M = ngJ\mu_B\left\{\frac{2J+1}{2J}\coth\left(\frac{2J+1}{2J}\right)a - \frac{1}{2J}\coth\left(\frac{a}{2J}\right)\right\} \tag{3.43}$$

where the saturation magnetization, $M_o = ngJ\mu_B$, $a = \frac{gJ\mu_B B}{k_B T} = \frac{\mu_m B}{k_B T}$. Using M_o in (3.43), we obtain

$$\frac{M}{M_o} = \frac{2J+1}{2J}\coth\left(\frac{2J+1}{2J}\right)a - \frac{1}{2J}\coth\left(\frac{a}{2J}\right) = B_J(a) \tag{3.44}$$

which is the Brillouin function.

A few special cases are

(i) $J \to \infty$, $B_J(a) \to L(a)$
(ii) $J = 1/2$, $B_J(a) = \tanh a$
(iii) $a \ll 1$, $B_J(a) = \frac{J+1}{3J}a$

Figure 3.7 shows the plot of the Brillouin function for various values of J. The Brillouin function turns to the Langevin function when the J value approaches the classical value ($J \to \infty$). Also, with very low B or high T, $L(a) = \frac{a}{3}$ whereas

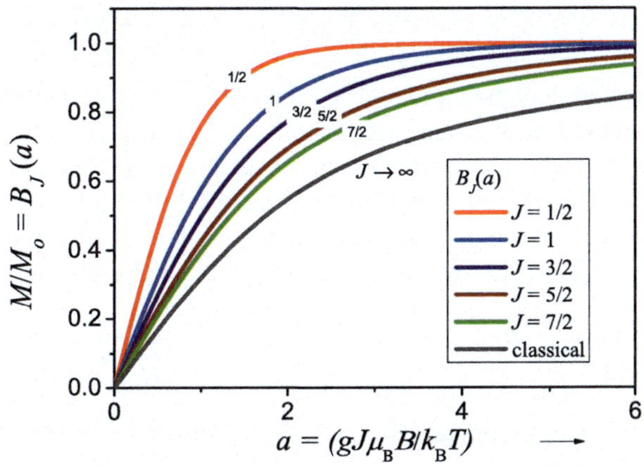

Fig. 3.7 The Brillouin function for various J values and the approach to the classical Langevin function as $J \to \infty$

$B_J(a) = \frac{J+1}{3J}a$. The quantum susceptibility of paramagnetic material from the Brillouin function is then,

$$\chi = \frac{M}{H} = \frac{ngJ\mu_B B_J(a)}{H} = \frac{ngJ\mu_B}{H}\left(\frac{(J+1)}{3J}\right)a$$
$$= \frac{ngJ\mu_B}{H}\left(\frac{(J+1)}{3J}\right)\frac{gJ\mu_B\mu_0 H}{k_B T} \tag{3.45}$$

$$\chi = \frac{n\mu_0 g^2\mu_B^2}{3k_B T}J(J+1) = \frac{n\mu_0\mu_{eff}^2}{3k_B T} = \frac{C}{T} \tag{3.46}$$

where

$$\mu_{eff} = g\mu_B\sqrt{J(J+1)} \tag{3.47}$$

$$C = \frac{n\mu_0 g^2\mu_B^2}{3k_B}J(J+1) \tag{3.48}$$

The values of a and C obtained here differ from those obtained using the classical theory.

3.7 Ferromagnetism

Ferromagnetism is a consequence of an interaction between spins that results in spontaneous magnetization even without an applied magnetic field. Weiss assumed that there is a molecular field that is proportional to the magnetization (Weiss 1906) given as

$$H_m = \gamma M \tag{3.49}$$

where γ is the molecular field constant. We use the H field here, and we can convert to the B field using (3.8) wherever necessary. It is also known that the ferromagnetic material becomes paramagnetic at the Curie temperature (T_c) (ISO symbol is T_C). The temperature-dependent susceptibility of a ferromagnetic material is not similar to that of a pure paramagnetic material, as shown below.

The paramagnetic susceptibility from Curie's law is

$$\chi = \frac{M}{H} = \frac{C}{T} \tag{3.50}$$

For the ferromagnetic case, H is replaced with $H + H_m$. Then the susceptibility becomes

$$\frac{M}{(H + \gamma M)} = \frac{C}{T} \tag{3.51}$$

$$MT - \gamma CM = CH$$
$$M(T - \gamma C) = CH$$
$$\chi = \frac{M}{H} = \frac{C}{T - \gamma C} = \frac{C}{T - \theta} \tag{3.52}$$

where $\theta = \gamma C$. It could be seen that θ is the Curie temperature, from (3.52), where the spontaneous magnetization vanishes (experimental T_c differ). The susceptibility temperature relation (3.52) is known as the Curie–Weiss law, valid above T_c. Therefore, from $\theta = \gamma C$ we obtain (cgs units)

$$C = \frac{ng^2\mu_B^2}{3k_B} J(J + 1) \tag{3.53}$$

$$\theta = T_c = \gamma \frac{ng^2\mu_B^2}{3k_B} J(J + 1) \tag{3.54}$$

In Fig. 3.8, curve (1) represents the Langevin function, straight line (2) corresponds to the molecular field H_m, (3) at the Curie temperature and (4) above the Curie temperature. However, if we consider a ferromagnetic material with an interaction due to the molecular field, the temperature-dependent susceptibility above T_c is given by the Curie–Weiss law. The slope of the straight lines increases with temperature and reaches a value of $a/3$ at $T = T_c$, where the spontaneous magnetization vanishes. From Fig. 3.8, it is noticeable that the magnitude of the spontaneous magnetization is maximum at very low temperature, corresponding to the intersection of the molecular field line with the $L(a)$ at M_s. The inset of Fig. 3.8 shows the temperature-dependent magnetization obtained from the intersection of the curve (1) with varying molecular field line (2) with temperature up to T_c which follows the law of corresponding states.

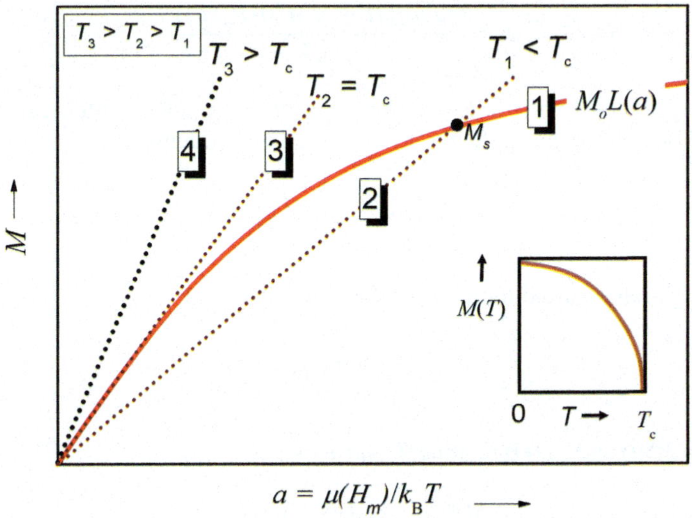

Fig. 3.8 Curve representing the Langevin function (1), straight lines corresponding to the molecular field at $T_1 < T_c$ (2), $T_2 = T_c$ (3) and $T_3 > T_c$ (4). The inset shows the temperature-dependent magnetization obtained from the intercept of the Langevin curve with the molecular field line (2) that shifts towards the T_c line (3) with increasing temperature

3.7.1 Law of Corresponding States

At a temperature close to absolute zero, the spontaneous magnetization is maximum, as is the saturation magnetization. The magnetization is a function of temperature with the ratio given as

$$\frac{M}{M_o} = \frac{M(T)}{M(0)} = B_J(a) \tag{3.55}$$

where

$$a = \frac{gJ\mu_B H_m}{k_B T} \tag{3.56}$$

For the temperature dependent molecular field, replacing H_m with $H + \gamma M(T)$ (SI unit), (3.56) becomes

$$a = \frac{gJ\mu_B(H + \gamma M(T))}{k_B T} \tag{3.57}$$

Using $M(T) = M(0)B_J(a)$,

$$a = \frac{gJ\mu_B(H + \gamma M(0)B_J(a))}{k_B T} \tag{3.58}$$

The expression (3.58) is a function of a on both sides, which can be solved numerically. Setting the applied field as zero and rearranging (3.55) and (3.58),

$$\frac{M(T)}{M(0)} = \frac{k_B T}{ng^2\mu_B^2 J^2 \gamma} a \tag{3.59}$$

For the high-temperature limit, using the molecular field,

$$T_c = \gamma \frac{ng^2\mu_B^2}{3k_B} J(J+1) \tag{3.60}$$

Using (3.59) and (3.60) relating T and T_c gives

$$\frac{M(T)}{M(0)} = \left(\frac{J+1}{3J}\right)\left(\frac{T}{T_c}\right) a \tag{3.61}$$

As $M(0)$ is the maximum possible saturation magnetization, the relation between M/M_o, which is proportional to T/T_c is known as the *law of corresponding states*. The universal behavior of M/M_o versus T/T_c for various J values is shown as Brillouin-Weiss curve in Fig. 3.9.

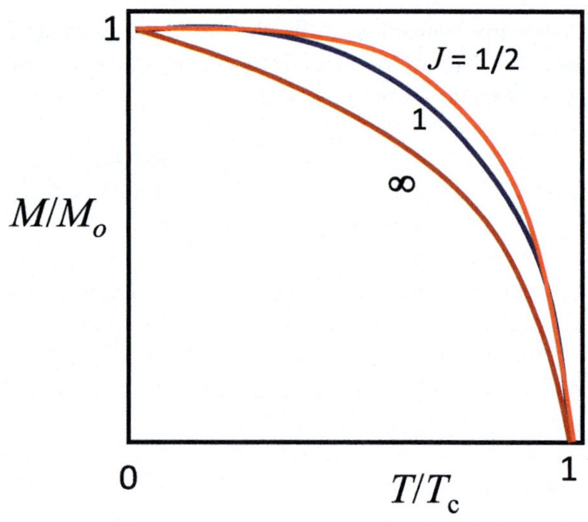

Fig. 3.9 The relative saturation magnetization (M/M_o) versus reduced temperature (T/T_c) shows universal behavior

3.8 Direct Exchange Interaction

The origin of a large molecular field resulting in spontaneous magnetization can be explained based on exchange interaction. Exchange interaction is responsible for ferromagnetism, ferrimagnetism and antiferromagnetism, resulting in an interaction between spins contrasting paramagnetism. Consider two electrons (labeled 1, 2) corresponding to atoms 'a' and 'b' as shown in Fig. 3.10. The Hamiltonian considering Coulombic repulsion is given as

$$H = -\frac{\hbar^2}{2m}(\nabla_a^2 + \nabla_b^2) - \left(\frac{Z_a}{r_{a1}} + \frac{Z_b}{r_{b2}}\right) \tag{3.62}$$

where m is the mass of the electron, Z is the nuclear charge and r is the distance between the electrons/nuclei (Dionne 2009), as mentioned in Fig. 3.10.

There can be two possible states, singlet and triplet, with the linear combination of wave functions given as (see Sect. 2.7)

$$\psi = \psi_{\text{symm}}\chi_0 = \frac{1}{\sqrt{2}}(\psi_a(r_1)\psi_b(r_2) + \psi_a(r_2)\psi_b(r_1))\chi_0 \tag{3.63}$$

$$\varphi = \psi_{\text{anti}}\chi_1 = \frac{1}{\sqrt{2}}(\psi_a(r_1)\psi_b(r_2) - \psi_a(r_2)\psi_b(r_1))\chi_1 \tag{3.64}$$

where (3.63) is for the singlet and (3.64) is for the triplet with $\frac{1}{\sqrt{2}}$ as the normalization constant. χ_0 and χ_1 are spin matrices corresponding to antiparallel and parallel spins, respectively, introduced to satisfy the fermionic characteristics of the wave function. The singlet and triplet energies are now

$$E_s = \int \psi^* \hat{H} \psi dr_1 dr_2 \tag{3.65}$$

$$E_T = \int \varphi^* \hat{H} \varphi dr_1 dr_2 \tag{3.66}$$

where the singlet and triplet wave functions are given as ψ and φ, respectively.

The total spin

$$S^2 = (s_1 + s_2)^2 = s_1^2 + s_2^2 + 2s_1 \cdot s_2 \tag{3.67}$$

$$2s_1 \cdot s_2 = S(S+1) - s_1(s_1+1) - s_2(s_2+1) \tag{3.68}$$

For $s_1 = s_2 = 1/2$ and $S = 0$, $s_1 \cdot s_2 = -\frac{3}{4}$ while that for $s_1 = s_2 = 1/2$ and $S = 1$, $s_1 \cdot s_2 = \frac{1}{4}$. The effective Hamiltonian will be a combination of spin-independent and spin-dependent terms

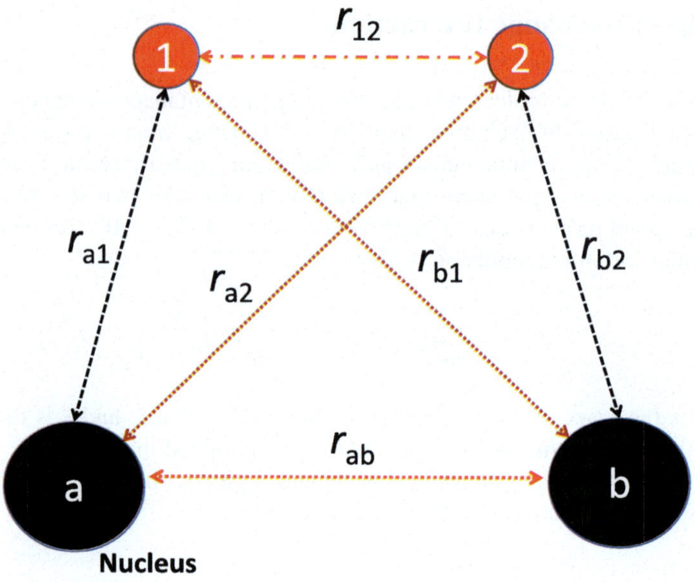

Fig. 3.10 Schematic representation of a pair of nuclei with electrons and the various interparticle distances used to evaluate the exchange interaction

$$H = \frac{1}{4}(E_s + 3E_T) - (E_s - E_T)s_1 \cdot s_2 \tag{3.69}$$

The energy eigenvalue will be

$$E = \Omega - 2\mathcal{J}s_1 \cdot s_2 \tag{3.70}$$

where Ω is the spin-independent part with the Coulomb integral also in it, and $\mathcal{J}$ is the exchange integral (often represented as J, similar to the notation for orbital angular momentum), which is spin-dependent.

The exchange energy between the ith and j^{th} spin pair is given as

$$E_{\text{ex}} = -\sum_{ij} \mathcal{J}_{ij} S_i \cdot S_j \tag{3.71}$$

For the simple two-electron exchange with near neighbors,

$$E_{\text{ex}} = -2\sum_{i>j} \mathcal{J} S_i \cdot S_j \tag{3.72}$$

$$E_{\text{ex}} = -2\mathcal{J} S_1 \cdot S_2 \tag{3.73}$$

The Heisenberg exchange interaction given by (3.73) determines the ferromagnetic or antiferromagnetic nature of the spins in the system. When $\mathcal{J} > 0$, minimum energy is achieved with S_1 and S_2 in parallel orientation (singlet state, $S = 1$), resulting in ferromagnetism. The ferromagnetic state, $S_1 \cdot S_2 = S_1 S_2 \cos(0)$ implies ↑↑ spin configuration. If $\mathcal{J} < 0$, then $S = 0$ entails ↑↓ configuration, as minimum energy could be achieved by $E_{\text{ex}} = -2(-\mathcal{J})S_1 S_2 \cos(180)$.

The positive values of $\mathcal{J}$ with large magnitude indicate strong ferromagnetism that depends on the ratio between the interatomic distance (r_{ab}) to 3d shell distance from the nucleus (r_{3d}). The dependence of the $\mathcal{J}$ value and hence on the ferro or antiferromagnetic interaction on the ratio of r_{ab}/r_{3d}, could be inferred from the Bethe-Slater curve for the d-block elements, as shown in Fig. 3.11. Ferromagnetic materials such as bcc-Fe, Co and Ni show positive $\mathcal{J}$ and lie in the upper portion of the Bethe-Slater curve. Antiferromagnetic materials such as fcc-Fe, Mn and Cr show negative $\mathcal{J}$ and closer interatomic separation. The approach towards paramagnetism is characterized by sizeable interatomic separation or negligible $\mathcal{J}$. The Curie temperature is proportional to $\mathcal{J}$ from the molecular field theory using the relation (Van Vleck 1941)

$$T_{\text{c}} = \frac{2|\mathcal{J}|zS(S+1)}{3k_{\text{B}}} \tag{3.74}$$

where z is the number of nearest neighbors (see Sect. 3.4. for relevance of S). It is thus apparent that Co with $\mathcal{J}$ values far larger than most of the elements in the Bethe-Slater diagram has the largest T_{c} of 1360–1390 K.

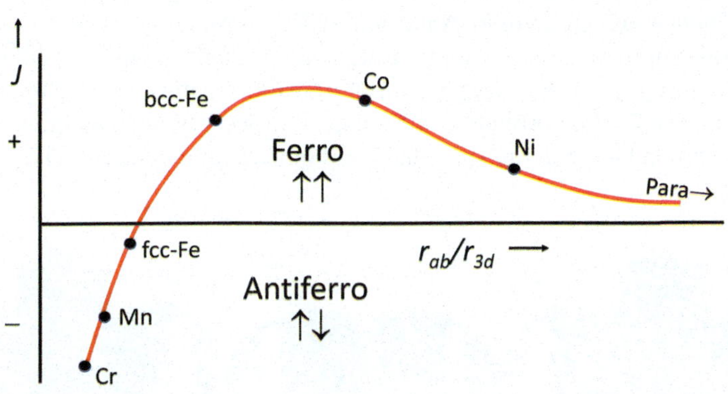

Fig. 3.11 Bethe-Slater diagram showing the dependence of exchange integral $\mathcal{J}$ on the ratio of interatomic distance to 3d shell distance from the nucleus (r_{ab}/r_{3d})

3.9 Superexchange Interaction

Metal ions separated by non-magnetic ions such as oxygen or fluorine exhibit another type of exchange interaction known as superexchange interaction. The 3d metal cation can have electrons in the d_{xy}, d_{yz}, d_{xz}, $d_{x^2-y^2}$ and d_{z^2} orbitals, and oxygen anion have p_x, p_y and p_z orbitals. Consider Mn^{2+} metal ion whose $d_{x^2-y^2}$ orbital is shared by p_x of oxygen ion, and electron hopping occurs between them. When one electron from p_x jumps to the d-state to form σ or σ* bond, another electron with the opposite spin left behind cannot be in the parallel state due to Pauli exclusion principle. The d-state spins thus align antiferromagnetically with another Mn^{2+} ion in the d-state. The interaction between Mn^{2+} ions on either side of oxygen gives $Mn^{2+}(5\mu_B\uparrow) - O^{2-} - Mn^{2+}(5\mu_B\downarrow)$ superexchange interaction resulting in antiferromagnetism. The Anderson's model (Anderson 1959) of superexchange is expressed in terms of an empirical Coulomb energy parameter U of d electrons and the transition matrix b, arising from the transition of the electrons from one atom to another. The b is related to the crystal field parameter Dq that determines the spread of the d wave function into the p wave function. The exchange integral is then expressed as

$$J_{\text{eff}} = \frac{2b^2}{U} \tag{3.75}$$

Superexchange interaction, and therefore strong antiferromagnetism, is possible with the mixing of orbitals $p_z - d_{z^2}$ or $p_x/p_y - d_{x^2-y^2}$. The p-d mixing of orbitals produces antiferromagnetic superexchange interaction through $p_z - d_{z^2}$ covalent bond, as illustrated in Fig. 3.12.

The Goodenough-Kanamori-Anderson (GKA) (Anderson 1963; Goodenough 1963) semi-empirical rules predict strong superexchange with half-filled orbitals at 180° resulting in antiferromagnetism in transition metal oxides. Superexchange interaction in half-filled orbitals at 90° ends up in weakly ferromagnetic structures such as a spin ladder, zig-zag spin chain and other exotic systems. The other types

Fig. 3.12 The $p_z - d_{z^2}$ overlap produces superexchange interaction, resulting in antiferromagnetic alignment of cations on either side of the anion

of indirect exchanges are double exchange as in mixed valent oxides and Ruderman-Kittel-Kasuya-Yosida (RKKY) (Ruderman and Kittel 1954; Kasuya 1956; Yosida 1957) exchange encountered between localized magnetic moments separated by non-magnetic metals.

3.10 Antiferromagnetism

Similar to ferromagnetism, antiferromagnetism is also characterized by interacting spins having a molecular field but in a negative direction, leading to the cancellation of the magnetic moment. The perfect cancellation of the antiparallel spins of the atom or ion pairs A and B, as shown in Fig. 3.13, is expected to show zero magnetization and susceptibility. However, the magnetization of an antiferromagnetic material is non-zero in an applied field depending on its orientation direction (Anderson 1950). Antiferromagnetic materials show paramagnetism when the temperature increases beyond the Néel temperature (T_N).

Consider A- and B- site identical lattices with magnetizations M_A and M_B, respectively. From the Weiss molecular field, the molecular field acting on the A-site (H_{mA}) is proportional to the magnetization of B-site. Similarly, the B-site molecular field is proportional to M_A. The molecular field constants for the A-A, A-B and B-B interactions are γ_{AA}, γ_{AB} and γ_{BB}, respectively. For an antiferromagnet, the magnitude

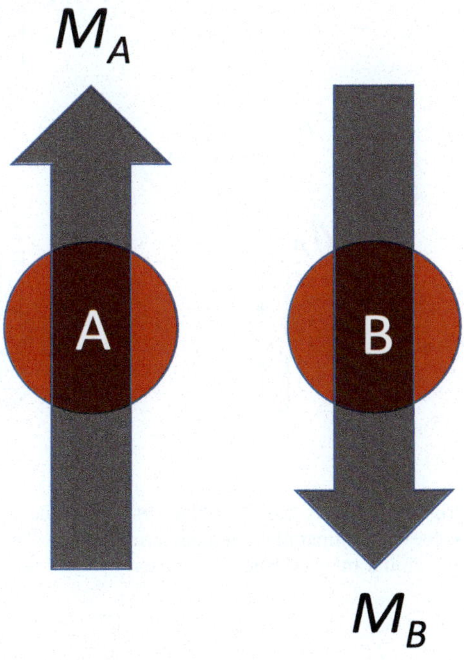

Fig. 3.13 A pair of atoms or ions having equal and opposite magnetic moments shall exhibit antiferromagnetism

of the magnetization of each lattice site is the same, while the directions are opposite for the dissimilar sites, and hence, the molecular field constant is negative.

The magnetization and molecular fields acting on the A- and B- sites are represented in Fig. 3.14 for various neighbors in an antiferromagnetic arrangement, from which we obtain the relations

$$H_{mA} = -\gamma_{AB} M_B + \gamma_{AA} M_A \tag{3.76}$$

$$H_{mB} = -\gamma_{BA} M_A + \gamma_{BB} M_B \tag{3.77}$$

Using $|\gamma_{AB}| = |\gamma_{BA}| = \gamma$, $|M_A| = |M_B|$ for a two sublattice and neglecting the second near-neighbor interaction, setting $|\gamma_{AA}| = |\gamma_{BB}| = \delta = 0$, we obtain

$$M_A T = C'(H + H_{mA}) = C'(H - \gamma M_B) \tag{3.78}$$

$$M_B T = C'(H + H_{mB}) = C'(H - \gamma M_A) \tag{3.79}$$

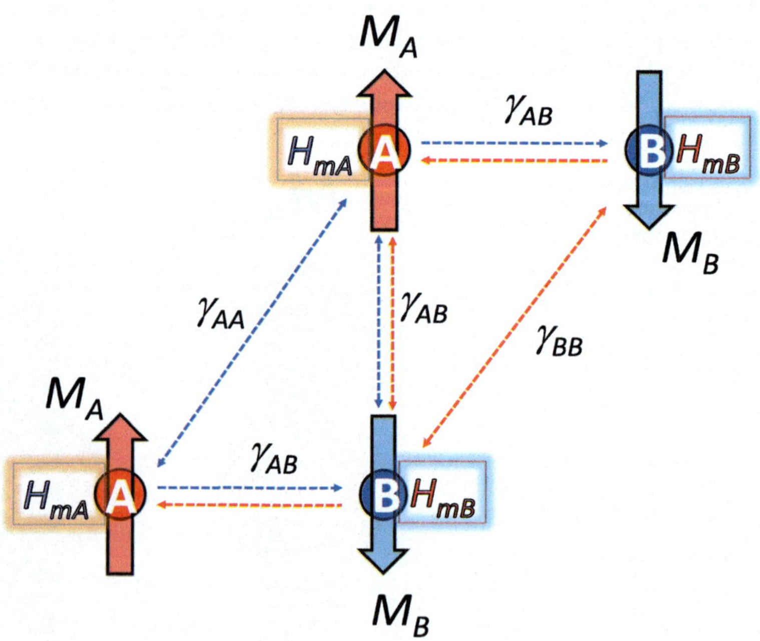

Fig. 3.14 The A- and B- sublattices experience near neighbor molecular field of γ_{AB} with negative values for antiparallel magnetic arrangement. The next nearest neighbor interactions are positive for similar magnetization directions

Adding (3.78) and (3.79),

$$M_AT + M_BT = C'\left[2H - \gamma(M_B + M_A)\right] \tag{3.80}$$

3.10.1 Magnetization Above T_N

The magnetization for the A- and B- sublattices beyond the Néel temperature is non-zero in an applied magnetic field and given by

$$M_A + M_B = M \tag{3.81}$$

$$MT = C'\left[2H - \gamma(M)\right] = 2C'H - \gamma C'M \tag{3.82}$$

$$M\left(T + C'\gamma\right) = 2C'H \tag{3.83}$$

$$\chi = \frac{M}{H} = \frac{2C'}{T + C'\gamma} = \frac{2C'}{T + \theta} \tag{3.84}$$

where $\theta = C'\gamma = T_{\mathrm{N}}$ is the antiferromagnetic Néel temperature. Although the sublattice magnetization is aligned along H, the negative molecular field reduces the overall M, reducing the χ at $T > T_{\mathrm{N}}$. The ideal value of $\frac{\theta}{T_{\mathrm{N}}} = 1$ may deviate in experiments due to the next neighbor contribution or changes in the molecular field constant.

3.10.2 Magnetization Below T_{N}

When $T < T_{\mathrm{N}}$, in the absence of field, spontaneous magnetization gives $M_A = -M_B$. The spontaneous magnetization and saturation magnetization at 0 K are represented as M_{sA}, M_{OA}; and M_{sB}, M_{OB} for the A- and B- sublattices, respectively. Using $H_{mA} = -\gamma M_{sB} = \gamma M_{sA}$, $H_{mB} = -\gamma M_{sA} = \gamma M_{sB}$ and $B = \mu_0(H + M)$ in the Brillouin function, we obtain

$$\frac{M_{sA}}{M_{0A}} = B_J\left(\frac{\mu_{mA} B_{sA}}{k_{\mathrm{B}} T}\right) \tag{3.85}$$

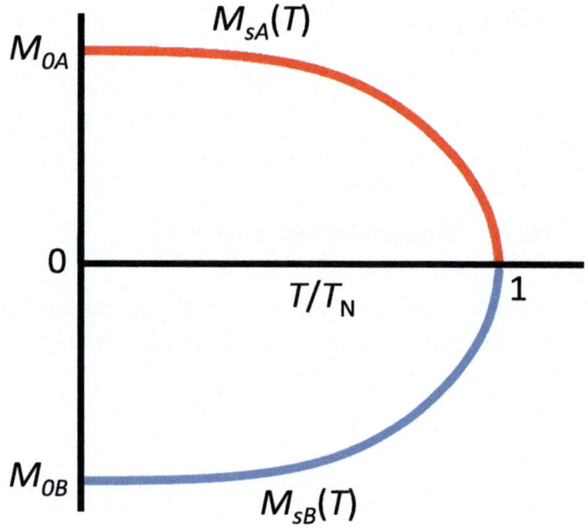

Fig. 3.15 The temperature-dependent spontaneous magnetization of the A and B antiferromagnetic sublattices

$$\frac{M_{sB}}{M_{0B}} = B_J\left(\frac{\mu_{mB}B_{sB}}{k_B T}\right) \tag{3.86}$$

where μ_{mA} and μ_{mB} are the quantized magnetic moments and B_{sA}, B_{sB} are the total magnetic fields, including the molecular field of the respective sites. Since $\mu_{mA} = \mu_{mB}$ and the molecular fields are opposite to each other, the graphical solution of (3.85) and (3.86) would produce a temperature-dependent magnetization for each sublattice, as shown in Fig. 3.15.

The temperature-dependent magnetization of the A- and B- sites obey the law of corresponding states but in the opposite way, leading to vanishing spontaneous magnetization at T_N. The condition leading to antiferromagnetism can be understood from the Heisenberg exchange interaction. The exchange integral becomes negative, leading to antiferromagnetism when the interatomic separation is shorter due to stronger Coulomb repulsion. This is mainly determined by the overlapping wave function lobes of the d and f shell electrons. Antiferromagnetism can also occur across a mediator, such as oxygen ions, through superexchange interaction.

3.10.3 Influence of External Magnetic Field

The effect of the external magnetic field on the magnetization of antiferromagnetic sublattice under perpendicular and parallel orientations is shown in Fig. 3.16. An applied field perpendicular to the spin-axis rotates the sublattice magnetization symmetrically along an angle, resulting in a balanced molecular field orientation.

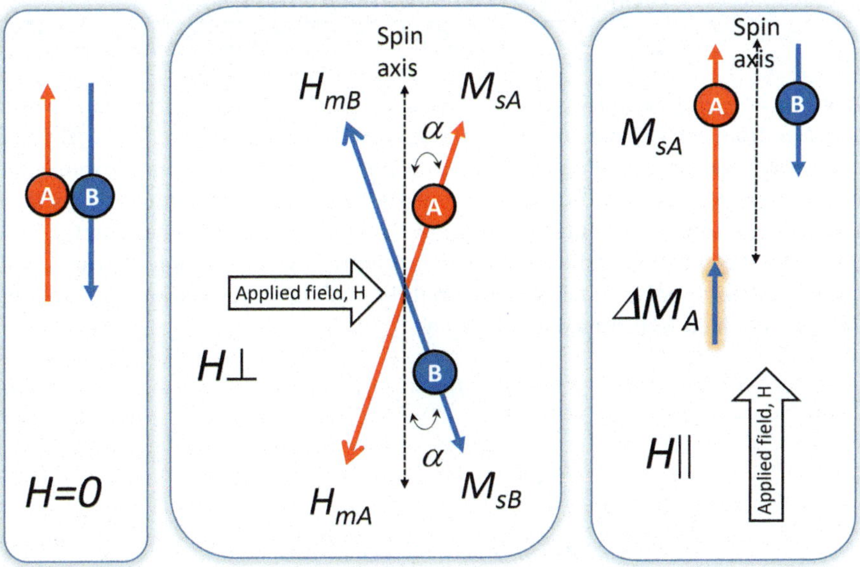

Fig. 3.16 The effect of sublattice magnetizations of A and B antiferromagnetic spins in zero applied field, field applied perpendicular ($H_{\perp}$) and parallel ($H_{||}$) to the spin-axis of A

Assuming smaller angle, α without changing the sublattice magnetization in an applied field gives

$$H_{\perp} = |H_{mA}| \sin\alpha + |H_{mB}| \sin\alpha = 2|H_{mA}| \sin\alpha \tag{3.87}$$

since $|H_{mA}| = |H_{mB}|$. Using the molecular field constant in (3.87),

$$H_{\perp} = 2\gamma M_{sA} \sin\alpha = 2\gamma M_{sB} \sin\alpha = \gamma M_{\perp} \tag{3.88}$$

$$\chi_{\perp} = \frac{M_{\perp}}{H_{\perp}} = \frac{2M_{sA}\sin\alpha}{H_{\perp}} = \frac{1}{\gamma} \tag{3.89}$$

Using $\theta = C'\gamma$ we obtain $\chi_{\perp} = \frac{1}{\gamma} = \frac{C'}{\theta}$, which gives the temperature-independent susceptibility below T_N, when the applied field is perpendicular to the spin direction. Once a magnetic field is applied parallel to any of the sublattice spins, say A, it increases the magnetization by ΔM_A and decreases the other by ΔM_B. Consequently, the molecular field in A increases and B drops while the net magnetization is $M = \Delta M_A + \Delta M_B$.

Assuming equal changes in the sublattice magnetization, the simplified susceptibility equation for the parallel applied field becomes

$$\chi_{\parallel} = \frac{\Delta M_A + \Delta M_B}{H} = \frac{2C'[(3J/(J+1)]B'_J(\eta)}{T - \gamma C'[(3J/(J+1)]B'_J(\eta)} \quad (3.90)$$

where $\eta = -\mu_0 \gamma M_p \big(g\mu_B \sqrt{J(J+1)}\big)/k_B T$, $M_p = M_{op} B_J(a)$ is the sublattice magnetization determined by the Brillouin function, and $B'_J(\eta)$ is the Brillouin function derivative. The temperature-dependent susceptibility of antiferromagnetic sublattice reduces as $T \to 0$ since $B'_J(\eta) \to 0$ whereas it is maximum at T_N as $B'_J(\eta)$ is in the upper limit. Susceptibility beyond T_N exhibits a paramagnetic trend.

The parallel and perpendicular sublattice orientations are possible only in a single crystal, and for a polycrystal, the susceptibility spreads over various angles α in a unit sphere which is averaged and given as

$$M_{\parallel} = \chi_{\parallel} H_{\parallel} = \chi_{\parallel} H \cos\alpha \quad (3.91)$$

$$M_{\perp} = \chi_{\perp} H_{\perp} = \chi_{\perp} H \sin\alpha \quad (3.92)$$

$$\chi_{pol} = \chi_{\parallel} \cos^2\alpha + \chi_{\perp} \sin^2\alpha \quad (3.93)$$

Since the orientations are proportional to sin$\alpha d\alpha$, we write (3.93) as

$$\chi_{pol} = \chi_{\parallel} \int_0^{\pi/2} \cos^2\alpha \sin\alpha d\alpha + \chi_{\perp} \int_0^{\pi/2} \sin^2\alpha \sin\alpha d\alpha = \chi_{\parallel}\frac{1}{3} + \chi_{\perp}\frac{2}{3} \quad (3.94)$$

$$\chi_{pol} = \chi_{\parallel} = \chi_{\perp} \quad \text{at } T = T_N$$
$$\chi_{pol} = \frac{2}{3}\chi_{\perp} \quad \text{at } T = 0\,\text{K} \quad (3.95)$$

The temperature-dependent susceptibilities of an antiferromagnet under various orientations and for a polycrystal are represented in Fig. 3.17. Maximum susceptibility could be achieved under a perpendicular orientation that remains constant up to T_N. The polycrystalline material exhibits temperature-dependent antiferromagnetic susceptibility due to the 1/3rd contributions from parallel orientation. Beyond T_N, the antiferromagnets show paramagnetic susceptibility, while the molecular field contribution results in higher susceptibility than pure paramagnets. Materials such as MnO, NiO, Cr_2O_3, FeO, rhombohedral FeF_3, $FeCl_2$ and hexagonal FeS are antiferromagnetic.

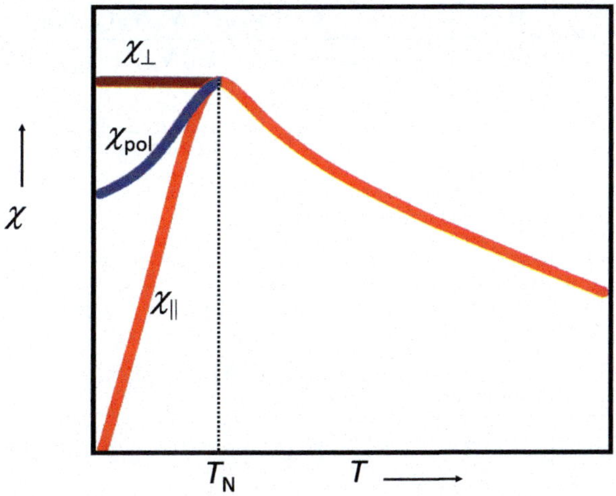

Fig. 3.17 The temperature-dependent antiferromagnetic susceptibility shows constant values for the spin axis perpendicular to the applied field, $\chi_\perp$. Temperature dependence for parallel orientation, $\chi_\parallel$ and polycrystalline materials, χ_{pol} show variation up to the Néel temperature, T_N; beyond which it exhibits a paramagnetic curve

3.11 Ferrimagnetism

A ferrimagnetic state is defined as the antiparallel alignment of spins of unequal magnitude. From Fig. 3.14, the molecular field γ_{AB} is negative, similar to an antiferromagnet, whereas the magnetizations of A- and B- sublattices are unequal for a ferrimagnet. Rewriting (3.76) and (3.77) for the ferrimagnetic lattice,

$$H_{mA} = -\gamma_{AB}M_B + \gamma_{AA}M_A \tag{3.96}$$

$$H_{mB} = -\gamma_{BA}M_A + \gamma_{BB}M_B \tag{3.97}$$

In ferrimagnets, $|M_A| \neq |M_B|$ and $|\gamma_{AA}| \neq |\gamma_{BB}|$. The ratio of molecular fields is defined as $\gamma_{12} = \frac{\gamma_{AA}}{\gamma_{AB}}$, $\gamma_{22} = \frac{\gamma_{BB}}{\gamma_{AB}}$.

Consider the fraction of A-site species as N_A and B-site species as N_B with the total number as N and $N_A + N_B = 1$. The A- and B- sublattice magnetizations are,

$$M_{sA} = NN_AM_A \tag{3.98}$$

$$M_{sB} = -NN_BM_B \tag{3.99}$$

and

$$M = M_{sA} + M_{sB} \tag{3.100}$$

If we consider sublattices with identical magnetic moments, then $M_A = M_B$ and the difference in the fraction N_A and N_B would give non-zero total magnetization. After extensive substitutions and simplifications (Néel 1948; Smart 1955), we get

$$\chi = \frac{M}{H} = \frac{C - \gamma_{AB} C^2 N_A N_B (2 + \gamma_{12} + \gamma_{22})/T}{T - \gamma_{AB} C(\gamma_{12} N_A + \gamma_{22} N_B) + \gamma_{AB}^2 C^2 N_A N_B (\gamma_{12}\gamma_{22} - 1)/T} \tag{3.101}$$

$$\frac{1}{\chi} = \frac{T}{C} + \frac{1}{\chi_o} - \frac{C_{ab}}{T - \theta} \tag{3.102}$$

where

$$\chi_o = \frac{1}{\gamma_{AB}\left(2N_A N_B - \gamma_{12} N_A^2 - \gamma_{22} N_B^2\right)} \tag{3.103}$$

$$C_{ab} = \gamma_{AB}^2 C N_A N_B \left[N_A(1 + \gamma_{12}) - N_B(1 + \gamma_{22})\right]^2 \tag{3.104}$$

$$\theta = \gamma_{AB} C N_A N_B (2 + \gamma_{12} + \gamma_{22}) \tag{3.105}$$

and C is the Curie constant determined by the magnetic moment of the sites.

The plot of $1/\chi$ versus T for a ferrimagnet as per (3.102) results in a hyperbolic curve, as shown in Fig. 3.18. At very large T, as the last term of (3.102) vanishes, the curve approximates to a straight line with a slope of $1/C$, and the y-intercept gives $1/\chi_o$. The asymptotic straight line emerges on approaching the Curie temperature from the high-temperature regime, and the hyperbolic curve falls to zero at the T_c. The temperature dependence of $1/\chi$ for paramagnetic and antiferromagnetic materials is also represented in Fig. 3.18. While the paramagnetic line passes through the origin, the temperature intercept is positive for ferromagnetic materials and negative for the asymptotic lines of antiferromagnetic and ferrimagnetic materials.

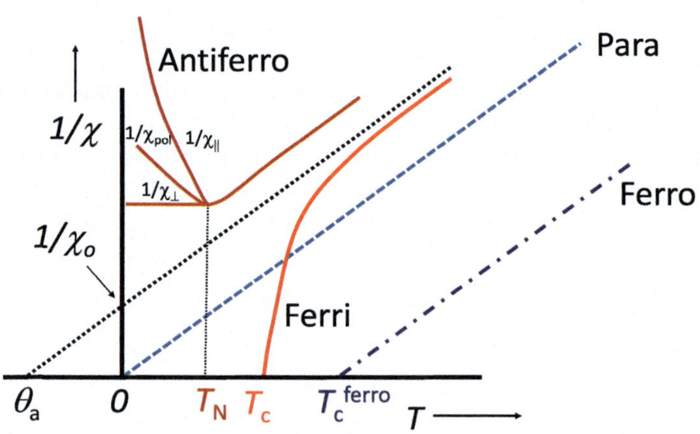

Fig. 3.18 The temperature-dependent $1/\chi$ for paramagnetic, ferromagnetic and antiferromagnetic materials. The asymptotic line that intercepts in the negative temperature axis branches off and falls to zero at the ferrimagnetic Curie temperature (T_c)

The asymptotic Curie temperature corresponds to the intercept in the negative temperature axis given as $\theta_a = -\frac{C}{\chi_o}$, that occurs when $T_c < 0$, where paramagnetism is observed down to 0 K. Setting $1/\chi \to 0$, the ferrimagnetic Curie temperature can be obtained as

$$T_c = \frac{C\gamma_{AB}}{2}\left[\gamma_{12}N_A + \gamma_{22}N_B \pm \left(\sqrt{(\gamma_{12}N_A - \gamma_{22}N_B)^2 + 4N_A N_B}\right)\right] \quad (3.106)$$

For $T_c > 0$ K, (3.106) is set to zero, from which we obtain

$$\gamma_{12}\gamma_{22} = 1 \quad (3.107)$$

The condition for ferrimagnetism is thus $|\gamma_{12}\gamma_{22}| < 1$ while $|\gamma_{12}\gamma_{22}| > 1$ would ensue antiferromagnetism.

In the low-temperature regime below T_c, the A- and B- sublattice magnetizations follow the Brillouin function given as

$$M_{sA}(T) = M_{0A}B_J(a_A) \quad (3.108)$$

$$M_{sB}(T) = M_{0B}B_J(a_B) \quad (3.109)$$

The net-magnetization is

$$M(T) = M_{sA}(T) - M_{sB}(T) \quad (3.110)$$

The temperature-dependent magnetization is the difference in the sublattice magnetizations obtained by the fractional site contributions determined by (3.110). Four types of temperature dependence could be observed based on the thermal response of the sublattice magnetizations (Chikazumi 1997), as shown in Fig. 3.19. Spinel ferrites such as γ-Fe_2O_3, Fe_3O_4 and $MnFe_2O_4$ of the type $AO{\cdot}B_2O_3$ show ferrimagnetic behavior contributed by A- and B- site cations.

The oxygen anions mediate the ferrimagnetic superexchange, and the temperature-dependent magnetic response depends on the relative fraction, magnetic moment and strength of the superexchange interaction. For example, if the A- and B- sites cancel each other or if both the sites have spin zero cations, then the spinel ferrite will be non-magnetic. On the other hand, if the A-site has a net magnetic moment larger and antiparallel to the B-site, the magnetization curve will be one among those shown in Fig. 3.19.

Curve (1): if the A- and B- sites are saturated, the net magnetization will be the difference between A and B, showing a temperature response following the law of corresponding states. Curve (2): if there is thermal fluctuation in the higher magnetic moment site; Curve (3): if the thermal fluctuation is in the lower magnetic moment site. Curve (4): exhibited by rare-earth iron garnets (RIG) with the formula $3R_2O_3{\cdot}5Fe_2O_3$ that are ferrimagnets, with the rare-earth (R) ion occupying the third

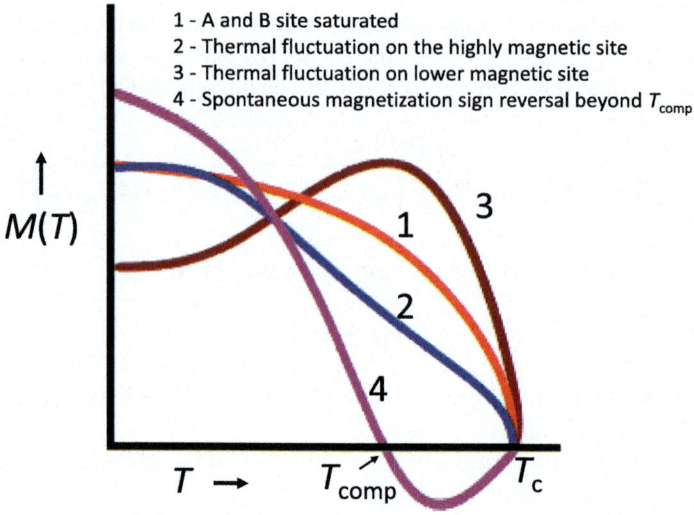

Fig. 3.19 The temperature-dependent magnetic behavior of ferrimagnetic materials with various sublattice magnetizations

site while two of the sites are occupied by ferric ions. The RIG ferrites such as GdIG show compensation temperature (T_{comp}) due to the drastic decrease of rare-earth ions magnetization with temperature compared to the ferric ions. At low temperatures, the R ions magnetization dominates while the sublattice magnetizations are compensated $\{M(R - Fe^{3+}) = 0\}$ at T_{comp}. The R ions contribution is negligible beyond T_{comp} exhibiting magnetization curve (4), as shown in Fig. 3.19. Although spontaneous magnetization sign changes beyond T_{comp}, bulk magnetic measurements shall give only absolute magnetization values resulting in an increase in magnetization beyond T_{comp} and then falling to zero at T_c.

3.12 Magnetism of Metals and Alloys

Metals and alloys have free or itinerant electrons and hence, the calculation of magnetic moment is not straightforward, unlike 4f elements or oxides, where the electrons are localized. The experimental magnetic moment per atom of Fe, Co and Ni are 2.2, 1.7 and 0.6 μ_B, respectively. The non-integer values of the magnetic moment could be explained using the itinerant electron forming bands near the Fermi level.

The ground state Fermi wave vector k_F, represented by a sphere of radius k_F where all the states are occupied within, and the Fermi energy E_F is related by the expression

$$k_F^3 = \left(\frac{2m_e^*}{\hbar^2} . E_F \right)^{3/2} \tag{3.111}$$

where m_e^* is the effective mass of the electron. The density of states (DOS) at 0 K where all the states are occupied up to k_F is given in terms of the electron number density N as follows

$$N = \int_0^{E_F} n(E) dE \tag{3.112}$$

The density of states for a 3D metal is given as

$$n(E) = \frac{1}{2\pi^2} \left(\frac{2m_e^*}{\hbar^2} \right)^{3/2} \sqrt{E} \tag{3.113}$$

The DOS is proportional to the square root of E, presenting parabolic dependence as a function of energy. With temperature, the occupation is determined by the Fermi–Dirac distribution function

$$f(E) = \frac{1}{1 + \exp\{(E - \mu)/k_B T\}} \tag{3.114}$$

where μ is the chemical potential. The electron number density above 0 K then becomes

$$N = \int_0^{E_F} n(E) f(E) dE \tag{3.115}$$

The temperature dependence of the chemical potential can be obtained as

$$\mu(T) \approx \mu(0) \left(1 - \frac{\pi^2}{12} \left(\frac{k_B T}{E_F} \right)^2 \right) \tag{3.116}$$

where $\mu(0) = E_F$ at 0 K. From (3.116), the chemical potential decreases with temperature due to the occupation of higher states from thermal excitation. Figure 3.20 shows the free electron DOS whose number density with temperature is determined by the Fermi–Dirac distribution function. At 0 K, all states below E_F are completely occupied, whereas it is spread out as the temperature increases.

Since each state can have two electrons, one with spin-up and another spin-down, the number densities are separate for the two spin states. For a paramagnetic itinerant electron system in an applied magnetic field, the Fermi energy level has to be the same, and therefore, the transfer of down-spin to up-spin states results in a splitting of the bands by $2\mu_B B$. The resulting susceptibility is known as the *Pauli susceptibility*,

Fig. 3.20 The free electron density of states follows the Fermi–Dirac distribution function

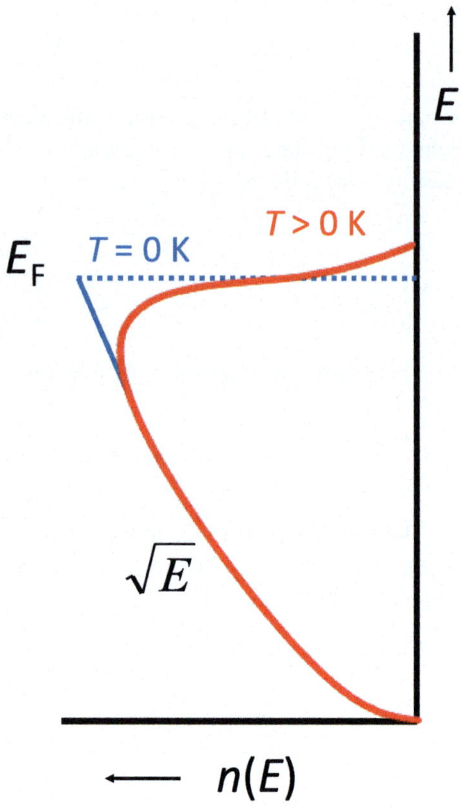

applicable for materials with large Fermi temperatures given as

$$\chi_{\text{Pauli}} = \frac{3N\mu_0{\mu_B}^2}{2E_F} \tag{3.117}$$

The Pauli paramagnets, such as Al, exhibit temperature-independent susceptibility. Considering the fact that free electrons move in a periodic potential, the Pauli susceptibility has a diamagnetic Landau-Peierls correction factor of $-\frac{1}{3}\left(\frac{m_e}{m_e^*}\right)^2$. The discrepancy in the susceptibility, even after the correction, arises due to the neglect of exchange interaction effects. The molecular field theory predicts spontaneous magnetization due to the internal field. Therefore, spontaneous spin splitting changes the Fermi energy by $E_F \pm \delta E$, and the total energy change is given as

$$\Delta E = \frac{1}{2}n(E_F)\delta E^2(1 - U \cdot n(E_F)) \tag{3.118}$$

where U is the Coulomb energy contribution known as the *Stoner exchange integral*, which could be calculated from density functional methods. The susceptibility is then

$$\chi = \frac{\chi_{\text{Pauli}}}{1 - U \cdot n(E_{\text{F}})} \tag{3.119}$$

Materials such as Fe, Co and Ni show significant paramagnetic susceptibility due to $U \cdot n(E_{\text{F}}) \geq 1$, known as the *Stoner criterion* (Stoner 1938). Large density of states at the Fermi energy combined with a large molecular field are responsible for ferromagnetism. Figure 3.21. shows the U, $n(E_{\text{F}})$ and their combined term as a function of various elements (Janak 1977). Fe, Co and Ni exhibit $U \cdot n(E_{\text{F}}) > 1$, fulfilling the Stoner criterion and ferromagnetism, while the large DOS for Pd compared with other elements explains their higher susceptibility. The ferromagnetism results from electrons localization at the top of the d-band producing large DOS for the 3d orbitals. For elements such as Cu, the DOS is very low at the E_{F}, causing diamagnetism.

The DOS for the paramagnetic, ferromagnetic and diamagnetic bands are shown in Fig. 3.22. The paramagnetic DOS is filled up to the Fermi level, while the spin transfer occurs for the ferromagnetic band, even in the absence of an external magnetic field, due to the molecular field. The diamagnetic bands are below the Fermi level, and thermal excitation does not alter the spin states.

In the case of alloys, the ground state magnetization proportional to the difference between the up-spin and down-spin electrons could be obtained from the Slater-Pauling diagram, with the largest magnetic moment for FeCo (Slater 1936; Pauling

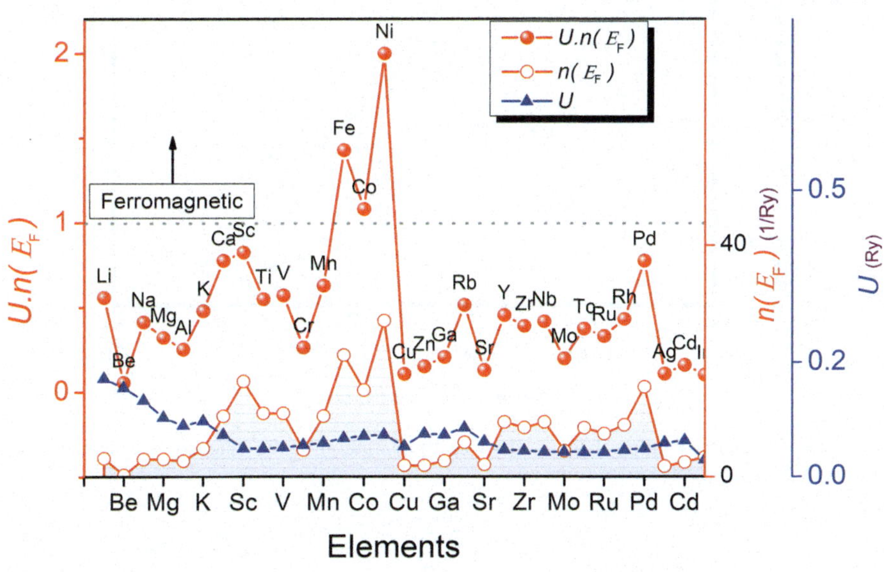

Fig. 3.21 The Stoner exchange integral (U) and density of states $n(E_{\text{F}})$ for various elements (Janak 1977). Elements are ferromagnetic if the Stoner criterion $U \cdot n(E_{\text{F}}) > 1$ is satisfied

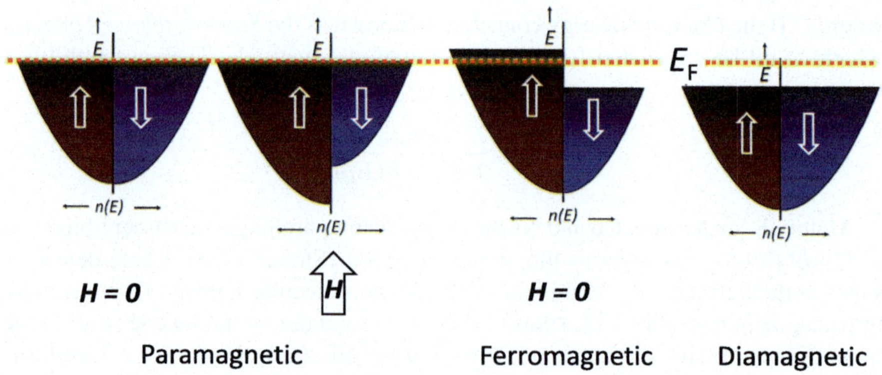

Fig. 3.22 The up and down -spin band structure for paramagnetic, ferromagnetic and diamagnetic states

1938), as shown in Fig. 3.23. The alloying of an element with a strongly ferromagnetic metal such as Co and Ni whose up-spin bands are full (up-spin valence electron number is constant) shows an increasing trend in the magnetization with decreasing total valence electron number Z. When down-spin electrons are constant, the magnetization decreases with decreasing total valence electron number as in bcc alloys of FeCo, FeCr and FeV, as shown in the left half of the Slater-Pauling diagram.

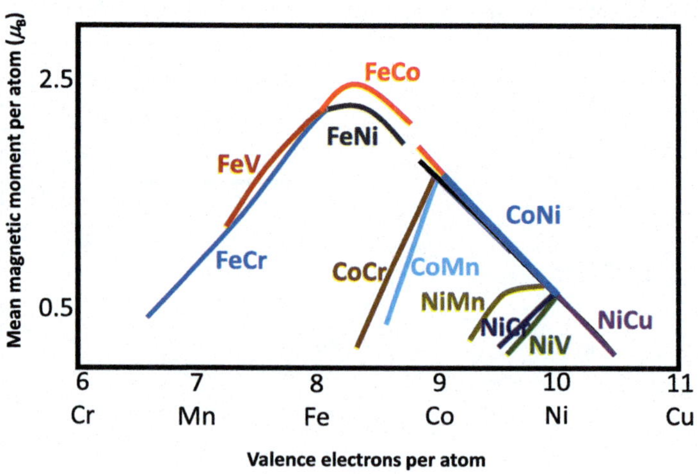

Fig. 3.23 The Slater-Pauling diagram showing the variation in the magnetic moment with the number of valence electrons for some alloys

3.13 Magnetic Anisotropy

Magnetic anisotropy influences the shape of the hysteresis loop and the coercivity of a magnetic material. Magnetic anisotropy determines the direction of spontaneous magnetization that may arise from crystal structure, shape, stress or other induced effects. They are thus classified into types such as magnetocrystalline anisotropy, shape anisotropy, stress anisotropy, induced anisotropy, surface anisotropy, random anisotropy and exchange anisotropy.

3.13.1 *Magnetocrystalline Anisotropy*

The magnetic anisotropy based on the material's crystal structure is known as magnetocrystalline anisotropy and may be cubic or uniaxial in origin. For a cubic crystal (Fig. 3.24a), the direction of easy magnetization is not symmetric as we see that bcc Fe has <100>, <111> and <110> as the easy, medium and hard directional sets of magnetizations, respectively. The hysteresis loop saturates at a lower field for the <100> direction set of bcc Fe than the <111>, as shown in Fig. 3.24b. Also, the crystal structure determines the easy direction of magnetization as we realize that bcc Fe has <100> as the easy direction while it is <111> for fcc Ni. The crystal anisotropy energy for a cubic crystal is given as

$$\begin{aligned} E_a &= K_0 + K_1\left(\alpha_1^2\alpha_2^2 + \alpha_2^2\alpha_3^2 + \alpha_3^2\alpha_1^2\right) \\ &\quad + K_2\left(\alpha_1^2\alpha_2^2\alpha_3^2\right) + K_3(\alpha_1^2\alpha_2^2 + \alpha_2^2\alpha_3^2 + \alpha_3^2\alpha_1^2)^2 + \cdots \end{aligned} \tag{3.120}$$

where the α terms are direction cosines, K_0 is direction independent and neglected, K_1 is the first order and K_2 is the second order magnetocrystalline anisotropy constant.

The magnetocrytalline anisotropy energy has the unit of J/m^3, and the sign of K_1 and K_2 mostly determines the direction of magnetization. Table 3.3 shows the magnetocrystalline anisotropy energy value neglecting K_0 when the magnetization vector lies along a particular $[hkl]$ direction. If $K_2 > -(9/4)K_1$ and K_1 is positive, then <100>, <110> and <111> are the easy, medium and hard directions, respectively (Cullity and Graham 2009). The K_1 for bcc Fe at room temperature is 48 kJ/m^3 while $K_2 = -1.3$–0.12 kJ/m^3 (Wijn 1991) and therefore, the easy direction is <100>. If K_1 is positive and $K_2 \to -\infty$ then the easy direction is <111>, medium is <100> and hard is <110>.

The magnetocrystalline anisotropy of hexagonal Co has easy direction along the c-axis with θ as the angle between the magnetization and c-axis. All other directions are equally hard. The anisotropy described as uniaxial is given by the energy relation

$$E_a = K_0 + K_1 \sin^2\theta + K_2 \sin^4\theta + \cdots \tag{3.121}$$

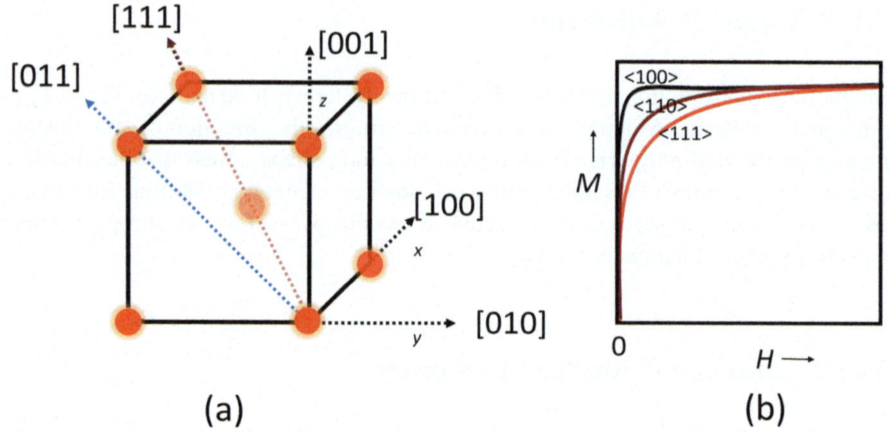

Fig. 3.24 **a** The [100], [001] and [010] directions of bcc Fe are easy magnetization directions and **b** the magnetization along the direction set <100> of bcc Fe could be easily saturated with a lower applied magnetic field than the other directions

Table 3.3 The magnetocrystalline anisotropy energy values and the direction of easy magnetization

[*hkl*]	$\alpha_1, \alpha_2, \alpha_3$	E_a	Direction of easy magnetization
[100]	1, 0, 0	0	<100> if $K_1 > 0$ and $K_2 > -9K_1$
[110]	$1/\sqrt{2}, 1/\sqrt{2}, 0$	$K_1/4$	<110> if $K_1 < 0$ and $K_2 > \frac{9}{4}\lvert K_1\rvert$
[111]	$1/\sqrt{3}, 1/\sqrt{3}, 1/\sqrt{3}$	$K_1/3 + K_2/27$	<111> if $K_1 > 0$ and $K_2 < \frac{9}{4}\lvert K_1\rvert$

The easy axis of magnetization along *c* is described by positive values for both K_1 and K_2. Conversely, if K_1 and K_2 are negative, the easy direction is along the basal plane. The magnetocrystalline anisotropy varies with temperature; therefore, the coercivity is also temperature-dependent.

The origin of magnetocrystalline anisotropy is spin–orbit interaction, where a non-zero $\xi L \cdot S$, would result in spin orientation in a particular preferred direction relative to *L*. When the spin–orbit interaction energy is larger than the crystal field energy, as in the case of rare-earth compounds, the magnetic moment is preferred along specific crystallographic directions due to the coupling of *L* with the lattice and *S*. Consequently, a large magnetic field is required to change the magnetic moment direction that results in large magnetocrystalline anisotropy. For spherically symmetric charge distribution, in the absence of spin–orbit coupling, the spin magnetic moment can point to any direction, and the orbital magnetic moment is quenched. In another model known as single-ion anisotropy, the magnetic ion in a crystalline electric field determines the magnitude of magnetic anisotropy. Magnetocrystalline anisotropy contributes mainly to a magnetic material's coercivity, which could be measured by the torque exerted in an applied magnetic field.

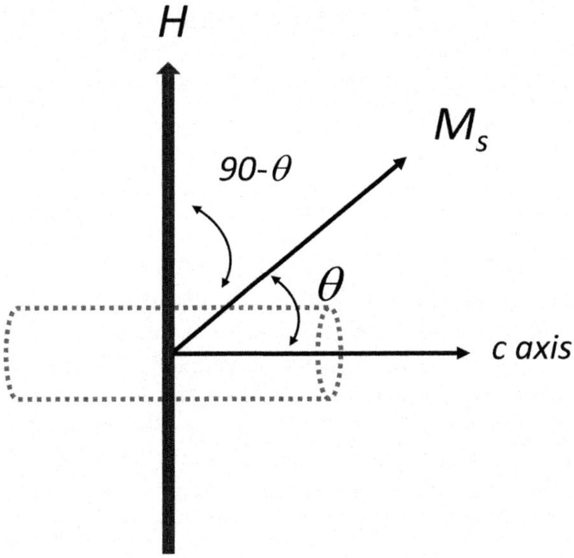

Fig. 3.25 The magnetization direction M_s orients at an angle θ from the c-axis of a uniaxial crystal where the applied field H is perpendicular to it

If M_s is oriented at an angle θ from the preferred easy direction (c-axis) of a uniaxial crystal and hence $90 - \theta$ with H as depicted in Fig. 3.25, the magnetocrystalline and magnetic potential energies are respectively given as

$$E_{\text{a}} = K_1 \sin^2 \theta + K_2 \sin^4 \theta \tag{3.122}$$

$$E_{\text{p}} = -\mu_0 M_s H \cos(90 - \theta) \tag{3.123}$$

From the minimization of energy,

$$\frac{dE_{\text{a}}}{d\theta} = 2K_1 \sin \theta \cos \theta + 4K_2 \sin^3 \theta \cos \theta \tag{3.124}$$

$$2K_1 \sin \theta \cos \theta + 4K_2 \sin^3 \theta \cos \theta = \mu_0 M_s H \cos \theta \tag{3.125}$$

$$2K_1 \sin \theta + 4K_2 \sin^3 \theta = \mu_0 M_s H \tag{3.126}$$

Since

$$M = M_s \cos(90 - \theta) = M_s \sin \theta \tag{3.127}$$

$$\mu_0 H_{\text{ani}} M_s = 2K_1(M/M_s) + 4K_2(M/M_s)^3 \quad (3.128)$$

Saturation will be achieved when $M/M_s = \sin\theta = 1$, and therefore the anisotropy field is given as

$$H_{\text{ani}} = \frac{(2K_1 + 4K_2)}{\mu_0 M_s} \quad (3.129)$$

Magnetic materials are classified as soft or hard magnetic materials based on the magnitude of the coercivity that is proportional to the magnetic anisotropy. Soft magnetic materials exhibit low coercivity (H_C) and a narrow hysteresis loop compared with hard magnetic materials, as shown in Fig. 3.26. The H_C of a typical soft magnetic material such as FeNi alloy is less than 800 A/m, while a hard or permanent magnet may show values of the order MA/m. The experimental values of the H_C in hard magnetic materials may be lower (Buschow and de Boer 2004). The magnetocrystalline anisotropy constant could be extracted from the hysteresis loops using the law-of-approach to saturation (Sivaranjani et al. 2022).

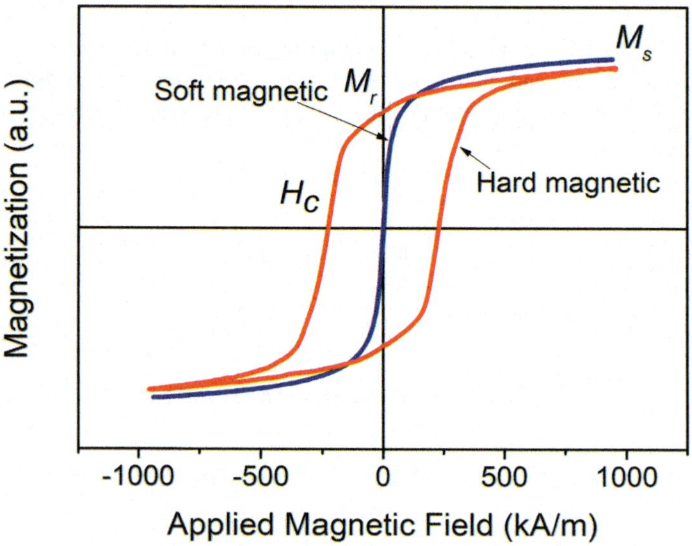

Fig. 3.26 A soft magnetic material has a narrow hysteresis loop, while a hard magnetic material has a broad hysteresis loop and high coercivity

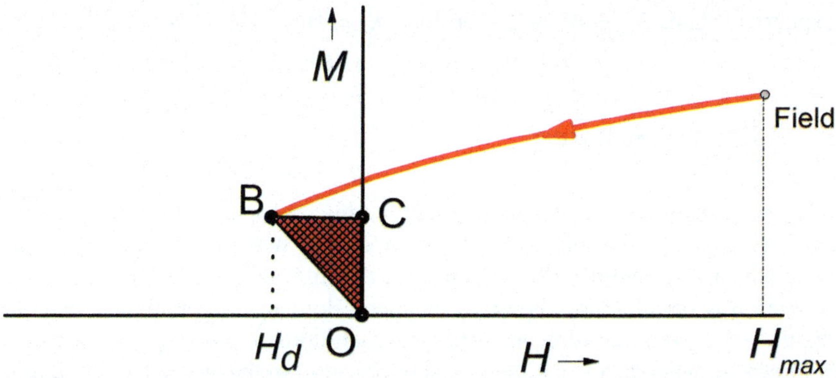

Fig. 3.27 A magnetic material is subjected to a maximum field of H_{max} and switched off. The demagnetization effect takes the magnetization to point B. The energy stored is given by the area of the triangle OBC. Point C corresponds to the remanent magnetization, determined by the demagnetizing field H_d at the point B

3.13.2 Shape Anisotropy

Shape anisotropy contribution to the coercivity arises from the shape of the material. If a magnetic material is magnetized, the demagnetizing field H_d takes the magnetization M to point B. The energy stored is then given by the area of the triangle OBC, as shown in Fig. 3.27. The magnetic potential energy from OBC is

$$E_{\text{ms}} = -\frac{1}{2}\mu_0 H_d \cdot M = \frac{1}{2}\mu_0 (N_d M) M = \frac{1}{2}\mu_0 N_d M^2 \tag{3.130}$$

where $H_d = -\mu_0 N_d M$ and N_d is the demagnetization factor.

Considering a prolate ellipsoid shape, the demagnetization factor along a and c-axis can be N_a and N_c, respectively, with M at an angle θ to the major axis (c). The magnetostatic energy is now

$$E_{\text{ms}} = \frac{1}{2}\mu_0 N_c (M\cos\theta)^2 + \frac{1}{2}\mu_0 N_a (M\sin\theta)^2 \tag{3.131}$$

$$E_{\text{ms}} = \frac{1}{2}\mu_0 N_c M^2 + \frac{1}{2}\mu_0 (N_a - N_c) M^2 \sin^2\theta \tag{3.132}$$

The second term in (3.132) has direction dependence, which is given in terms of shape anisotropy constant (K_s) as

$$K_s = \frac{1}{2}\mu_0 (N_a - N_c) M^2 \tag{3.133}$$

For a spherical shape $N_a = N_c$ and, therefore, $K_s = 0$.

3.14 Magnetic Domains

The direction of spontaneous magnetization in a microscopic scale is along certain preferred directions, determined by the competition between the exchange and magnetic anisotropy energies. Such regions of definite size, where the direction of magnetization is parallel, are known as magnetic domains, and they are separated by domain walls. In a multidomain structure, each domain has a specific magnetization direction, whereas the net magnetization is zero in the absence of an applied magnetic field, as illustrated in Fig. 3.28. Saturation magnetization is reached with a sufficient applied magnetic field such that all the domains point to the field direction. There are two types of domain walls: Bloch wall and Néel wall. The Bloch wall appears in bulk materials, whereas the in-plane domain wall direction changes in thin films, categorized as Néel wall. The thickness of the domain wall is obtained by minimizing the total energies.

Let us assume a 180° domain wall where the domain directions are opposite to each other on either side of the wall, made of N atomic layers of a cubic lattice. The exchange energy between similar spins (S) at a small angle (θ) is

$$E_{\text{ex}} = -2\mathcal{J}S^2\cos\theta = -2\mathcal{J}S^2\left(1 - \frac{\theta^2}{2!} + \frac{\theta^4}{4!} + \cdots\right) \tag{3.134}$$

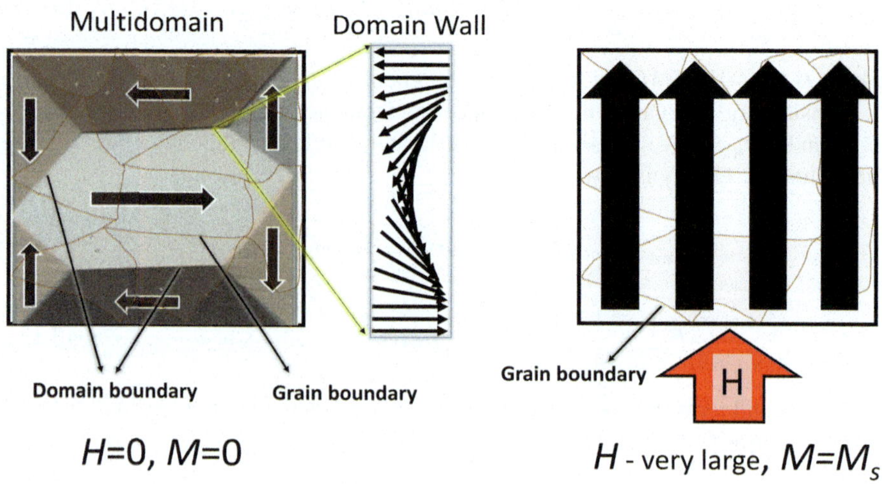

Fig. 3.28 The multidomain state is configured so that the net magnetization is zero. The domains are separated by domain walls where there is a gradual change in the magnetization direction. In a sufficient applied magnetic field, the domain walls vanish, and the magnetization reaches saturation magnetization

Neglecting the higher order terms for smaller θ and considering only the angle-dependent part gives

$$E_{\text{ex}} = \mathcal{J}S^2\theta^2 \tag{3.135}$$

Since E_{ex} is spread over N layers of N spins

$$E_{\text{ex}} = \mathcal{J}S^2\theta^2 N \tag{3.136}$$

The angle between neighboring spins is $\theta = \pi/N$

$$E_{\text{ex}}^A = \mathcal{J}S^2\left(\frac{\pi}{N}\right)^2 N = \mathcal{J}S^2\frac{\pi^2}{N} \tag{3.137}$$

If a is the lattice constant, the exchange energy per unit area is

$$E_{\text{ex}}^A = \frac{\mathcal{J}S^2}{a^2}\frac{\pi^2}{N} \tag{3.138}$$

The wall thickness (δ) for the N layer with lattice constant a is then $\delta = Na$. The exchange energy in terms of wall thickness is given as

$$E_{\text{ex}}^A = \frac{\mathcal{J}S^2\pi^2}{\delta a} \tag{3.139}$$

The anisotropy energy per unit area is

$$E_{\text{a}}^A = KNa = K\delta \tag{3.140}$$

where K is the anisotropy energy density in J/m^3 and the factor Na gives the anisotropy energy per unit area. From (3.139) and (3.140), the total energy is

$$E^A = \frac{\mathcal{J}S^2\pi^2}{\delta a} + K\delta \tag{3.141}$$

$$\frac{dE^A}{d\delta} = -\frac{\mathcal{J}S^2\pi^2}{\delta^2 a} + K = 0 \tag{3.142}$$

$$\delta^2 = \frac{\mathcal{J}S^2\pi^2}{Ka} \tag{3.143}$$

The 180° domain wall thickness is then

$$\delta = \pi\sqrt{\frac{\mathcal{J}S^2}{Ka}} = \pi\sqrt{\frac{A}{K}} \tag{3.144}$$

Here, A is the exchange stiffness constant given as

$$A = \frac{n\mathcal{J}S^2}{a} \tag{3.145}$$

and n is the number of atoms per unit cell where $n = 1$ in (3.144). For bcc Fe, considering $n = 2$, $\mathcal{J} = 2.16 \times 10^{-21}$ J, $S = 1$ and $a = 2.856$ Å, the exchange stiffness constant $A = 1.513 \times 10^{-11}$ J/m ≈ 15 pJ/m.

The strength of the exchange integral is directly proportional to the Curie temperature and best correlated by the Weiss expression given as (Weiss 1948)

$$\mathcal{J} = N_n k_B T_c \tag{3.146}$$

N_n is 0.54 for simple cubic spin ½, 0.34 for bcc spin ½ and 0.15 for bcc spin 1 lattices. The energy of the wall for uniaxial and cubic anisotropy is

$$E_w = 4\sqrt{AK} \tag{3.147}$$

The wall thickness and wall energy of bcc Fe are $\delta = \pi\sqrt{\frac{15\times10^{-12}}{48\times10^3}} \approx 60$ nm and 3.4×10^{-3} J/m^2, respectively.

The domain magnetization takes possible directions according to the easy axes of the crystals and for cubic Fe, the <100> set has six directions. For the fcc Ni, the <111> easy axis set has possible domain wall angles 180°, 109° and 71°. In addition to 180° domain walls, all the other types are categorized as 90°. The wall thickness for a 90° domain wall is given as

$$\delta = \frac{\pi}{2}\sqrt{\frac{A}{K}} \tag{3.148}$$

where the wall energy is also reduced by half. Considering strain effects, 90° domain walls are prevalent when the magnetostriction is small. The Bloch wall energy in thin films is higher when the thickness is small, whereas the Néel wall energy increases with thickness, and therefore, the Néel wall is prevalent in films < 50 nm thick.

Assuming a spherical shape, the uniaxial crystals will show a critical single-domain radius

$$R_c = C_o\frac{\sqrt{AK}}{\mu_0 M_s^2} \tag{3.149}$$

where C_o is a constant, while for weak anisotropy materials, the critical single domain radius is (O'Handley 2000)

$$R_c = 3\sqrt{\frac{A}{\mu_0 M_s^2}}\left(\sqrt{\ln\left(\frac{2R_c}{A}\right)} - 1\right) \tag{3.150}$$

The critical single domain radius for the low anisotropy materials would be difficult to estimate using (3.150) and therefore (3.149) is used conveniently. The critical single domain size for bcc Fe is ~ 50 nm, which has been obtained experimentally for cubic particles (Sivaranjani et al. 2020).

3.15 Magnetic Frustration

An important feature in describing an unusual magnetic structure with a non-collinear spin arrangement may require the concept of frustration, which was first introduced by G. Toulouse in 1977 (Toulouse 1977). Magnetic frustration is manifested if magnetic interactions in all the spin systems in the geometric system are not satisfied simultaneously. Such a situation results from the topology of the magnetic lattice combined with the nature and strength of magnetic exchange interactions. As examples, two typical configurations are schematized in Fig. 3.29, where only first nearest neighbor magnetic interactions are considered: in Fig. 3.29 (left), the presence of both ferromagnetic ($\mathcal{J}_1 > 0$) and antiferromagnetic ($\mathcal{J}_1 < 0$) interactions between moments located on a square lattice. In Fig. 3.29 (right), the presence of only antiferromagnetic interactions between moments located on a triangular lattice lead to unsatisfied magnetic interactions. But in some cases, magnetic frustration also arises from the mixture of longer-range ferromagnetic and antiferromagnetic interactions as encountered in metallic and ionic systems, where Ruderman-Kittel-Kasuya-Yoshida (RKKY) interactions, superexchange (SE) and supersuperexchange (SSE) interactions, respectively, are present.

The magnetic frustration, initially reserved for triangular and pyrochlore structures and Kagomé lattice, has been widely applied to describe the magnetic structure of certain crystalline and amorphous fluorides. These structures result from the particular nature of the cationic topology and the presence of antiferromagnetic interactions, to which zero-field and in-field ^{57}Fe Mössbauer spectrometry has made a significant contribution (Ferey et al. 1985; Greneche and Varret 1993). In addition, the frustration concept helps to explain the magnetic disorder at the surface of nanoparticles, which is dominated by the presence of structural disorder and dangling chemical bonds.

In a triangular lattice, consider Heisenberg model spins occupying the three vertices. If two spins are parallel or antiparallel to each other, the third spin is in a state of frustration. The energy of the system can be written as

$$E = \mathcal{J}(S_1 \cdot S_2 + S_2 \cdot S_3 \cdot + S_3 \cdot S_1) = -\frac{3}{2}\mathcal{J}S^2 + \frac{\mathcal{J}}{2}(S_1 + S_2 + S_3)^2 \tag{3.151}$$

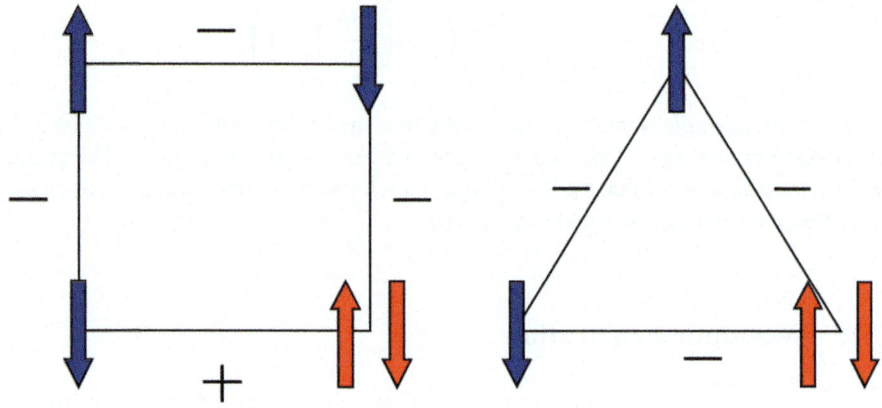

Fig. 3.29 The illustrations of frustrated magnetic structures, (left) a square lattice and (right) a triangular lattice

The minimum energy then corresponds to an angle of 120° to each other such that $S_1 + S_2 + S_3 = 0$ in a triangular antiferromagnet. Magnetic frustrations in antiferromagnetic systems are predominant when the Curie–Weiss temperature, $|\Theta_C|$ is greater than the T_N. Another type of frustration arises due to the competition between the near neighbor and next near neighbor exchange interactions in a helimagnet such as Ho, Dy and Tb through the long-range RKKY interaction. The RKKY interaction is an oscillatory exchange with distance given by

$$\mathcal{J}(R) = J_o \frac{\cos(2k_F R)}{(k_F R)^3} \tag{3.152}$$

where J_o is a constants, k_F is the Fermi wave number and R is the distance where $R >> 1/k_F$. Alternate ferromagnetic and antiferromagnetic interactions are encountered whose magnitude decreases with distance. A non-frustrated helical state arises due to the Dzyaloshinskii-Moriya interaction (DMI). Exotic states such as spin liquid, valence bond crystal and nematic phase can occur in frustrated systems.

The Dzyaloshinskii-Moriya interaction, arising due to spin–orbit interactions, is given by the Hamiltonian,

$$\mathcal{H} = D_{ij} \cdot (S_i \times S_i) \tag{3.153}$$

The DMI is antisymmetric, and the magnitude of the vector D_{ij} is two orders smaller than the Heisenberg exchange integral. The noncollinear spin results in spin canting and chiral magnetic structures. Randomized spin freezing below the spin freezing temperature, T_f, leads to another magnetic structure known as spin glass, as represented in Fig. 3.30a. The local spontaneous magnetization in such a case is determined by $< S_i > \neq 0$ while the k vector magnetization approaches zero as,

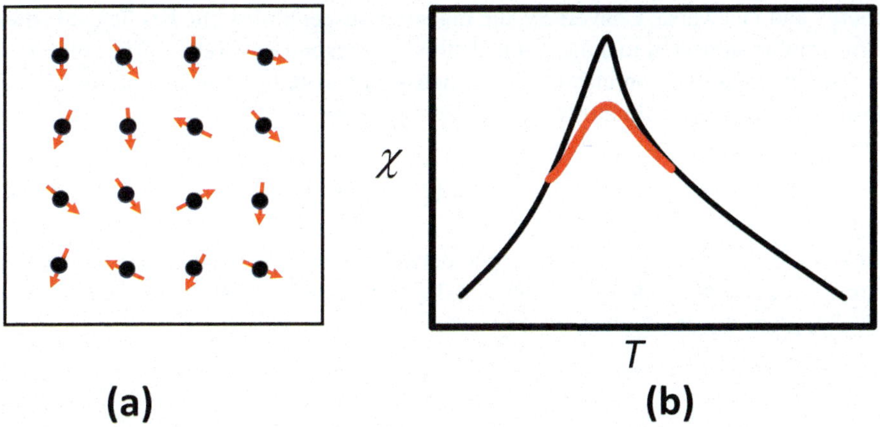

Fig. 3.30 **a** The spin glass model that shows randomized spins producing non-zero magnetization at low temperature and **b** the temperature-dependent ac susceptibility of spin glass. The cusp in **b** is spread out with increasing applied field strength

$\frac{1}{N}\sum_i \langle S_i \rangle \exp(ik.R) = 0$ as the number of spins $N \to \infty$ where R is the distance for $T < T_f$. The temperature-dependent ac susceptibility displays a cusp at the T_f (Fig. 3.30b), influenced by the ac frequency. With increasing applied field strength, the cusp broadens and shifts towards low temperature, and therefore, T_f has to be determined as $H \to 0$. The spin glass exhibits multiple metastable free energy valleys, with each valley corresponding to the up and down magnetization states. The order parameter as per Edwards and Anderson model (EA) in terms of thermal and configurational averages is given by (Edwards and Anderson 1975) $q_{EA} = \langle \langle S_i \rangle^2 \rangle$. The q_{EA} is a local nonequilibrium single valley parameter measured through ac susceptibility.

3.16 Magnetostriction

The magnetostrictive effect, discovered in 1842 by J. Joule is the characteristics of a magnetic material leading to a change in the dimension (strain) in the presence of an external magnetic field. The linear saturation magnetostriction is defined as $\lambda_s = \pm\frac{\Delta l}{L}$ where Δl is the change in length and L is the initial length. The strain produced ranges from $\pm$ 10^{-3} to 10^{-7} in materials such as Fe, Ni, FeNi, Tb and yttrium iron garnet. It is important to note that the highest value of 0.002 m/m is observed with Terfenol-D, $Tb_xDy_{1-x}Fe_2$ (x ~ 0.3) alloy. It is well established that the variations in dimensions induced by magnetostriction are associated with changes in the orientation of the magnetic domains. Only phenomenological expressions were proposed to describe the magnetostrictive coefficient in high symmetry systems. Certain amorphous alloys and Co ferrites can possess magnetostrictive properties and can be used as actuators and sensors after converting magnetic energy into kinetic energy and vice versa. Conversely, the magnetic susceptibility can be changed when a material is subjected to a mechanical stress, corresponding to the Villari effect.

The magnetostriction elastic energy density contribution for an isotopic crystal from the strain and magnetostriction tensors is given by

$$E_{\mathrm{ms}} = \frac{1}{2}\lambda_s\sigma\left(1 - 3\cos^2\theta\right) \tag{3.154}$$

where σ is the stress and θ is the angle between the magnetization and the applied stress direction. Considering the angular dependence of (3.154) and comparing with (3.122) gives the uniaxial anisotropy term

$$K_{\mathrm{u}} = \frac{3}{2}\lambda_s\sigma \tag{3.155}$$

For a cubic polycrystal, the magnetostriction averaged over various crystallographic directions is given by

$$\lambda_s = \frac{2}{5}\lambda_{\{100\}} + \frac{3}{5}\lambda_{\{111\}} \tag{3.156}$$

3.17 Superparamagnetism

The thermal fluctuation of isolated single-domain particles, due to the reduction in the anisotropy energy compared with the thermal energy, would produce magnetic effects similar to that given by Langevin's paramagnetism and known as superparamagnetism (Bean and Livingston 1959). Consider a spherical particle magnetized to a remanent state M_a and a final demagnetized state M_b. Depending on the thermal energy, the particle may change its state within a time period, and the rate of decrease is given by

$$-\int_{M_a}^{M_b} dM/dt = \int_{M_a}^{M_b} M/\tau \tag{3.157}$$

$$M_b = M_a \exp(-t/\tau) \tag{3.158}$$

Upon saturation, $M_b = M_a$ and using the Boltzmann function $\exp(-KV/k_{\mathrm{B}}T)$,

$$\frac{1}{\tau} = \frac{1}{\tau_{\mathrm{o}}}\exp(-KV/k_{\mathrm{B}}T) \tag{3.159}$$

where τ is the relaxation time and τ_o is the measurement-dependent relaxation time constant.

The anisotropy energy is

$$KV = 25k_BT \quad (3.160)$$

where the factor 25 arises considering the relaxation time for a typical experiment. The critical superparamagnetic volume and blocking temperature could be obtained from (3.160). It is important to note that the blocking temperature is not a phase transition but a continuous evolution of $\tau(T)$, while its estimation depends on the characteristic time measurement associated with the experimental technique. The anisotropy energy keeps the magnetization direction along one of the easy axes for large particles, whereas smaller particles below the thermal energy flip their magnetizations between the minimum energy states, as illustrated in Fig. 3.31. The presence of an external magnetic field favors the blocking of magnetic structures within nanoparticles, which makes in-field low-temperature ^{57}Fe Mössbauer spectrometry particularly interesting for studying the dynamics of Fe-containing nanoparticles.

The coercivity of superparamagnetic particles is zero due to the easy reversal of magnetization direction. Since the anisotropy energy is larger for materials with substantial magnetocrystalline anisotropy, permanent magnetic materials show very small critical superparamagnetic sizes. The coercivity is maximum for the critical single domain size where each particle has a single domain, and an applied field rotates the domain direction. On increasing the particle size from the single domain

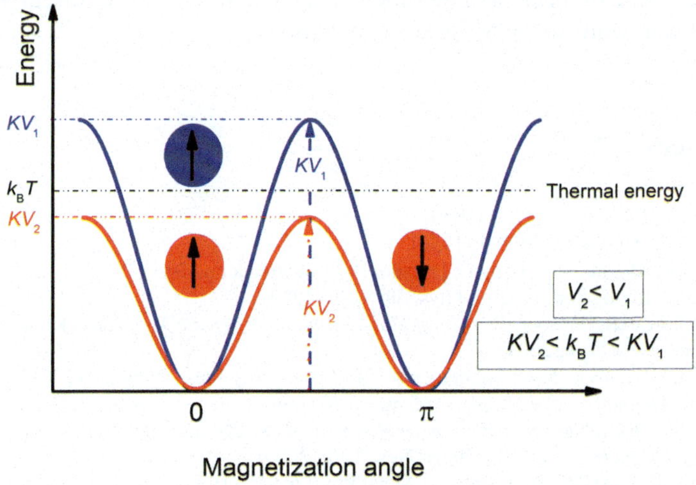

Fig. 3.31 The anisotropy energy for large particles (KV_1) exceeds the thermal energy (k_BT), and therefore, the magnetization angle is stable along one of the easy directions. The magnetization angle flips between their energy minima for smaller particles whose anisotropy energy (KV_2) is lower than the thermal energy

state to the multidomain particles, the coercivity decreases due to domain wall motion, which requires less energy than domain rotation. However, domain wall pinning due to defects or impurities may increase the coercivity as more energy has to be spent to overcome the energy barrier. Consequently, the study of an assembly of nanoparticles or nanoarchitectures requires firstly modeling the size distribution and the distribution of morphologies, giving rise to a distribution of relaxation times and then modeling the distribution of distances or the nature of grain boundaries to better understand the role of dipolar interactions in the interparticle interactions or coupling between magnetic grains. Numerous theoretical models have been established to describe ideal homogeneous assemblies of nanoparticles, but they are not capable of strictly describing artificial and natural nanostructures, despite the recent enormous progress made in the synthesis, control, homogeneity and reproducibility of samples. However, these theoretical models are very useful for better interpreting experimental data, provided they are obtained using complementary instrumental techniques, and even at different temperatures and magnetic fields.

3.18 Summary

The essential magnetic characteristics of materials such as paramagnetism, ferromagnetism, ferrimagnetism, antiferromagnetism and superparamagnetism are elaborated. The coercivity contribution from magnetic anisotropy and domain aspects helps to understand the classification of magnetic materials. The temperature-dependent magnetic behaviors of various magnetic materials are discussed, which shall hold well while using Mössbauer spectrometry.

References

P.W. Anderson, Phys. Rev. **115**, 2 (1959)
P.W. Anderson, Phys. Rev. **79**, 705 (1950)
P.W. Anderson, *Magnetism*, ed. by G.T. Rado, H. Suhl (Academic Press, New York, 1963)
C.P. Bean, J.D. Livingston, J. Appl. Phys. **30**, S120 (1959)
K.H.J. Buschow, F.R. de Boer, *Physics of Magnetism and Magnetic Materials* (Kluwer Academic Publishers, New York, 2004)
S. Chikazumi, *Physics of Ferromagnetism* (Oxford University Press, New York, 1997)
J.M.D. Coey, *Magnetism and Magnetic Materials* (Cambridge University Press, Cambridge, 2009)
B.D. Cullity, C.D. Graham, *Introduction to Magnetic Materials*, 2nd edn. (Wiley, NJ, 2009)
G.F. Dionne, *Magnetic Oxides* (Springer, New York, 2009)
S.F. Edwards, P.W. Anderson, J. Phys. F: Metal Phys. **6**, 965 (1975)
G. Ferey, M. Leblanc, R. de Pape, J. Pannetier, in *Inorganic Solid Fluorides*, ed. by P. Hagenmuller (Academic Press, New York, 1985), p. 395.
J.B. Goodenough, *Magnetism and the Chemical Bond* (Interscience Publisher, New York, 1963)
J.-M. Greneche, F. Varret, in *Mössbauer Spectroscopy Applied to Magnetism and Magnetic Materials*, ed. by G. Long, F. Grandjean (Plenum, New York, 1993).

J.F. Janak, Phys. Rev. B **16**, 255 (1977)
J. James, Ann. Electr. Magn. Chem. **8**, 219 (1842)
T. Kasuya, Prog. Theor. Phys. **16**, 45 (1956)
L. Néel, Ann. Phys. **3**, 137 (1948)
R.C. O'Handley, *Modern Magnetic Materials: Principles and Applications* (Wiley, New York, 2000)
L. Pauling, Phys. Rev. **54**, 899 (1938)
M.A. Ruderman, C. Kittel. Phys. Rev. **96**, 99 (1954)
K.S. Sivaranjani, G. Antilen Jacob, R. Justin Joseyphus, Phys. Status Solidi B, 2200050 (2022)
K.S. Sivaranjani, G. Antilen Jacob, R. Justin Joseyphus, J. Magn. Magn. Mater. **513**, 167228 (2020)
J.C. Slater, Phys. Rev. **49**, 931 (1936)
J.S. Smart, Am. J. Phys. **23**, 356 (1955)
E.C. Stoner, Proc. R. Soc. Lond. A **165**, 372 (1938)
G. Toulouse, Commun. Phys. **2**, 115 (1977)
J.H. Van Vleck, J. Chem. Phys. **9**, 85 (1941)
P. Weiss, Compt. Rend. **143**, 1136 (1906)
P.R. Weiss, Phys. Rev. **74**, 1493 (1948)
H.P.J. Wijn (ed.), *Magnetic Properties of Metals: d-Elements, Alloys and Compounds* (Springer, Heidelberg, 1991)
K. Yosida, Phys. Rev. **106**, 893 (1957)

Chapter 4
The Mössbauer Effect

4.1 The Phenomenon of Nuclear Recoil

The phenomenon of recoilless emission and resonant absorption of nuclear gamma rays in solids was discovered by Rudolf L. Mössbauer during his experiments with the 129 keV transition of ^{191}Os to ^{191}Ir (Mössbauer 1958). The discovery was facilitated by (Mössbauer 2000) (i) the choice of the NaI detector, which is close to 100% efficiency for the 129 keV γ-ray, (ii) the choice of low-temperature experiments with a custom-built cryostat, (iii) the experimental proof of nuclear resonance fluorescence using the Doppler effect with the 411 keV decay of ^{198}Hg to ^{198}Au by Moon (1951) and (iv) the paper on neutron resonance absorption by Lamb (1939).

The factors affecting the absorption and emission lines were discussed by Kuhn (1929), corresponding to thorium decay through β and γ emission. The γ-ray recoil broadening has an energy shift due to recoil energy loss, and it is challenging to achieve absorption if the energy distribution is shifted to a large extent. Figure 4.1 illustrates the absorption or emission lines with (a) natural line broadening, (b) the Doppler effect line broadening and (c) the Doppler effect line broadening accompanied by an energy shift. The energy shift due to the recoil energy loss occurs in both the absorber and the emitter. Ideally, the absorption and emission should be of the same energies and with similar line broadening. However, it is seen that there is not only a line broadening (due to the natural and Doppler effects) but also an energy shift (due to the recoil effect).

Let us consider a free nucleus emitting a γ-ray photon from an excited state, E_f, to a lower energy state, E_i. This emitted photon must be absorbed by another nucleus having the same energy difference, $E_f - E_i$. Figure 4.2 shows the free nucleus with a mass M emitting γ-ray of energy E_γ. The nucleus has to recoil to conserve its momentum, resulting in a change in velocity after the gamma ray is emitted.

According to energy conservation, the energy equation can be written as

R. J. Joseyphus and J.-M. Greneche, *Fundamentals of ^{57}Fe Mössbauer Spectrometry*,
https://doi.org/10.1007/978-981-99-8653-8_4

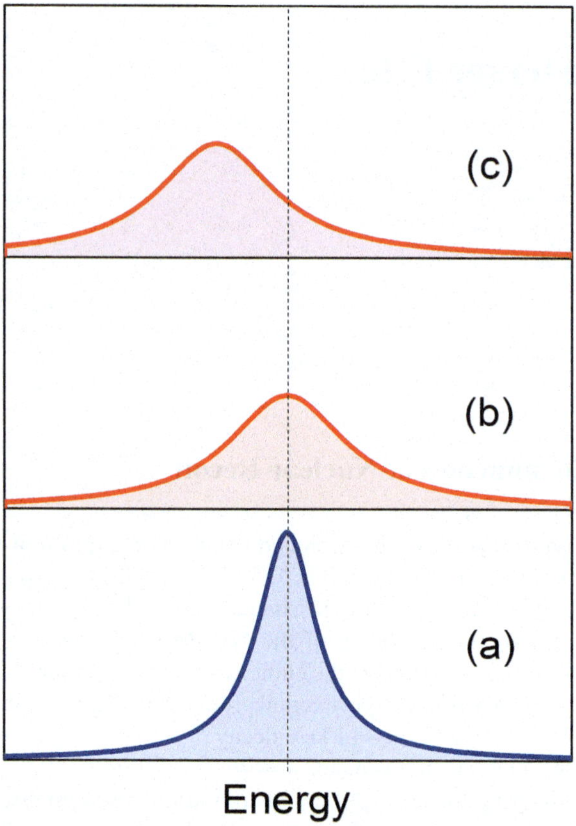

Fig. 4.1 Absorption/emission lines showing **a** natural line broadening **b** the Doppler effect line broadening, and **c** the Doppler effect line broadening along with an energy shift

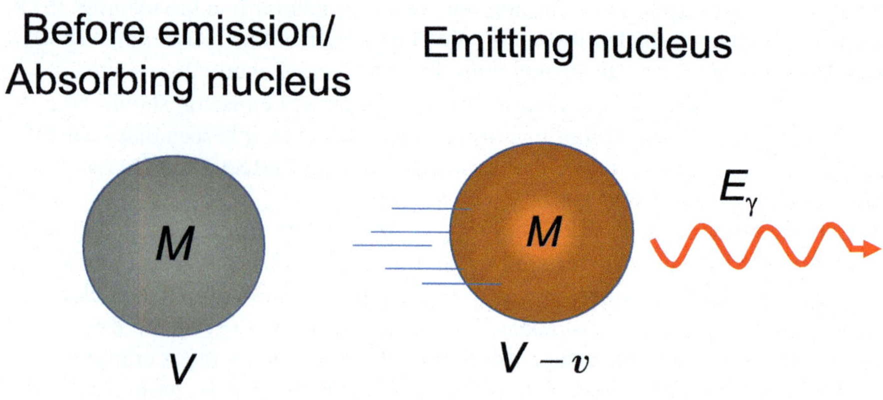

Fig. 4.2 A free nucleus of mass, M emitting a γ-ray of energy, E_γ. The velocity of the nucleus before emission is V; after emission, it is $V - v$ due to recoil

$$E_t + \frac{1}{2}MV^2 = E_\gamma + \frac{1}{2}M(V - v)^2 \tag{4.1}$$

where $E_t = E_f - E_i$, E_γ, M, V and v are the transition energy of the nucleus before emission, the energy of the emitted gamma ray, the mass of the nucleus, the velocity of the nucleus before emission and velocity of the nucleus after emission, respectively. Note that $E_t = E_\gamma$ under ideal conditions and that there is no change in the nuclear mass since it is only γ-ray emission without any change in the number of nucleons. The v after emission is negative because the nucleus recoils just after emission. The energy difference is then given by

$$\Delta E = E_t - E_\gamma = \frac{1}{2}Mv^2 - \frac{1}{2}(2MVv) = E_R - E_D \tag{4.2}$$

From (4.2), we have obtained the recoil energy $E_R = \frac{1}{2}Mv^2$ and the Doppler effect energy $E_D = MVv$. The velocity, V, is affected by the thermal motion of the nucleus, which contributes to the Doppler energy term. The negative sign implies that E_R and E_D are opposite to each other. The energy levels corresponding to E_t with similar excited and ground state levels for the absorber and emitter are schematized in Fig. 4.3.

The momentum before and after emission must be conserved, and the energy is written as

$$\frac{p^2}{2M} = \frac{p_\gamma^2}{2M} = \frac{E_\gamma^2}{2Mc^2} \tag{4.3}$$

where p is the momentum of the nucleus before emission while the momentum after emission is the γ-ray momentum given as, $p_\gamma = -\frac{E_\gamma}{c}$. The negative sign is introduced to take care of the opposite momentum due to recoil. From (4.3), we have the recoil energy given by

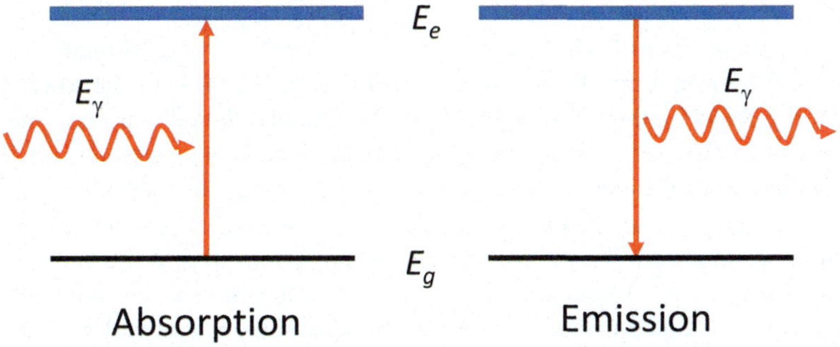

Fig. 4.3 The absorption and emission of γ-rays is feasible if $E_\gamma = E_e - E_g$

Table 4.1 Few gamma ray emitting isotopes, their E_γ and the calculated recoil energy losses using (4.4)

Isotope	E_γ (keV)	E_R (eV)
^{198}Au	411	0.46
^{192}Pt	316.5	0.28
^{192}Os	205.7	0.19
^{176}Yb	82.2	0.02
^{161}Dy	25.7	2.20×10^{-3}
^{149}Sm	22.5	1.82×10^{-3}
^{119}Sn	23.9	2.57×10^{-3}
^{83}Kr	9.3	5.59×10^{-4}
^{61}Ni	67.4	0.04
^{57}Fe	14.4	1.95×10^{-3}
^{57}Fe	122.1	0.14
^{40}K	29.4	0.01

$$E_R = \frac{E_\gamma^2}{2Mc^2} \tag{4.4}$$

suggesting that the recoil energy loss depends on the energy of the emitting gamma ray and the mass of the nucleus. Table 4.1 shows the E_R values for some γ emitting nuclei.

The E_R is of the order 0.1–100 meV for most light and heavier isotopes (Table 4.1) and increases drastically with E_γ. Thus, ^{83}Kr, which has a smaller E_γ, but lighter than ^{119}Sn, has a lower E_R. Similarly, ^{57}Fe, whose transition energy of the first excited level is 14.4 keV, has a lower E_R than the second excited state energy of 136.3 keV. Nuclei with high E_γ, although heavier, show very high E_R, as in the cases of ^{198}Au and ^{192}Pt. The recoil energy loss is higher than the width of the absorption lines, which has disadvantages that are discussed further.

The resonant absorption occurs when there is an overlap in the line-broadened gamma ray emission energy equal to the transition energy ($E_\gamma = E_t$). As there is a loss of recoil energy in both the emission and absorption, full resonance can only be achieved by removing the recoil, as illustrated in Fig. 4.4. In the lower part of Fig. 4.4, the peaks on the left side correspond to the broadened emission line which has a lack of energy, $-E_R$, due to recoil. In the broadened absorption line on the right side, we have a similar energy loss of $+E_R$ due to the recoil energy loss, resulting in a non-overlapping spectrum. There is a partial overlap when the recoil energy is minimized, as shown by the absorption and emission lines in the central part of Fig. 4.4. However, we should note that if the linewidth is narrow, as indicated by the striped absorption and emission Lorentzian profiles, there is no overlap at all. When the recoil energy loss is removed, the emission and absorption line profiles overlap entirely, as shown in the upper part of Fig. 4.4.

Due to the large γ-ray energies compared to the photons in atomic transitions, the E_R is significant in the case of nuclear transitions. The overlap in emission

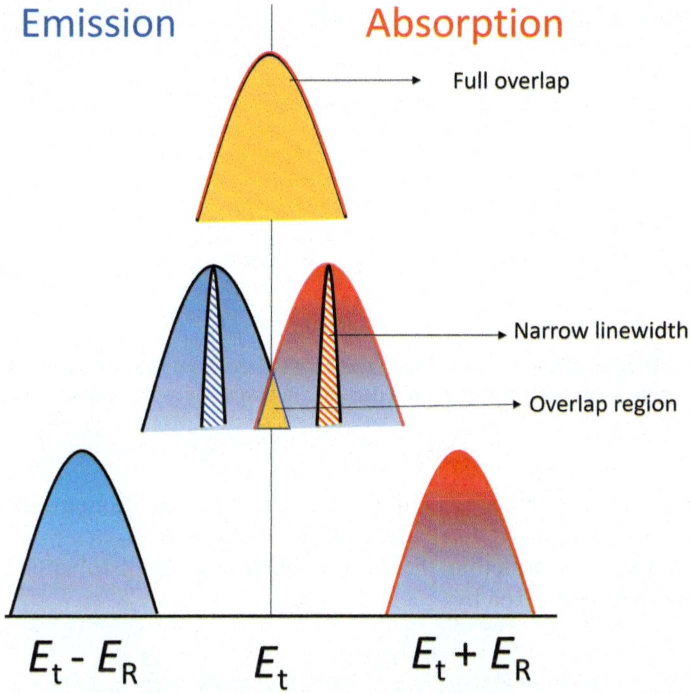

Fig. 4.4 The emission (left) and absorption (right) line profiles without overlap due to recoil are shown at the bottom. Partial overlap when energy loss due to recoil is minimized or no overlap with narrow linewidth (middle). The emission and absorption line profiles overlap completely when there is no energy loss due to recoil (top)

and absorption line profiles is, therefore, minimal. For resonant absorption, $\Gamma/2E_R$ should be greater than one. The Γ/E_R is 10^{-5} and 10^3 for nuclear and atomic systems, respectively. Therefore, optical resonance fluorescence is readily observed compared to nuclear resonance fluorescence.

4.2 Thermal Velocity Distribution

As the atoms move towards the absorber due to thermal motion, they encounter an increase in frequency due to the Doppler effect. The Doppler effect is the increase or decrease in the frequency of the wave as the source approaches or moves away from the observer, as illustrated in Fig. 4.5. If the original frequency is υ_o, measured when the source and observer are in the same position, then when a source approaches an observer with a velocity v, the frequency is increased by a factor of $(1 + v/c)$ with respect to the speed of the wave in vacuum, c. The frequency is then said to

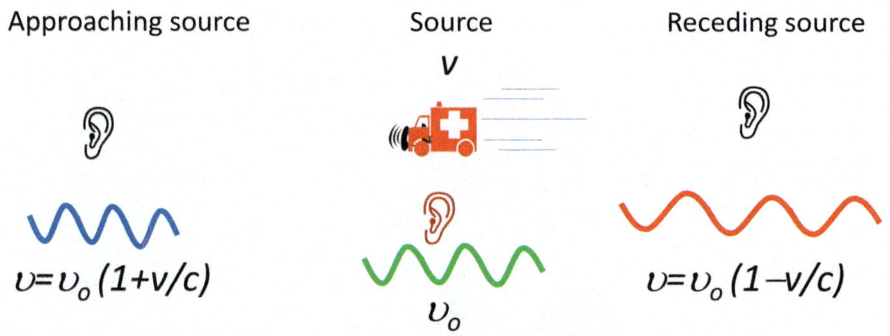

Fig. 4.5 The Doppler effect of sound waves could be noticed through an increase in frequency as the source approaches an observer (left) or a decrease in frequency as the source moves away from the observer (right)

be blue-shifted. On the other hand, if the source moves away from the observer, the frequency is red-shifted and decreases by a factor of $(1 - v/c)$.

From Maxwell's distribution, the most probable velocity distribution of a gas in thermal equilibrium is given by

$$f(v)dv = A\exp\left(-\frac{E_K}{k_B T}\right)dv = A\exp\left(-\frac{Mv^2}{2k_B T}\right)dv \tag{4.5}$$

Using Maxwell's relation for a free particle undergoing thermal motion along the x-direction, the kinetic energy is given by

$$\frac{1}{2}M\langle v^2\rangle \tag{4.6}$$

where $\langle v^2\rangle$ is the mean square velocity given as

$$\langle v^2\rangle = \frac{\int_0^\infty v^2 f(v)dv}{\int_0^\infty f(v)dv} \tag{4.7}$$

The normalization constant A in (4.5) is given for the 1D case as

$$A = \sqrt{\frac{M}{2\pi k_B T}} \tag{4.8}$$

Combining (4.5), (4.7) and (4.8), we get the velocity distribution in 1D as

$$f(v)dv = \sqrt{\frac{M}{2\pi k_B T}} \exp\left(-\frac{M\langle v^2\rangle}{2k_B T}\right) dv \tag{4.9}$$

The most probable speed (v_p) is given by the peak in the distribution function given by (4.9), while the mean speed, $\langle v \rangle$, and the root mean square (rms) speed are less than v_p.

4.3 Resonance Using Doppler Energy

The motion of the nucleus, parallel or antiparallel to the gamma rays, induces the Doppler energy. In order to match the energy ΔE of the emitting nucleus with the energy of the absorbing nucleus, the E_R and E_D have to be now taken into account. From (4.2), we have $E_t - E_\gamma = E_R - E_D$ and we obtain the shift in the energy due to the Doppler effect and recoil energy as $E_s = E_t + E_D = E_R + E_\gamma$. Figure 4.6 shows the source emitting a gamma ray of energy, E_γ, which has a deficit of energy by $E_t - E_R$ where; (i) E_γ is not equal to the transition energy in an absorber (bottom) and (ii) the deficient energy is compensated by a Doppler energy, E_D, and therefore E_γ is absorbed by the absorber (top).

The transition energy, E_t, is not sharp and has a finite width in addition to the Doppler energy broadening, E_D. Let us assume that the source, emitting gamma rays, approaches the absorbing nucleus. Moon (Moon 1951) proposed that resonant absorption of the gamma ray is possible under three conditions: (i) the emitting and absorbing nuclei are identical, (ii) the gamma ray transition goes to the ground state and (iii) a relative velocity can be given to the source and absorber such that the Doppler effect restores the energy lost due to nuclear recoil.

The cross section of the resonant scattering is given by the Breit-Wigner formula (see Sect. 2.14)

$$\sigma = \frac{\pi}{k^2} \frac{\Gamma_i \Gamma_f}{(E - E_o)^2 + \frac{\Gamma^2}{4}} \tag{4.10}$$

where Γ_i, Γ_f are the partial widths of the resonance at decay, Γ is the full width at half maximum (FWHM) at resonance, E is the peak energy, E_o is the rest mass energy and k is the wave number of the projectile. From (4.10) considering $\Gamma_i = \Gamma_f$ and using $k = 2\pi/\lambda$, we get the cross section for resonance scattering of gamma rays as

$$\sigma = C\lambda^2 \frac{\Gamma^2}{(E_\gamma - E_t)^2 + \frac{\Gamma^2}{4}} = C\frac{(hc)^2}{E^2} \frac{\Gamma^2}{(E_\gamma - E_t)^2 + \frac{\Gamma^2}{4}} \tag{4.11}$$

where $\lambda = hc/E$, C is a constant and Γ is the FWHM in eV.

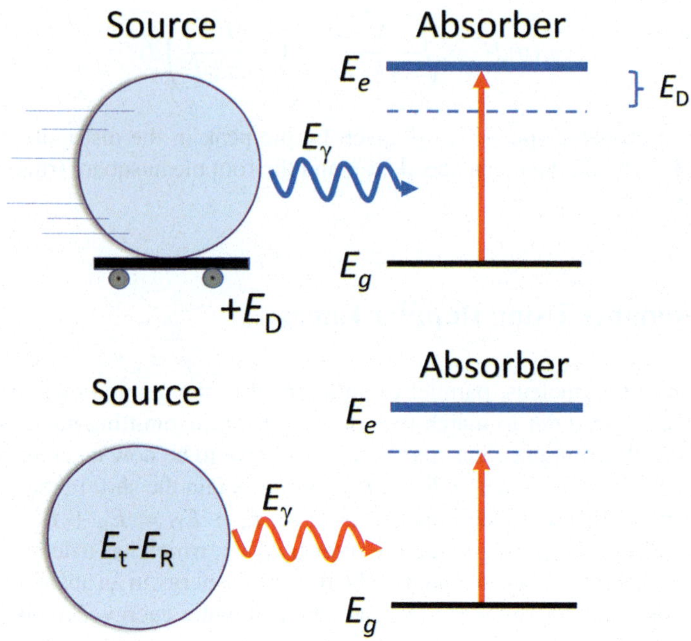

Fig. 4.6 The source emitting a gamma ray of energy, E_γ which has a deficit $E_t - E_R$ in the absorber (bottom). Doppler energy, E_D compensates for the energy loss in the source and absorber, resulting in a transition (top). Note that the absorber will also have recoil energy loss

The velocity distribution of the source and the absorber undergoing thermal collision in the direction of γ-ray is given in terms of the relative velocity V by

$$f(V)dV = \sqrt{\frac{M}{4\pi k_B T}} \exp\left(-\frac{MV^2}{4k_B T}\right) dV = \frac{1}{\pi \Gamma_T} \exp\left(-\frac{MV^2}{4k_B T}\right) dV \quad (4.12)$$

where Γ_T is the thermal full width at half maximum that has the unit of velocity. Now, a Doppler energy change is given in the form of a velocity change that gets added to the transition energies of the moving nucleus. Combining the cross section equation (4.11) in the velocity distribution (4.12) and adding a Doppler velocity energy

$$E_D = \frac{E_\gamma V_D}{c} \quad (4.13)$$

where V_D is the Doppler velocity, we get the thermal resonance cross section as

$$\begin{aligned}\sigma_T &= \int_{-\infty}^{+\infty} \sigma f(V) dV \\ &= \int_{-\infty}^{+\infty} \frac{\lambda^2}{8\pi}\left(\frac{\Gamma^2}{\left(E_\gamma - E_t + E_D\right)^2 + \frac{\Gamma^2}{4}}\right)\sqrt{\frac{M}{4\pi k_B T}}\exp\left(-\frac{MV^2}{4k_B T}\right) dV \\ &= \sqrt{\frac{M}{4\pi k_B T}}\exp\left(-\frac{M(V_t - V_D)^2}{4k_B T}\right)\frac{\lambda^2}{8\pi}\int_{-\infty}^{+\infty}\left(\frac{\Gamma^2}{\frac{E_\gamma^2}{c^2}(V - V_t + V_D)^2 + \frac{\Gamma^2}{4}}\right) dV\end{aligned} \tag{4.14}$$

where $V = V_t - V_D$, $V_t = \frac{E_\gamma}{Mc}$ and V_D are the velocity of the source (initial nucleus), velocity of the final nucleus that undergoes a transition and the Doppler velocity, respectively. The thermal resonance cross section can be simplified using (4.15) as

$$\int_{-\infty}^{+\infty}\left(\frac{\Gamma^2}{\frac{E_\gamma^2}{c^2}(V - V_t + V_D)^2 + \frac{\Gamma^2}{4}}\right) dV = \frac{2\pi\Gamma c}{E_\gamma} \tag{4.15}$$

$$\sigma_T = \sqrt{\frac{M}{4\pi k_B T}}\exp\left(-\frac{M(V_t - V_D)^2}{4k_B T}\right)\frac{\lambda^2}{4}\frac{c\Gamma}{E_\gamma} \tag{4.16}$$

Maximum nuclear resonant cross section occurs when $V_t = V_D$, i.e., the absorber nuclear velocity is matched by the Doppler velocity. Moon (Moon 1951) used 411 keV γ-ray using ^{198}Hg source, which required a very high Doppler velocity of 600 m/s (using a rotor) to observe partial resonance. The Doppler energy from (4.13) is 0.82 eV, and complete resonance could not be achieved as the recoil energy from (4.4) is much larger. Mössbauer, in his experiments, solved this problem by eliminating recoil energy loss and choosing the correct isotope.

4.4 Recoilless Emission and Resonant Absorption

Mössbauer found that recoilless emission energy is possible for bound nuclei in solids under certain conditions. One condition is the replacement of nuclear mass by the lattice mass, and the other is to prevent the energy transfer by phonon excitation (see Sect. 2.3.3 and Sect. 2.15). Replacing the mass of the nucleus with the lattice mass M_L in (4.4), the recoil energy is given as

$$E_R = \frac{E_\gamma^2}{2M_L c^2} \tag{4.17}$$

where the mass of the lattice is of the order of Avogadro number 6.023×10^{23}. Then, the recoil energy loss will be lesser by an order of 10^{-24} eV. Now, the energy E_R is available to be transferred to the whole lattice.

From the Debye temperature relationship, we have $\hbar\omega_D = k_B\Theta_D$ where ω_D is the phonon cut-off frequency and Θ_D is the Debye temperature. If $E_R > k_B\Theta_D$ then the energy shall be transferred to the lattice, whereas recoilless emission without energy loss is possible if $E_R < k_B\Theta_D$. This is illustrated in Fig. 4.7 for the condition of $E_R < \hbar\omega_D$ where there is no energy transfer to the lattice and $E_R > \hbar\omega_D$, where there is energy transfer to the lattice.

The role of phonon vibrations on the recoil energy transfer requires that we choose materials with large Θ_D. The fraction of transition that occurs without energy transfer to the lattice is known as the Debye–Waller factor or Lamb-Mössbauer factor (f). It can be shown that the recoilless fraction for an Einstein solid is equal to

$$f = 1 - \frac{E_R}{\hbar\omega} \tag{4.18}$$

where $\hbar\omega$ is the phonon energy. The Einstein solid assumes that each atom is an independent quantum harmonic oscillator that oscillates with the same frequency, unlike the Debye solid. The temperature dependence of f could be accurately predicted using the Debye model.

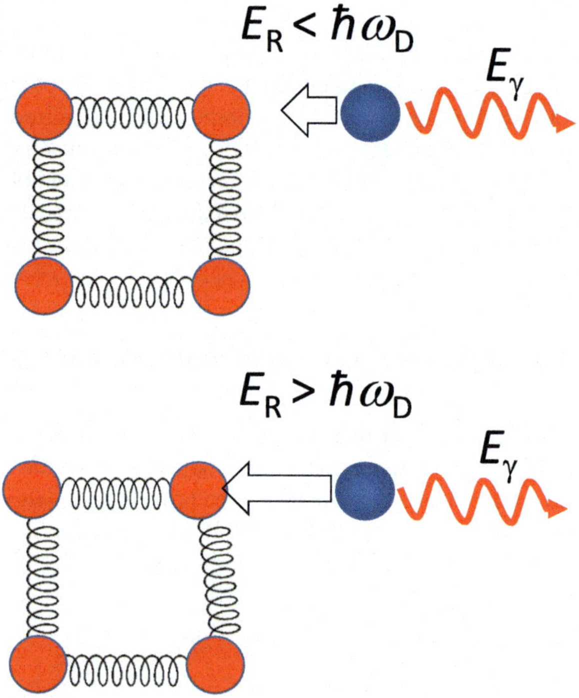

Fig. 4.7 Schematic representation of $E_R < \hbar\omega_D$ where there is no energy transfer to the lattice (zero phonon transition) and $E_R > \hbar\omega_D$ when there is phonon excitation

4.5 Lamb-Mössbauer Factor

The Debye–Waller factor is a measure of the zero phonon transition, as this term includes the mean square displacement of a lattice resulting from a scattering event. If the relative probability of gamma ray emission or absorption is accompanied by a transition from the initial energy state i to the final energy state f, then

$$P = |\langle i|\exp(ik \cdot r)|f\rangle|^2 \tag{4.19}$$

where $\hbar k$ is the momentum vector of the gamma ray and r is the position of the atom. For a zero phonon transition, i and f are identical, giving

$$P = |\langle i|\exp(ik \cdot r)|i\rangle|^2 \tag{4.20}$$

and the thermal averaging gives

$$P = |\langle i|\langle\exp(ik \cdot r)\rangle|i\rangle|^2 \tag{4.21}$$

By replacing r with x, the recoil free fraction is now

$$f = \langle\exp(ik \cdot x)\rangle^2 = \exp\left\langle -k^2 x^2 \right\rangle \tag{4.22}$$

which depends on the mean square displacement of the lattice site $\left\langle x^2 \right\rangle$.

The recoil free fraction is written as

$$f = \exp(-2W) \tag{4.23}$$

which is also known as the Debye–Waller factor. Here $\left\langle x^2 \right\rangle$ is proportional to the potential energy of a harmonic oscillator. The Debye–Waller parameter is written as follows (Mössbauer 1958, 1962)

$$f = \exp(-2W) = \exp\left\{ -\int_0^{\omega_D} (ku_e)^2 \langle x^2 \rangle \rho(\omega) d\omega \right\} \tag{4.24}$$

where u_e is the polarization vector and $\int_0^{\omega_D} \rho(\omega) d\omega$ is the density of oscillators within the interval 0 to ω_D. The mean value of the quantized potential energy of the oscillator is

$$\left(\overline{n} + \frac{1}{2}\right)\hbar\omega = NM\omega^2 \left\langle x^2 \right\rangle \tag{4.25}$$

where N is the number of nuclei and $\overline{n}$ is the occupancy of phonon wave vector in the thermal equilibrium.

The mean square displacement is

$$\langle x^2 \rangle = \left(\overline{n} + \frac{1}{2}\right)\frac{\hbar}{NM\omega} \tag{4.26}$$

Using (4.26) in (4.24), the Debye–Waller factor becomes

$$f = \exp(-2W) = \exp\left\{-\int\limits_0^{\omega_D} \frac{(ku_e)^2}{MN\omega}\left(\overline{n} + \frac{1}{2}\right)\hbar\rho(\omega)d\omega\right\} \tag{4.27}$$

For a cubic crystal $(ku_e)^2 = \frac{k^2}{3}$. Using $\hbar\omega_D = k_B\Theta_D$, $E_R = \frac{p^2}{2M}$, $k = \frac{p}{\hbar}$ and

$$\int\limits_0^{\omega_D} \rho(\omega)d\omega = \int\limits_0^{\omega_D} \frac{9N\omega^2}{\omega_D^3}d\omega \tag{4.28}$$

the Debye–Waller factor (4.27) becomes

$$f = \exp\left\{-\frac{k^2}{3}\frac{1}{MN\omega}\int\limits_0^{\omega_D} \frac{9N\omega^2}{\omega_D^3}\hbar\left(\overline{n} + \frac{1}{2}\right)d\omega\right\} \tag{4.29a}$$

$$f = \exp\left\{-3\frac{p^2}{\hbar^2}\frac{1}{M\omega_D^3}\int\limits_0^{\omega_D}\left(\overline{n} + \frac{1}{2}\right)\hbar\omega d\omega\right\} \tag{4.29b}$$

$$f = \exp\left\{-6\frac{E_R}{\hbar^2\omega_D^3}\int\limits_0^{\omega_D}\left(\overline{n} + \frac{1}{2}\right)\hbar\omega d\omega\right\} \tag{4.29c}$$

$$f = \exp\left\{-6\hbar\frac{E_R}{(\hbar\omega_D)^3}\int\limits_0^{\omega_D}\left(\overline{n} + \frac{1}{2}\right)\hbar\omega d\omega\right\} \tag{4.29d}$$

Replacing $\overline{n} = \frac{1}{e^{\frac{\hbar\omega}{k_B T}} - 1}$, the energy of the oscillator with a distribution given by $\int \frac{\hbar\omega}{e^{\frac{\hbar\omega}{k_B T}} - 1} d\omega$ for a Debye solid in the limits from 0 to Θ_D/T, (4.29d) becomes

$$f = \exp\left\{-6\hbar\frac{E_R}{(\hbar\omega_D)^3}\int\limits_0^{\omega_D}\left[\frac{1}{e^{\frac{\hbar\omega}{k_B T}} - 1} + \frac{1}{2}\right](\hbar\omega)d\omega\right\} \tag{4.30}$$

Using $x = \frac{\hbar\omega}{k_B T}$, (4.30) is given as

$$f = \exp\left\{-6\frac{E_R}{k_B\Theta_D}\left(\frac{T}{\Theta_D}\right)^2\int_0^{\frac{\Theta_D}{T}}\left[\frac{1}{e^x-1}+\frac{1}{2}\right]xdx\right\} \tag{4.31}$$

$$f = \exp\left\{-\frac{3}{2}\frac{E_R}{k_B\Theta_D}\left(1+4\left(\frac{T}{\Theta_D}\right)^2\int_0^{\frac{\Theta_D}{T}}\frac{xdx}{e^x-1}\right)\right\} \tag{4.32}$$

where Θ_D is the Debye temperature of the solid and T is the measurement temperature. The evaluation of the integral gives a value of $\pi^2/6$ in the extreme limits.

For the low-temperature limit when $T << \Theta_D$

$$f = \exp\left\{-\frac{E_R}{k_B\Theta_D}\left(\frac{3}{2}+\frac{\pi^2T^2}{\Theta_D^2}\right)\right\} \tag{4.33}$$

that gives

$$f = \exp\left\{-\frac{E_R}{k_B\Theta_D}\left(\frac{3}{2}\right)\right\} \text{ when } T = 0 \tag{4.34}$$

At high temperatures $T > \Theta_D$,

$$f = \exp\left\{-6\frac{E_R T}{k_B\Theta_D^2}\right\} \tag{4.35}$$

As the Lamb-Mössbauer factor, also known as the f-factor, should be as large as possible, the maximum being unity, the Mössbauer effect absorption will be well pronounced if:

(i) The recoil energy is low by choosing a lower gamma ray transition energy and integrating the emitting source in a solid matrix.
(ii) The measurements are performed at the lowest temperature.
(iii) The Debye temperature of the material is much higher than the measurement temperature.

This could be understood by the fact that α-Fe, with a Debye temperature of 420–480 K, has a large f-factor that is non-vanishing at high temperatures, whereas the f-factor of the Fe impurity in Au ($\Theta_D = 164$ K) diminishes to a large extent, as shown in Fig. 4.8. The f-factor was obtained corresponding to the γ-ray energy of 14.4 keV. Although the f-factor is maximum (> 0.9) at 0 K, it does not reach unity. If the energy corresponding to the γ-ray transition of 122 keV is used, then the f-factor is negligible with a value of 3×10^{-3} even at 0 K.

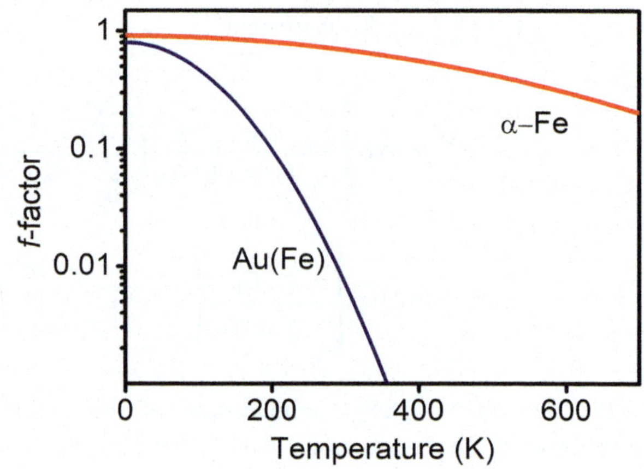

Fig. 4.8 The temperature-dependent f-factors (log scale) of α-Fe ($\Theta_D = 420$ K) and Fe impurity in Au host ($\Theta_D = 164$ K)

Other factors that affect the Mössbauer resonant absorption are the internal conversion coefficient, the absorption process such as the photoelectric effect, the natural abundance, the thickness of the absorber, the absorption by heavy elements, the activity of the source, the natural linewidth that depends on the lifetime of the states, the background noise and other instrumentation effects.

4.6 The Mössbauer Spectrum

The resonant absorption of the recoilless emission of the gamma ray from the source gives rise to Lorentzian peaks according to the energy levels of the absorber. For a γ-ray from the source transmitted through the sample (absorber), the spectrum consists of a singlet, a doublet, a sextet or a combination of these. The origin of the singlet/doublet/sextet is discussed in the next chapter. Figure 4.9 shows the typical singlet, doublet and sextet spectra of the transmitted γ-beam where the area inside the inverted Lorentzian indicates the energy absorbed by the sample. The velocity scale in the x-axis represents the Doppler velocity given to achieve the resonant spectrum.

4.6.1 Linewidth and Cross Section

The uncertainty in the energy and time is given by

$$\Delta E \Delta t \geq \frac{\hbar}{2} \tag{4.36}$$

The Breit-Wigner relationship gives a Lorentzian line shape for the nuclear transitions. If the full width at half maximum (FWHM) is Γ, then the uncertainty in

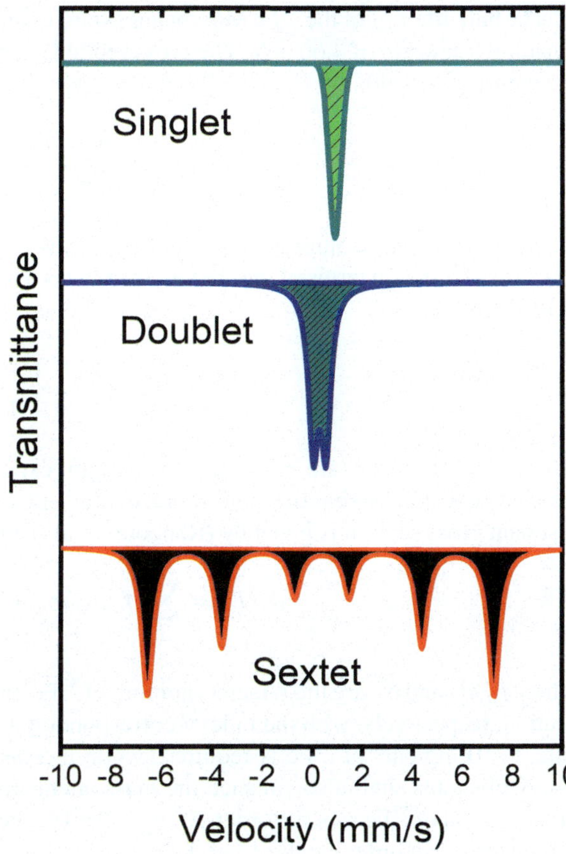

Fig. 4.9 The typical Mössbauer spectra representing transmitted signal with respect to velocity. The line broadened inverted Lorentzian may be composed of a single (singlet), double (doublet) or six (sextet) peaks. A combination of these is also exhibited according to the nature of the sample

energy is given by

$$\Delta E = \frac{\Gamma}{2} = \frac{\hbar}{2\tau} \tag{4.37}$$

where τ is the lifetime of the nuclear excited state and Γ is the natural linewidth. Hence, the relationship becomes

$$\Gamma = \frac{\hbar}{\tau} \tag{4.38}$$

In terms of the half-life ($\tau_{1/2}$), the relationship becomes

$$\Gamma = 0.693\frac{\hbar}{\tau_{1/2}} \tag{4.39}$$

The half-life $\tau_{1/2}$ of the ^{57}Fe Mössbauer excited state at 14.4 keV is 97.9 ns, giving a natural linewidth of 4.66 neV. The cross section of the Lorentzian line is related to the natural linewidth by

$$\sigma = \sigma_o \frac{\Gamma^2}{\Gamma^2 + 4(E - E_o^2)} \tag{4.40}$$

where σ_o is the maximum cross section and E_o is the resonant gamma ray energy. The cross section in terms of the nuclear spin states and the internal conversion factor is given as

$$\sigma_o = \frac{\lambda^2}{2\pi} \frac{2I_e + 1}{2I_g + 1} \left[\frac{1}{\alpha + 1} \right] \tag{4.41}$$

where λ, I_e, I_g and α are the γ-ray wavelength, excited state nuclear spin, ground state nuclear spin and internal conversion coefficient, respectively. The total absorber resonant cross section is given by (Rancourt et al. 1993).

$$\sigma_a(E) = \sum_i f_{a,i} n_{a,i} \sigma_{a,i}(E) \tag{4.42}$$

where f_a, n_a and σ_a are the f-factor, number of ^{57}Fe/cm^2 and intrinsic absorber cross section, respectively, with the index i corresponding to various Fe sites. As a thumb rule, 5–10 mg/cm^2 of ^{57}Fe is required, while increasing the sample quantity with heavy elements shall also enhance the mass attenuation. The resonance absorption cross section of ^{57}Fe is of the order 10^6 b for the 14.4 keV gamma ray energy whereas it reduces by an order for the 136.5 keV.

The effective thickness of a Mössbauer sample, t_a, is optimum when the experimental linewidth is as close as possible to the linewidth of the source and the signal-to-noise ratio is extremely high. The t_a can be expressed as

$$t_a = f_a n_a \sigma_a e,$$

where e is the physical thickness of the sample. When considering a thin sample of about 15 μm, the optimal quantity can be estimated to be 8×10^{17} nuclei of ^{57}Fe/cm^2. Considering that the natural abundance of ^{57}Fe is about 2.2%, the optimum natural content of Fe is estimated to be 2.35×10^{18} Fe nuclei/cm^2, equivalent to 5 mg Fe/cm^2.

4.7 Transmission Mössbauer Spectrometer Instrumentation

The Mössbauer spectrometer is used to tune the energy spectrum to the energy levels of the absorber by giving a Doppler energy. The γ-ray energy, resonantly tuned to the energy levels of the absorber, is then recorded using an electronic package that can be analyzed later. The essential components of the transmission Mössbauer spectrometer are shown as a block diagram in Fig. 4.10.

A typical transmission Mössbauer spectrometer consists of components connected to the source and detector parts. The radioactive source is attached to one end of a constant velocity or constant acceleration electromagnetic transducer. An electrical signal in a sine or triangular waveform from a function generator is transmitted to the transducer in the constant acceleration mode. This gives the required Doppler velocity to the source. The components mentioned in the block diagram of Fig. 4.10 are described below.

4.7.1 Mössbauer Velocity Transducer

Mössbauer velocity transducers (MVT) can produce typical velocities up to 1000 mm/s and a resonant frequency in the range 10–25 Hz. The transducer operates on the principle of generating mechanical motion through a coil in an applied magnetic field generated by an AlNiCo/SmCo/NdFeB permanent magnet. The Lorentz force creates a mechanical motion similar to that of a loudspeaker by giving the appropriate waveform. Another coil acts as a pickup coil to sense whether the motion is linear.

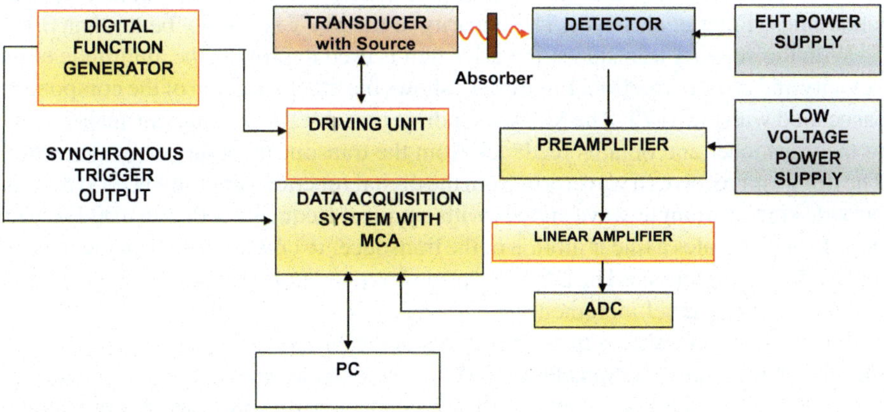

Fig. 4.10 Block diagram of the Mössbauer spectrometer with its typical components

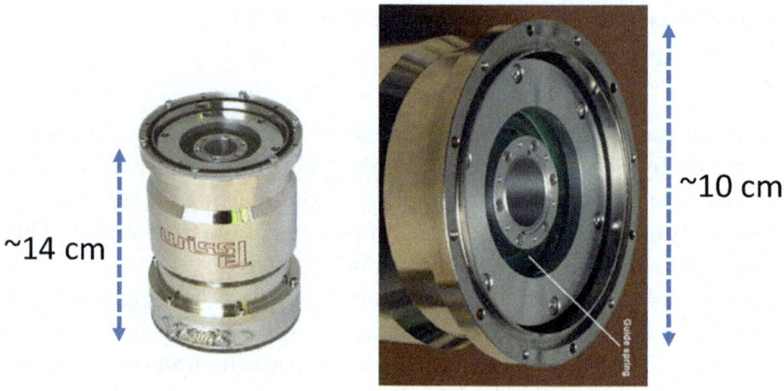

Fig. 4.11 Transducer used to provide a Doppler velocity (Courtesy: WissEl GmbH, Germany)

The MVT is a crucial component of a Mössbauer spectrometer that must produce a perfectly linear velocity. This can be achieved by multiple electronic and mechanical components associated with the Mössbauer velocity transducer, as shown in Fig. 4.11. The velocity transducer has a high impedance and is powered by a current amplifier that drives the transducer for positive and negative velocities. The velocity feedback pickup coil senses the actual velocity of the transducer in the form of an electromotive force (emf) that has an amplitude and a phase. If this emf signal is proportional to the input signal to the velocity transducer, it implies that there is no error in the linear velocity that could be detected from the feedback given to the differential amplifier. The nonlinearity can be corrected to some extent by providing a difference signal to the input of the velocity transducer. The reference input signal is the original waveform given to the velocity transducer and compared in the difference amplifier. Another device that can be attached to the velocity transducer is the position regulator made of photoresistors that corrects slow velocity drifts induced by the op-amp electronics (Filter et al. 1996). The linearity of the MVT is usually better than 0.1%.

With the velocity transducer, the drive unit is used to provide the amplitude of the velocity and error correction. Figure 4.12 shows the block diagram of the components associated with the MVT. The Mössbauer drive unit (MDU) provides an analog signal to the transducer and obtains feedback from the transducer for any error correction. The drive unit receives a waveform from the digital function generator (DFG) that can provide sine or triangle wave signals with a typical frequency of up to 100 Hz. This waveform generates a linear motion of the transducer to create velocity in the forward and backward directions. The DFG also provides a start pulse for multichannel scaling and pulses for channel incrementing.

Figure 4.13 (left) shows the components of the Mössbauer spectrometer, namely the Mössbauer velocity transducer (MVT), the detector and the preamplifier. The right part shows the components that are connected to a nuclear instrumentation module (NIM) bin, namely the Mössbauer drive unit (MDU), the digital function generator (DFG), the high voltage power supply (HV) and the linear amplifier

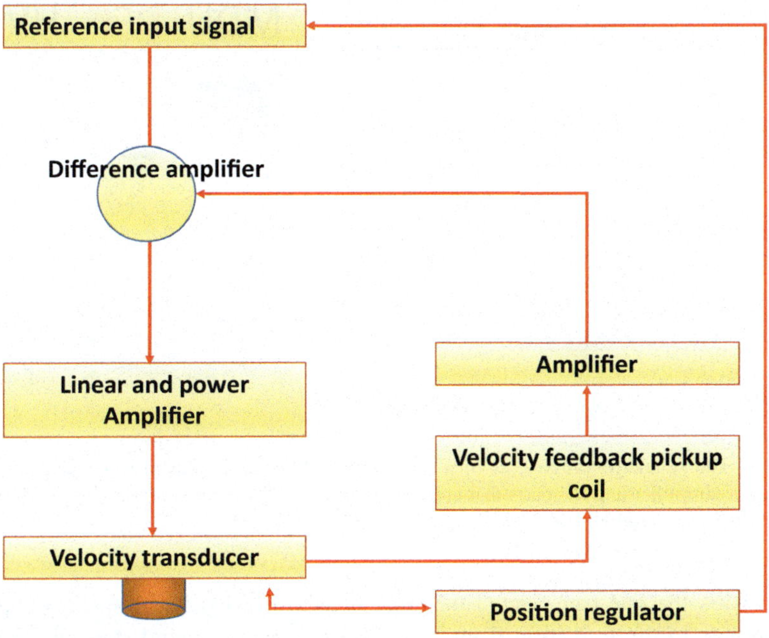

Fig. 4.12 Block diagram showing the various components associated with the Mössbauer velocity transducer

(LA). Apart from these components, the data acquisition system, which includes the multichannel analyzer, is usually integrated into the computer as a separate module.

4.7.2 *^{57}Co Radioactive Source*

The Mössbauer radioactive source for the study of Fe-based systems is ^{57}Co, which decays by electron capture to the excited states of ^{57}Fe and finally reaches the ground state by emission of γ-ray. ^{57}Co can be produced in a cyclotron by bombarding ^{60}Ni with a proton or ^{56}Fe with a deuteron, giving the following reactions.

$$^{60}\mathrm{Ni} + \mathrm{p} \rightarrow {}^{57}\mathrm{Co} + \alpha \tag{4.43a}$$

$$^{56}\mathrm{Fe} + {}^{2}\mathrm{H} \rightarrow {}^{57}\mathrm{Co} + \mathrm{n} \tag{4.43b}$$

The half-life of ^{57}Co is 271.81 days, and the decay scheme (Bhat 1992) with the energy levels are shown in Fig. 4.14. The ^{57}Co isotope decays to the excited states of ^{57}Fe, and the most probable decay is to the $5/2^-$ state of energy 136.4743 keV by electron capture. The E2 multipolarity being negligible, the M1 transition (see

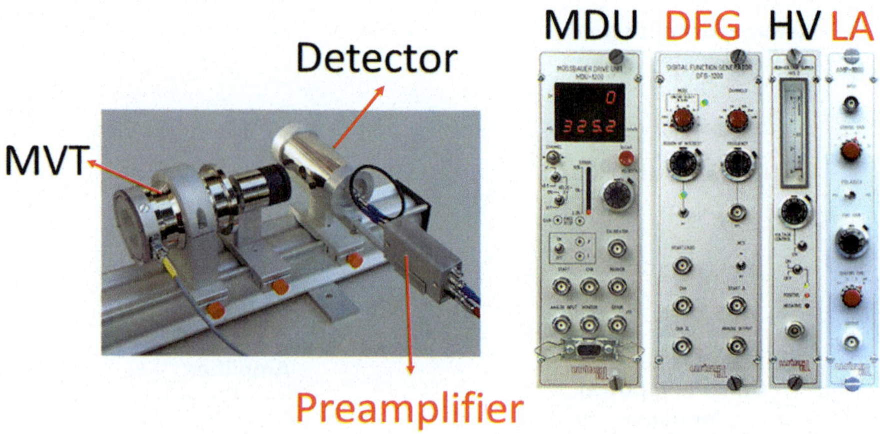

Fig. 4.13 The standard components of a transmission Mössbauer spectrometer showing MVT, detector, preamplifier and the parts that are in the NIM bin that include MDU, DFG, HV and LA (Courtesy: WissEl GmbH, Germany)

Sect. 1.9) is the one (see Sect. 1.9.3) that is used in Mössbauer spectroscopy and corresponds to the transition from 14.4129 keV to the ground state of ^{57}Fe. The lifetimes of the energy levels are also shown in Fig. 4.14: that of the 14.4129 keV energy state is 97.9 ns, and the intensity of the $3/2^-$ to $1/2^-$ transition is about 9.16%. The internal conversion coefficient is 8.56. Two other γ-ray transitions corresponding to E_γ of 122.0614 and 136.4743 keV (E2) are also shown in Fig. 4.14 as ($5/2^- \rightarrow 3/2^-$) and ($5/2^- \rightarrow 1/2^-$) transitions, respectively.

The ^{57}Co source is usually prepared by thermal diffusion of ^{57}Co into a non-magnetic matrix such as stainless steel (SS) or Rh. The diffusion process is carried out under high vacuum conditions ($>10^{-5}$ mmHg) at an elevated temperature of ~ 1373 K by electrodeposition of ^{57}Co onto the matrix material for a few hours (Dezsi and Molnar 1967). This foil or plate is then cut into the required shape, sealed and fixed in a Ti alloy capsule with a thread such as M4 on one side and a Be, Al or Kapton window on the other. The schematic diagrams of the disc type and point type source holders are depicted in Fig. 4.15. These sources usually have an activity range from 5–150 mCi (1 mCi = 37 MBq). The typical linewidth of the sources is 0.20 mm/s. For thick samples and high-pressure studies, a synchrotron-based Mössbauer source is useful, as discussed in the later sections.

4.7.3 Detectors

Mössbauer gamma ray detectors are mainly proportional counter detectors, scintillation detectors and semiconductor detectors. The proportional counter is a gas detector that detects radiation based on the ionization of the gas by the radiation.

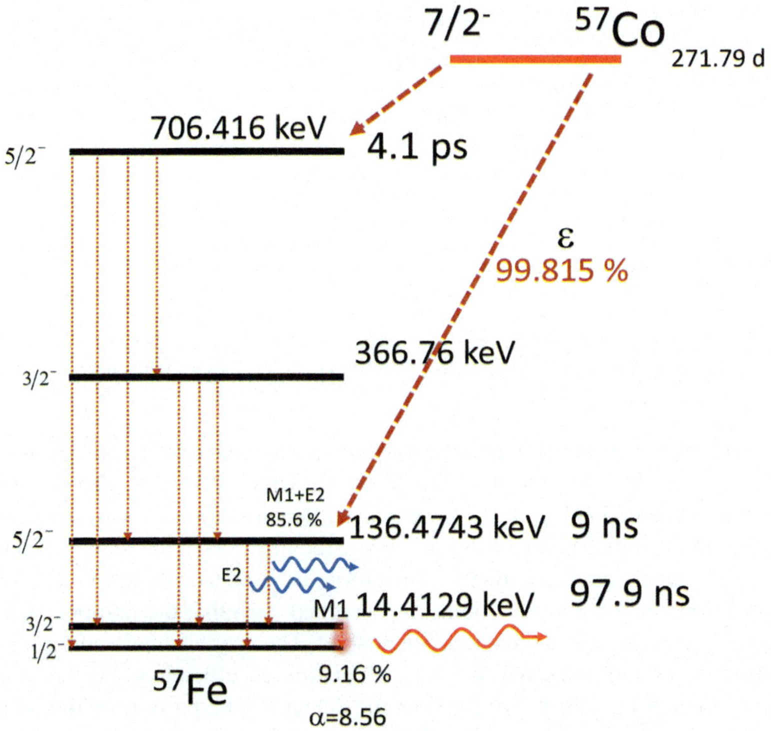

Fig. 4.14 The energy level diagram of ^{57}Co that decays to ^{57}Fe

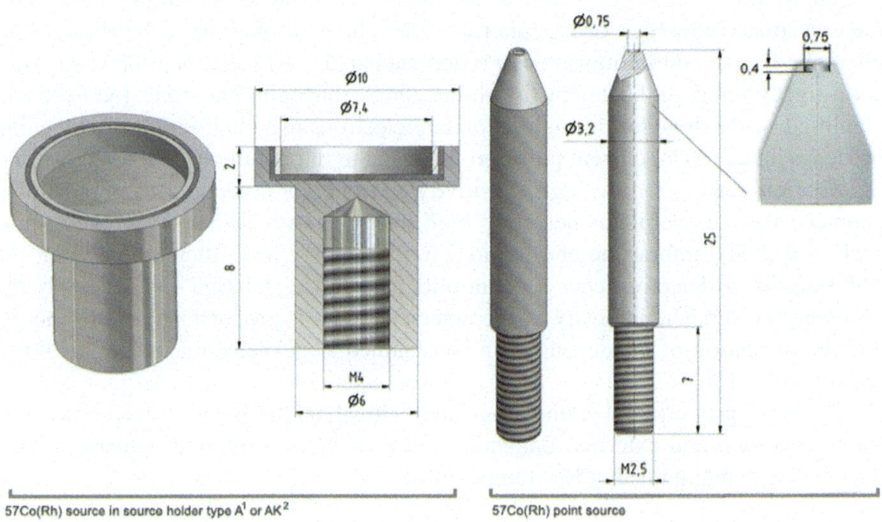

Fig. 4.15 The ^{57}Co sealed source holders and their typical dimensions (Courtesy: WissEl GmbH, Germany)

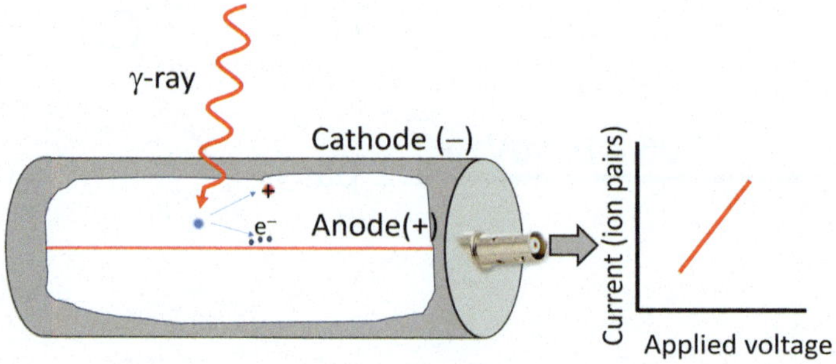

Fig. 4.16 The proportional counters are gas detectors that work on the principle of ionization of the gas by the gamma ray, and the ion pairs produced are proportional to the energy

The pulse height is proportional to the voltage, hence the name. The detector usually has a mixture of 90% Ar and 10% methane, known as P-10 gas. The proportional counter, as illustrated in Fig. 4.16, has a thin anode wire mounted within an insulated metallic cathode tube. When a gamma ray hits the inert gas in the tube, it ionizes and produces electrons that cause further ionization. Beyond a threshold voltage, charge multiplication occurs, and the counter is designed such that the number of original ion pairs is proportional to the energy deposited. The proportional counter has good resolution at low energies below 75 keV but limited efficiency. Furthermore, the efficiency decreases with the lifetime of the detector. The proportional counter can be calibrated with an isotope such as ^{59}Ni ($t_{1/2} = 7.5 \times 10^5$ y) that emits photons of 6.83 keV (Crouthamel et al. 1970).

Scintillation detectors emit light through fluorescence when the luminescent material is irradiated with high-energy photons. The schematic diagram of the photomultiplier tube-based scintillation detector is depicted in Fig. 4.17. The NaI(Tl) scintillator emits light photons, collected by the photocathode that emits electrons, focussed and amplified by the dynodes through secondary electrons in a photomultiplier tube that produces an electrical current pulse proportional to the number of photons. NaI(Tl) scintillation detectors have high sensitivity to radiation compared to proportional counters and are useful for detecting high energy γ-rays. Other detector crystals, such as CsI(Tl), anthracene and liquid scintillators, are less efficient than NaI(Tl). The scintillator detector needs a preamplifier for noise isolation and to supply the necessary power. The intensity of the detected signal is proportional to the energy, and the resolution of the detector can be obtained as (FWHM/maximum height of the signal) × 100%.

The use of proportional counters is easier with all additional Mössbauer facilities for temperature and external magnetic field than the scintillation detector, which requires to be maintained at low temperature.

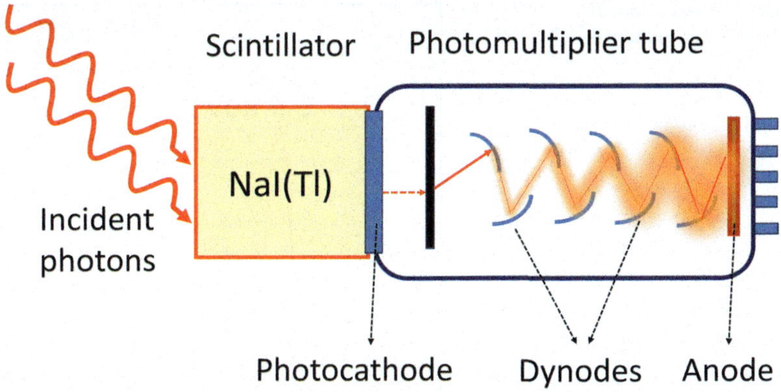

Fig. 4.17 The scintillation detector converts the photons to amplified electrons through the dynodes in the scintillator-photomultiplier tube assembly

4.7.4 Multichannel Analyzer

The multichannel analyzer allows data to be acquired in either multichannel scaling (MCS) or pulse height (PHA) mode. The required pulses corresponding to the 14.4 keV gamma ray energy alone can be selected by rejecting the other pulses using a lower and upper discriminator. The number of channels can be 512, 1024, 2048, or 4096 before folding: while higher the number of channels, the longer the registration time and the better the resolution. The signals are allocated to the channels according to the pulse amplitude in PHA mode and their time of arrival in MCS mode.

Figure 4.18 shows the pulse height spectrum of ^{57}Co in the Rh matrix, which has the 14.4 keV peak required for the Mössbauer spectrum that could be selected using Upper Level Discriminator (ULD) and Lower Level Discriminator (LLD). The 6.4 keV X-ray Kα peak is visible on the left side of the main peak, and Rh X-ray (unresolved 20 and 23 keV) peaks emanating from the source matrix to its right. The low energy X-ray intensities could be reduced using X-ray filters.

4.7.5 Experimental Features

Some basic rules must be considered to optimize the registration time of Mössbauer spectra. As concluded above, the Fe content is optimized with 5 mg Fe/cm^2 and must be homogeneously distributed in the case of a powder sample. Such a rule cannot be strictly applied when the sample contains a certain amount of heavy elements, which can absorb the γ-photons: tests should be done, but a lower quantity of Fe should be considered. In general, exceeding this optimal thickness changes the shape of Mössbauer spectra, in particular, the relative intensities of the magnetic sextet lines

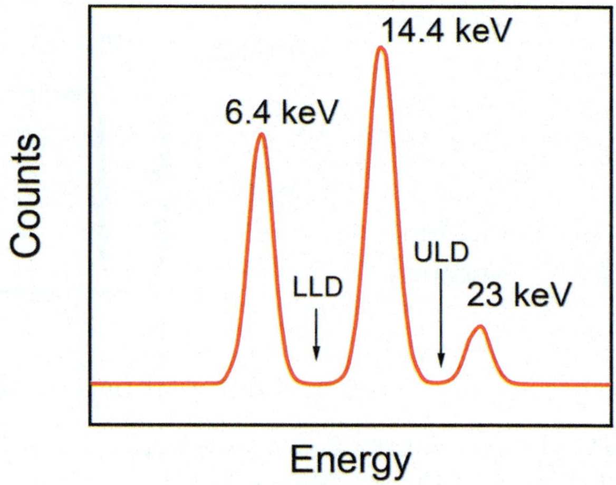

Fig. 4.18 The pulse height spectrum of ^{57}Co in the Rh matrix

and/or the widths of the different lines of the quadrupolar and magnetic components, due to polarization and thickness effects. Corrections can be made in this case, but reference spectra must also be established. On the contrary, when the amount of Fe in the material to be studied is too low, the effective surface of the samples must be appropriately reduced by focusing with lead masks. A second alternative is to enrich the sample with ^{57}Fe. When the composition of Fe content is poorly known (natural samples in geology, biology, and heritage culture), different samples must be tested to optimize the registration time.

The second recommendation concerns the source-sample-detector distances, which depend on the activity of the source and the speed of the acquisition electronics: the transducer on which the source is mounted and the detector are usually placed on a bench and can move around the sample. The minimum distance between source-detector must not lead to a too large solid angle for the γ-photon radiation, thus avoiding polarization effects. It is finally possible to carefully select the 14.4 keV energy peak to make the transmission as efficient as possible.

The source is subjected to a periodic motion, a constant acceleration triangular or sinusoidal mode. The first mode is the most common one, while the second one is mainly used to increase the resolution around the zero velocity, i.e., to study accurately the values of isomer shift.

One additional crucial recommendation is the frequent calibration of the spectrometer: the first objective is to establish the exact value of the velocity of the source and to follow the stability with time of the electromechanical part of the transducer. It is, therefore, convenient to use an α-Fe foil with an appropriate thickness to obtain linewidths as small as possible, allowing also (second objective) to estimate the value of its isomer shift, to be quoted as a correction for the other experiments. In practice, the calibration spectrum can be performed simultaneously with the registration of spectra, or before and/or after each experiment (series of experiments) given for a velocity.

4.7.6 Low Temperature Cryostat

Since the use of Mössbauer spectrometry as a characterization spectroscopic technique, bath cryostats, flow cryostats, dilution cryostats have been specifically developed by some companies, and in-house cryostats have been designed for particular applications by some laboratories, allowing measurements in a large temperature range. It is important to mechanically isolate the cryostat from the pumping units to break the transmission of mechanical vibrations, thus avoiding an inhomogeneous broadening of Mössbauer lines. A more practical design is to hold the sample on a tube with appropriate probes to regulate the temperature, located in a chamber with helium gas or vacuum according to the expected measurement temperature. Mylar, beryllium, kapton, quartz, sapphire, spectrosil or high-purity aluminum windows are used to facilitate the transmission of gamma rays. More recently, a new generation cryostat has been developed by several companies around a cryogenerator operating with helium gas in a closed circuit. Applied to Mössbauer measurements, the main challenge is to decouple the cryogenic system and the sample holder to avoid the vibrations caused by the helium compression/expansion cycle.

Based on the working principle, cryocoolers are classified as Stirling, pulse-tube, Brayton, Joule–Thomson and Gifford–McMahon (GM). The cryogenic refrigerators use liquid nitrogen (77 K) or liquid He (4.2 K) and work in a vibration-free environment, whereas the replenished cryogens are expensive. The closed-cycle cryofree systems mostly use high-pressure helium as the working fluid, and the major equipment components are the compressor and cold head. The heat exchange between the warm and cold fluids is regulated by valves between the compressor and cold head, housing a regenerator, such as Gadolinium oxysulphide, in a GM cryocooler. The cold head is connected to the He gas inlet and outlet on one side and the sample chamber on the other, with the gamma ray transparent window (Fig. 4.19). The high-pressure He gas passes through the regenerator, followed by expansion, cooling and heat removal by He gas through a low-pressure path, regulated by a rotating valve that operates at a frequency of 1–2 Hz. The vibrations in the 10–15 μm range could be further reduced below 3 nm in recent models that offer temperature down to 3 K. Further low temperatures below 2 mK could be obtained using dilution refrigerators.

The pulse tube refrigerator is a valveless cryocooler driven by oscillating pressure through an orifice, achieving adiabatic compression and expansion in a pulse tube. The dilution refrigerators generally use pulse-tube cryocooling for precooling a ^{3}He/^{4}He mixture. At temperatures below 0.87 K, the ^{3}He, a low-density Fermi fluid, has higher vapor pressure than bosonic ^{4}He, phase-separates into a concentrated phase (100%) that stays on the top and a dilute phase (6.6%) below, in the ^{4}He superfluid. Cooling is achieved due to the latent heat of the evaporated ^{3}He liquid diffusing across the phase boundary.

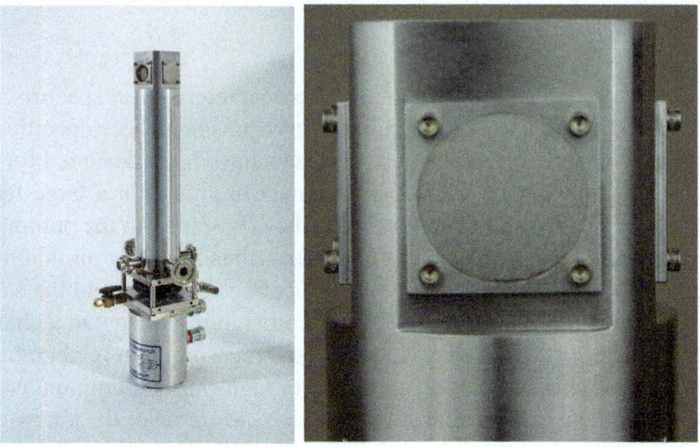

Fig. 4.19 The GM closed cycle cryofree cold head chamber (left) and the gamma ray transparent window (right). (Courtesy: Advanced research systems, USA)

4.7.7 High Temperature Furnace

For high-temperature Mössbauer experiments, the sample has to be in an inert/vacuum atmosphere, and the sample can be heated using a heater attached to a temperature controller. Figure 4.20 shows the typical diagram of a high-temperature furnace that could be used to measure samples at temperatures up to 1373 K. Mylar windows are provided through which the γ-ray enters and leaves the furnace. A thermocouple senses the sample temperature, and the furnace has thermal insulation. Gas/vacuum inlet and outlet provisions further help to prevent sample oxidation.

4.7.8 Cryofurnace

The combination of low and high-temperature measurements can be achieved by using a cryofurnace: its great advantage is to perform in situ measurements in the same atmosphere to follow, for example, (i) the transformation resulting in the departure of water molecules at high temperature and (ii) the transformation at high temperature of amorphous materials sensitive to atmosphere and to check the consequences on the hyperfine structure at low temperature in order to model any structural and/or magnetic change.

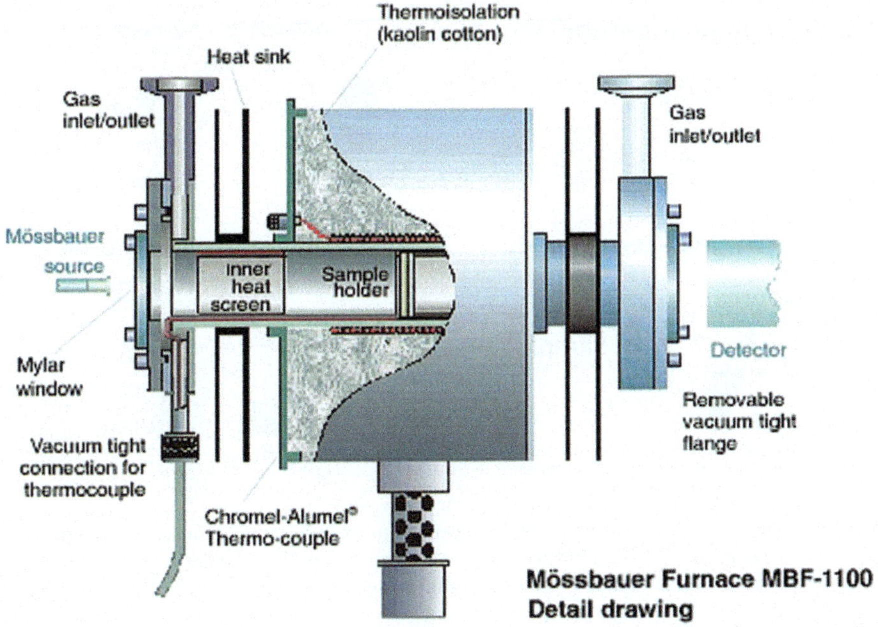

Fig. 4.20 High-temperature furnace used in Mössbauer spectrometry for temperatures up to 1373 K. (Courtesy: WissEl GmbH, Germany)

4.8 Other Modes of Mössbauer Spectrometry

The Mössbauer spectrometry is not limited to the availability of isotopes alone, as other energy sources can also be used in the investigation. The geometrical arrangement of the Mössbauer spectrometer shall be determined by the type of radioactive source and the energy used for the experiments, as depicted in Fig. 4.21. The conventional spectrometer uses a sealed ^{57}Co radioactive source in transmission geometry, while the same in reflection geometry may use either the scattered gamma ray or the conversion electrons. The transmission of the gamma ray through the sample is achieved by a synchrotron source, while nuclear resonant inelastic X-ray experiments are performed in scattering geometry.

4.8.1 Conversion Electron Mössbauer Spectrometry

When the 14.4 keV γ-ray strikes the absorber, it undergoes resonant absorption, and the absorbed energies are detected in transmission geometry in conventional Mössbauer spectrometry. It is also seen that there is energy transfer to the orbital electron emitting X-rays and electrons due to internal conversion and Auger electrons, as

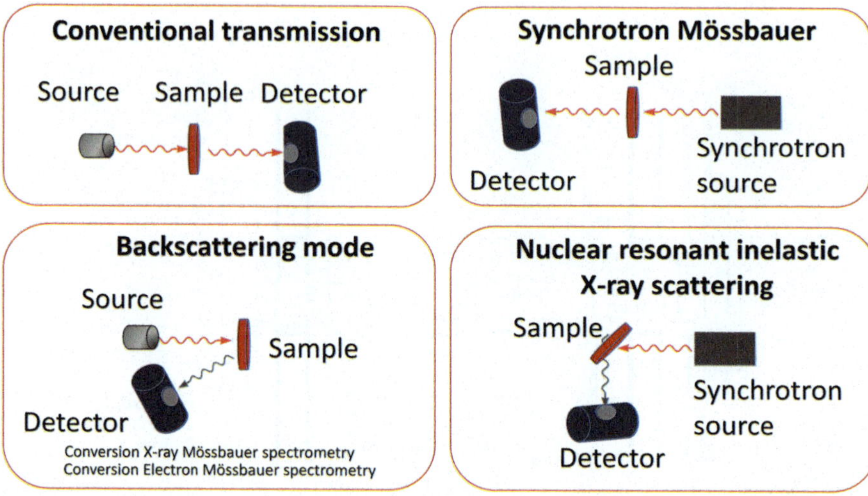

Fig. 4.21 The various modes of experiments apart from the conventional transmission type Mössbauer experiments are CXMS, CEMS that operate in backscattering mode using ^{57}Co radioactive source, transmission and nuclear resonant inelastic X-ray scattering using a synchrotron source

illustrated in Fig. 4.22. The electrons and X-rays can be detected in the backscattering mode, known as conversion electron Mössbauer spectrometry (CEMS) and conversion X-ray Mössbauer spectrometry (CXMS). In addition, the backscattered γ-ray can also be used to record the spectrum under reflection geometry. In conversion electron Mössbauer spectrometry, either all electron energy could be analyzed, forming the basis for integral CEMS or a specific energy is chosen that is known as a depth selective/differential CEMS. The energy of the K shell electron corresponds to the photoemission energy (E_γ − 7.1 keV binding energy) of 7.3 keV. If the gamma ray knocks the K shell electron, then the L shell electron fills the K shell, giving an X-ray emission energy of 6.4 keV. Instead of X-ray emission, if the energy is transferred to another electron, it is known as an Auger electron. In the ^{57}Fe transitions, the energies of conversion photoelectrons are 7.3, 13.6 and 14.3 keV, corresponding to K, L and M lines, respectively. The Auger electron energies are 5.4 and 0.6 keV.

The electron emission energies are detected by CEMS in 2π geometry and an X-ray energy of 6.4 keV is detected by CXMS. The backscattered electron has the most probability of which 80% for the K electrons, 53% for the KLL Auger electrons and 27% for the K X-ray, while the γ-ray energy of 14.4 keV has a chance of 10%. Figure 4.23 represents the typical energy spectrum showing Auger and conversion electron energies for the ^{57}Co isotope. The Auger LXY (X, Y are subshell M, N transitions), KLL and KLM energy peaks are observed at 0.6, 5.4 and 6.2 keV, respectively (Itoh et al. 1983). The K, L and M conversion electron peaks appear at higher energies beyond 6.5 keV.

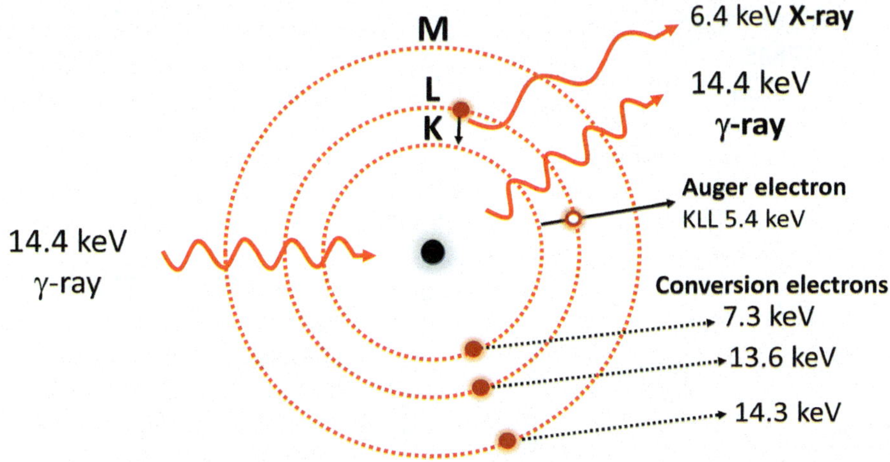

Fig. 4.22 The 14.4 keV gamma ray can knock out orbital electrons, resulting in the generation of X-rays and Auger electrons apart from nuclear electromagnetic energy transfer ejecting conversion electrons

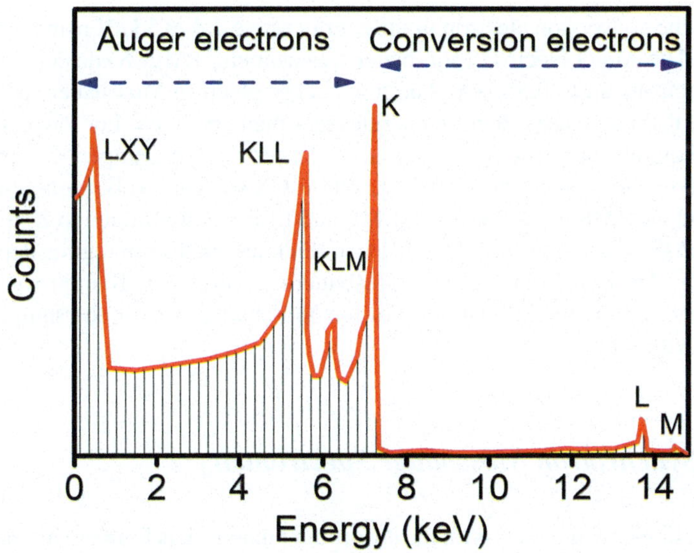

Fig. 4.23 Auger and conversion electron energy spectrum for the ^{57}Co isotope

The different Mössbauer spectrometry techniques in backscattering geometry are illustrated in Fig. 4.24. The integral conversion electron Mössbauer spectrometry (ICEMS) is based on the analysis of scattered electrons of all energies. The depth selective conversion electron Mössbauer spectrometry (DCEMS) is based on

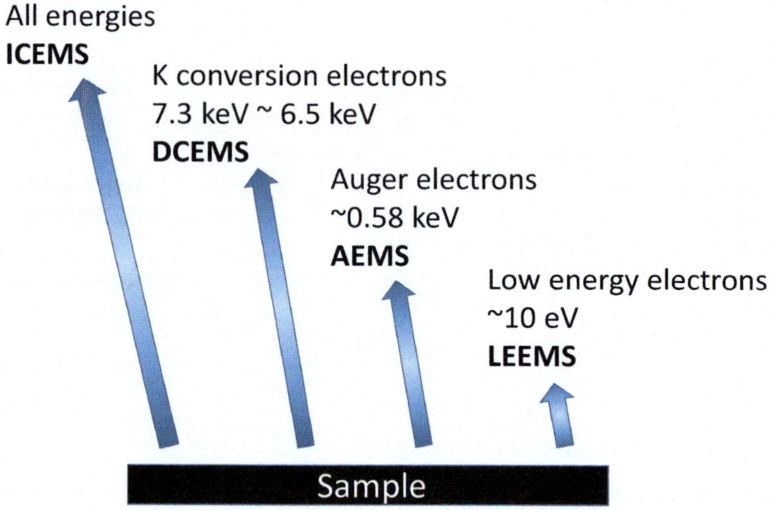

Fig. 4.24 Various Mössbauer spectrometry techniques in backscattering geometry

detecting the conversion electron in the energy range 7.3–6.5 keV using a beta spectrometer. The Auger electron Mössbauer spectrometry (AEMS) detects low energy Auger electrons up to 0.58 keV. The low energy electron Mössbauer spectrometry (LEEMS) technique uses electron energies less than 10–20 eV. For low-energy electrons, the quality of the scattered signals is larger for thin samples compared with transmission Mössbauer spectrometry. While CXMS can probe sample depths of 20 μm and DCEMS up to 500 nm, AEMS and LEEMS are limited to 20 nm. Therefore, LEEMS is useful to study thin films of thickness in the nm regime deposited on a substrate preventing transmission Mössbauer measurement. To improve the statistics and shorten the time registration of the Mössbauer spectra, the samples can be enriched with ^{57}Fe.

4.8.2 Synchrotron Mössbauer Spectrometry

When a gamma ray interacts with the absorbing nucleus, it is found to interact in such a way that it is classified as coherent or incoherent scattering, which could be elastic or inelastic (Champeney 1979), as illustrated in Fig. 4.25. If the incoming and outgoing energy after the scattering is the same, then it is known as coherent elastic scattering. This could be due to a recoilless event, as there is no energy transfer after the scattering. On the other hand, coherent inelastic scattering changes the energy levels of the ground state. Incoherent scattering arises from an absorbing nucleus having an energy level different from the source nucleus, producing elastic and inelastic scattering. In addition to Mössbauer nucleus scattering (resonant scattering), there can be

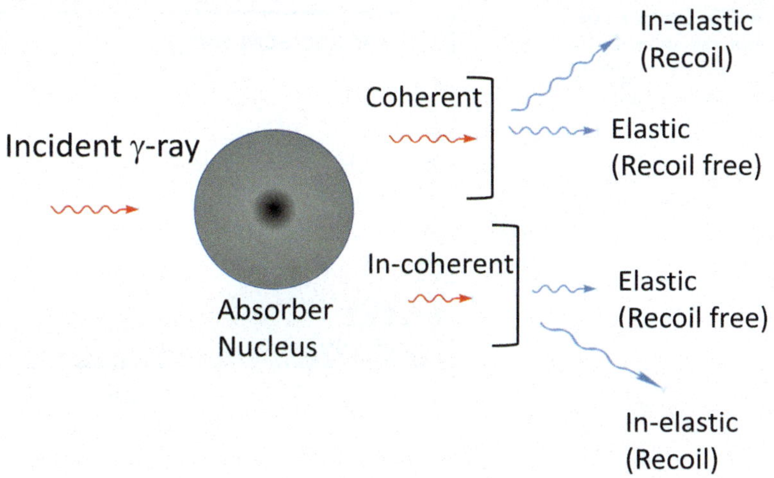

Fig. 4.25 The interaction of the gamma ray with an absorbing nucleus can be empirically classified as coherent or in-coherent scattering; further, it can be elastic or inelastic

Compton and non-resonant nuclear scattering, which are always incoherent. It is to be noted here that resonance can be achieved using Doppler velocity in conventional experiments, whereas another way is to use a synchrotron energy source.

Synchrotron radiation (SR) has high brilliance (~10^{22} photons/s/mrad^2/mm^2/0.1% energy bandwidth), energy tunability, smaller beam size, 100% polarizability and pulse structure, whereas the energy resolution is only a few meV. The interaction of synchrotron radiation with the nuclei is accompanied by prompt electronic scattering within a fraction of ns followed by delayed nuclear scattering that could be hundreds of ns, as illustrated in Fig. 4.26. The use of synchrotron radiation for Mössbauer spectrometry was proposed by Ruby (1974) based on using a timed resonant detector to distinguish electron scattering from nuclear scattering. Nuclear resonant scattering using synchrotron radiation was experimentally demonstrated by Cohen et al. (1978). The experimental arrangement comprises the synchrotron radiation directed through Si monochromators on a ^{57}Fe target. The resonant nuclear energy was recorded using a detector gated for 100 ns by varying the energy range close to 14.41 keV using the monochromator. A broad resonance with a width of 5×10^{-7} eV was observed due to magnetic hyperfine interaction for the resonant peak of 14.41303 keV.

Nuclear Bragg scattering using yttrium iron garnet (YIG) single crystal films to enhance the nuclear scattering cross section was utilized by Gerdau et al. (1985), and an energy resolution of 10^{-8} eV was achieved. Subsequently, energy resolution up to 10^{-14} eV could be realized by nuclear monochromators and used for Mössbauer spectrometry based on nuclear Bragg reflection of synchrotron radiation. The polarization and monochromatization could be accomplished by Si single crystals, and the nuclear Bragg reflector was aligned along the electronically forbidden planes of a crystal such as $FeBO_3$, YIG or α-Fe_2O_3.

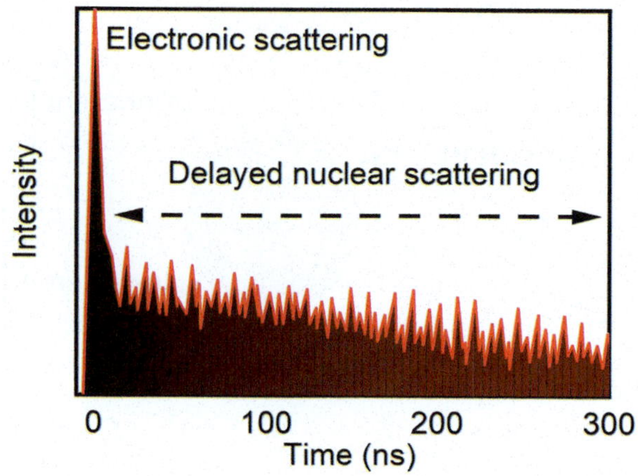

Fig. 4.26 The scattering of synchrotron radiation on Mössbauer nuclei comprises prompt electronic scattering and delayed nuclear scattering components that can be recorded over a time period by a suitable detector

Since the bandwidth of the synchrotron radiation is large compared with the Mössbauer linewidth, the detection of time-differential nuclear forward scattering technique is used in SR-based Mössbauer spectroscopy. If the synchrotron radiation pulses (<100 ps) excite the Mössbauer nuclei, the nuclear scattering will be delayed according to the lifetime of the excited state. The ^{57}Fe resonance timescale is 141 ns for a width Γ of 4.66 neV, whereas the electron scattering timescale is of the order of femtoseconds, which could be easily discriminated.

4.8.3 *Time Domain Nuclear Resonant Scattering*

4.8.3.1 Nuclear Resonant Forward Scattering

The process of coherent elastic scattering, where nuclear transitions can be excited simultaneously by short photon pulses of sufficient bandwidth, forms the basis for synchrotron Mössbauer spectroscopy. The pulses then create constructive and destructive interference during de-excitation, leading to forward scattered recoilless radiation oscillations. Thus, SR-based Mössbauer spectroscopy uses the time-resolved technique in contrast to the conventional energy-resolved technique. The time domain signal has components due to the coherent interference forming quantum beats, dynamic beats due to multiple scattering and speedup of the decay due to the enhancement of the radiative channel.

The Fourier expansion for the time-dependent electric field amplitude of recoilless γ-ray is given as

$$E(t) \propto \int_{-\infty}^{+\infty} \varepsilon_\omega \exp(i\omega t) d\omega \tag{4.44}$$

where ε_{ω} is the amplitude of the wave at a frequency ω. The time-dependent electrical field of the transmitted SR wave packet for a depth, z, is given by (Smirnov 1999).

$$E_t(z,t) = \varepsilon_{\omega_o} e^{-\frac{\mu_e z}{2}} \left\{ \delta(t) - \frac{1}{2t_o} T \exp\left(i\omega_o t - \frac{q\tau}{2} \right) \sigma(z,t) \right\} \tag{4.45}$$

where $\sigma(z,t) = \frac{J_1(\sqrt{T\tau})}{\sqrt{T\tau}}$, $T = \sigma_o f \kappa n z = \mu z$, $\tau = t/t_o$, $q = 1 + \frac{\Delta}{\Gamma}$

t_o is the characteristic time defined as $t_o = \hbar/\Gamma$, t is the decay time, Γ is the natural width of the excited level, ω_o is the resonant frequency for nuclear transition, μ_e is the absorption length, I_o is the stationary intensity of the synchrotron beam, J_1 is the first order Bessel function, T is the effective thickness of the resonant absorber, σ_o is the resonance cross section determined by the internal conversion coefficient, f is the Lamb-Mössbauer factor, κ is the isotope enrichment, n is the concentration of the resonant nuclei and q is the broadening parameter known as q-factor, determined by the width Δ.

The time integral δ function gives the function to be zero everywhere except as $t \to 0$ given by

$$\delta(t) = \frac{1}{2\pi} \int_{+\infty}^{-\infty} \exp(i\omega t) d\omega \tag{4.46}$$

The time-dependent forward scattering intensity for the energy band ΔE is given by

$$I(T,\tau) = \frac{I_o}{t_o} \frac{\Gamma}{\Delta E} e^{-(\mu_e z + q\tau)} \frac{T^2}{4} \sigma^2(T,\tau) \tag{4.47}$$

Equation (4.47) has the following factors: (i) T^2 corresponding to the enhancement of scattering according to the sample characteristics, (ii) the nuclear decay exponential term that includes the thickness of the target, resonance broadening (Δ) and (iii) $\sigma^2(T,\tau)$ term that produces dynamic beats as it propagates. The polarization of the scattered radiation with hyperfine splitting causes interference, forming quantum beats with time.

The incident field from the synchrotron radiation source can be $E(t) = p_\sigma a(t) \exp(ik \cdot y - i\overline{\omega} t)$ where p_σ is the time-independent polarization component, $a(t)$ is the function that describes the pulse structure and energy spectrum, and k is the wave vector with average energy $\overline{\omega} = c|k|$. The intensity per time of the SR pulse is given as (Sturhahn, 2004)

$$\frac{dI}{dt} = I_o \frac{\eta^2}{16\tau} \exp\left(-\frac{t}{\tau}\right) \left\{ \sum_{ij} p_\sigma W_l^* W_j p_\sigma \exp(-i[\omega_j - \omega_l]t) \right\} \tag{4.48}$$

where η is the effective thickness of the platelet, τ is the lifetime of the excited nuclear state with j and l as the energy level index for energy ω and W is the normalization weight tensor. We get oscillations only when all values of ω_j are not the same or else natural decay is observed.

4.8.3.2 Time-Integrated Nuclear Resonant Forward Scattering

In the nuclear resonant forward scattering experiments, the time interval between adjacent synchrotron radiation pulses has to be longer than the mean lifetime of the nuclear excited state. This can be overcome using time-integrated nuclear resonant forward scattering proposed by Coussement et al. (1996). The setup uses the sample to be measured and a single line reference mounted on a Mössbauer velocity transducer. This interference process between the unknown sample and the reference is also known as heterodyne detection. The superposition of hyperfine split excitations with the incident photon gives quantum beats. The frequencies of the quantum beats are correlated to the hyperfine split energy levels. The interference of electric fields with the reference amplitude E_o and sample amplitude E_j is given as

$$E_{\mathrm{fs}}(\omega) = E_o(\omega) + \sum_j E_j(\omega) \tag{4.49}$$

The total amplitude of the scattered wave with probability amplitude A can be obtained through quantum electrodynamics (Odeurs et al. 1998). If a particular nucleus i has excitation level j, the amplitude is then

$$A_{ij} = A_j e^{-i\omega_j t} e^{i(k-k')\cdot r_i} \tag{4.50}$$

where ω_j corresponds to the frequency of excitation of ith nucleus at position r_i, k is the wave vector of the incoming photon, k' is the wave vector of the scattered photon and A_j is the amplitude constant. If N is the total number of nuclei and n is the number of nuclear excitations by the photon, then the total scattering probability is

$$P \propto N \sum_{j=1}^{n} \left|A_j\right|^2 + \sum e^{-i(\omega_j-\omega_{j'})t} \sum A_j A_{j'}^* e^{i(k-k')\cdot(r_i-r_{i'})} \tag{4.51}$$

If the nuclear energies of the sample and reference are given as $\hbar\omega_i$ and $\hbar\omega_r$, respectively, having an isomer shift, then the intensity for a thin sample is given by

$$I(t) = \left(N_s^2 + N_r^2\right)e^{-\tau} + 2N_s N_r e^{-\tau}\cos(\omega_r - \omega_i)t \tag{4.52}$$

where $\tau = t/t_o$, N_s and N_r are the number of resonant nuclei for sample and reference, respectively. The integrated intensity is given by the Lorentzian (L'abbé et al. 2001).

$$\overline{I}_{\text{fs}} = N_s^2 + N_r^2 + 2N_s N_r \sum_i \frac{1}{\tau^2(\omega_i - \omega_r)^2 + 1} \tag{4.53}$$

From (4.53), we can understand that the integrated intensity will be maximum when the sample and the absorber are in resonance ($\omega_i = \omega_r$). Thus, the time domain signal is converted to the energy domain, and we obtain a plot of Doppler velocity versus counts. For thick samples, the width increases, resulting in thickness broadening.

Apart from extracting energy information by the analysis of the time decay pattern, other information, such as phonon density of states, could be obtained using SR experiments based on inelastic scattering.

4.9 Analysis of the Mössbauer Spectrum

The Mössbauer spectra for the crystalline samples are least-squares fitted with Lorentzian lines for discrete fields. A Mössbauer spectrum, in general, consists of a sum of Lorentzian lines if the relaxation effects are not present.

The theoretical function which is fitted to the experimental spectrum depends on a number of input parameters. These are varied in the fitting procedure to obtain a minimum in the value of Q^2, which is a statistically weighted sum over all data points of the squares of the difference between the postulated function and the experimental data. Its explicit form is,

$$Q^2 = \sum_{k=1}^{n} \frac{1}{y_k} \left[F\left(x_k; \alpha_1, \ldots \alpha_j, \ldots \alpha_p\right) - y_k \right]^2 \tag{4.54}$$

$F(x_k, \alpha)$ is the value of the postulated function evaluated at k^{th} data point and y_k is the number of counts in the same velocity channel. The function is determined by the variable parameters α_j and the spectrum consists of n data points. The velocity at each point is denoted by x_k. The function is linearized by truncating its Taylor expansion written in the abbreviated form as

$$F_k = F_k^0 + \sum_j \left[\frac{\partial F_k^\alpha}{\partial \alpha_j} \right] \Delta\alpha_j \tag{4.55}$$

The fit is started by estimating the value of α_j^0 from the experimental spectrum. Then the adjustments, $\Delta\alpha_j$, required for Q^2 to be minimum are approximated by the solution of the normal equations,

$$\sum_{i,k} \frac{1}{y_k} \frac{\partial F_k^0}{\partial \alpha_i} \frac{\partial F_k^0}{\partial \alpha_j} \Delta\alpha_j = \sum_k \left[\frac{F_k - y_k}{y_k} \right] \frac{\partial F_k^0}{\partial \alpha_i} \tag{4.56}$$

These equations are more conveniently expressed in the matrix form,

$$\sum_j (AM)_{ij} \Delta\alpha_j = (BM)_i \tag{4.57}$$

where *AM* is the measurement matrix and *BM* is the data vector.

The function F_k designed to fit the Mössbauer data is not linear in α_j and therefore, the solution of the above equation will not immediately yield the minimum Q^2. In general, the new parameter values $\alpha_j^0 + \Delta\alpha_j$ will give a value of Q^2, which is smaller than the previous value. If Q^2 decreases, the iterations are repeated until the changes, $\Delta\alpha_j$ becomes negligibly small. The final smallest value of Q^2 is normalized by dividing the number of degrees of freedom to give the value of χ^2 for the fit. The function $F(x_k, \alpha)$ is given by

$$F(x_k,\alpha) = B\left[1 + \left(\frac{\sigma x_k}{100V}\right) - \left(\frac{2A}{100\pi}\right) \sum_{j=1}^{N} \frac{h_j}{w_j + (4/w_j)(x_k - e_j)^2}\right] \tag{4.58}$$

where B is the background level of the spectrum, σ is a measure of the slope of the background, V is the peak-to-peak velocity used in the experiment and A is the total area of the spectrum measured in units of velocity times the percentage of the background. The quantities h_j, w_j and e_j represent the relative intensity, width and velocity position, respectively, of the jth line. α represents the parameter varied in the fitting procedure to get minimum Q^2. The quantities x_k and N do not vary and are determined by the input. The above least-squares program has many flexible options, such as varying line intensities, calculating the intensities internally, or prefixing certain input parameters.

The widths of all spectral lines of a sextet or the width of any one spectral line of a sextet can be varied independently or prefixed. Figure 4.27 shows the raw data spectrum (a) with the channel number and (b) raw data converted to velocity scale in the x-axis along with the fitted line. The spectrum shows only a singlet indicating isomer shift, whereas the conversion to velocity scale specifies zero isomer shift with reference to Fe. Although various fitting programs are available that follow different algorithms, the objective of the fitting is to obtain goodness of fit close to 1 with acceptable fit parameters. For example, the FWHM cannot be below the natural linewidth and instrumental line broadening. However, error limits, much smaller are generally reported from the fitting results.

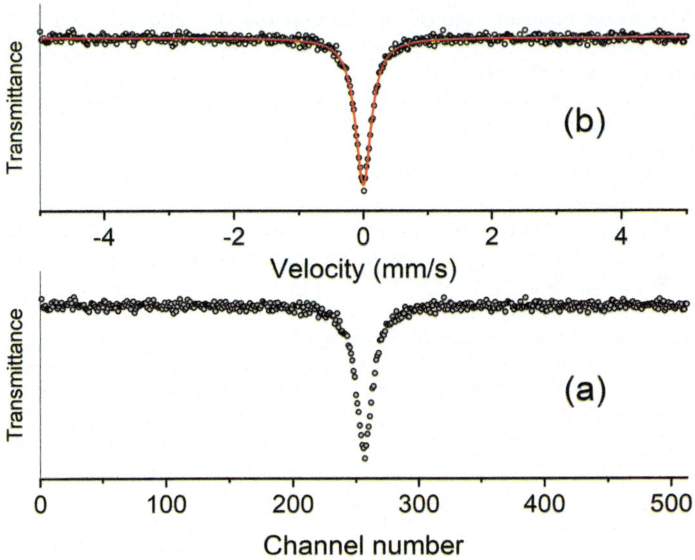

Fig. 4.27 Single line ^{57}Fe Mössbauer spectrum raw data with **a** channel number scale indicating the presence of only isomer shift and **b** converted to velocity scale. The continuous line is the fitted Lorentzian, indicating zero isomer shift with reference to Fe. The fitting gives reduced chi square 0.00188, FWHM = 0.2953 mm/s and coefficient of determination $R^2 = 0.981$. The statistical errors in the isomer shift is 0.0012 and the FWHM is 0.0037 (in mm/s), obtained from the fitting, that are different from experimental errors

4.10 Summary

The basic concept of Mössbauer effect that deals with the nuclear recoilless resonant absorption of gamma rays has been discussed. The instrumentation related to the Mössbauer spectrometer and various modes of measurements are presented.

References

M.R. Bhat, Nucl. Data Sheets **67**, 195 (1992)
D.C. Champeney, Rep. Prog. Phys. **42**, 1017 (1979)
R.L. Cohen, G.L. Miller, K.W. West, Phys. Rev. Lett. **41**, 381 (1978)
R. Coussement, S. Cottenier, C. L'abbé, Phys. Rev. B **54**, 16003 (1996)
C.E. Crouthamel, F. Adams, R. Dams, in ed. by R. Belcher, H. Freiser, *Applied Gamma Ray Spectrometry*, (Pergamon Press, Surrey, 1970)
I. Dézsi, B. Molnár, Nucl. Instrum. Methods **54**, 105 (1967)
W.F. Filter, R.H. Sands, W.R. Dunham, Nucl. Instrum. Methods Phys. Res., Sect. B **119**, 565 (1996)
E. Gerdau, R. Rüffer, H. Winkler, W. Tolksdorf, C.P. Klages, J.P. Hannon, Phys. Rev. Lett. **54**, 835 (1985)
J. Itoh, T. Toriyama, K. Saneyoshi, K. Hisatake, Nucl. Instrum. Methods Phys. Res. **205**, 279 (1983)

W. Kuhn, The London, Edinburgh and Dublin Philosophical Magazine and J. Sci. **8**, 625 (1929)
C. L'abbé, R. Callens, J. Odeurs, Hyperfine Interact. **135**, 275 (2001)
W.E. Lamb Jr., Phys. Rev. **55**, 190 (1939)
P.B. Moon, Proc. Phys. Soc. **64**, 76 (1951)
R.L. Mössbauer, Z. Phys. **151**, 124 (1958)
R.L. Mössbauer, Rev. Nuc. Sci. **12**, 123 (1962)
R.L. Mössbauer, Hyperfine Interact. **126**, 1 (2000)
J. Odeurs, R. Coussement, C. L'abbé, G. Neyens, G.R. Hoy, E.E. Alp, W. Sturhanhn, T. Toellner, C. Johnson, Hyperfine Interact. **113**, 455 (1998)
D.G. Rancourt, A.M. McDonald, A.E. Lalonde, J.Y. Ping, Am. Mineral. **78**, 1 (1993)
S.L. Ruby, J. Phys. Colloques **35**, C6-209 (1974)
G.V. Smirnov, Hyperfine Interact. **123/124**, 31 (1999)
W. Sturhahn, J. Phys. Condens. Matter **16**, S497 (2004)

Chapter 5
The Mössbauer Hyperfine Parameters

5.1 Mössbauer Hyperfine Parameters

Hyperfine interactions, i.e., interactions between a nuclear quantity and appropriate electronic or atomic quantity, are of great significance and provide details of the electronic and spin structure. There are three main hyperfine interactions leading to (three observables):

(1) the electric monopole interaction (isomer shift),
(2) the electric quadrupole interaction (quadrupole splitting) and
(3) the magnetic dipole interaction (nuclear Zeeman effect).

Apart from the Mössbauer effect, the hyperfinc interaction parameters could be deduced from perturbed angular correlation, nuclear magnetic resonance (NMR) and electron spin resonance. But, the discovery of the Mössbauer effect has made it possible to study the hyperfine interactions precisely and directly, and identify mixed magnetic phases thanks to a wealth of new information. Since most highly magnetic materials are Fe-based, we focus on ^{57}Fe Mössbauer spectrometry.

5.2 Isomer Shift (Electric Monopole Interaction)

The isomer shift is a measure of the s electron density at the nucleus relative to the source and absorber. The difference in effective charge radii between the ground and the excited (isomeric) levels then shifts the gamma ray energy levels, first observed by Kistner and Sunyar (1960). It is due to the Coulomb interaction between the charge distribution of the nucleus and the electrons. This interaction causes a shift in energy levels, and the peaks are also shifted for an absorber having an isomer shift, as shown in Fig. 5.1. If the source and absorber have similar energy levels or if the shift is equal for the source and absorber, no isomer shift will be observed in the

R. J. Joseyphus and J.-M. Greneche, *Fundamentals of ^{57}Fe Mössbauer Spectrometry*,
https://doi.org/10.1007/978-981-99-8653-8_5

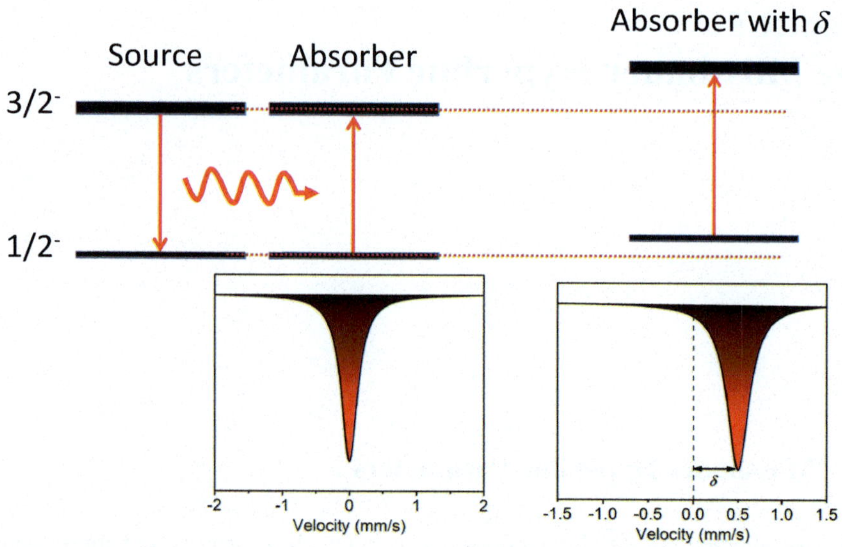

Fig. 5.1 The interaction with the electric monopole alters the nuclear energy levels, resulting in a shift in the spectrum. This shift, known as the isomer shift, can be positive or negative, as represented on the velocity scale

spectrum. For a positive isomer shift, the energy level in the absorber corresponds to a positive Doppler velocity and vice-versa. This shift is known as the isomer shift or central shift, and the position of the center of gravity of the spectrum relative to the zero velocity is a measure of the isomer shift (δ).

The nuclear charge distribution is altered by changing the nuclear volume in the ground state relative to the excited state, both in the source and in the absorber.

The Coulomb energy due to continuous charge distribution, with a charge density $\rho(r)$ and electric potential $V(r)$ is

$$E_c = \frac{1}{2}\int_0^{\infty} \rho(r)V(r)4\pi r^2 \mathrm{d}r \tag{5.1}$$

The charge density is given by $\rho(r) = -e|\psi(0)|^2$. (We can now call this ψ as the Mössbauer wave function). Assuming a constant wave function for a nucleus of radius *R*, the electrostatic interaction energy of the nuclear charge (*Ze*) with the orbital electron density for the ground state is given as

$$E_r = -\int_0^{R_r} V(r)e|\psi(0)|^2 2\pi r^2 dr = -2\pi e|\psi(0)|^2 \int_0^{R_r} V(r)r^2 \mathrm{d}r \tag{5.2}$$

where the limit corresponds to the ground state nuclear radius R_g and $V(r) = kV_{r\to 0} = Ze/R_g$ for a point nucleus. The Coulomb constant in SI unit is $k = 1/4\pi\varepsilon_0$ while it is unity in cgs electrostatic unit. Assuming a uniformly charged sphere, the electric potential for $r \leq R_g$ is

$$V(r) = \frac{Ze}{2R_g}\left(3 - \frac{r^2}{R_g^2}\right) \quad (5.3)$$

Using (5.3) in (5.2),

$$E_g = -\pi Ze^2|\psi(0)|^2 \int_0^{R_g} \left(\frac{3}{R_g} - \frac{r^2}{R_g^3}\right) r^2 \mathrm{d}r \quad (5.4a)$$

$$E_g = -\pi Ze^2|\psi(0)|^2 \left| \frac{r^3}{R_g} - \frac{r^5}{5R_g^3} \right|_0^{R_g} = -\frac{4\pi}{5} Ze^2 R_g^2 |\psi(0)|^2 \quad (5.4b)$$

A similar relation is obtained for the excited state E_e with excited state radius R_e.

The energy shift due to the finite nuclear size for an s electron over the nuclear volume with respect to the point nucleus from (5.2) and (5.4) is given as (Wertheim 1964; Shirley 1964),

$$\delta E = \frac{2}{5}\pi Ze^2 R^2 |\psi(0)|^2 \quad (5.5)$$

The energy shift between the excited and ground states is

$$(\delta E)_e - (\delta E)_g = \frac{2}{5}\pi Ze^2 |\psi(0)|^2 (R_e^2 - R_g^2) \quad (5.6)$$

The difference in energy of the gamma ray relative to the source (S) and the absorber (A) during a nuclear transition between ground and excited states is given by

$$\delta = \left[(\delta E)_e - (\delta E)_g\right]_\mathrm{A} - \left[(\delta E)_e - (\delta E)_g\right]_\mathrm{S} = \Delta E_\mathrm{A} - \Delta E_\mathrm{S} \quad (5.7a)$$

Since the s electron charge densities of the source and absorber are different,

$$\Delta E_\mathrm{A} = \frac{2}{5}\pi Ze^2 |\psi(0)_\mathrm{A}|^2 (R_e^2 - R_g^2) \quad (5.7b)$$

$$\Delta E_\mathrm{S} = \frac{2}{5}\pi Ze^2 |\psi(0)_\mathrm{S}|^2 (R_e^2 - R_g^2) \quad (5.7c)$$

$$\delta = \frac{2}{5}\pi Ze^2 \left\{|\psi(0)_\mathrm{A}|^2 - |\psi(0)_\mathrm{S}|^2\right\} (R_e^2 - R_g^2) \quad (5.7d)$$

$$\delta = \frac{2}{5}\pi Ze^2\{|\psi(0)_A|^2 - |\psi(0)_S|^2\}\left\{2R^2\frac{\delta R}{R}\right\} \tag{5.7e}$$

$$\delta = \left(\frac{4\pi Ze^2}{5}\right)R^2\{|\psi(0)_A|^2 - |\psi(0)_S|^2\}\left(\frac{\delta R}{R}\right) \tag{5.8}$$

with $R = \frac{R_e+R_g}{2}$ and $\delta R = R_e - R_g$

where Ze is the nuclear charge, R_e and R_g are nuclear radii in its excited and ground states, respectively.

5.2.1 Relativistic Correction

The contribution to the s electron density from the valence electrons is affected by the chemical bond and the charge state. The $|\psi_{ns}(0)|^2$ is found using the Fermi-Segré formula (Fermi and Segré 1933)

$$|\psi_{ns}(0)|^2 = \frac{ZZ_o^2}{\pi(a_o n_e)^3}\left(1 - \frac{d\sigma}{dn}\right) = \frac{Z(1+m)^2}{\pi a_o^3(n-\sigma)^3}\left(1 - \frac{d\sigma}{dn}\right) \tag{5.9}$$

where Z is the nuclear charge, m is the charge on the atom or ion, n_e is the effective quantum number, a_o is the Bohr radius and σ is the quantum defect. The above formula is for free atoms or ions and does not take into account of shielding effects. It must also be modified for relativistic effects in solids (Shirley 1964).

Considering relativistic effects, the electron density near the nucleus for a point nucleus is given as

$$P(r) = \frac{2(1+\rho)}{\Gamma^2(2\rho+1)}\left(\frac{2Zr}{a_o}\right)^{2(\rho-1)}|\psi(0)|^2 \tag{5.10}$$

where $\rho = \sqrt{1-\alpha^2 Z^2}$, α is the fine structure constant, Γ is the gamma function and $|\psi(0)|^2$ is the non-relativistic electron density.

As Z is neglected, in particular for light elements, $\rho \to 1$ and $P(r) \to |\psi(0)|^2$. The energy shift for s electron for a finite nuclear volume using first-order perturbation theory becomes

$$\Delta E = \frac{6\pi R_\infty a_o^3(1+\rho)}{Z\rho(2\rho+1)(2\rho+3)\Gamma^2(2\rho+1)}\left(\frac{2ZR}{a_o}\right)^{2\rho}|\psi(0)|^2 \tag{5.11}$$

where $r = R$ and on solving further gives the isomer shift as

$$\delta = \left(\frac{4\pi Ze^2}{5}\right)R^2\left(\frac{30(1+\rho)}{(2\rho+1)(2\rho+3)\Gamma^2(2\rho+1)}\left(\frac{2ZR}{a_o}\right)^{2(\rho-1)}\right)|\psi(0)|^2\left(\frac{\delta R}{R}\right) \tag{5.12}$$

The second term in the bracket of (5.12) is the relativistic term using perturbation theory, whereas it is modified into a relativity factor due to wave function distortions at the origin.

In terms of Doppler velocity relative to the source and the absorber, (5.12) becomes

$$v = \left(\frac{4\pi Ze^2c}{5E_\gamma}\right)R^2(\Lambda(Z))\left(|\psi(0)_A|^2 - |\psi(0)_S|^2\right)\left(\frac{\delta R}{R}\right) \tag{5.13}$$

where

$$\Lambda(Z) = \frac{30(1+\rho)}{(2\rho+1)(2\rho+3)\Gamma^2(2\rho+1)}\left(\frac{2ZR}{a_o}\right)^{2(\rho-1)} \tag{5.14}$$

Since $\Lambda(Z) \to 1$ in the non-relativistic limit, we get back (5.8) for δ.

Electrons close to the nucleus have higher momentum than valence electrons. Theoretical calculations show that the Hartree–Fock orbital energies for the 4s orbital of neutral Fe have an energy difference of – 0.1 eV between the relativistic and non-relativistic approximations. This energy difference increases to – 62.4 eV for the 1s orbital and further increases for heavier atoms. Thus, non-relativistic calculations will likely show a larger deviation from the experimental values.

5.2.2 *Isomer Shift Calibration Constant*

The non-relativistic Doppler velocity is given as

$$v = \left(\frac{4\pi Ze^2R^2c}{5E_\gamma}\right)\left(|\psi(0)_A|^2 - |\psi(0)_S|^2\right)\left(\frac{\delta R}{R}\right) \tag{5.15a}$$

The isomer shift relating the Doppler velocity using $\delta = \frac{v}{c}E_\gamma$, (5.8) is given as,

$$\delta = \left(\frac{4\pi Ze^2R^2}{5}\right)\left(\frac{\delta R}{R}\right)\left(|\psi(0)_A|^2 - |\psi(0)_S|^2\right) = \text{const}\left(|\psi(0)_A|^2 - |\psi(0)_S|^2\right) \tag{5.15b}$$

The $\delta R/R$ is negative with a value in the range $-(0.9 \sim 1.3) \times 10^{-3}$ (Filatov, 2009) for ^{57}Fe, implying a decrease in the nuclear charge density in the excited state from the s electron contribution.

The Eq. (5.15b) can be further written in terms of relative velocity as

$$\delta = \alpha\left(|\psi(0)_A|^2 - |\psi(0)_S|^2\right) \quad (5.16)$$

where

$$\alpha = \left(\frac{4\pi c Z e^2 R^2}{5E_\gamma}\right)\frac{\delta R}{R} \quad (5.17)$$

The isomer shift, δ, therefore, depends on the electron density at the nucleus of the source $|\psi(0)_S|^2$ and absorber $|\psi(0)_A|^2$. In a much simpler form, for ^{57}Fe, the isomer shift contribution can be written as a linear equation

$$\delta = \alpha\rho_A(0) - b \quad (5.18)$$

where α is known as the calibration constant as defined by (5.17), which was previously reported as -0.6 to $-0.1\ a_o^3$mm/s. $\rho_A(0)$ and b are the s electron densities at the nucleus of the absorber and reference, respectively. A more rational value was found as $\alpha \approx -0.3\ a_o^3$mm/s (Wdowik and Ruebenbauer 2007; Filatov 2009; Kurian and Filatov 2010).

5.2.3 *Factors Affecting Isomer Shift*

The electron density is provided by an internal factor, written as $\rho_{ns} = |\psi_{ns}(R)|^2$ (see Sect. 2.4) and there are other external factors, such as the bonding given as ρ_{bond} that alter the s electron density. Among the internal contribution of the electron density, the 3d, 3s and 4s electrons only leads to changes in the charge density at the nucleus because they are affected by the bonding, and therefore, the 1s and 2s contributions are neglected. ρ_{bond} influences the 3s + 4s electron density and, therefore, the s electron density at the nucleus indirectly by either allowing or preventing them from participating in the bonding. Indeed, most of the s electron density at the nucleus is affected by (i) valence shell s electron, i.e., the 4s electron, (ii) the screening by d electrons, i.e., the 3d electrons, (iii) the oxidation state of Fe, (iv) the covalency charge transfer effect, and (v) the σ donor and π acceptor of the surrounding ligands. Effects such as bonding and oxidation states influence the s and d electron density, which is reflected in the δ values.

5.2.4 *Overlap Distortion*

A solid containing A and B ions can have a local wave function and a non-local overlap wave function contributing to the electron density at the nucleus (Sawatzky and van der Woude 1974). The overlap distortion increases the electronic charge at the nucleus, decreasing the effective nuclear charge. The ion A will rearrange by moving slightly outwards, thus decreasing the charge density at the nucleus, which could influence the δ.

The change in charge density is given in terms of the change in the effective nuclear charge for the jth Slater orbital ΔZ_j as

$$\Delta Z_j = -\int_0^{R_j} \Delta\rho(r) d^3 r \tag{5.19}$$

where $\Delta\rho(r)$ is the total change in charge density due to overlap distortion and rearrangement of ion. The most contributor is given as

$$\Delta Z_1 \approx \frac{4\pi}{3} R_1^3 \Delta\rho(0) \tag{5.20}$$

for $j = 1$ and the radius R_1 corresponding to it. The wave function now becomes

$$\psi'_{ns} \approx \psi_{ns} + \sum_j \frac{\delta\psi_{ns}}{\delta Z_j} \delta Z_j \tag{5.21}$$

The first-order change in charge density is then

$$\Delta\rho_r(0) \approx -3\frac{\rho(0)}{Z_1}\left(\frac{4}{3}\pi R_1^3\right)\Delta\rho(0) \tag{5.22}$$

The total change in charge density is the sum of the overlap charge density $\Delta\rho_{\text{ov}}(0)$ and the first-order term given as

$$\Delta\rho(0) = \Delta\rho_{\text{ov}}(0) + \Delta\rho_r(0) \tag{5.23}$$

$$\Delta\rho(0) = \frac{\Delta\rho_{\text{ov}}(0)}{\left(1 + \frac{4\pi R_1^3 \rho(0)}{Z_1}\right)} \tag{5.24}$$

The expression (5.24) gives the variation in nuclear charge density due to wave function overlap and the ionic displacement for occupied orbitals. The overlap charge density with interatomic distance in the octahedral coordination exhibits a trend, as shown in Fig. 5.2, with the highest overlap charge density (in a.u. or a_o) for chloride ligands followed by oxides and fluorides (Sawatzky and van der Woude 1974). The change in isomer shift values also shows a similar trend from a more negative value with a shorter Fe–O distance. In contrast to the overlap distortion, the crystalline potential decreases the electron density.

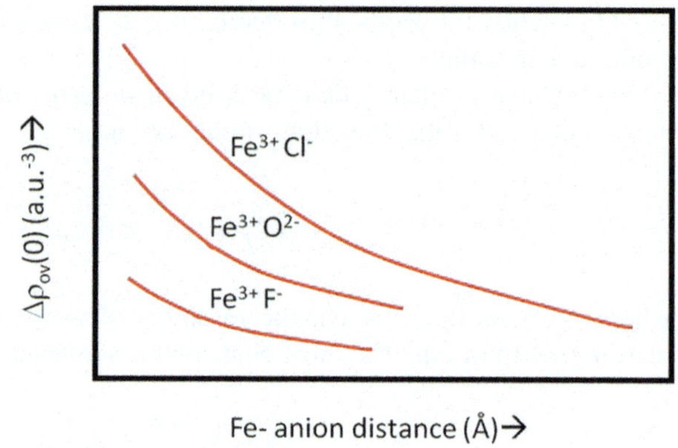

Fig. 5.2 The change in overlap charge density at the nucleus with interatomic distance in octahedral coordination for various ligands (Sawatzky and van der Woude 1974). The Fe-anion distance ranges from 1.9 to 2.6 Å, and $\Delta\rho_{ov}(0)$ is from 1 to 9

5.2.5 Electronegativity

The charge transfer to the 3d and 4s orbital could affect the isomer shift values, which can be observed from the δ dependence on electronegativity for the Fe-halides, as shown in Fig. 5.3. The linear combination of atomic orbitals-molecular orbitals (LCAO-MO) calculations have shown effective charges of 0.09 for FeF_2, which turns into a more negative value of – 0.4 for FeI_2 (Reschke and Trautwein 1976). The calculations also suggest the lowest 4s population for FeF_2 and the largest for FeI_2. This is also reflected in the total charge density, which gave a difference of $2a_o^{-3}$ between them, with the highest for FeI_2 showing the lowest δ. The plot of the isomer shift as a function of Pauling electronegativity and relativistic charge density for different halides is demonstrated in Fig. 5.3. The relativistic charge density varies from 15,069 to 15,066.5 a_o^{-3} between the halide FeI_2 and FeF_2, respectively.

The s electron density at the nucleus increases when the ionic to covalency state is changed, and the isomer shift then follows (Danon 1963)

$$\delta\left(Fe^{2+}(\text{ionic})\right) > \delta\left(Fe^{3+}(\text{ionic})\right) > \delta\left(Fe^{2+}(\text{covalent})\right) > \delta\left(Fe^{3+}(\text{covalent})\right).$$

The decrease in the 3d electrons lowers the atomic shielding of s electrons, resulting in a higher s electron density at the nucleus. Therefore, δ is higher for Fe^{2+} than for Fe^{3+}.

For iron-nitrosyl complexes

$$\delta\left(CN^-\right) > \delta(CO) > \delta\left(NO^+\right)$$

where the increase in s electron density for NO in Fe nitrosyl is due to the extensive delocalization of the d electrons in the π bond.

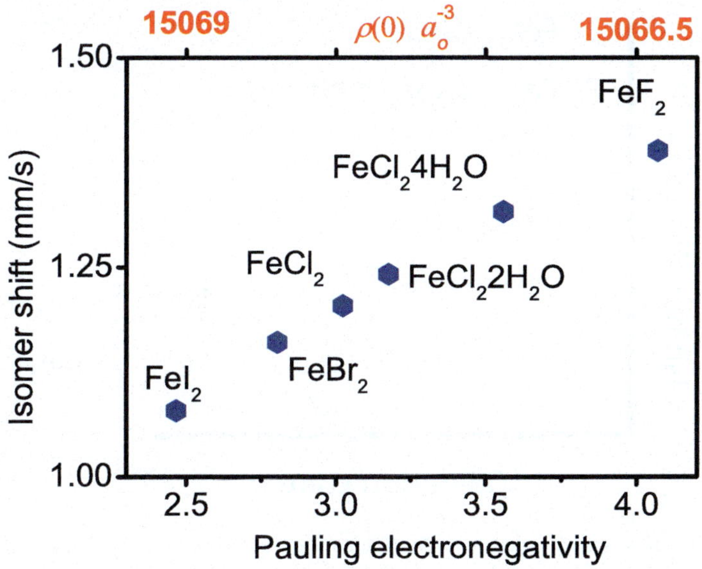

Fig. 5.3 The isomer shift values for various Fe halides show a linear variation with electronegativity and relativistic electron density (Sawatzky and van der Woude 1974; Reschke and Trautwein 1976)

5.2.6 Oxidation State

The isomer shift decreases with the oxidation state as the d electrons are reduced, resulting in lower shielding of s electrons with a higher oxidation state. Figure 5.4 shows the general trend of the isomer shift values with the oxidation state of Fe in the high-spin and low-spin complexes. It can be seen that the isomer shift is the largest for Fe with the lowest oxidation number and smallest for Fe^{3+}, where it could also be negative. In addition, the isomer shift increases with the coordination number for a particular oxidation state. Therefore, the tetrahedral site with four oxygen coordination anions shows a smaller isomer shift than the octahedral site with six oxygen ions. The isomer shift ranges for the various low-spin and high-spin oxidation states of Fe at 300 K are depicted in Fig. 5.5. Fe with the highest oxidation state (Fe^{6+}) could show more negative values, i.e. $\delta < -1$ mm/s while the least one (Fe^{1+}) could show $\delta > 1$ mm/s at room temperature. The high-spin complexes show a larger isomer shift than their low-spin counterparts. It follows that

$$\delta\left(Fe^{1+}\right) > \delta\left(Fe^{2+}\right) > \delta\left(Fe^{3+}\right) > \delta\left(Fe^{4+}\right) > \delta\left(Fe^{5+}\right) > \delta\left(Fe^{6+}\right)$$

and the experimental isomer shift also obeys a general trend as a function of the coordination number for a given anionic species, (Menil 1985)

$$\delta(c) > \delta(o) > \delta(tb) > \delta(t) > \delta(sp)$$

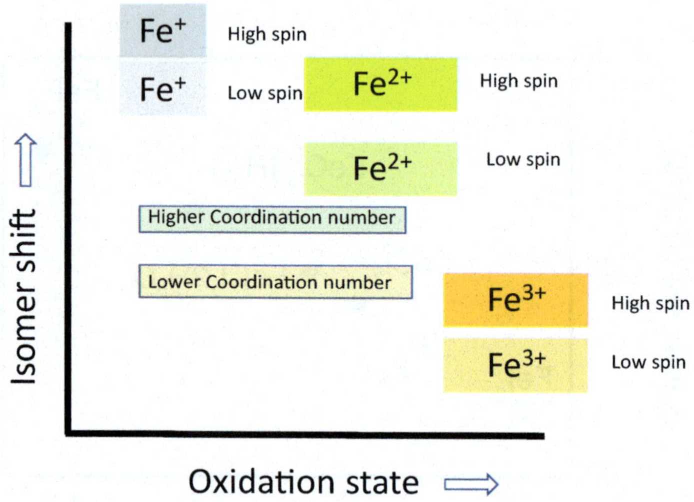

Fig. 5.4 The influence of various coordination numbers and spin states of Fe ions on the isomer shift

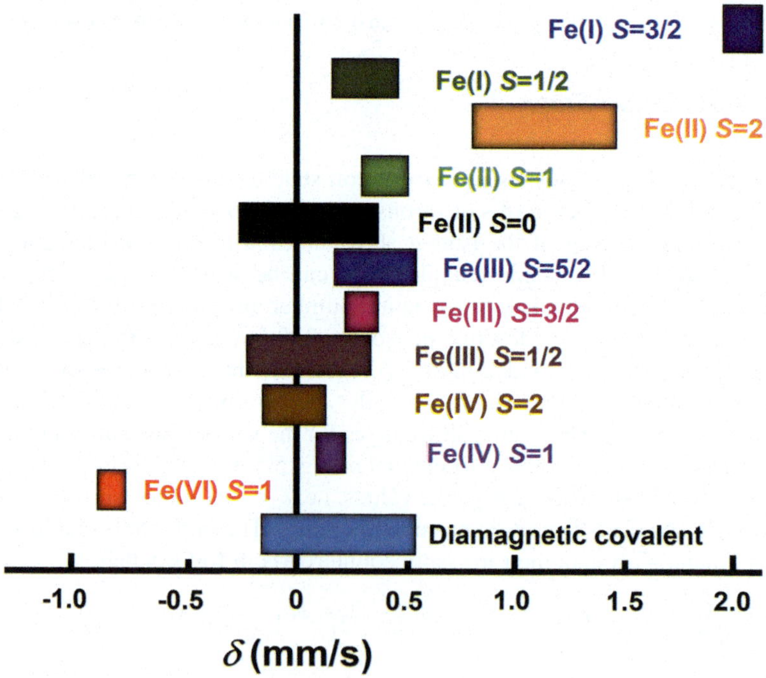

Fig. 5.5 Ranges of isomer shift values at 300 K for various Fe oxidation states relative to α-Fe (Greenwood and Gibb 1971)

where *c*, *o*, *tb*, *t* and *sp* represent cubic, octahedron, trigonal bipyramid, tetrahedron and square plane coordination, respectively.

5.2.7 *Atomic Volume*

The difference in isomer shift between Fe^{2+} and Fe^{3+} at 295 K is $\delta(Fe^{2+}) - \delta(Fe^{3+}) = 0.9$ mm/s and that of $\delta(Fe^{2+}) - \delta(\text{Fe metal}) = 1.28$ mm/s while there is an anomalous change in δ between paramagnetic and ferromagnetic states of metallic Fe with a difference of – 0.01 mm/s, due to experimental uncertainties.The correlation between isomer shift and electron density was predicted by theoretical calculations (Ingalls 1967) using $\alpha \approx -0.4\ a_o^3$mm/s. The calculated differences in electron densities between the different free ions and that of Fe metal are

$$\left|\psi_{Fe^{2+}}(0)\right|^2 - \left|\psi_{Fe^{3+}}(0)\right|^2 = -1.9\ a_o^{-3} \tag{5.25}$$

$$\left|\psi_{Fe^{2+}}(0)\right|^2 - \left|\psi_{Fe^{1+}}(0)\right|^2 = 0.5\ a_o^{-3} \tag{5.26}$$

$$\left|\psi_{Fe^{2+}}(0)\right|^2 - \left|\psi_{Fe^{0}}(0)\right|^2 = -3\ a_o^{-3} \tag{5.27}$$

For the Γ_1 in k space of ferromagnetic Fe at room temperature and for an atomic volume Ω of 80 a_o^{-3},

$$\left|\psi_{\Gamma_1}(0)\right|^2 = 7.1\ a_o^{-3} \tag{5.28}$$

$$|\psi_{4s}(0)|^2 = 3.8\ a_o^{-3} \tag{5.29}$$

The total contribution to the electron density due to volume change can be given by

$$|\psi(0)|^2 = n_s\left|\psi_{\Gamma_1}(0)\right|^2 + \beta\langle n u_{\max}^2\rangle + C \tag{5.30}$$

where n_s is the number of s states per atom, $\left|\psi_{\Gamma_1}(0)\right|^2$ is the electron density corresponding to Γ_1, $u_{\max}^2$ is the maximum of the squared 3d radial wave function, n is the number of 3d electrons, β is the correlation coefficient relating 3d and 3s charge density and C is a constant.

With pressure, there will be a change in atomic volume and a transfer of electrons from the s-band to the d-band, so that

$$\Omega \frac{\partial\left(|\psi(0)|^2\right)}{\partial\Omega} \approx -3.7 \tag{5.31}$$

giving

$$\Omega \frac{\partial(\delta)}{\partial \Omega} \approx 1.37 \text{ mm/s} \tag{5.32}$$

which is close to the experimental value of 1.34 mm/s for Fe at 295 K. At the Curie temperature corresponding to the ferromagnetic to the paramagnetic transformation of Fe, the atomic volume of the unit cell, $\Omega \approx 83\, a_o^{-3}$ and thus, $\left|\psi_{\Gamma_1}(0)\right|^2 = 6.8\, a_o^{-3}$. The $\langle nu_{\text{max}}^2 \rangle = 5.84\, a_o^{-3}$ compared with a value of 5.64 a_o^{-3} at room temperature. This gives $\left|\psi_{\text{Fe}^0}(0)\right|_{\text{para}}^2 - \left|\psi_{\text{Fe}^0}(0)\right|_{\text{ferro}}^2 \approx 0.13\, a_o^{-3}$ that could result in an isomer shift difference of – 0.02 mm/s. However, the charge density calculations and their experimental agreement are greatly influenced by the calibration constant α. It has been shown that $\alpha \approx -(0.29 - 0.23)\ a_o^3$mm/s is more reasonable (Duff and Dass 1971; Duff 1974). This increases $\left|\psi_{\Gamma_1}(0)\right|^2$ to $8.35\, a_o^{-3}$ and $|\psi(0)|^2$ by 25%. The α value is strongly influenced by the choice of density function calculation using an appropriate basis set, resulting in deviation among various reports.

5.2.8 *Metals and Alloys*

The widely accepted relation between the charge density change and number changes due to valence and screening electrons in metallic alloys is given by

$$\Delta\rho(0) \cong C\rho_s(0)(\Delta n_s + R\Delta n_d) \tag{5.33}$$

where C is a proportionality constant, $\rho_s(0)$ is the free atom s electron density, Δn_s is the number change in valence s electron contribution with alloying, Δn_d is the screening of the s electron by the d electron and the screening constant, R is in the range of 0 to – 1.0. In dilute alloys with various impurity elements ($Z = 0$ (vacancy) to $Z = 56$(Ba)) in Fe, good agreement was reached between the theoretical and experimental isomer shift of the first near neighbor Fe atoms, as shown in Fig. 5.6. (Akai et al. 1986). It has been shown that the change in s charges is more important than the d electron screening effect. The angular momentum of 3sp, 4sp and 5sp impurities influences the change in charge of the Fe site.

The isomer shift value must be corrected for the Fe impurity volume in the host volume that gives (Dezsi et al. 1989)

$$\Delta\delta = \left(\frac{0.615 B_M}{0.615 B_M + B_{\text{Fe}}}\right)\left(\frac{V_{\text{M}} - V_{\text{Fe}}}{V_{\text{Fe}}}\right)\frac{\partial\delta}{\partial \ln V} \tag{5.34}$$

where B_M, V_{M}, V_{Fe} are the bulk modulus of the host matrix, the molar volume of the host matrix and the molar volume of Fe, respectively and $\frac{\partial\delta}{\partial \ln V}$ is the volume derivative of isomer shift (1.37 mm/s) obtained from high-pressure experiments. The

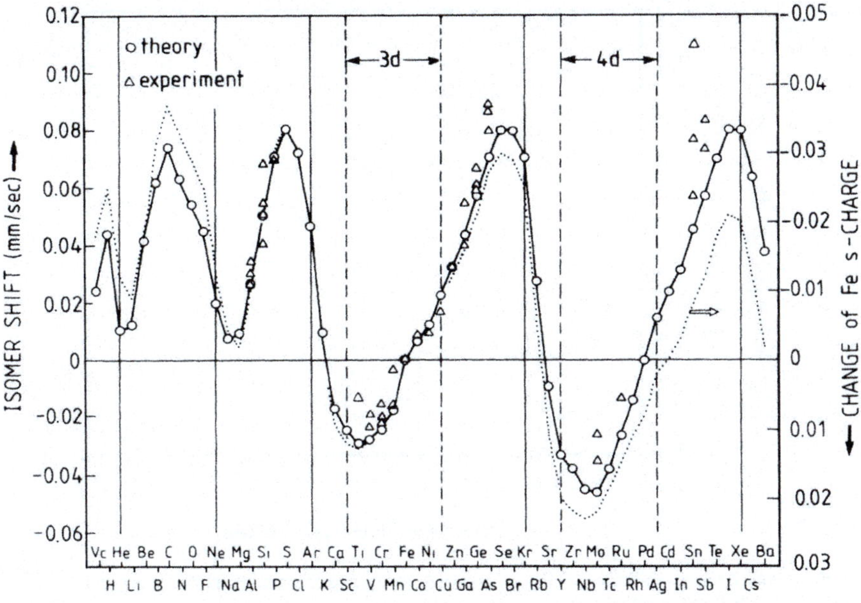

Fig. 5.6 The theoretical and experimental isomer shift of nearest neighbor Fe atoms as a function of the nuclear charge of the impurity atoms. Reprinted with permission (Akai et al. 1986), © American Physical Society

corrected δ value is close to the experimental deviation of 0.05 mm/s in amorphous and crystalline intermetallic alloys. At very low d electron density of the host matrix, a slight decrease in charge transfer to Fe occurs. In addition, s electrons are transferred from more electropositive atoms, and the empty d-band reduces charge transfer. However, a correlation of the electronegativity of the host atoms with δ is required.

In another study, the first-principles linear muffin-tin orbital-atomic sphere approximation-real space approach was employed to obtain a correlation of isomer shift with the volume effect of Fe impurities in the metallic hosts (Terrazos and F-Pessôa 1997). The calculated value of isomer shift was obtained in terms of the impurity charge density, $\rho_{\text{imp}}(0)$ and the charge density of the Fe site, $\rho_{\text{Fe}}(0)$ in the bcc-Fe using the following relation

$$\delta = \alpha\left(\rho_{\text{imp}}(0) - \rho_{\text{Fe}}(0)\right) \tag{5.35}$$

with $\alpha \approx -0.24\ a_o^3$mm/s. The increasing trend in the isomer shift values with Wigner-Seitz (WS) radius is in agreement with the theoretical and experimental values, but there are discrepancies in the magnitude. The values of δ vary from −0.2 to 0.4 mm/s for WS radius, 2.6 to 4.2 a.u. for the hosts on the left side of Fe in the periodic table. For the other side, the values are positive and reach a maximum of 0.5 mm/s, corresponding to a WS radius of 3.4 a.u. The probability of finding the 4s electron at the impurity nucleus was highest, with the smallest WS radius

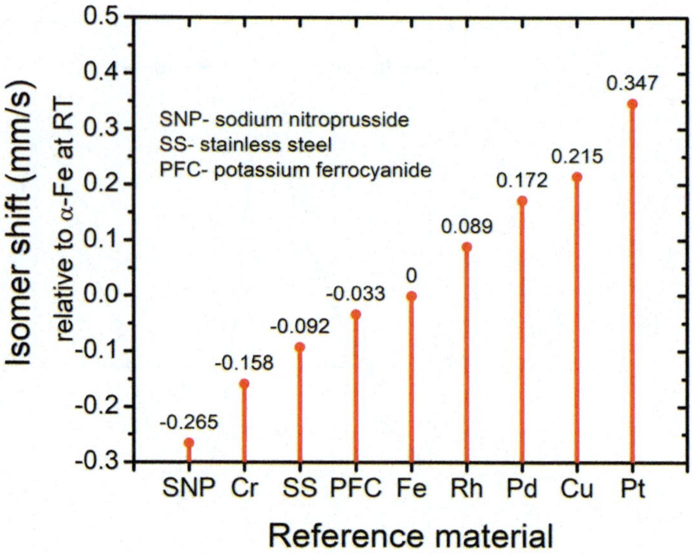

Fig. 5.7 The isomer shifts of ^{57}Fe in various reference materials relative to α-Fe at room temperature (Stevens 1983)

corresponding to hosts such as Ni and Co. A good correlation was obtained between the 4s electrons of the impurity and those of Fe, $\frac{n_{4s}^{imp}}{n_{4s}^{Fe}}$ with δ at the impurity site in terms of WS radius. The study showed that the main contribution to δ comes from the 4s valence electrons of Fe. Size effects are associated with the probability of finding 4s electrons close to the nucleus. The overall charge transfer at 4s impurity increases with WS radius, while competing size effect increases δ. The charge transfer increases with the WS radius of the host, suggesting the role of electronegativity in determining δ for Fe impurities in metallic hosts.

The isomer shift of ^{57}Fe in various materials such as sodium nitroprusside, stainless steel, Cr, Pd and Pt show variation in the values with respect to bcc Fe at room temperature (Fig. 5.7). If the isomer shift value is given for the source in the matrix material, a correction can be undertaken by adding the value shown in Fig. 5.7 to obtain the isomer shift relative to α-Fe.

5.2.9 Second Order Doppler Shift

The second-order Doppler effect is due to the relativistic effect between the laboratory time scale and the atomic time scale which causes time dilation. Due to this time dilation, the emission and absorption lines are not identical for an identical source and absorber at different temperatures.

The time dilation is given by

$$t' = \frac{t}{\sqrt{1 - \frac{v^2}{c^2}}} \tag{5.36}$$

where v is the velocity of the moving object, c is the velocity of light in vacuum, t' is the time in a moving frame of reference and t is the time in a stationary frame of reference. In terms of Doppler wavelength for a source moving towards the observer

$$\lambda' = \frac{(c - v)t}{\sqrt{1 - \frac{v^2}{c^2}}} \tag{5.37}$$

For a source moving away from the observer

$$\lambda' = \frac{(c + v)t}{\sqrt{1 - \frac{v^2}{c^2}}} \tag{5.38}$$

Replacing the proper time in terms of λ

$$\lambda' = \frac{(c + v)\frac{\lambda}{c}}{\sqrt{1 - \frac{v^2}{c^2}}} = \frac{\left(1 + \frac{v}{c}\right)\lambda}{\sqrt{1 - \frac{v^2}{c^2}}} \tag{5.39}$$

In terms of frequency

$$\nu = \nu' \frac{\left(1 + \frac{v}{c}\right)}{\sqrt{1 - \frac{v^2}{c^2}}} \tag{5.40}$$

or

$$\nu' = \nu \frac{\sqrt{1 - \frac{v^2}{c^2}}}{\left(1 + \frac{v}{c}\right)} \tag{5.41}$$

where ν' is the frequency of the moving frame of reference and ν is the frequency of the observer.

In the low-velocity limit, no matter of the source approaching or moving away, the denominator term vanishes and therefore,

$$\nu' = \nu\sqrt{1 - \frac{v^2}{c^2}} \tag{5.42}$$

and the binomial approximation gives

$$\nu' \approx \nu\left(1 - \frac{\upsilon^2}{2c^2}\right) \tag{5.43}$$

$$\Delta\nu = \nu' - \nu \approx -\nu\left(\frac{\upsilon^2}{2c^2}\right) \tag{5.44}$$

The corresponding energy change is given by

$$\Delta E \approx -E\left(\frac{\upsilon^2}{2c^2}\right) \tag{5.45}$$

In terms of Mössbauer transition energy E_γ, the shift in the energy relative to source and absorber due to the second order Doppler shift is

$$\Delta E_{\text{SOD}} \approx -E_\gamma\left(\frac{\upsilon^2}{2c^2}\right) \tag{5.46}$$

Thus, the Doppler shift decreases the observed energy level by taking into account the second order of velocity. The average velocity of the atoms in the first order is considered to be zero, whereas it is non-zero when the second order is considered within a factor of $\frac{\upsilon^2}{c^2} \approx \frac{k_B T}{Mc^2} \approx 10^{-13}$. For a Mössbauer nucleus, the mean square velocity is $\langle \upsilon^2 \rangle$ due to thermal averaging. The velocity scale isomer shift is

$$\delta_{\text{SOD}} = -\frac{\langle \upsilon^2 \rangle}{2c} \tag{5.47}$$

where $\delta_{\text{SOD}} \to 0$ as $\langle \upsilon^2 \rangle \to 0$ and $t' \to t$.

In the Debye approximation (Menil 1985)

$$\langle \upsilon^2 \rangle = \frac{9}{8}\frac{k_B}{M}\Theta_D + \frac{9}{8}\frac{k_B}{M}T\left(\frac{T}{\Theta_D}\right)^3 \int_0^{\frac{\Theta_D}{T}} \frac{x^3 dx}{e^x - 1} \tag{5.48}$$

Now, the isomer shift has a correction term given by

$$\delta = \delta_{\text{obs}} + \delta_{\text{SOD}} \tag{5.49}$$

where δ_{obs} is the first-order isomer shift when the source and absorber are at the same temperature.

5.3 Quadrupole Interaction

The interaction of nuclear quadrupole moment with the electric field gradient (EFG) due to other charges in the crystal is called quadrupole interaction. The EFG comes from the charges in neighboring atoms/ions, lone pair electrons, d electrons and bonding electrons. A non-zero nuclear quadrupole moment is possible only if there is a difference in electron densities between the various axes. The excited state $3/2^-$ of Fe nucleus is split into two components of m_I; $\pm$ 3/2 and $\pm$ 1/2, as shown in Fig. 5.8. This would result in a transition from the ground state to the two excited states and, consequently, two absorption peaks. The magnitude of the quadrupole splitting (Δ or ΔE_Q) is determined by the distance between the two peaks, measured in terms of Doppler velocity difference with its unit in mm/s. The quadrupole splitting is a measure of the deviation from the spherical charge distribution of the nucleus.

The Cartesian multipole expansion of the position coordinates given in terms of $\frac{1}{|r-r'|}$, where $r >> r'$ in the expression of the electric potential at a point separated by r in the z-axis from the charge distribution can be written as

$$\frac{1}{|r-r'|} = \frac{1}{r} + \frac{r \cdot r'}{r^3} + \frac{1}{2r^5} \sum_{i,j=1}^{3} x_i x_j \left(3x'_i x'_j - \delta_{ij} r'^2\right) + \cdots \tag{5.50}$$

$$\begin{aligned} V(r) &= \frac{1}{4\pi\varepsilon_0} \left(\begin{array}{c} \frac{\int \rho(r') d^3r'}{r} + \frac{r \cdot \int \rho(r') r' d^3r'}{r^3} \\ + \frac{1}{2r^5} \sum_{i,j=1}^{3} x_i x_j \int \rho(r') \left(3x'_i x'_j - \delta_{ij} r'^2\right) d^3r' + \cdots \end{array} \right) \\ &= k \left(\frac{q}{r} + \frac{r.p}{r^3} + \frac{\sum x_i x_j Q_{ij}}{2r^5} + \cdots \right) \end{aligned} \tag{5.51}$$

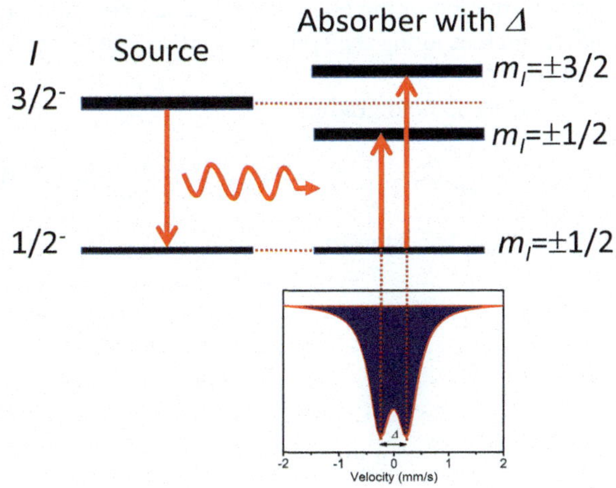

Fig. 5.8 The effect of electric quadrupole interaction on nuclear energy levels of ^{57}Fe

The first term in the equation, the monopole moment, is a scalar. The dipole moment is a vector given by

$$p = \int \rho(r')r'd^3r' \tag{5.52}$$

The quadrupole moment Q_{ij} associated with the continuous charge distribution is a second rank tensor

$$Q_{ij} = \int \rho(r')(3x_i'x_j' - r'^2\delta_{ij})d^3r' \tag{5.53}$$

The sum of diagonal elements (trace of the matrix) disappears so that

$$Q_{xx} + Q_{yy} + Q_{zz} = 0 \tag{5.54}$$

The classical multipole expansion terms (5.51) can be extended to the nuclear charge distribution case, and the nuclear quadrupole interaction can be expressed by the Hamiltonian

$$H = Q_{ij} \cdot \nabla E \tag{5.55}$$

with

$$Q_{ij} = \int \rho_n(r)(3x_ix_j - r^2\delta_{ij})d^3r \tag{5.56}$$

where ρ_n is the nuclear charge density; x_i, x_j are the Cartesian coordinates of r, δ_{ij} is the Kronecker delta and $r^2 = x^2 + y^2 + z^2$. The sign of Q_{ij} depends on the shape of the deformation of the nucleus with a negative quadrupole moment representing an oblate nucleus along the spin axis and a positive moment for a prolate or elongated nucleus.

The electric field gradient,

$$\nabla E = \nabla(-\nabla V) = -\nabla^2 V \tag{5.57}$$

is given by a 3×3 tensor

$$\begin{vmatrix} V_{xx} & V_{xy} & V_{xz} \\ V_{yx} & V_{yy} & V_{yz} \\ V_{zx} & V_{zy} & V_{zz} \end{vmatrix} \tag{5.58}$$

where $V_{ij} = \frac{\partial^2 V}{\partial_i \partial_j}$ that gives

$$\frac{\partial^2 V}{\partial x^2}, \frac{\partial^2 V}{\partial y^2}, \frac{\partial^2 V}{\partial x \partial y}, \frac{\partial^2 V}{\partial x \partial z}, \frac{\partial^2 V}{\partial y \partial x}, \frac{\partial^2 V}{\partial y \partial z}, \frac{\partial^2 V}{\partial z \partial x}, \frac{\partial^2 V}{\partial z \partial y} \text{and} \frac{\partial^2 V}{\partial z^2} \tag{5.59}$$

Only five of the nine EFG components are independent parameters. Due to symmetry, three off-diagonal elements $V_{xy} = V_{yx}$, $V_{xz} = V_{zx}$ and $V_{yz} = V_{zy}$ are dependent.

The electron densities for p, d and f are zero at the nucleus, while the EFG is contributed only by them due to the non-spherical symmetry. At the nucleus, it is thus, $\nabla^2 V = 0$ since $\rho(r) = 0$. Hence,

$$V_{xx} + V_{yy} + V_{zz} = 0 \tag{5.60}$$

A unique axis system of the Mössbauer atom called the principal axes of the EFG tensor, can be defined such that the off-diagonal elements disappear and the diagonal elements are ordered as

$$\begin{vmatrix} V_{xx} & 0 & 0 \\ 0 & V_{yy} & 0 \\ 0 & 0 & V_{zz} \end{vmatrix} \tag{5.61}$$

The axes are chosen such that

$$|V_{zz}| \geq |V_{yy}| \geq |V_{xx}| \tag{5.62}$$

With respect to the principal axes, the EFG tensor is described by only two independent parameters, usually chosen as $V_{zz} = eq$ and the asymmetry parameter η, defined as

$$\eta = \frac{V_{xx} - V_{yy}}{V_{zz}} \text{ where } 0 \leq \eta \leq 1 \tag{5.63}$$

For a spherically symmetric charge distribution, $\eta = 0$. The EFG elements in spherical polar coordinates are obtained further.

The potential due to the point charge q at r from the origin is $V = q/r$ (the constant $1/4\pi\varepsilon_0$ is included in SI unit), and the electric field component at the nucleus is $E_i = -\frac{\partial V}{\partial x_i} = qx_i/r^3$ where $x_1 = x, x_2 = y, x_3 = z$. The divergence of E is given by the Laplacian of V in (5.57).

The components of V are

$$V_{xx} = q(3x^2 - r^2)/r^5 \tag{5.64a}$$

$$V_{yy} = q(3y^2 - r^2)/r^5 \tag{5.64b}$$

$$V_{zz} = q(3z^2 - r^2)/r^5 \tag{5.64c}$$

$$V_{xy} = V_{yx} = 3qxy/r^5 \tag{5.64d}$$

$$V_{xz} = V_{zx} = 3qxz/r^5 \tag{5.64e}$$

$$V_{yz} = V_{zy} = 3qyz/r^5 \tag{5.64f}$$

that can be written as

$$V_{ij} = q\left(3x_i x_j - r^2\delta_{ij}\right)/r^5 \tag{5.65}$$

Using

$$\begin{aligned}
r^2 &= x^2 + y^2 + z^2 \\
\theta &= \cos^{-1}(z/r) \\
\varphi &= \tan^{-1}(y/x) \\
x &= r\sin\theta\cos\varphi \\
y &= r\sin\theta\sin\varphi \\
z &= r\cos\theta
\end{aligned}$$

we obtain

$$V_{xx} = q\left(3\sin^2\theta\cos^2\theta - 1\right)/r^3 \tag{5.66a}$$

$$V_{yy} = q\left(3\sin^2\theta\sin^2\varphi - 1\right)/r^3 \tag{5.66b}$$

$$V_{zz} = q\left(3\cos^2\theta - 1\right)/r^3 \tag{5.66c}$$

$$V_{xy} = V_{yx} = 3q\sin^2\theta\sin\varphi\cos\varphi/r^3 \tag{5.66d}$$

$$V_{xz} = V_{zx} = 3q\sin\theta\cos\theta\cos\varphi/r^3 \tag{5.66e}$$

$$V_{yz} = V_{zy} = 3q\sin\theta\cos\theta\sin\varphi/r^3 \tag{5.66f}$$

The off-diagonal elements will be zero by letting the principal z-axis pass through the charge, and the EFG elements are now

$$V_{xx} = V_{yy} = -q/r^3 \tag{5.67}$$

$$V_{zz} = 2q/r^3 \tag{5.68}$$

and $\eta = 0$. The Wigner-Eckart theorem (WET) reduces the number of independent matrix elements consisting of scalar, vector, and tensor operators, and it can be written in terms of CG coefficient and a reduced matrix element.

The eigenvalue equation of the interaction Hamiltonian, solving the matrix elements of the quadrupole operator Q_{ij} using WET, can be written as

$$H = \frac{eQ}{4I(2I-1)}\left\{V_{zz}\left[3I_z^2 - I(I+1)\right] + \left(V_{xx} - V_{yy}\right)\left(I_x^2 - I_y^2\right)\right\} \tag{5.69}$$

$$E_Q = \frac{e^2qQ}{4I(2I-1)}\left[3m_I^2 - I(I+1)\right]\left[1 + \frac{\eta^2}{3}\right]^{\frac{1}{2}} \tag{5.70}$$

where the spin quantum number $m_I = I, I - 1, \ldots, -I$ and η is the asymmetry parameter. The vector spin operator components are given as (I_x, I_y, I_z), and Q is a scalar in (5.69). A nuclear state with $I > 1/2$ is $(2I + 1)$ fold degenerate, and it is split into substates $|I, \pm m_I>$ without shifting the center of gravity. In the case of ^{57}Fe, the nuclear ground state with $I = 1/2$ is not split because of spherical symmetry, whereas the excited state with $I = 3/2$ is split into two substates $|\pm 3/2, \pm 1/2\rangle$ and $|\pm 3/2, \pm 3/2\rangle$, as shown in Fig. 5.8. Hence, two resonance lines are observed, and the separation between these two lines is called quadrupole splitting.

The quadrupole splitting (Δ) is given by

$$\Delta = \frac{1}{2}e^2qQ\left[1 + \frac{\eta^2}{3}\right]^{\frac{1}{2}} \tag{5.71}$$

The quadrupole moment, Q, (Q_{ij} in eb) has been shown to have a value of 0.15–0.16 b for the ^{57}Fe excited state nucleus (Dufek et al. 1995; Martinez-Pinedo et al. 2001). Most of the older literature used the Q value of 0.2 b, which would lead to an error since the deviation is about 25%. The asymmetry parameter in (5.71) can enhance the quadrupole splitting by a maximum of 15.5%. $V_{zz} = eq$; hence, q can be either positive or negative due to various factors, as discussed below.

The resonance lines are equal in intensity for an isotropic system, i.e., an isotropic powdered sample, resulting in symmetric quadrupole doublets. Anisotropic vibrations of the lattice (Goldanskii-Karyagin effect) give rise to asymmetric quadrupole doublets that depend on temperature. The texture effect, which results from a preferential crystalline orientation, for example, of a layered system like some clays, also gives rise to asymmetric quadrupole doublets. To verify the presence of texture effect, a rotation of the sample with respect to the γ-beam using the magic angle (54.7°) configuration should show a symmetric quadrupole doublet.

Since the quadrupole moment is constant for a given Mössbauer nucleus, the quadrupole splitting depends on the EFG. The various contributions to EFG can be represented as

$$V_{zz} = (1 - \gamma_\infty)(V_{zz})_{\text{LAT}} + (1 - R)(V_{zz})_{\text{val}} \tag{5.72}$$

$$V_{zz} = (1 - \gamma_\infty)(V_{zz})_{\text{LAT}} + (1 - R)\{(V_{zz})_{\text{CF}} + (V_{zz})_{\text{MO}}\} \tag{5.73}$$

where γ_∞ is called the Sternheimer anti-shielding factor and R is the Sternheimer shielding factor, which represents the values resulting from the anti-shielding and shielding of the nucleus by the core electrons, respectively. $(V_{zz})_{\text{LAT}}$, $(V_{zz})_{\text{val}}$, $(V_{zz})_{\text{CF}}$ and $(V_{zz})_{\text{MO}}$ are the contributions of the lattice or ligand, valence electron, crystal field and molecular orbitals, respectively. The contribution of the valence electron is further affected by the crystal fields and molecular orbitals. Thus, the quadrupole splitting would give information about the electronic structure, the bonding and the molecular symmetry.

5.3.1 *Valence Electron Contribution*

The EFG elements is expressed as

$$V_{ij} = \int \rho(x, y, z)\left(3x_i x_j - r^2\delta_{ij}\right)\mathrm{d}^3r \tag{5.74}$$

The z-component in polar coordinates is

$$V_{zz} = \int \rho(r, \theta, \varphi)\left(3\cos^2\theta - 1\right)r^{-3}\mathrm{d}^3r \tag{5.75}$$

The z-component in terms of the wave function is

$$V_{zz} = \int \psi_{lm}^*\left(3\cos^2\theta - 1\right)r^{-3}\psi_{lm}\mathrm{d}^3r \tag{5.76}$$

In spherical harmonics,

$$V_{zz} = \int Y_{lm}^*\left(3\cos^2\theta - 1\right)r^{-3}Y_{lm}\mathrm{d}^3r \tag{5.77}$$

In the absence of crystal field and molecular orbitals, the free atom's valence contribution due to the different sub-shells can be obtained. The z-components are

$$V_{zz} = (1 - \gamma_\infty)(V_{zz})_{\text{LAT}} + (1 - R)(V_{zz})_{\text{val}} \tag{5.78}$$

$$\eta V_{zz} = (1 - \gamma_\infty)\eta_{\text{LAT}}(V_{zz})_{\text{LAT}} + (1 - R)\eta_{\text{val}}(V_{zz})_{\text{val}} \tag{5.79}$$

Taking the expectation value, the charge of the electron as $q = (-e)$, the valence electron contribution can be written in Dirac notation form as

Table 5.1 The EFG elements for the p and d orbitals

Orbital	$\frac{(V_{zz})_{val}}{q\langle r^{-3}\rangle}$	$\frac{(V_{xx})_{val}}{q\langle r^{-3}\rangle}$	$\frac{(V_{yy})_{val}}{q\langle r^{-3}\rangle}$	η
d_{xy}	− 4/7	+ 2/7	+ 2/7	0
$d_{x^2-y^2}$	− 4/7	+ 2/7	+ 2/7	0
d_{z^2}	+ 4/7	− 2/7	− 2/7	0
p_z	+ 4/5	− 2/5	− 2/5	0
p_y	− 2/5	− 2/5	+ 4/5	+ 3
p_x	− 2/5	+ 4/5	− 2/5	− 3
d_{xz}	+ 2/7	+ 2/7	− 4/7	+ 3
d_{yz}	+ 2/7	− 4/7	+ 2/7	− 3

Note that $q = -e$ that would lead to sign changes

$$(V_{zz})_{\text{val}} = q\sum_i \langle l_i m_i|3\cos^2\theta - 1|l_i m_i\rangle \langle r_i^{-3}\rangle \tag{5.80}$$

where i is the number of possible valence electron states. The other EFG components are

$$(V_{xx})_{\text{val}} = q\sum_i \langle l_i m_i|3\sin^2\theta\cos^2\varphi - 1|l_i m_i\rangle \langle r_i^{-3}\rangle \tag{5.81}$$

$$(V_{yy})_{\text{val}} = q\sum_i \langle l_i m_i|3\sin^2\theta\sin^2\varphi - 1|l_i m_i\rangle \langle r_i^{-3}\rangle \tag{5.82}$$

As the s electrons are spherically symmetric, the contributions come only from p and d electrons which have the magnitudes as given in Table 5.1.

5.3.2 *Crystal Field Contribution*

The crystalline field lifts the degeneracy of the d orbital wave functions of the ferrous ion, ψ_n corresponding to $|x^2 - y^2\rangle$, $|3z^2 - r^2\rangle$, $|xy\rangle$, $|yz\rangle$ and $|xz\rangle$. The crystal field perturbation can be modeled as equivalent to a distorted octahedron given by (Ingalls 1964)

$$V = V_{\text{cubic}} + V_{\text{axial}} + V_{\text{rhombic}} \tag{5.83}$$

where the crystalline field axes are allowed to pass through the neighboring ions, making the V_{axial} corresponding to tetragonal symmetry.

The crystal field terms induce a splitting of the energy levels, as illustrated in Fig. 5.9. The splitting is represented further as

$$V = V_{\text{cubic}} - \frac{2}{21} B_2^o \langle r^2 \rangle \left(3L_z^2 - L(L+1)\right) - \frac{2}{21} B_2^2 \langle r^2 \rangle \left(3L_x^2 - 3L_y^2\right) \quad (5.84)$$

where B_2^o and B_2^2 are the axial and rhombic fields in Hutchings notation, respectively. The EFG components for the orbital states of the ferrous ion are shown in Table 5.2.

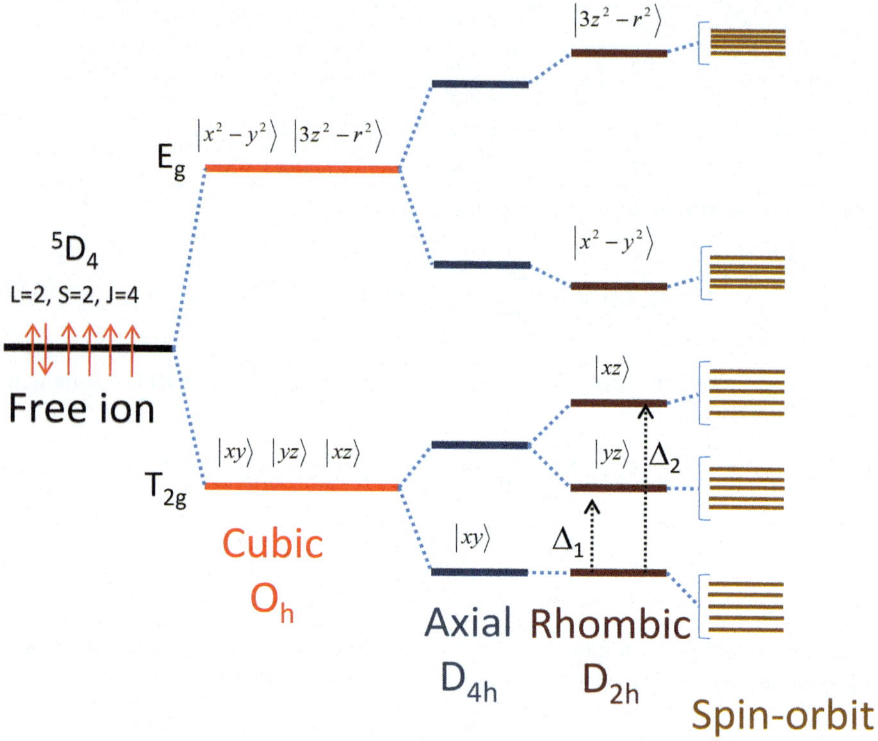

Fig. 5.9 The splitting of energy levels under different crystal field symmetries and spin–orbit interaction for ferrous ion. Adapted from (Ingalls 1964)

Table 5.2 EFG components for the orbital states of the ferrous ion

Wave function	$\langle V_{\text{axial}} \rangle_n$	$\frac{(V_{zz})_n}{q\langle r^{-3}\rangle}$	$\langle V_{\text{rhombic}} \rangle_n$	$\frac{(V_{xx}-V_{yy})_n}{q\langle r^{-3}\rangle}$
$\lvert x^2 - y^2\rangle$	$-\frac{4}{7} B_2^o \langle r^2\rangle$	$-\frac{4}{7}$	0	0
$\lvert 3z^2 - r^2\rangle$	$\frac{4}{7} B_2^o \langle r^2\rangle$	$+\frac{4}{7}$	0	0
$\lvert xy\rangle$	$-\frac{4}{7} B_2^o \langle r^2\rangle$	$-\frac{4}{7}$	0	0
$\lvert yz\rangle$	$\frac{2}{7} B_2^o \langle r^2\rangle$	$+\frac{2}{7}$	$-\frac{6}{7} B_2^2 \langle r^2\rangle$	$-\frac{6}{7}$
$\lvert xz\rangle$	$\frac{2}{7} B_2^o \langle r^2\rangle$	$+\frac{2}{7}$	$\frac{6}{7} B_2^2 \langle r^2\rangle$	$+\frac{6}{7}$

The temperature dependence of the EFG was evaluated by associating the Boltzmann distribution function, $\exp(-E_n/k_B T)$ with the EFG components. The EFG components disappear at higher temperatures and at any temperature when the $|xy\rangle$, $|yz\rangle$ and $|xz\rangle$ states are degenerate. The maximum EFG contribution appears at low temperatures with a non-degenerate single-occupied state. The quadrupole energy splitting in terms of the valence contribution can be written in terms of the absolute value of $\frac{4}{7}\langle r^{-3}\rangle$ for the free ion as

$$\Delta E = \frac{e^2 q Q}{2}(1 - R)\left(\frac{4}{7}\langle r^{-3}\rangle\right) \tag{5.85}$$

The temperature dependence of the energy splitting is characterized by a reduction factor, F that reduces with temperature and is given as

$$\Delta E = \frac{e^2 q Q}{2}(1 - R)\left(\frac{4}{7}\langle r^{-3}\rangle\right)F(T) \tag{5.86}$$

The spin–orbit coupling further causes a spread in the quadrupole splitting at low temperatures. The spin–orbit interaction that lifts the spin degeneracy of the orbital state is given by

$$V_{so} = -\lambda L \cdot S \tag{5.87}$$

where λ is the coupling constant. The five-fold degeneracy is now lifted, giving a wave function as

$$\psi_n = \sum_{i,j=1}^{5} A_{ij}^n \psi_i \chi_j \tag{5.88}$$

The separation energy

$$\Delta(|xy\rangle \sim |yz\rangle) = \Delta_1 = \frac{6}{7}\langle r^2\rangle\left(B_2^0 - B_2^2\right) \tag{5.89}$$

$$\Delta(|xy\rangle \sim |xz\rangle) = \Delta_2 = \frac{6}{7}\langle r^2\rangle\left(B_2^0 + B_2^2\right) \tag{5.90}$$

Numerical calculations have shown that the reduction factor, F, becomes less than 0.2 when $\frac{\Delta_1}{\lambda}$ or $\frac{\Delta_2}{\lambda} = 1$ and reaches more than 0.9 when the ratio is 20.

The energy splitting can be written as

$$\Delta E = \frac{e^2 q Q}{2}(1 - R)\left(\frac{4}{7}\langle r^{-3}\rangle\right)F(\Delta_1, \Delta_2) \tag{5.91}$$

Incorporating the radial expansion due to screening that gives the covalency factor, the equation changes further as

$$\Delta E = \frac{e^2 qQ}{2}(1 - R)\left(\frac{4}{7}\langle r^{-3}\rangle\right)\alpha^2 F\left(\alpha^2\lambda_o\right) \tag{5.92}$$

where α^2 has values between 0.6 and 0.9. Thus, the energy splitting is

$$\Delta E = \frac{e^2 qQ}{2}(1 - R)\left(\frac{4}{7}\langle r^{-3}\rangle\right)\alpha^2 F(\text{CF}, \text{COV}, T) \tag{5.93}$$

where $F(\text{CF}, \text{COV}, T)$ is the reduction factor, which is a function of crystal field, covalency and temperature, respectively.

5.4 Nuclear Zeeman Interaction

The interaction of the nuclear magnetic moment (μ) with the internal magnetic field (H) at the nucleus due to the electrons of the atom is called the nuclear Zeeman effect or magnetic hyperfine interaction (see Sect. 2.10 and Sect. 2.11). The Hamiltonian of this interaction can be written as

$$H_m = -\mu \cdot H = -g_N \mu_N I \cdot H \tag{5.94}$$

where g_N and μ_N are the nuclear g-factor and the nuclear magneton, respectively.

The energy levels are given by

$$E_{mI} = -g_N \mu_N H m_I \tag{5.95}$$

where $m_I = I, I - 1, \ldots, -I$. From the eigenvalue equation, one can understand that there are $(2I + 1)$ equidistant energy levels and that the splitting between adjacent level is $g_N \mu_N H$.

The transition between the excited state and ground state sublevels is governed by selection rules. In the case of the magnetic dipole interaction, (for ^{57}Fe) $\Delta I = 1$; and $\Delta m_I = 0, \pm 1$ are allowed, resulting in six transitions, as shown in Fig. 5.10, which can be observed as a six-line pattern (bottom). The relative intensities for a polycrystalline material where the Debye–Waller factor is assumed to be isotropic are given by the square of the Clebsch-Gordon coefficients and are in the ratio 3:2:1:1:2:3, as shown in the simulated spectrum of Fig. 5.11 with and without isomer shift.

The energy separation in terms of Doppler velocity (given by the Mössbauer velocity transducer) using (4.13) is

$$\Delta V_D = \frac{c\Delta E_m}{E_\gamma} \tag{5.96}$$

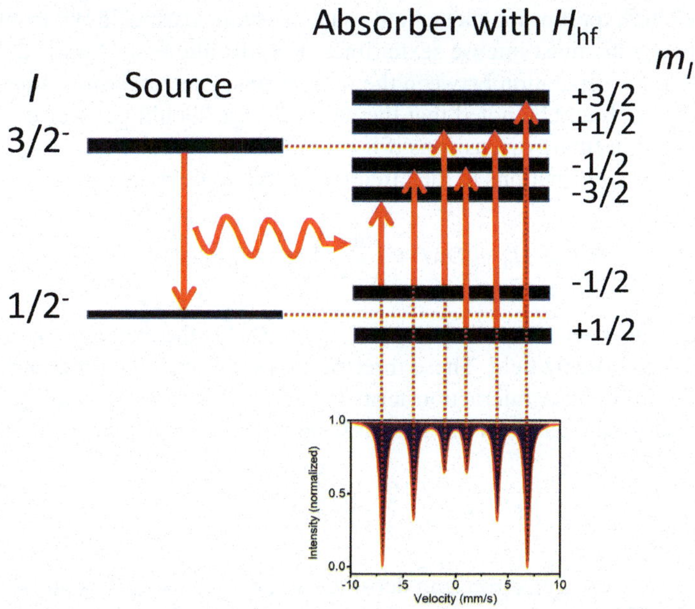

Fig. 5.10 The nuclear Zeeman interaction results in splitting the ground state ($1/2^-$) energy level into two and the excited state ($3/2^-$) into four. The allowed transitions give a magnetic sextet

Fig. 5.11 The simulated Mössbauer hyperfine spectra with and without monopolar electric interaction (isomer shift)

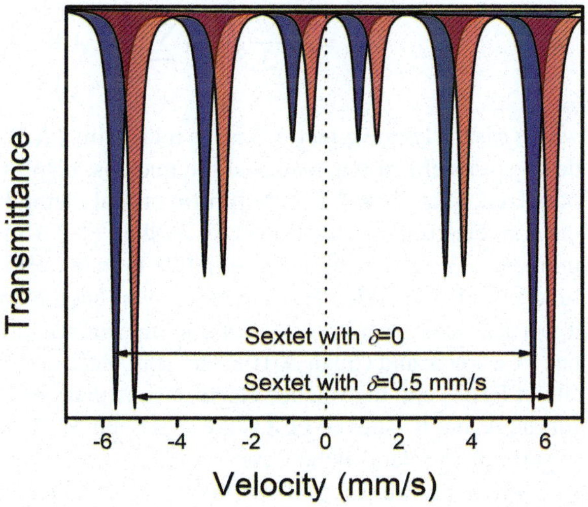

from which the hyperfine magnetic field could be obtained using (5.95) from the separation between the most intense sextet lines. For metallic Fe of hyperfine magnetic field − 33 T, the separation between the most intense, i.e., outermost lines is nearly 10.66 mm/s. It should be noted that the negative sign should be inferred for the Fe hyperfine field in future discussions if it is omitted.

The various contributions to the effective magnetic field are given by

$$H_{\mathrm{eff}} = H_a - DM + \left(\frac{4\pi}{3}\right)M + H_S + H_L + H_D \tag{5.97}$$

where H_a is the external applied magnetic field, DM is the demagnetizing field and $(4\pi/3)M$ is the Lorentz field. These three fields would be of the order of a few tesla (T). The Fermi contact interaction term H_S arises due to a direct coupling of the nucleus with an imbalance in the s electron spin density at the nucleus given by

$$H_S = -\left(\frac{16\pi}{3}\right)\mu_{\mathrm{B}} \sum \langle S_{\uparrow} - S_{\downarrow}\rangle \tag{5.98a}$$

where $S_{\uparrow}$ and $S_{\downarrow}$ are s electron spin densities at the nucleus with up and down spins, respectively. This contribution would be of the order of 10 T for the iron group elements.

The contribution to the hyperfine field due to all the core s electrons is given by

$$H_{\mathrm{c}} = -\frac{8\pi}{3} g\mu_{\mathrm{B}} S \sum_n \left[|\psi_{ns}^{\uparrow}(0)|^2 - |\psi_{ns}^{\downarrow}(0)|^2 \right] \tag{5.98b}$$

where g is the Lande g-factor, S is the total spin, $\psi_{ns}^{\uparrow}(0)$ is the up spin and $\psi_{ns}^{\downarrow}(0)$ is the down spin s electron densities at the nucleus, respectively. The first-principle spin-polarized, relativistic linear muffin-tin orbital method has shown that the majority of the core electron contribution is by 2s shell (− 53 T), while the 3s contributes next in magnitude with an opposite sign (29 T) in bcc Fe (Guo and Ebert 1996). Density functional theory calculations have shown that the core electron contribution to the hyperfine field is nearly − 29 T while the valence electron contribution is − 5 T for bcc Fe (Novak and Chlan 2010). The core electron contribution is − 50 T for the B-site of Fe_3O_4 and the valence electron contribution is 16 T. The Fermi contact fields for the A and B -sites of Fe_3O_4 are − 34 and 38 T, respectively. The magnitudes of the orbital, dipolar, valence, core and effective Fermi contact contributions for most Fe compounds are 0.5–5, 0.005, 5–20, 29–56 and 33–44 T, respectively. In the case of FeCo alloys, the change in the sign of the conduction band contribution results in a decreasing hyperfine field on the Co-rich side.

The H_L refers to the contribution arising from non-zero orbital magnetic moments of the parent atom and is given by

$$H_L = -2\mu_B\langle r^{-3}\rangle\langle L\rangle \tag{5.99}$$

This term becomes zero whenever orbital angular momentum is quenched. Since the g factor is then close to 2, the orbital term is expressed as (Marshall and Johnson 1962)

$$H_L = -2\mu_B(g-2)\langle r^{-3}\rangle\langle S\rangle \tag{5.100}$$

The dipolar interaction of the nucleus with the spin moment of the parent atom is given by

$$H_D = -2\mu_B\left\langle 3r\frac{(S.r)}{r^5} - \frac{S}{r^3}\right\rangle \tag{5.101}$$

This contribution is estimated to be the order of 1 T, vanishes in cubic symmetry and the absence of spin–orbit coupling.

The hyperfine field lines could not be resolved as the mean field decreases or when the FWHM (Γ) increases, as represented in the simulated patterns (Fig. 5.12). It could be visualized that the unresolved sextet resembles a paramagnetic doublet, and the ferromagnetic or antiferromagnetic nature is hard to be ascertained without an external magnetic field. If the hyperfine field term in (5.94) is zero due to rapid spin relaxation, a doublet ensues as in paramagnetism and superparamagnetism. The hyperfine field, if mentioned as H_{hf}, would be in kOe, while the SI unit shall be represented as B_{hf} in T.

5.5 Combined Hyperfine Parameters

In the case of the combined quadrupole and magnetic hyperfine interaction effects, there may be changes in the resonant line positions or linewidths of the spectrum. Normally, one interaction is found to be small compared to the other when they are simultaneously present. The term e^2qQ is known as the quadrupole coupling constant, with the factor 1/2 included by a few authors. Accompanying quadrupole and magnetic hyperfine interactions, if $e^2qQ << g_N\mu_N H$, then the eigenvalues for arbitrary orientations are

$$E = -g_N\mu_N H m_I + (-1)^{|m_I|+\frac{1}{2}}\frac{e^2qQ}{4}\left[\frac{3\cos^2\theta - 1 + \eta\sin^2\theta\cos 2\phi}{2}\right] \tag{5.102}$$

where θ and ϕ are the polar and azimuthal angles with the magnetic axis and principal EFG axis, respectively.

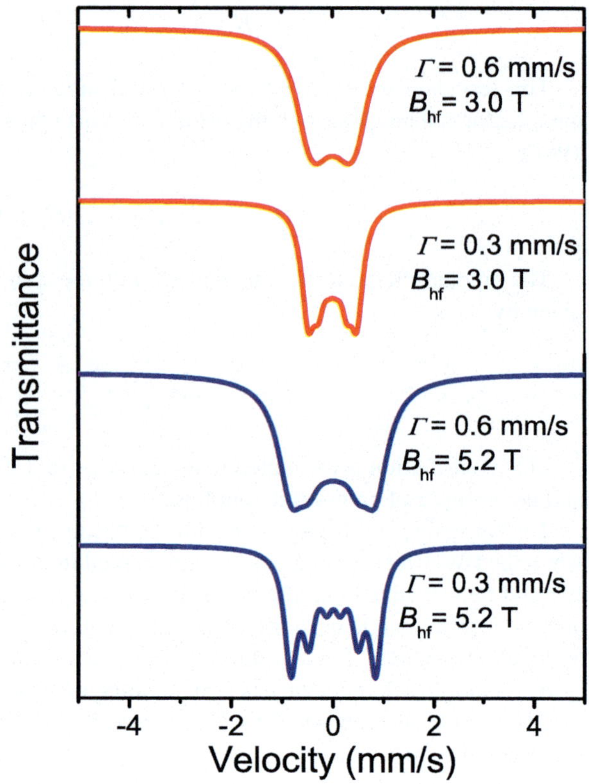

Fig. 5.12 The simulated Mössbauer spectra with various hyperfine fields (B_{hf}) and FWHM (Γ). Line broadening and reducing hyperfine fields result in the sextets collapsing to a doublet-like pattern

For axially symmetric EFG and principal axis along H is θ, then (5.102) can be written as

$$E = -g_{\mathrm{N}}\mu_{\mathrm{N}}Hm_I + (-1)^{|m_I|+\frac{1}{2}}\frac{e^2qQ}{4}\left[\frac{3\cos^2\theta - 1}{2}\right] \tag{5.103}$$

since $\eta = 0$. The $m_I = \pm 3/2$ energy levels are shifted up while $m_I = \pm 1/2$ are shifted down in the presence of quadrupole interaction. The magnetic hyperfine splitting with electric quadrupole interaction, producing shifted energy levels in the $I = 3/2$ without affecting the $I = 1/2$ level, is represented in Fig. 5.13. The quadrupole interaction term vanishes when $\theta = 54.7°$ as

$$E = -g_{\mathrm{N}}\mu_{\mathrm{N}}Hm_I + (-1)^{|m_I|+\frac{1}{2}}\frac{e^2qQ}{4}\left[\frac{3\frac{1}{3} - 1}{2}\right] = -g_{\mathrm{N}}\mu_{\mathrm{N}}Hm_I \tag{5.104}$$

The quadrupole interaction term is positive for $\theta = 0$ and $V_{zz} > 0$. When the EFG tensor is not axially symmetric, but the magnetic axis lies along one of its principal axes, then the eigenvalues of the four excited state splittings for $I = 3/2$ are given as (Greenwood and Gibb 1971)

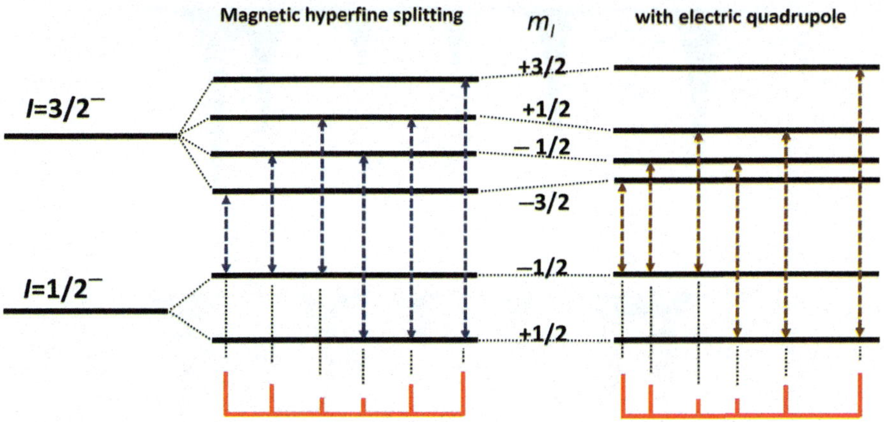

Fig. 5.13 The magnetic hyperfine splitting without (left) and with electric quadrupole perturbation (right)

$$(1/2)g_N\mu_N H \pm (e^2qQ/4)\left[(1 + 4g_N\mu_N H/e^2qQ)^2 + \eta^2/3\right]^{1/2} \tag{5.105}$$

and

$$(-1/2)g_N\mu_N H \pm (e^2qQ/4)\left[(1 - 4g_N\mu_N H/e^2qQ)^2 + \eta^2/3\right]^{1/2} \tag{5.106}$$

The solutions are obtained not analytically but by a full mathematical analysis using a computer. The main use of the effect of combined interactions of this type is in the determination of the sign of the EFG in polycrystalline absorbers exhibiting purely quadrupole doublets by applying an external magnetic field.

The asymmetric splitting of the energy levels presents unequal separation between the lines in the sextet (Fig. 5.14) as the V_{zz} is positive or negative. Considering the peak positions from the outermost lines from the negative velocity side as P1 to P6, the hyperfine magnetic field could be obtained from the difference between the outermost lines, P1 − P6. The quadrupole shift is obtained as

$$2\varepsilon = \frac{1}{2}\{(\mathrm{P6} - \mathrm{P5}) - (\mathrm{P2} - \mathrm{P1})\} = \frac{eQV_{zz}}{4}\{3\cos^2\theta - 1 + \eta\sin^2\theta\cos 2\phi\} \tag{5.107}$$

where θ and ϕ are the polar and azimuthal angles between the hyperfine field and the principal axis of the electric field gradient.

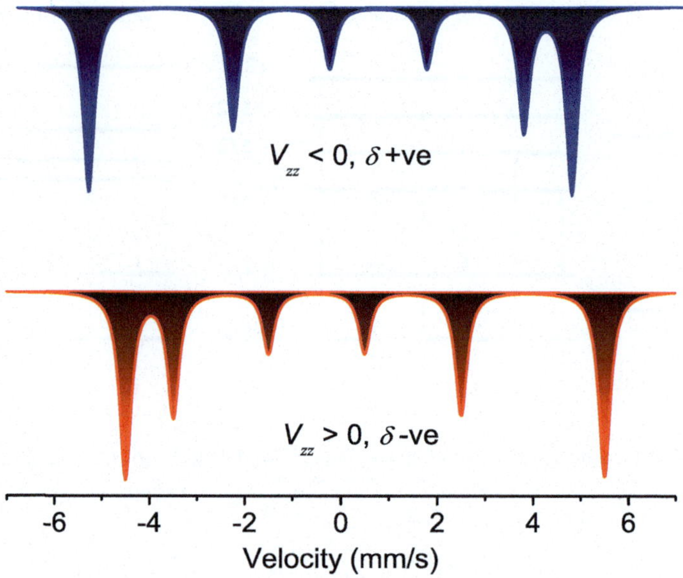

Fig. 5.14 The simulated spectra of hyperfine splitting with isomer shift and quadrupole interaction

5.6 Hyperfine Parameters in an External Magnetic Field

In the diamagnetic spectrum ($S = 0$) with a singlet, the internal magnetic field is zero and there will not be quadrupole splitting due to the charge-symmetric environment. Quadrupole doublet shall arise if there is charge asymmetry due to covalency. A paramagnetic doublet can become a sextet below the Curie/Néel temperature. The relative intensities of the sextet is given by the ratio

$$3 : \frac{4\sin^2\theta}{1+\cos^2\theta} : 1 : 1 : \frac{4\sin^2\theta}{1+\cos^2\theta} : 3 \tag{5.108}$$

For a polycrystalline sample, this ratio becomes 3:2:1:1:2:3 as $\overline{sin^2\theta} = 2/3$.

The application of an external magnetic field will not induce any net magnetic field in pure diamagnets, and therefore, the effective field (H_{eff}) will be equal to the external magnetic field (H_{app}). Applying a strong magnetic field on a magnetic single crystal shall result in the hyperfine field splitting according to the direction of the applied field and the γ-ray. In paramagnetic ($S \neq 0$) non-Kramer doublets, an internal magnetic field (H_{int}) can be induced with an external magnetic field. The magnitude of the internal field can be determined from the effective magnetic field as $H_{\text{int}} = H_{\text{eff}} \pm H_{\text{app}}$.

In antiferromagnetic polycrystals showing a sextet, a spread in angles would broaden the lines. It is to be noted here that the antiferromagnets show equal magnitude sextets that are merged but in opposite directions. This is similar to ferromagnetic

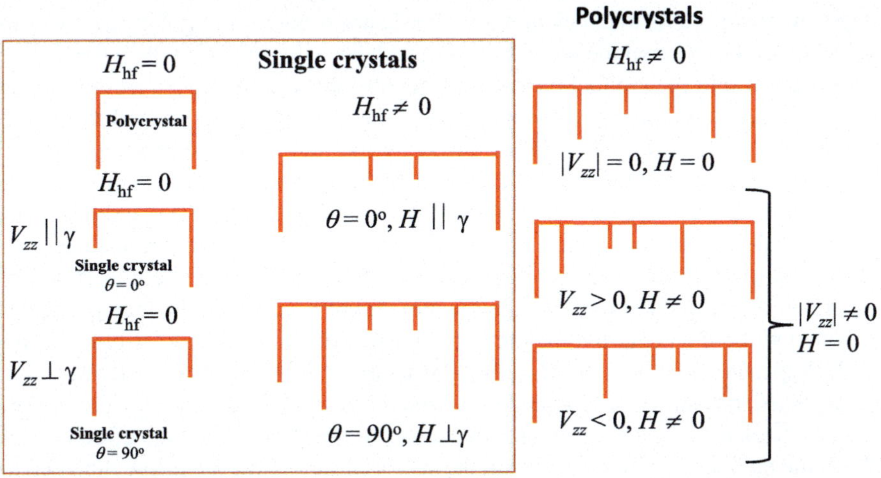

Fig. 5.15 Comparison of the intensity of the doublet and sextet lines under different orientation directions in single crystals and polycrystals. *H*-applied field, θ- the angle between principle quantization *z*-axis and γ-ray direction

sextets with equal magnitude sextets but in the same direction. On the other hand, applying an external magnetic field to the ferromagnet would result in narrower lines. If it is a ferrimagnetic sextet (composed of two sextets of unequal magnitude), the lines may be resolved as the applied field will add with one sextet component and subtract with the other. The effects of the applied magnetic field *(H)* on the doublet and sextet line patterns in single crystals and polycrystals are illustrated in Fig. 5.15. When an external magnetic field is applied parallel to the γ-ray direction in a polycrystalline ferromagnet, the relative intensities are 3:0:1:1:0:3 (Chappert et al. 1979). If the antiferromagnetic hyperfine fields are perpendicular to the γ-rays, the relative intensities shall be 3:4:1:1:4:3.

Mössbauer measurements with polarized γ-radiation can lead to a significant simplification of the spectra, and help to interpret them, particularly when their hyperfine structures are complex. In addition, they allow the quadrupolar or magnetic components from different polarizations to be studied separately. In particular, $\Delta m = \pm 1$ and $\Delta m = 0$ transitions can be extracted using a linearly polarized beam. Moreover, the polarized beam can be used to determine texture parameters related to the spatial arrangements of the spins in the plane perpendicular to the radiation beam (Szymanski 2006, 2018). The high-field Mössbauer spectrometry was developed for a long time, a few years after the Mössbauer effect discovery (Chappert 1974; Chappert et al. 1979; Greneche 1995, 2013; Reissner 2021).

In addition to conventional Mössbauer spectrometry, external magnetic fields produced by either a magnet or electromagnet can be applied to the sample to obtain complementary information on soft magnetic materials, but their effects are not sufficient on hard magnetic samples and magnetic nanostructures. Superconducting cryomagnetic devices must be used to carry out useful Mössbauer studies. At this stage, the first point is to check that the Mössbauer source is not polarized under the influence of the external magnetic field, i.e., to have a distance for which the leakage fields cannot split the monochromatic character of the source into six emission lines. In the case of cryomagnets, a superconducting coil is placed around the source to compensate for the leak field produced by the main superconducting coil (see Fig. 5.16). The second basic point concerns the orientation of the magnetic applied field with respect to the direction of propagation of the gamma ray, either a priori parallel or perpendicular, giving rise to different intensities for the intermediate lines of the magnetic sextet. The first objective is to find out whether the hyperfine field is parallel or antiparallel to the magnetic moment of Fe. Indeed, the magnetic sextet obtained without an external magnetic field does not allow us to determine the sign of the hyperfine field, which is highly dependent on the Fermi contact term.

The hyperfine structure, which depends on the nature of the magnetic interactions, can be correlated to the Fe moments that allow us to estimate the total effective field B_{eff} at the nucleus, which results from the vectorial sum of the hyperfine field and the applied magnetic field. Figure 5.17 shows the ideal shape of the Mössbauer spectra under a magnetic field when the magnetic field is oriented parallel to the gamma radiation for: (a) isotropic magnetic powder out of the magnetic field, (b) ferromagnetic structure for which the hyperfine field is opposite to the magnetic moment, (c) ferromagnetic structure for which the hyperfine field is parallel to the magnetic moment,

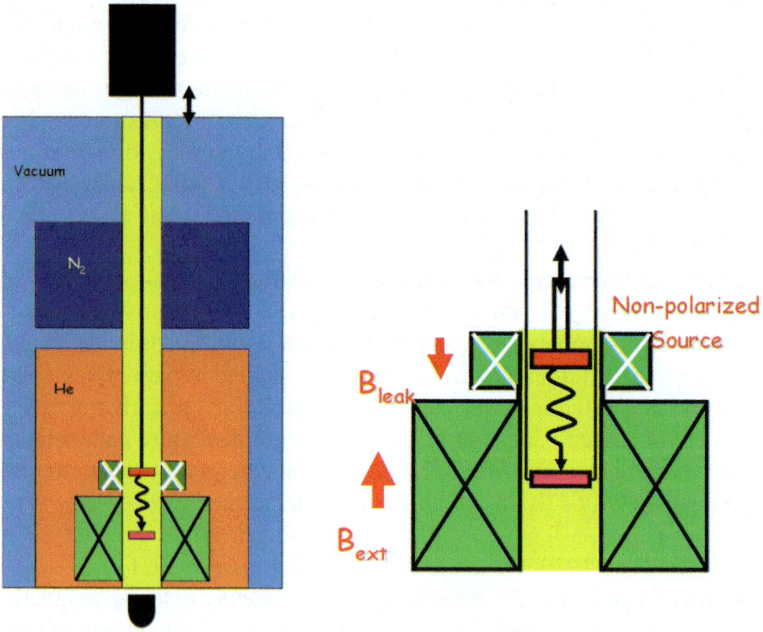

Fig. 5.16 Schematic description of a superconducting cryomagnetic device (left) and local view of the main and compensating coils to obtain an unpolarized source (right)

(d) antiferromagnetic, (e) ferrimagnetic, (f) speromagnetic, (g) sperimagnetic and (h) asperimagnetic structures. Let us note that the last three structures are concerned with non-collinear magnetic structures in the blocked state, i.e., below the freezing temperature, resulting from the presence of frustrated magnetic interactions and/or the presence of strong local anisotropies preventing collinear magnetic structures.

As shown in Fig. 5.17, the intensities of the intermediate lines of a Mössbauer magnetic sextet are correlated to the Fe moment configuration with respect to the γ-beam direction. Indeed, the relative area ratios are given by 3:p:1:1:p:3, where $p = 4\sin^2\theta/(2 - \sin^2\theta)$ where θ is the angle defined between the hyperfine field held by the nucleus probe and the propagation direction of the γ-beam as given in (5.108). In the case of a typical magnetic powdered sample, i.e., random distribution of the orientations of the Fe magnetic moments, one observes the spectrum with relative ratio 3:2:1:1:2:3 (see Fig. 5.17a). Some disagreement with this relative ratio must be attributed to a lack of random homogeneity of the powder, i.e., the presence of some preferential orientation or magnetic texture.

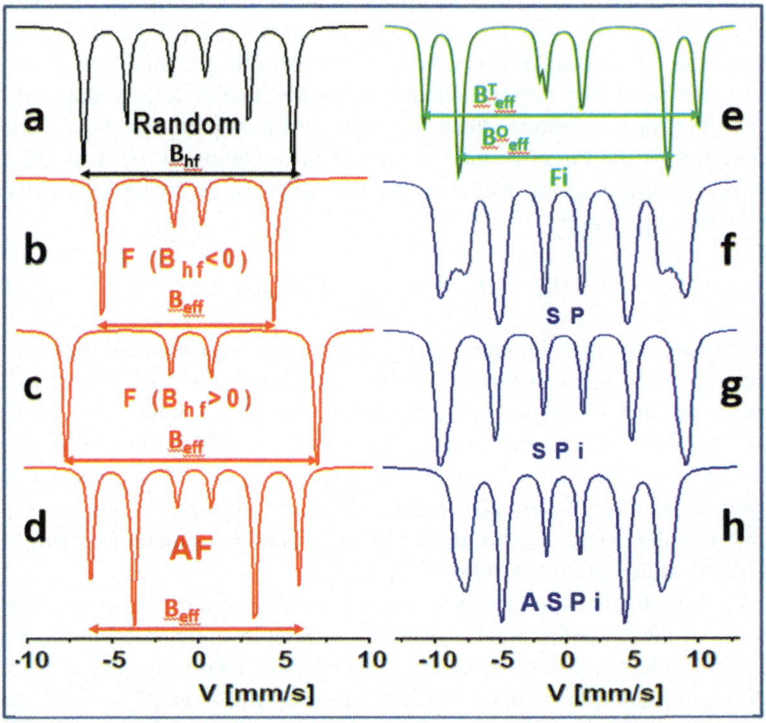

Fig. 5.17 Theoretical Mössbauer spectra on linear and non-collinear magnetic structures in the presence of external field applied parallel to the γ-beam: **a** random powder in zero-field, ferromagnetic with hyperfine field antiparallel and parallel to the magnetic moment **b** and **c**, respectively, **d** antiferromagnetic, **e** ferrimagnetic structures and non-collinear magnetic structures (speromagnetic, sperimagnetic and asperimagnetic: **f**, **g**, **h**, respectively)

When the magnetic moments are oriented parallel to the γ-beam, one expects 3:0:1:1:0:3, i.e., the intermediate lines disappear. Such a situation occurs in a perfect ferromagnetic or ferrimagnetic material subjected to an applied field large enough to align the Fe moments. We must obviously consider two magnetic components in the case of ferrimagnetic structures, as observed in the case of spinel ferrites, for which the values of isomer shift allow us to distinguish unambiguously the Fe located in tetrahedral and octahedral sites. When the effective field is smaller or larger than the hyperfine field, it can be concluded that the hyperfine field is opposite or parallel to the magnetic moment, respectively, because the magnetic moment is forced to be aligned parallel to the external magnetic field. Therefore, the main contribution to the hyperfine field corresponding to the Fermi or contact term is negative or positive (see Fig. 5.17b, c). In the case of an antiferromagnet, the magnetic moments are rather oriented perpendicular to the external magnetic field because of the dominant antiferromagnetic interaction: consequently, the shape of the Zeeman sextet is 3:4:1:1:4:3 (maximal intensities of intermediate lines) when the magnetic field is applied parallel to the γ-beam (see Fig. 5.17d). When the external magnetic field is applied perpendicular to the γ-beam, the situation is reversed, i.e., the respective relative ratios must be inverted (3:4:1:1:4:3 becomes 3:0:1:1:0:3 and vice versa).

In practice, these ideal configurations are not strictly satisfied: the intensities of intermediate lines differ from 0 and 4 due to the presence of defects, the contribution of anisotropies or the surface effects in the case of nanoparticles. Nevertheless, the refined values of the effective field(s) (B_{eff}) and the angle(s) (θ) from the in-field Mössbauer spectrum, allow us to estimate the values of the hyperfine field(s) using the following expression

$$B_{\text{hf}}^2 = B_{\text{eff}}^2 + B_{\text{app}}^2 - 2B_{\text{eff}}B_{\text{app}}\cos\theta \tag{5.109}$$

For low-temperature measurements using a transverse magnetic field, electromagnets with tapered pole caps can produce fields of nearly 2.5 T. The typical axial hole diameter is around 0.6 cm for low-temperature transverse field experiments (Petitt and Walter 1973). The source and the detector sit inside the electromagnet bore, while the pole gap is nearly 5 cm. Much larger fields over 8 T could be obtained using superconducting solenoids. Recent cryostats have optical access for parallel and perpendicular magnetic fields up to 10 T using split pair superconducting magnets and at temperatures of 20 mK–800 K (Fig. 5.18).

Finally, Mössbauer experiments performed under an external magnetic field allow us to know if the hyperfine field is oriented parallel or antiparallel to the magnetic moment, to understand the magnetic configuration of Fe moments in the case of crystalline and amorphous materials and to verify the dynamics of the magnetic structure in the case of nanostructured materials. It is obvious that these results must be compared with those obtained by field-cooled and zero-field-cooled magnetic measurements (FC-ZFC), X-ray magnetic circular dichroism (XMCD) and nuclear resonant scattering of synchrotron radiation.

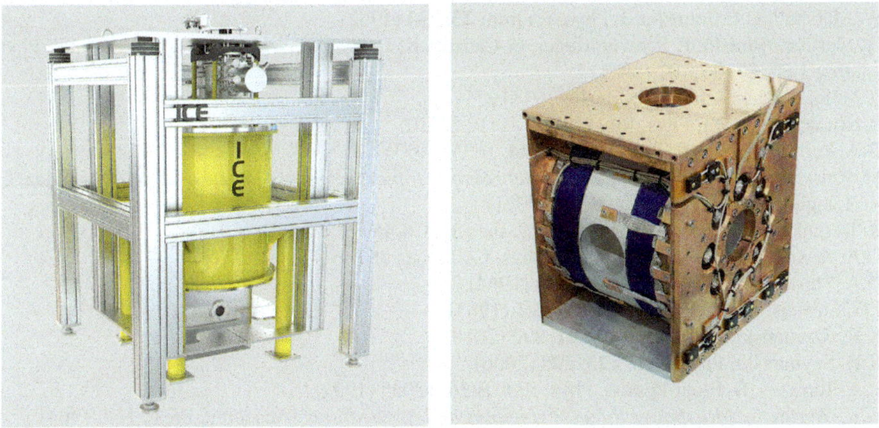

Fig. 5.18 The low-temperature cryogen free cryostat with the antivibration mount (left) and the split pair Nb-Ti superconducting magnet (right). (Courtesy: ICEoxford, UK)

5.7 Summary

This chapter has introduced the fundamentals of Mössbauer hyperfine parameters such as isomer shift, quadrupole splitting and nuclear Zeeman effect. The intrinsic and extrinsic factors that affect the hyperfine parameters are discussed, useful to interpret the Mössbauer spectra.

References

H. Akai, S. Blugel, R. Zeller, P.H. Dederichs, Phys. Rev. Lett. **56**, 2407 (1986)
J. Chappert, J. Phys. Colloques **35**, C6-71 (1974)
J. Chappert, J. Teillet, F. Varret, J. Magn. Magn. Mater. **11**, 200 (1979)
J. Danon, J. Chem. Phys. **39**, 236 (1963)
I. Dezsi, U. Gonser, G. Langouche, Phys. Rev. Lett. **62**, 1659 (1989)
P. Dufek, P. Blaha, K. Schawarz, Phys. Rev. Lett. **75**, 3545 (1995)
K.J. Duff, Phys. Rev. B **9**, 66 (1974)
K.J. Duff, T.P. Das, Phys. Rev. B **3**, 2294 (1971)
E. Fermi, E. Segré, Z. Physik **82**, 729 (1933)
M. Filatov, Coord. Chem. Rev. **253**, 594 (2009)
N.N. Greenwood, T.C. Gibb, *Mössbauer Spectroscopy* (Chapman and Hall Ltd., London, 1971)
J.M. Greneche, Acta Phys. Slovaca **45**, 45 (1995)
J.M. Greneche, in *Mössbauer Spectroscopy*, ed. by Y. Yoshida, G. Langouche, (Springer, Berlin, Heidelberg, 2013)
G.Y. Guo, H. Ebert, Phys. Rev. B **53**, 2492 (1996)
R. Ingalls, Phys. Rev. **133**, A787 (1964)
R. Ingalls, Phys. Rev. **155**, 157 (1967)
O.C. Kistner, A.W. Sunyar, Phys. Rev. Lett. **4**, 412 (1960)
R. Kurian, M. Filatov, Phys. Chem. Chem. Phys. **12**, 2758 (2010)

W. Marshall, C.E. Johnson, J. Phys. Radium **23**, 733 (1962)
G. Martinez-Pinedo, P. Schwerdtfeger, E. Caurier, K. Langanke, W. Nazarewicz, T. Söhnel, Phys. Rev. Lett. **87**, 062701 (2001)
F. Menil, J. Phys. Chem. Solids **46**, 763 (1985)
P. Novak, V, Chalan, Phys. Rev. B **81**, 174412 (2010)
G.A. Petitt, R.J. Walter, Rev. Sci. Inst. **44**, 1635 (1973)
M. Reissner, in *Modern Mössbauer Spectroscopy, Topics in Applied Physics*, ed. by Y. Yoshida, G. Langouche, (Springer, Singapore, 2021)
R. Reschke, A. Trautwein, J. Phys. Colloques **37**, C6-459 (1976)
G.A. Sawatzky, F. van der Woude, J. Phys. Colloques **35**, C6-47 (1974)
D.A. Shirley, Rev. Mod. Phys. **36**, 340 (1964)
J.G. Stevens, Hyperfine Interact. **13**, 221 (1983)
K.R. Szymanski, Eur. Phys. J. B. **91**, 292 (2018)
K.R. Szymanski, Phys. Rep. **423**, 295 (2006)
L.A. Terrazos, S. Frota-Pessôa, Phys. Rev. B **56**, 13035 (1997)
G.K. Wertheim, *Mössbauer Effect: Principles and Applications* (Academic Press, NY, 1964)
U.D. Wdowik, K. Ruebenbauer, Phys. Rev. B **76**, 155118 (2007)

Chapter 6
Mössbauer Spectrometry of Fe and Its Alloys

6.1 Metallic Body-Centered Cubic Fe

Body-centered cubic (bcc, α) iron is a well-known ferromagnet exhibiting a large saturation magnetization next to FeCo alloys. The Mössbauer effect in metallic α-Fe was studied in the early 1960s (Hanna et al. 1960) with the source and absorber magnetized perpendicular and parallel to each other and also at temperatures from 4 to 1300 K (Preston et al. 1962). The temperature-dependent Mössbauer experiments used ^{57}Fe enriched up to 53.6% as the absorber and a 1 mCi ^{57}Co source embedded in Cu, maintained at liquid nitrogen temperature (77 K). The room temperature spectrum of α-Fe is a sextet with a magnetic hyperfine field of 33.0 T and transforms into a paramagnetic single line at 1042 K corresponding to the Curie temperature (T_c). The variation in the hyperfine field with temperature for the pure bcc iron is shown in Fig. 6.1. The internal magnetic hyperfine field has a value of 33.8 T at ~ 4 K, and the $B_{hf}(T)/B_o(T)$ is proportional to T/T_c obeying the law of corresponding states (see Sect. 3.7.1). The α → γ (bcc to fcc) phase transformation manifests at a temperature higher than T_c, indicated by a discontinuity in the temperature-dependent isomer shift.

The pressure-dependent phase diagram of the Fe system is shown in Fig. 6.2. The bcc (α) phase is ferromagnetic and transforms into the fcc-Fe (γ) phase at high temperature and hcp-Fe (ε) phase under very high pressure. Experimental studies require pressure beyond 13 GPa (1 GPa = 10 kbar) to observe the high-pressure ε phase. The pressure-dependent phase transformation has been verified by Mössbauer spectrometry, where the internal hyperfine field of α-Fe vanishes with pressures above 13 GPa (Pipkorn et al. 1964), suggesting the transformation to the non-magnetic hcp phase. A linear decrease in the hyperfine field with pressure and a correlative volume change in α-Fe was observed with increasing pressure at room temperature. The isomer shift variation with pressure is related to the ratio of the volume change (ΔV) to the original volume (V) given by

R. J. Joseyphus and J.-M. Greneche, *Fundamentals of ^{57}Fe Mössbauer Spectrometry*,
https://doi.org/10.1007/978-981-99-8653-8_6

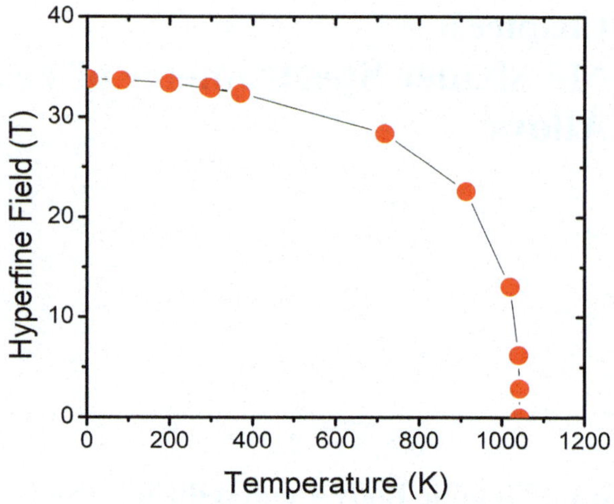

Fig. 6.1 The hyperfine field with temperature for α-Fe. Data from (Preston et al. 1962)

$$\Delta\delta = -0.14\frac{\Delta V}{V}\ \text{cm/s} \tag{6.1}$$

A linear decrease of the isomer shift with pressure is expected due to the volume change according to (6.1) up to 13 GPa, where the phase remains α-Fe. Deviation from the linearity of the isomer shift for α-Fe with pressure near 6 GPa (Moyzis and Drickamer 1968) may arise from differing calibrations (Williamson et al. 1972). A sudden drop in isomer shift was observed near 15 GPa using synchrotron Mössbauer

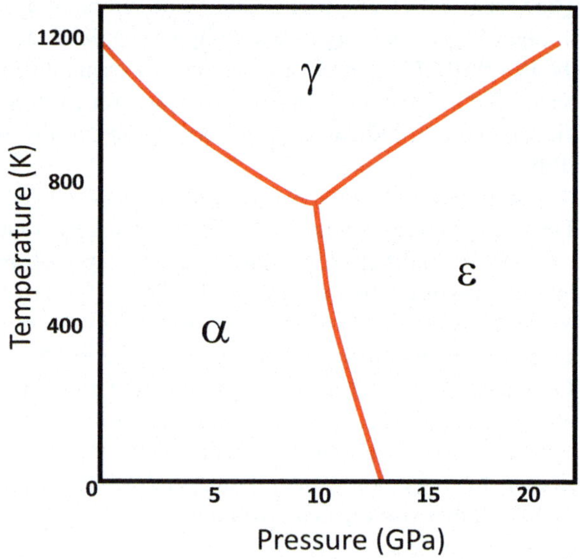

Fig. 6.2 The phase diagram of pure Fe shows the hcp phase under high pressure

spectrometry corresponding to bcc to hcp transformation, supported by theoretical studies (Desmarais 2021).

The 4s and 3s electron wave function contributions that influence the isomer shift are given as

$$|\psi_{4s}(0)|^2 = 0.53|\psi_\Gamma(0)|^2 \tag{6.2}$$

$$|\psi_{3s}(0)|^2 = \beta\langle nR_m^2\rangle \tag{6.3}$$

$|\psi_\Gamma(0)|^2$ is the 4s contribution at Γ ($k = 0$) and the factor 0.53 in (6.2) arises due to the number of s electron per Fe while $\beta = 5.5\ a_o^{-2}$, where a_o is the Bohr radius and R_m is the maximum of radial wave function. The 4s wave function is affected to a large extent by pressure and the 3s next, influenced by 3d.

6.2 Hexagonal Fe

The transformation of α-Fe to ε-Fe is characterized by the appearance of a central line in the hyperfine field sextet as the pressure increases beyond 14 GPa (Nicol and Jura 1963). The paramagnetic nature of ε-Fe was confirmed from experiments at 20 K and high pressure near 15 GPa, which showed a broadened doublet with a quadrupole splitting of 0.17 mm/s (Williamson et al. 1972). Hyperfine splitting could not be observed in Mössbauer studies undertaken under high pressure up to 21.5 GPa and at temperatures as low as 30 mK (Cort et al. 1982). Experiments at 20 GPa and 4.5 K using external magnetic fields up to 7 T excluded any induced hyperfine magnetic field (Nasu et al. 2002). Figure 6.3 shows the Mössbauer spectra under different longitudinal magnetic fields applied to ε-Fe under a pressure of 20 GPa at 4.5 K. A non-resolved quadrupole doublet (0.14 mm/s) was observed in the absence of an external magnetic field at 4.2 K corresponding to hexagonal ε-Fe whose value was close to that at 300 K. In external magnetic fields up to 7 T, the magnitudes of the hyperfine fields were nearly equal to the external fields, and the experimental conditions suggested a local magnetic moment of less than 0.05 μ_B.

Density function theory predicts an antiferromagnetic order for the ε-Fe phase, while Mössbauer spectrometry results show that it is paramagnetic (Neumann et al. 2004). However, antiferromagnetic fluctuations cannot be excluded as they cannot be detected in the time scale of Mössbauer experiments (Monza et al. 2011). Synchrotron radiation perturbed angular correlation spectroscopy experiments at 30 GPa suggested an electric field gradient *eq* of 1.2×10^{16} V/cm^2 (equivalent to 0.07 mm/s) for the quadrupole interaction and a magnetic hyperfine field of 0.4 T (Bessas et al. 2020). This is why these hyperfine parameters cannot be estimated from a conventional Mössbauer spectrum because of the lack of resolution. The isomer shift variation with pressure reported in the literature decreases non-linearly (Fig. 6.4), exhibiting a significant deviation of − 0.23 mm/s on transforming to the

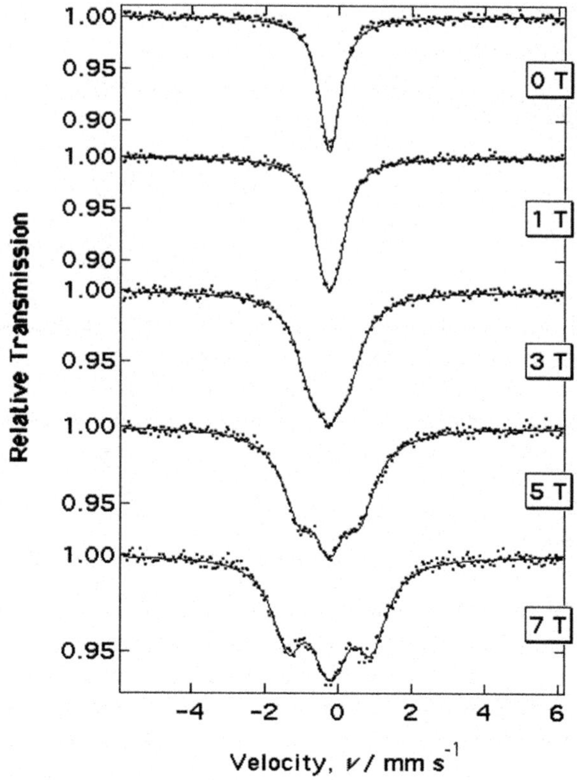

Fig. 6.3 Mössbauer spectra of ε-Fe with a pressure of 20 GPa at 4.5 K under different longitudinal applied magnetic fields. Reprinted with permission (Nasu et al. 2002), ©IOP Publishing

hcp phase reaching – 0.8 mm/s at 240 GPa with respect to bcc Fe at room temperature, as studied by synchrotron Mössbauer spectrometry (Gavriliuk 2023).

6.3 Face-Centered Cubic Fe

The α-Fe transforms to γ-Fe at high temperatures above ~ 1183 K and remains paramagnetic at this temperature. The fcc phase is antiferromagnetic, but the Néel temperature (T_N) could not be determined for the bulk as the phase transformation temperature is far above the T_N. On the other hand, theoretical and various experimental studies on γ-Fe precipitates and thin films indicate a wide range of T_N from 38 to 200 K (Gonser et al. 1963; Li et al. 1994; Zhou et al. 1999).

Thin films of ~ 18 Å Fe on Cu(001) substrate prepared by vapor deposition showed an antiferromagnetic nature (Keune et al. 1977). The sample was subjected to a longitudinal external magnetic field of 4.9 T at 13 K. The effective hyperfine field determined from the spectral pattern in an applied magnetic field is 5.2 T. The antiferromagnetic saturation hyperfine field is around 2.4 T, and the sextet could not

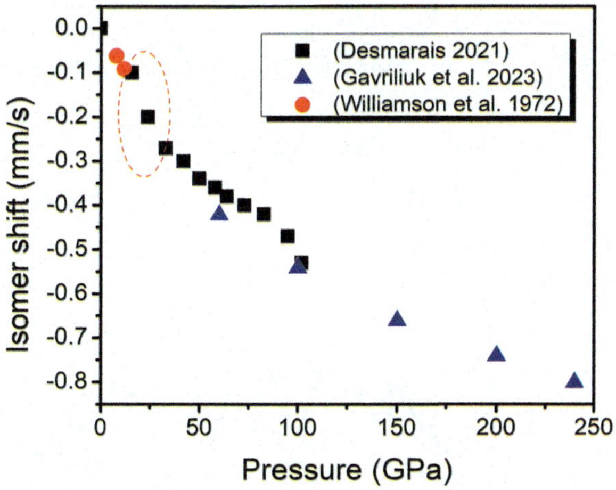

Fig. 6.4 The isomer shift variation with pressure for Fe at room temperature. The steep drop in the isomer shift near 15 GPa corresponds to the α → ε transformation

be resolved even with the natural linewidth. Ferromagnetic order was excluded since the hyperfine field was unresolved at low temperatures and an expected effective field of 3 or 7 T with the external field was absent. The effective field obtained from the hyperfine field distribution showed almost similar values when fitted with fixed intensity ratios as well as for the simulated spectrum considering random spins. At room temperature, the paramagnetic peak gives a linewidth of 0.31 mm/s that broadens on approaching the T_N of 80 K due to relaxation effects and distribution of T_N in the sample. The paramagnetic fcc Fe exhibits an isomer shift of − 0.09 mm/s at 300 K and a linewidth of 0.3 mm/s.

The antiferromagnetic transition could be observed only from line broadening, even at 4.2 K (Halbauer and Gonser 1983). Theoretical studies using local density discrete variational method have shown that the antiferromagnetic Fe hyperfine field contribution from conduction electrons is + 4.4 T while the core electron contribution is − 15.7 T, resulting in a net contribution of − 11.3 T (Guenzburger and Ellis 1995). However, these values are affected by the presence of other neighbors, and the net hyperfine fields are shown to be − 13.6 and − 4.4 T with Al and Cu, respectively. The net hyperfine field for the ferromagnetic Fe with Al and Cu neighbors, on the other hand, is nearly − 30 T.

6.4 Face-Centered Tetragonal Fe

Mössbauer spectrometry was effectively used to identify the magnetic properties of the non-equilibrium phases of Fe. Molecular beam epitaxy was used to obtain the face-centered tetragonal (fct) phase of Fe by depositing 7 Fe monolayers on the substrate Cu(001) at ~ 90 K (Ellerbrock et al. 1995). The conversion electron Mössbauer spectra (CEMS) at 160 and 296 K for the fct sample are shown in Fig. 6.5,

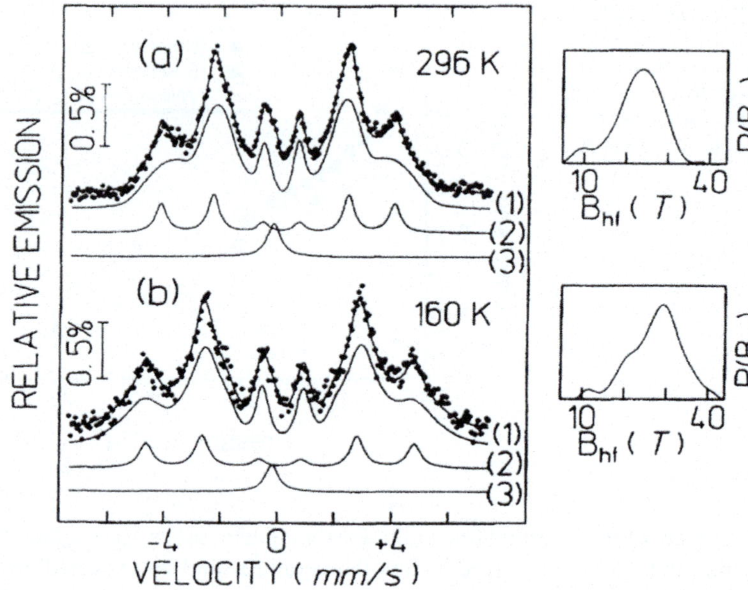

Fig. 6.5 CEMS of 7 Fe monolayers deposited on Cu(001) substrate at ~ 90 K and the spectra recorded at **a** 296 and **b** 160 K. The figures in the right panel are the hyperfine field distribution plots. Reprinted with permission (Ellerbrock et al. 1995), ©American Physical Society

along with the hyperfine field distribution plots. The fct-Fe phase of the sample is characterized by a peak hyperfine field of 31 T at 160 K, which decreases further to 29 T at 296 K. The distorted fcc structure shows features such as a broad hyperfine field distribution, non-zero quadrupole and isomer shifts. The $eQV_{zz}/4$ values of + 0.18 and 0.07 mm/s are obtained at 160 and 296 K, respectively. The V_{zz} value of $+ 1.2 \times 10^{17}$ V/cm^2 is obtained from the low-temperature CEMS, and the lattice parameter ratio, *c/a* is 1.035. The studies suggest a ferromagnetic nature for fct Fe in contrast to the antiferromagnetic fcc Fe. The 3 ML Fe thin film in Pd(110) substrate resulted in a peak hyperfine field of 39.2 T at 25 K due to electronic 3d-4d hybridization (Cuenya et al. 2005). *Ab initio* Monte Carlo simulations suggest that the ferromagnetic to the antiferromagnetic state occurs as the *c/a* becomes larger than 1.4 (Wang et al. 2006).

6.5 Impurity Studies

6.5.1 Fe Impurity in Metal Hosts

The dilute concentration of Fe impurities in metal hosts renders the ^{57}Fe fraction insufficient for resonant absorption, and therefore, ^{57}Co is diffused into the metal host for most impurity studies. The Mössbauer effect has been studied for the dilute

Fe impurities (< 0.1 at. %) in Cu, Ag, Au, V, Nb, Ta, Mo, W, Rh and Pt with applied fields and temperatures up to 6.2 T and 0.5 K, respectively (Kitchens et al. 1965). The internal fields of V, Nb and Ta are close to zero, independent of temperature, indicating the absence of a localized magnetic moment. Using the Brillouin function, the internal field (B_{hf}) and saturation hyperfine field, $B_{hf}(0)$, are related as

$$B_{hf} = B_{hf}(0) B_J(x) \tag{6.4}$$

$B_{hf}(0)$ has been obtained as – 8, – 31.5, – 29.5, – 19.5, – 11.5, – 10.3, – 7.6 and – 4.0 T for Cu, Pt, Pd, Au, Mo, Rh, W and Ag, respectively, all showing negative values.

Fe impurities in the Cu system have been extensively studied because of their Kondo effect. Mössbauer spectrometry of Fe impurities in various concentrations from 600 ppm to less than 3 ppm in Cu have been studied at low temperatures (0.05 K) and with an applied field up to 7 T (Tholence et al. 1970). It was concluded that an Fe atom could be considered isolated if it has no other Fe neighbors in a sphere of about 1 nm. The isolated impurities show antiferromagnetic Curie–Weiss behavior with $T_N \approx 29$ K, while the others are almost magnetic. It has also been shown that isolated Fe atoms exhibit a hyperfine field of magnitude 8 T while clustering occurs with higher concentrations, studied with up to 1.1 at.% Fe in Cu (Window 1970). In another Fe:Cu system where Fe is almost 10 ppm, $B_{hf}(0)$ was observed as – 11.1 T and an additional antiferromagnetic polarization was suggested at low temperatures (Steiner et al. 1973). Alternatively, a temperature-independent orbital contribution to Knight shift may also exist. Increasing Fe to 0.6 and 1.6 at.% in Cu resulted in internal hyperfine fields of – 14 and – 29 T, respectively, at 5 K (Gonser et al. 1965). The Cu at such concentration levels can exist as a solid solution, γ-Fe or α-Fe precipitates, depending on the synthesis conditions.

Experiments with Fe impurities (< 1 at.%) in various host metals have also concluded that the local magnetic moment emerges at an electron concentration of 5.5 and disappears at 7. It reappears at 8.25, reaching a maximum value for the 4d metal corresponding to Pd (11.3 μ_B at 100 K) (Clogston et al. 1962). Fe impurity studies in Tc, Ru and Ir with outer electrons 7, 8 and 9 showed saturation hyperfine fields of – 0.73, – 0.1 and – 0.93 T, respectively, (Takabatake et al. 1980) with an applied field of 5 T. The lower saturation hyperfine field in the Ru host is thought to be due to the spin fluctuation temperature exceeding 700 K. The study suggests no difference between systems with localized spin fluctuations and Kondo systems at low temperatures. The ground state of the Fe impurity in Pd at 0 K has been reported to be non-magnetic, and the giant magnetic moment is due to spin fluctuations beyond 250 mK (Swieca et al. 1997). However, it has been pointed out that the magnetic moment of Fe in d-metal hosts is determined by the host d-band structure rather than by spin fluctuations or Kondo effects (Kapoor et al. 1996). Experiments with Zn host have shown a $B_{hf}(0)$ of 4.5 T with an external field of 7 T and Kondo behavior below 1.5 at.% Fe (Munayco et al. 2006).

6.5.2 Metal Impurities in Fe Host

Figure 6.6 shows the hyperfine fields (sign omitted) at 4.2 K of various impurity elements in the Fe host (Rao 1985), where the impurity concentration is less than 4 at.%. Isotopes other than ^{57}Fe are used in most studies if the impurity elements are Mössbauer active. A large induced magnetic moment at the local site produces a large hyperfine field, as for Dy and Er. For the doping of Dy in an α-Fe host, the ^{161}Tb isotope was implanted at an energy of 130 keV in a high-purity Fe foil using an isotope separator (Wit et al. 1978). The ^{161}Dy Mössbauer spectrum at 5 K showed outer line sextet at a velocity of ± 25 cm/s and 4f hyperfine fields in the range of –600 T. The rare-earth metal behaves similar to the rare-earth + 3 ion in the metal host. The exchange interaction between the rare-earth spin and the polarized conduction electron spin determines the direction and magnitude of the magnetic moment. The polarization is due to the 4d, 4f and s-electron moments, and the 4f contributes the most while core polarization can be given as

$$B_{cp} = -9(g_J - 1)J(\mathrm{T}) \tag{6.5}$$

where g_J is the g-factor and J is determined by Hund's rule. It is also interesting to note the large hyperfine field of the gold impurity nucleus, which is otherwise diamagnetic in the elemental form. This is due to the polarization of the diluent by the ferromagnetic host, and in the case of Au, the contribution is from the core electron polarization and the lone 6s electron (Roberts and Thomson 1963).

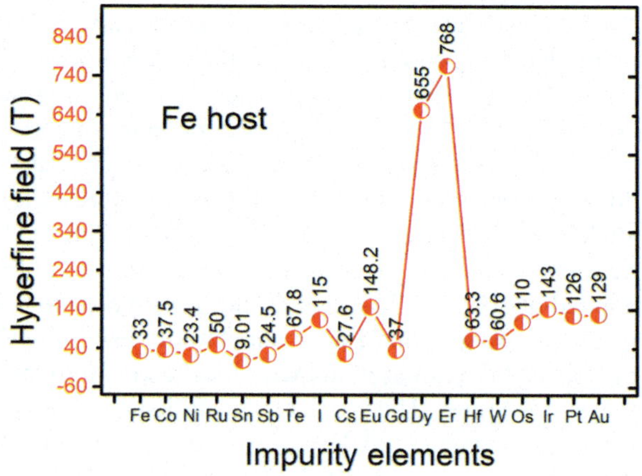

Fig. 6.6 The hyperfine fields (magnitude) of various impurities (< 4 at.%) in the Fe host at 4.2 K (Rao 1985)

6.6 FeCo Alloys

The phase diagram of FeCo presenting the different phases along with the Curie temperature as a function of Co concentration is shown in Fig. 6.7. FeCo alloys have a composition-dependent Curie temperature, which is higher than the T_c of 1043 K corresponding to α-Fe. The bcc FeCo phase has two variants: an ordered phase (α′) and a disordered phase (α). The disordered phase is bcc with Fe and Co atoms occupying random positions in the lattice with the A2 structure, while the ordered phase has Co occupying the body center position and Fe the cube corners or vice versa in a B2 structure, as represented in Fig. 6.8. The bcc phase transforms to fcc at a high temperature depending on the composition of the alloy. The disordered alloy is ductile while the ordered one is brittle.

The hyperfine fields of the FeCo alloys with composition under different states are represented in Fig. 6.9. The room temperature hyperfine field of the FeCo alloys varies with the composition exhibiting Slater-Pauling behavior with values larger than 36 T in the composition range 20–35 at.% Co (Johnson et al. 1963). The maximum value of 36.5 T for the hyperfine field corresponds to $Fe_{75}Co_{25}$. The ordered alloy has a reduced hyperfine field compared to the disordered alloy, with a maximum difference of 1 T for the $Fe_{50}Co_{50}$ (deMayo et al. 1970). The Fe atom in a Co-host shows a hyperfine field of 32.9 T, while that of Co in a Fe-host is 28.6 T (Shirley and Westenbarger 1965).

The isomer shift ranges from 0.04 to 0.01 mm/s, and the disordered alloys show higher values than the ordered alloys. The hyperfine field varies with Co neighbors in the FeCo alloys and can be modeled as (deMayo 1981)

$$B = B_o + B_{NL} + \sum_i n_i B_i \tag{6.6}$$

where B_o is the effective field of Fe (33.0 T), B_{NL} is the non-local contribution, n_i and B_i are the number of Co neighbors and their hyperfine field contributions,

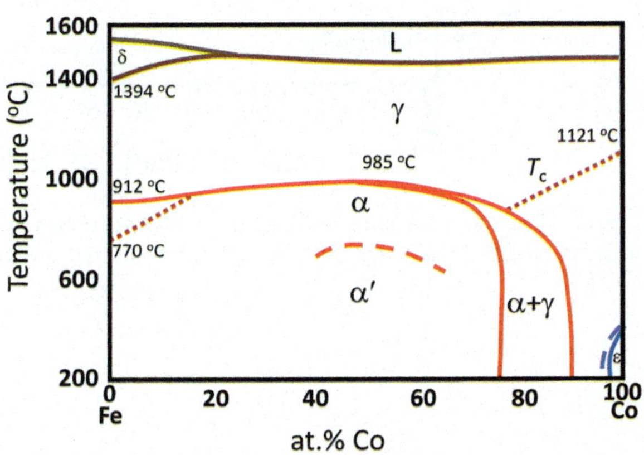

Fig. 6.7 The phase diagram of FeCo with the Curie temperature shown as a dashed line coincides with the α → γ transformation temperature in the intermediate compositions. The order (α′)-disorder (α) transformation temperature is maximum close to the equiatomic composition

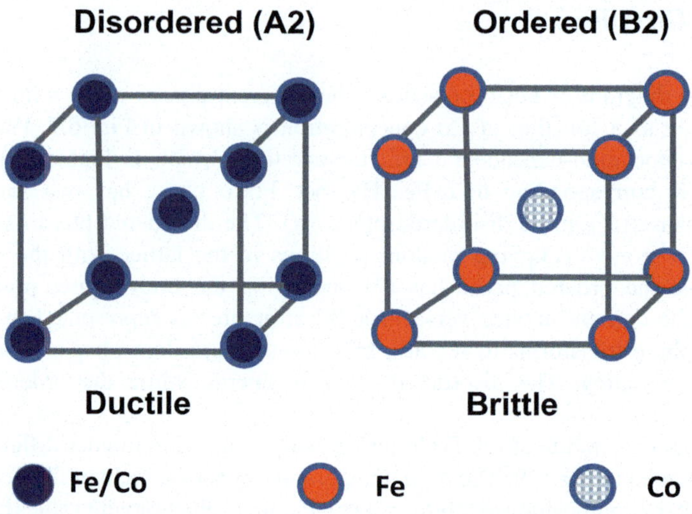

Fig. 6.8 The Fe and Co atoms occupy random positions in the bcc structure of the disordered (left) FeCo, while the body center is occupied by one type of atom and the corners by the other in the ordered alloy (right)

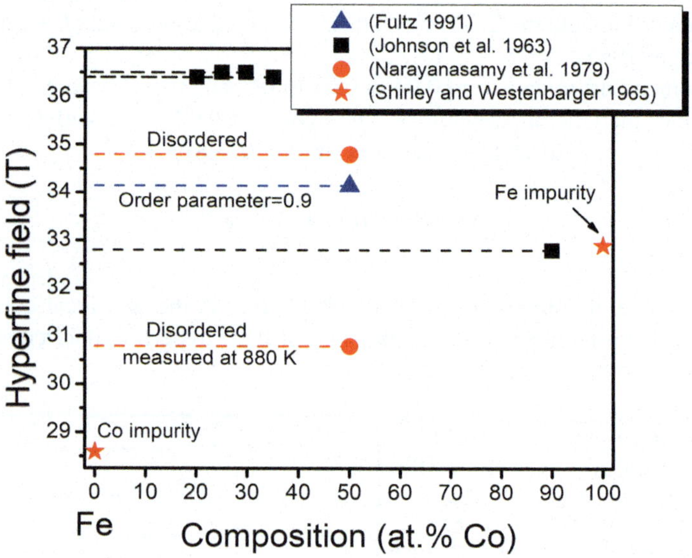

Fig. 6.9 The room temperature hyperfine fields with cobalt concentration in the FeCo alloys. The hyperfine field of equiatomic disordered FeCo measured at 880 K is also shown

respectively. For the first and second neighbors, the i values are 1 and 2, respectively. A continuous decrease of the hyperfine field with the short-range order parameter was obtained in the FeCo alloys, implying the decrease of B_{hf} with ordering. The equiatomic disordered FeCo showed a room temperature hyperfine field of 34.8 T that decreased to 30.8 T at 880 K (Narayanasamy et al. 1979). The hyperfine field contribution from the near neighbor is 0.73 T while the next near neighbor contributes 0.65 T in equiatomic FeCo (Frąckowiak 1985). The B_{hf} of disordered equiatomic FeCo reduces below 34.2 T with significant ordering (Fultz 1991).

Mössbauer spectrometry for ordered and disordered FeCo alloys was studied at 77 and 295 K (Hamdeh et al. 1989). The disordered alloys were obtained by quenching at a cooling rate of 10^5–10^6 K/s while slow cooling to room temperature for two days resulted in an ordered alloy. For the disordered alloy, it was presumed that the 3d↑ states fill continuously up to 30 at.% Co while the 3d↓ and 4s states decrease continuously with Co. The hyperfine field contribution to the disordered alloy was modeled by assuming contributions from (i) the local component due to exchange polarization of the 3d electrons with the core and conduction 4s electrons and (ii) the non-local component due to spin polarization of the 4s electron from the neighboring atoms by exchange interaction. The non-local component has contributions from the solute atoms, and the Fe magnetic moment perturbed by Co atoms.

The simulated hyperfine fields for the ordered and disordered alloys as a function of composition, together with the experimental values at 77 K, are shown in Fig. 6.10. The reduction in the experimental values is suggested to be due to the loss of 4s↓ electrons with Co concentration. The loss of 3d↓ electrons could explain the changes in the hyperfine field and the isomer shift with ordering. The non-local contribution reduces the average hyperfine field ($\overline{B}_{hf}$)in the ordered alloys. The local Fe environment, even up to 9 Å, plays a significant role in determining the Mössbauer parameters.

In contrast to the bulk, the chemically synthesized FeCo was found to be disordered, as shown by Mössbauer spectrometry (Kodama et al. 2006; Karipoth et al. 2016). The composition-dependent hyperfine field and isomer shifts for the FeCo alloys obtained by the polyol process are shown in Fig. 6.11 (Kodama et al. 2006). The average hyperfine field values are larger than 34.0 T for compositions close to the equiatomic region, establishing the disordered nature of the alloy. In addition, the composition-dependent isomer shifts and hyperfine fields imply the variation of Fe environments with the method of synthesis compared to the bulk alloy.

6.7 FeNi Alloys

The FeNi alloys show a bcc phase in the Fe-rich region and an fcc phase in a wide range of compositions from 5 to 70 at.% Fe as per the phase diagram shown in Fig. 6.12. The invar composition near 65 at.% Fe has been well studied. Alloys with less than 30 at.% Ni show a paramagnetic fcc phase, and the hyperfine field increases with Ni content, reaches a maximum value near 30 T, and gradually decreases with

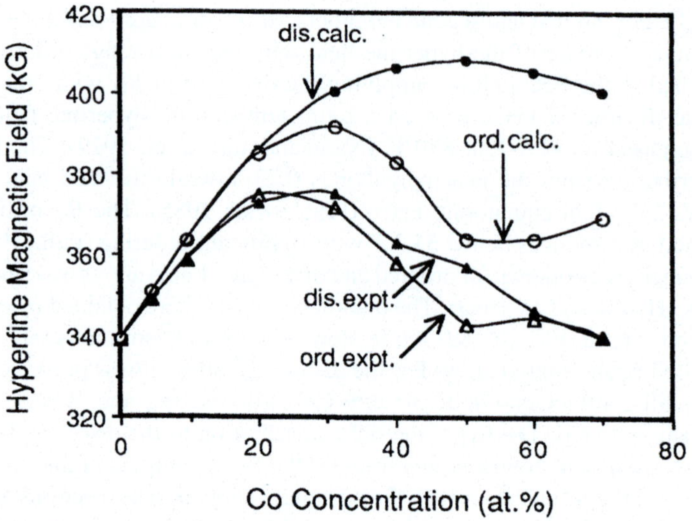

Fig. 6.10 Experimental and simulated average hyperfine fields at 77 K for ordered and disordered FeCo alloys with Co concentration. Reprinted with permission (Hamdeh et al. 1989), © The American Physical Society

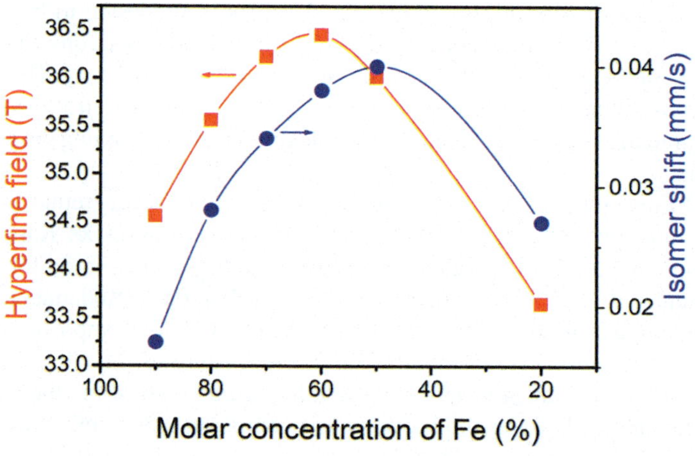

Fig. 6.11 The hyperfine field and isomer shift as a function of the molar concentration of Fe in FeCo synthesized by the polyol process. Reproduced with permission (Kodama et al. 2006), ©Wiley

increasing Ni (Johnson et al. 1963). The temperature-dependent Mössbauer spectra for FeNi compositions with 31.8–65.6 at.% Ni have been reported (Tomiyoshi et al. 1971). The hyperfine field for fcc $Fe_{34.4}Ni_{65.6}$ was observed as 30 T at 88 K with a broad outer line, and collapse of the sextet occurs at 875 K corresponding to the T_c. As the Fe concentration increases, the line broadening is more dominant at

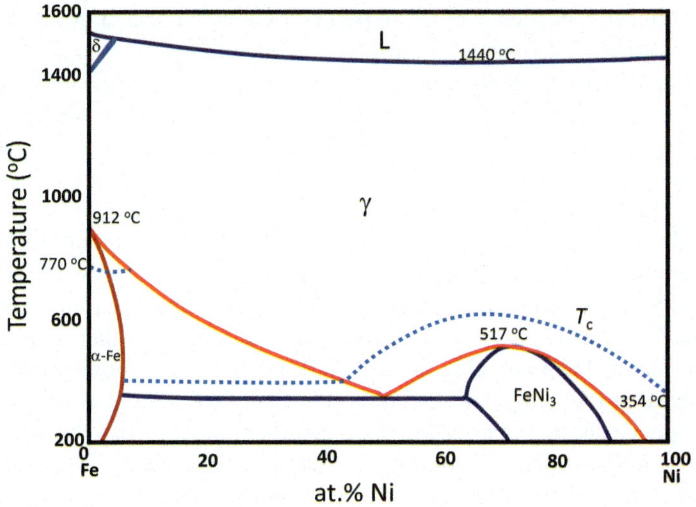

Fig. 6.12 The phase diagram of FeNi with Ni composition

low temperatures due to the distribution of hyperfine fields. For the 31.8 at.% Ni composition, the hyperfine field at 226 K varies continuously from 0 to 35 T with a peak at 25 T.

The investigation of FeNi alloy with 32 at.% Ni resulted in both ferromagnetic and paramagnetic absorption lines at 77 K while the 34 at.% Ni was ferromagnetic (Nakamura et al. 1964). In the critical invar composition, the coexistence of ferromagnetic and antiferromagnetic fcc FeNi phases or spin glass like structures are found to be common (Gonser et al 1979; Hesse 1989). The average hyperfine field for the $Fe_{65}Ni_{35}$ invar alloy decreases from 31 T at 30 K to less than 30 T near 0 K (Abd-Elmeguid et al. 1987). With an external field of 2 T, the average effective hyperfine field reaches beyond 31 T. It was concluded that the antiferromagnetic spins in the ferromagnetic matrix at 4.2 K are spin-flipped back to the ferromagnetic state above 30 K.

The disordered bcc FeNi alloy in the Ni composition range 15–25 at.% showed hyperfine fields close to 34.2 T at 293 K (Vincze et al. 1974). The hyperfine field gradually increases from 34.0 to 35.5 T in the 15–25 at.% Ni regime at 80 K for the disordered alloy. The average hyperfine field in solid solution is expressed by the empirical relation

$$\overline{B}_{hf} = a\overline{\mu}_{\mathrm{A}} + b\overline{\mu} \tag{6.7}$$

where $\overline{\mu}_{\mathrm{A}}$ is the average magnetic moment of the solute atom A in the host alloy with an average magnetic moment $\overline{\mu}$. The proportionality constants a and b depend on the type of alloy. Using Fe as solute atom A, the values a and b are obtained as 7 and 8 T/μ_{B}, respectively. However, this relationship is more appropriate for alloys that show a linear dependence of the hyperfine field with concentration. In addition, short-range ordering also leads to considerable uncertainty.

The average hyperfine field of the $FeNi_3$ alloy is 29.6 T when the long-range order parameter, S, becomes zero, corresponding to complete disorder and decreases to 27.7 T when S becomes 0.89 (Drijver et al. 1977b). The linewidth (Γ) also decreases from a value greater than 0.7 mm/s corresponding to $S = 0$ to less than 0.4 mm/s for $S > 0.9$. The $\overline{B}_{hf}$ for the $FeNi_3$ alloy with an order parameter 0.88 showed a gradual decrease with temperature (Drijver et al. 1977b). It should be noted here that temperature-dependent studies resulted in the order to disorder transformation just above 780 K, resulting in a decrease in $\overline{B}_{hf}$. On cooling, the transition occurred just below 780 K. The T_c of the ordered phase (940 K) is higher than that of the disordered phase. The linewidth of the disordered phase is almost twice that of the ordered phase at low temperatures due to the hyperfine field distribution resulting from different Fe environments. The order–disorder transition in FeNi alloy has been proven to be a first-order transition by Mössbauer spectrometry. The square of the long-range order parameter showed a linear relation with the average hyperfine field in $FeNi_3$ alloy given by $\overline{B}_{hf} = B_d - KS^2$ where B_d is the mean hyperfine field of the disordered alloy, K is a constant and S is the long-range order parameter (Frąckowiak 1985). A fit of the data from Frąckowiak (1985) and Drijver et al. (1974) shows a good agreement with the linear relationship (Fig. 6.13), yielding $B_d = 29.45(4)$ T and K = 2.04(8) T.

Mössbauer spectra of the disordered FeNi and $FeNi_3$ produced by rapid cooling in liquid nitrogen have been reported (Narayanasamy et al. 1979). Temperature-dependent hyperfine field studies showed a distribution width of 4 T for the $FeNi_3$ over the entire temperature range. For disordered FeNi, the distribution width fluctuates around 4.4–4.8 T. This wide distribution is due to the local fluctuation of the short-range exchange field at different lattice sites in the disordered alloy. The average hyperfine fields as a function of temperature for disordered FeNi and $FeNi_3$ alloys

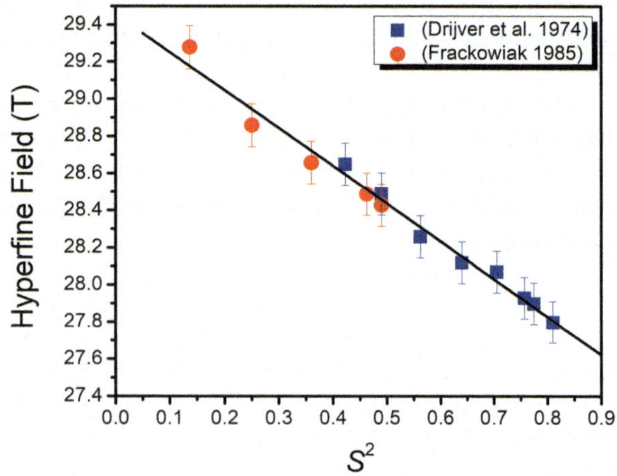

Fig. 6.13 The average hyperfine field shows a linear relationship with the square of the order parameter in $FeNi_3$ alloy. The hyperfine field of the disordered alloy obtained from the fit is 29.45(4) T

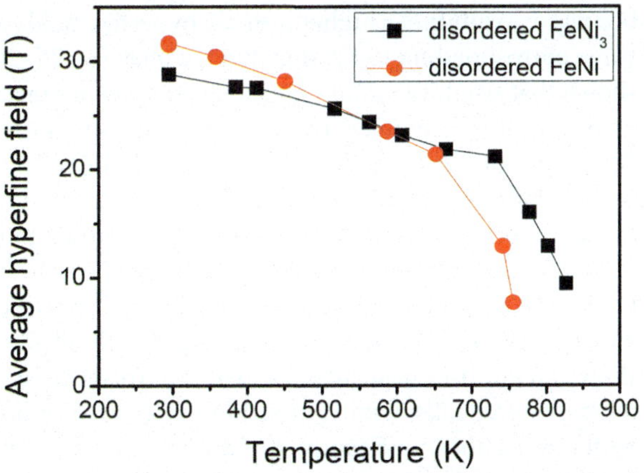

Fig. 6.14 The average hyperfine fields of disordered $FeNi_3$ and FeNi as a function of the temperature. Data from (Narayanasamy et al. 1979)

are shown in Fig. 6.14. The Curie temperatures are 860 and 770 K for $FeNi_3$ and FeNi, respectively. The average hyperfine field of the FeNi alloy is 31.6 T upon ordering, whereas the disordered alloy is larger by 0.35 T at room temperature (Narayanasamy et al. 1982). The ordered FeNi exhibits a larger isomer shift (0.09 mm/s) than the disordered alloy (0.02 mm/s), implying the transfer of electrons to the Fe 3d-band and also the reduction of the core polarization, inferred from the decrease in the hyperfine field.

Significant ordering could not be achieved to obtain $L1_0$ FeNi phase in the laboratory, whereas the Mössbauer spectra of the phase are recorded with meteorite samples. The room temperature (RT) hyperfine field of the $L1_0$ equiatomic ordered alloy was found to be 28.5 T that increased to 31.0 T on complete disorder (Larsen et al. 1982). The quadrupole shift for the ordered alloy was close to 0.2 mm/s that decreased with disordering.

6.8 FeCu Alloys

FeCu alloys have a low solubility, and therefore, non-equilibrium process techniques have been used to improve the solubility. Low temperature (4.2–100 K) high field (8 T) Mössbauer studies up to 4.6 at.% Cu in the Fe-Cu system have shown that below 1.5 at.% Fe, it exhibits a Kondo state, reducing the saturation hyperfine field while α-Fe precipitates at the upper limit (Campbell et al. 1976).

The $Fe_{1-x}Cu_x$ alloys obtained by radio frequency (rf) sputtering showed solubility throughout the composition with a bcc structure when $x < 0.4$ and an fcc structure when $x > 0.6$ (Sumiyama et al. 1985). Both bcc and fcc phases are ferromagnetic with a narrow sextet for the Fe-rich (bcc) composition, while it becomes wider and paramagnetic at room temperature for the Cu-rich (fcc) composition. The average hyperfine field at 4.2 K also showed a linear increase with Fe concentration for the bcc phase exhibiting a higher average hyperfine field (> 25 T), reaching the maximum value corresponding to Fe, than the fcc phase (~ 25–20 T). Subsequently, it was also shown that solubility could be achieved by using mechanical alloying.

FeCu alloys obtained by dc magnetron sputtering showed bcc and fcc phases depending on the composition, similar to samples obtained by rf sputtering (Chien et al. 1986). At intermediate compositions, a mixture of bcc and fcc phases was evident. The T_c obtained by Mössbauer spectrometry is below 50 K for the fcc phase up to 20 at.% Fe and then substantially increases to nearly 500 K for the bcc phase in 60 at.% Fe. The bcc FeCu alloys at higher Fe concentrations show a much higher T_c of ~ 900 K. The ratio of B_{hf} at 4.2 K with different compositions to that of Fe shows an almost linear increase with Fe irrespective of the bcc or fcc phases. The δ increases gradually with Cu reaching up to 0.18 mm/s relative to Fe. Thus, alloying with Cu has not significantly affected the ferromagnetic nature of Fe, while it favors filled majority band spins.

6.9 FeZn Alloys

FeZn has a limited solubility for Zn (< 3 at.%) in bcc Fe at room temperature. However, the zinc solubility in bcc Fe extends up to 41 at.% Zn during the peritectic reaction. With increasing Zn concentration, other phases such as $\Gamma(Fe_3Zn_{10})$, $\Gamma_1(Fe_5Zn_{21})$, $\delta(FeZn_{10})$ and ζ $(FeZn_{13})$ appear. The hyperfine parameters, structure and composition related to the FeZn phases are listed in Table 6.1. All phases, with the exception of the α phase, are non-magnetic and composed of one or more quadrupolar doublets corresponding to various sites.

The temperature-independent quadrupole splitting of 0.5 mm/s was obtained in the 4.2–300 K range with 0.3 at.% enriched Fe in Zn (Williams et al. 1976). An applied field of 6 T failed to resolve the hyperfine structure. Studies have shown that Fe concentrations up to 1.5 at.% in Zn is a Kondo system (Munayco et al. 2006). Thin films with 0.6 at.% Fe show asymmetric quadrupolar doublets at 300 and 4.3 K, suggesting texture effects. Also, the $V_{zz} < 0$ was found to be in the *c*-axis plane of the film. The saturation hyperfine field is 4.5 T, determined in an applied field of 7 T at 4.3 K.

FeZn obtained by rf sputtering showed a bcc phase up to 30 at.% Zn exhibiting a magnetic sextet at room temperature. The sextet broadens with increasing Zn content accompanied by a single line corresponding to the Γ phase (Laggoun et al. 1990). The $\overline{B}_{hf}$ gradually decreases from the value corresponding to Fe and reaches ~ 24 T with 45 at.% Zn. The decrease in $\overline{B}_{hf}$ and the linear increase in $\overline{\delta}$ to 0.21 mm/s up to 45 at.% Zn can be explained by the increase in the number of Zn neighbors up to 6.

Table 6.1 The different phases of the FeZn alloy and its hyperfine parameters at room temperature (Grant and Cook 1994; Laggoun et al. 1990)

Phase	Structure	Composition range (at.% Fe)	Hyperfine parameters Δ: quadrupole splitting, δ: isomer shift
Zeta (ζ)	Monoclinic, C2/m	5.9–7.1	$\Delta \approx 0.15$ mm/s $\delta \approx 0.49$ mm/s
Delta (δ)	Hexagonal P6₃mc	8.1–13.2	$\Delta \approx 0.51$ mm/s, $\delta \approx 0.47$ mm/s (A) $\Delta \approx 0.22$ mm/s, $\delta \approx 0.47$ mm/s (B) $\Delta \approx 0.37$ mm/s, $\delta \approx 0.28$ mm/s (C)
Gamma-1 (Γ_1)	Cubic $F\overline{4}3m$	18.9–24	$\Delta \approx 0.08$ mm/s, $\delta \approx 0.47$ mm/s (A) $\Delta \approx 0.56$ mm/s, $\delta \approx -0.04$ mm/s (B) $\Delta \approx 0.22$ mm/s, $\delta \approx 0.04$ mm/s (C)
Gamma (Γ)	Cubic $I\overline{4}3m$	18–31	$\Delta \approx 0.16$ mm/s, $\delta \approx 0.37$ mm/s (A) $\Delta \approx 0.50$ mm/s, $\delta \approx 0.13$ mm/s (B)
Alpha (α)	$Im\overline{3}m$	> 58	$\overline{B}_{hf} = 33$–24 T $\delta = 0.0$–0.21 mm/s

6.10 FeMn Alloys

FeMn alloys have a bcc (α) phase up to 5 at.% Mn beyond which it solidifies into the fcc (γ) phase. In the 15–18 at.% Mn range, the hcp (ε) phase coexists by martensite transformation. Early experiments near equiatomic FeMn compositions showed an antiferromagnetic nature with unresolved hyperfine fields in the range 3.3–2.2 T for the 60–43 at.% Mn alloys (Kimball et al. 1963). The hcp phase has a T_N of 240 K, obtained from the broadened hyperfine field for the 17 at.% Mn alloy, while the other compositions have a T_N of 500 K. Other studies with the 50–30 at.% Mn compositions showed hyperfine fields in the range 3.7–4.1 T at 0 K and T_N from 519 to 424 K with decreasing Mn content (Ishikawa and Endoh 1967). Studies have shown that polycrystalline samples exhibit an antiferromagnetic disordered γ phase.

FeMn fcc ingots with compositions 17.6, 25.6 and 28.6 at.% Mn were cold rolled to obtain the hcp phase (Ohno and Mekata 1971). The extrapolated hyperfine field is 1.6 T at 0 K for all the hcp compositions. On the other hand, the T_N for fcc FeMn phase with 25.9 and 28.6 at.% Mn are 400 and 420 K, respectively. The B_{hf} at 0 K for the fcc compositions are higher than their hcp counterparts by almost 2.2–2.5 T. In the 95 at.% Mn alloy, which is a cubic antiferromagnet, the B_{hf} is less than 1.6 T at 4 K (Kimball et al. 1966).

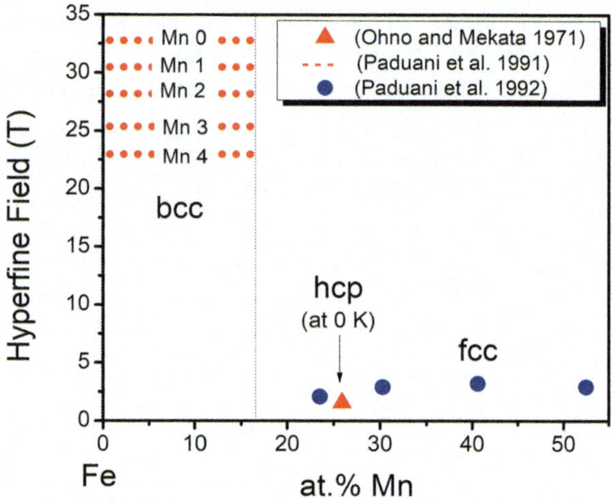

Fig. 6.15 Room temperature hyperfine fields for FeMn with Mn composition show a decreasing trend with Mn neighbors for the bcc phase. The fcc and hcp phases are antiferromagnetic

The bcc phase of FeMn could be stabilized by adding 3 at.% Ti beyond 5 up to 17.5 at.% Mn (Paduani et al. 1991). The transformation of hyperfine split sextets at room temperature to pure singlet is complete at 19.9 at.% Mn. The hyperfine field shows a distribution corresponding to different Mn near neighbors with a peak of 32.8 T for one Mn, linearly decreasing to 23.0 T for 4 Mn near neighbors, depicted in Fig. 6.15, along with the hyperfine fields of hcp and fcc antiferromagnetic phases (Ohno and Mekata 1971; Paduani et al. 1991, 1992). Earlier studies have shown a linear reduction in the hyperfine field to 27 T with 3 Mn near neighbor with a concomitant reduction in Fe atoms due to polarization of conduction electrons (Kang et al. 1974). The T_c obtained from the thermomagnetic analysis was close to 1050 K for the starting compositions and decreased with Mn content to reach a value close to 850 K for 17.5 at.% Mn. It should be noted here that the α extended phase using cold working techniques showed a T_c lower than 600 K for 10 at.% Mn (Yamauchi et al. 1974).

6.11 FeCr Alloys

FeCr alloy at equilibrium has a bcc structure and exhibits Fe-rich (α_1) and Cr-rich (α_2) bcc phases. Slow cooling from a high temperature in the equiatomic composition range results in a tetragonal σ phase which can then transform to $\alpha_1 + \alpha_2$. A linear dependence of the average hyperfine field ($\overline{B}_{hf}$) and the average isomer shift ($\overline{\delta}$) has been reported in the bcc FeCr alloys (Dubiel et al. 1981). The $\overline{B}_{hf}$ decreases from 33 to 23.5 T, for FeCr annealed at 1173 K, when the Cr content increases from 1.3 to 45.5 at.%. The $\overline{\delta}$ varies from a value close to zero to − 0.07 mm/s with increasing Cr concentration up to 45.5 at.%. A broad distribution of the hyperfine field is obtained with the 45.5 at.% Cr sample annealed at temperatures above 1073 K,

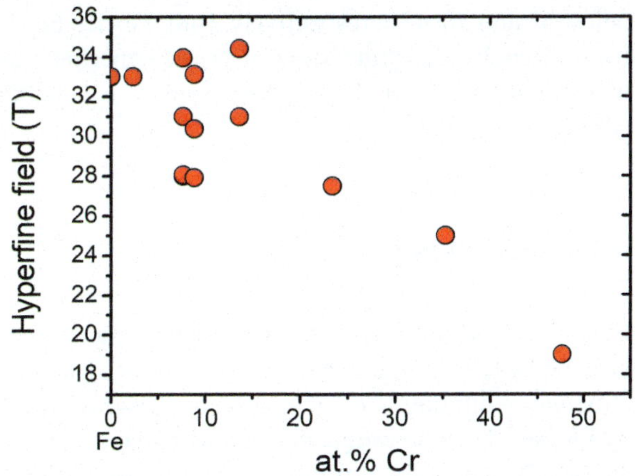

Fig. 6.16 The hyperfine field values of bulk FeCr alloys with Cr concentration up to 50 at.%. Data from (Petrov et al. 2002)

whereas it exhibits only a singlet for the sample annealed at 973 K corresponding to the paramagnetic σ phase. However, such thermal treatments rarely affect the state of the samples below 15 at.% Cr.

The local changes in spin and charge densities of Fe nuclei by Cr neighbors are related to the modifications of hyperfine parameters. The narrow sextet at low Cr concentration broadens with increasing Cr, consisting of various hyperfine field components, and the B_{hf} reduces further to ~ 18 T as it approaches the equiatomic composition, as shown in Fig. 6.16 (Petrov et al. 2002). For the 60.4 at.% Cr in FeCr, a singlet corresponding to the paramagnetic bcc phase is obtained. This transformation pattern of the spectra from narrow to broad and to a single line is similar to the FeMn alloys but for the transformation from ferro to para of the bcc phase with concentration. The magnetic, thermodynamic and electronic properties of FeCr alloys were studied in depth by Mössbauer spectrometry (Dubiel 2024).

6.12 FeV Alloys

FeV alloys exhibit bcc phase in the entire composition range with a tetragonal σ phase near the equiatomic region. FeV alloys with Fe up to 50 at.% show disordered nature exhibiting a singlet, transforming to a doublet with large linewidth as Fe concentration increases (Preston et al. 1966). Mössbauer spectra recorded at 77 K for the σ-FeV in the range 36–46 at.% V exhibit a broad spectrum composed of a doublet or singlet and three sextets. The B_{hf} ranges from 1.6 to 13.5 T, originating from four different Fe environments (Sumitomo et al. 1973).

Disordered and ordered bcc FeV alloys have been obtained in the composition range of 40–65 at.% V through sudden cooling and annealing of the samples, respectively (Shiga and Nakamura 1978). The Mössbauer spectra at 5 K exhibit broad hyperfine sextet for the disordered and ordered alloys as the V concentration is below 65 and 50 at.%, respectively. Bulk Fe-rich bcc FeV obtained in the disordered form through quenching showed broad hyperfine sextets for the $Fe_{60}V_{40}$ at room temperature while the heat treated sample was paramagnetic with a T_c of 143 K (Krause et al. 2000).

6.13 FeSi Alloys

FeSi alloys exhibit a disordered (A2) and ordered (B2) bcc solid solution below 11 at.% Si, $DO_3(Fe_3Si)$ and $\eta(Fe_5Si_3)$ phases at higher Si. Ordered Fe_3Si with the cubic structure exhibiting a T_c of 823 K, and Fe_5Si_3 with the hexagonal structure of T_c 373 K have been investigated using Mössbauer spectrometry (Shinjo and Nakamura 1963). The spectrum at room temperature for the Fe_3Si is fitted with two sextets of B_{hf} 30.5 and 19.5 T and that of Fe_5Si_3 at 77 K with 23.0 and 13.0 T. The extrapolated B_{hf} at 0 K are 32.5 and 21.0 T for the Fe_3Si; 23.5 and 13.5 T for the Fe_5Si_3.

Since Fe_3Si exhibits DO_3 structure, three site positions A-, B- and D- with respective full, half and random Fe occupancies, are assumed later. The ordered FeSi alloy up to 25 at.% Si has two phases in the 10–20 at.% range, with one showing DO_3 structure (Häggström et al. 1973). The B_{hf} reaches a value close to Fe when the near neighbors are greater than 8 Fe. With 4 Fe/Si near neighbor (nn), the B_{hf} reduces to 19.0 T, whereas δ increases to 0.25 mm/s as the Si composition reaches 18.4 at.%.

Mössbauer experiments with FeSi (up to 26 at.% Si) were performed at 300 K (Rixecker et al. 1993) and then in the temperature range from 77 to 1000 K (Randrianantoandro et al. 1999). The resolved sextets became broader as they reached the Curie temperature and finally collapsed to a singlet. The room temperature spectra for the FeSi alloys with Si up to 26 at.% are shown in Fig. 6.17. The spectra are composed of sextets arising from various Fe environments. The temperature-dependent average hyperfine fields for various compositions (Fig. 6.18) indicate that the T_c gradually decreases with Si and reaches a value of 825 K corresponding to 26 at.% Si. The critical exponent (β) is 0.36 for all the compositions.

FeSi alloys prepared through quenching are amorphous on increasing the Si concentration beyond 25 at.% (Mangin and Marchal 1978). While the mean hyperfine field at 0 K is 24.6 T for the Fe_3Si, it reduces to 16.8 T for the Fe_5Si_3 phase; 15.8 and 4.2 T for the 35 and 45 at.% Si amorphous alloys, respectively. The T_c for the 50 at.% Si amorphous alloy reaches below 200 K.

Molecular beam epitaxy has been utilized to obtain β-$FeSi_2$ orthorhombic phase that showed a doublet from the conversion electron Mössbauer spectrometry (Fanciulli et al. 1995). The doublet was fitted with two Fe site environments, resulting in quadrupole splitting of 0.53 and − 0.32 mm/s with δ of 0.08 and 0.09 mm/s, respectively. The electric field gradients of the two Fe sites were found to be antiparallel. Metastable fluorite type cubic γ-$FeSi_2$ phase has been obtained as precipitates in Si(100) (Desimoni et al. 1995). The room temperature CEMS spectrum suggested the disordered cubic structure and the doublet showed an average quadrupole splitting, Δ_{avg}, of 0.58 mm/s and δ of 0.20 mm/s. The Δ_{avg} decreased to 0.50 mm/s

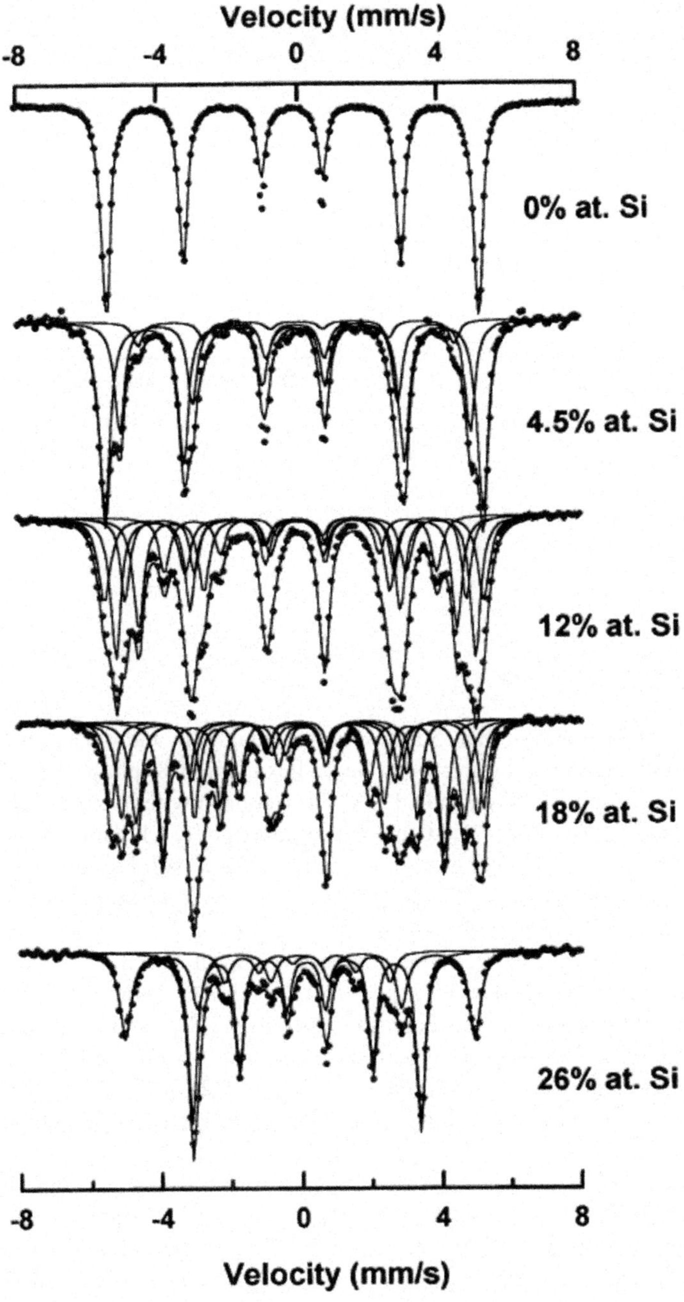

Fig. 6.17 The room temperature Mössbauer spectra of FeSi with various compositions. Reproduced with permission (Randrianantoandro et al. 1999), ©Elsevier

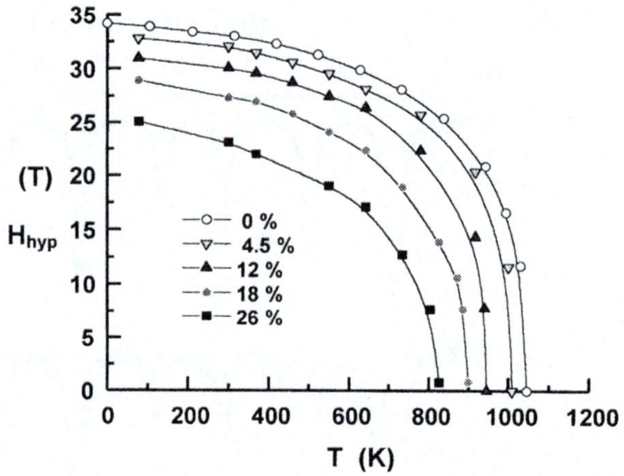

Fig. 6.18 The temperature-dependent average hyperfine fields of FeSi alloys up to 26 at.% Si. Reproduced with permission (Randrianantoandro et al. 1999), ©Elsevier

with a longer annealing duration at 733 K. High-temperature annealing at 1073 K transforms γ-$FeSi_2$ to β-$FeSi_2$.

6.14 FeGe Alloys

FeGe alloys with Ge less than 50 at.% is ferromagnetic and exists in the α phase solid solution. Other phases are: β-Fe_5Ge_3, a hexagonal $B8_2$ phase; ε-Fe_3Ge, a high-temperature hexagonal phase; and $L1_2$-Fe_3Ge, a cubic low-temperature phase (Kanematsu and Ohoyama 1965). The Curie temperatures of ε, $L1_2$ and β phases are 650, 760 and 480 K, respectively. Apart from the α-FeGe solid solution at lower Ge concentration, various other phases such as η, cubic B20, hexagonal ζ, monoclinic and δ-$FeGe_2$ are also reported.

The ε and $L1_2$ Fe_3Ge showed room temperature B_{hf} of 23.9 and 24.7 T, respectively (Yamamoto 1965). The Mössbauer spectra of the $Fe_{2-x}Ge_x$ alloys with x = 0.32–0.5 display broad hyperfine sextets corresponding to the 2(a) and 2(d) Fe lattice sites (Germagnoli et al. 1966). The hyperfine fields from the broadened lines for the x = 0.5 are 23.3 and 16.8 T, while they are 25.7 and 20.7 T for x = 0.32. The site assignment is unclear, whereas the relative intensity increases for the site showing the highest B_{hf}.

The Mössbauer parameters reported for the various phases are listed in Table 6.2. The β phase with the composition $Fe_{1.67}Ge$ at 84 K and RT showed distinct 2a, 2d and 2c site hyperfine fields in increasing order (Bhargava and Iyengar 1976; Schurer et al. 1978). The magnitude of the B_{hf} from the 2d and 2c sites of the $Fe_{1.55}Ge$ at RT are lower by 3.5 T compared to the $Fe_{1.67}Ge$, while the 2a site is completely occupied by Fe atoms. The hexagonal FeGe is antiferromagnetic and presents a T_N close to 410 K, transforming to a paramagnetic doublet (Tomiyoshi et al. 1966). The spectrum at 79 K shows a hyperfine field of 15.5 T, and its variation with temperature is close to the Brillouin curves with $J = 1/2$ and 1. The cubic B20 phase exhibits a conical spin structure with an angle of 57–61° between V_{zz} and B_{hf} (Ericsson et al. 1981).

The temperature-dependent hyperfine fields of β, cubic (B20) and hexagonal FeGe reported in the literature (Ericsson et al. 1981; Häggström et al. 1975; Schurer et al. 1978) are shown in Fig. 6.19. The largest average hyperfine field was exhibited by the β-$Fe_{1.95}$Ge whose contribution from the 2(a), 2(d) and 2(c) sites are 12.1, 17.5 and 21.8 T at 300 K, respectively, with an outer line broadening of 0.88 mm/s (Schurer et al. 1978). The spectra with temperature for the monoclinic phase are composed of three magnetic sextets that undergo a transition to a singlet at 335 K (Wäppling et al. 1970; Wäppling and Häggström 1968). The spectra display varying

Table 6.2 Various phases of Fe-Ge alloys and their Mössbauer parameters

Phase	T	B_{hf}(T)	$2\varepsilon/\Delta$ (mm/s)	References
β-Fe_5Ge_3	84 K	13.7	− 0.06	Bhargava and Iyengar (1976)
		20.2	0.16	
		24.5	0.18	
β-Fe_5Ge_3	RT	12.2 (2a)	− 0.1	Schurer et al. (1978)
		17.6 (2d)	0.38	
		21.1 (2c)	0.29	
β-Fe_5Ge_3 ($T_c = 480$ K)	520 K	–	0.21 (2a)	
			0.37 (2d)	
			0.39 (2c)	
η-FeGe	RT	–	0.42	Eliezer et al. (1975)
α-FeGe	RT	33.0–29.7	< 0.03	Brossard et al. (1971)
0–18 at.% Ge				
Amorphous FeGe	RT	–	~ 0.5	Hamdeh et al. (1992)
(51–89 at.% Ge)	22 K	5.0–25.0	~ 0.5	
	(< 53 at.% Ge)			
$FeGe_2$ ($T_N = 287$ K)	77 K	14.8	0.05	Forsyth et al. (1964)
	294 K	–	~ 0.1	
FeGe	80 K	11.1	0.16	Ericsson et al. (1981)
(cubic B20)	274 K	2.2	0.01	
($T_c = 275.5$ K)	700 K	–	0.07	
ς-FeGe (hexagonal B35)	0 K	15.2	0.01 (20 K)	Wäppling et al. (1970)
$T_N = 410$ K	RT	11.9	0.35	
FeGe	0 K	20.1 (FeI)	~ 0.1	
(monoclinic)		9.6 (FeII)	~ 0.05	
($T_N = 335$ K)		12.0 (FeIII)	~ −0.02	

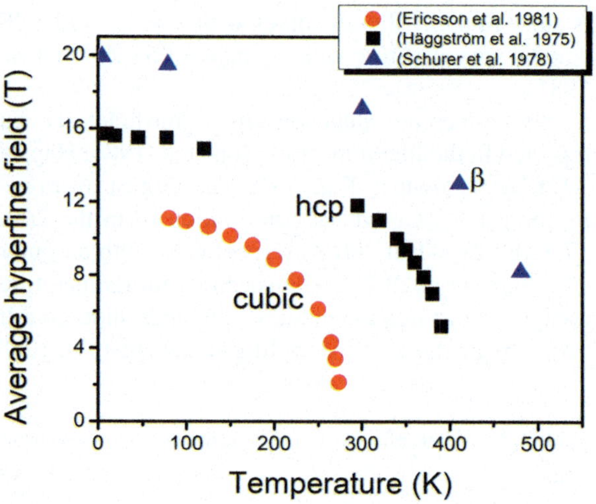

Fig. 6.19 The temperature-dependent average hyperfine fields of cubic FeGe, hexagonal FeGe and β-$Fe_{1.95}Ge$. The average hyperfine field of β-$Fe_{1.95}Ge$ was obtained from three Fe sites

shapes with temperature originating from Fe environments in 4(i) (FeI) and 2(a)/2(c) sites (FeII) and another transition from the third sextet (FeIII) at 110 K. The FeI and FeII sextets of the monoclinic FeGe show a transition near 110 K and an antiferromagnetic to paramagnetic transition at 335 K. The temperature-dependent quadrupole splitting shows a positive increase for FeI, a decrease but still positive for another sextet FeIII, whereas it decreases further with negative values for FeII. On the other hand, the cubic FeGe is a simple ferromagnet with a T_c just above 273 K. The saturation hyperfine field for the cubic FeGe is 11.5 T, and the broad sextet components decrease, reaching nearly 1.6 T at 273 K. However, below 180 K, the spectra of the monoclinic phase exhibit complex behavior.

Literature reveals that FeGe alloys exhibit several phases and magnetic behavior ranging from para, ferro, antiferro, helical, conical, skrymionic and chiral bobber states. The ferromagnetic behavior in the cubic FeGe is accomplished through a conical spin structure. The helical magnetic transition in FeGe is predicted to be of first order caused by the Dzyaloshinskii-Moriya (DM) interaction (Bak and Jensen 1980). Bulk non-centrosymmetric FeGe has been shown to exhibit helimagnetic, spiral and skyrmionic states due to the competition between exchange and DM interactions. Nuclear forward scattering experiments in B20 FeGe were carried out at high pressure and low temperatures (Barla et al. 2015). An inhomogeneous quasi-static magnetic (IQM) state with pressure up to 19 GPa in the helimagnetic to paramagnetic transition region was revealed. The transition temperature from the helimagnetic to the IQM decreases with pressure, and the average hyperfine field in the IQM region is 4 T, which persists up to 28.5 GPa.

6.15 FeSn Alloys

FeSn alloys exhibit a solid solution for less than 8 at.% Sn, DO_{19}-Fe_3Sn, hexagonal Fe_5Sn_3, monoclinic Fe_3Sn_2, B35-FeSn and Sn rich C16-$FeSn_2$ phases. Among these phases, only FeSn and $FeSn_2$ are antiferromagnetic, and all the phases have transition temperatures above room temperature (Trumpy et al. 1970). The FeSn solid solution has a T_c of nearly 1033 K, and the hyperfine field decreases with Sn neighbors reaching 29.8 T at 77 K for 8 at.% Sn corresponding to 2 nn Sn. However, the average magnetic moment is the largest (2.41 μ_B) for the 8 at.% Sn solid solution and the Fe site B_{hf} reaches a maximum of 34.3 T. Fe_3Sn is ferromagnetic with a T_c of 743 K, and the B_{hf} at 77, 293 and 473 K are 26.2, 25.4 and 23.1 T, respectively (Djéga-Mariadassou et al. 1966).

The Fe_5Sn_3 phase has two Fe sites, 2(a) and 2(d), which display sextet components due to various Fe-Sn neighbors. The 2(a) site exhibits an average B_{hf} of 17.1 T, and the 2(d) site shows 23.1 and 18.5 T at 297 K (Yamamoto 1966). The sextet with a B_{hf} of 15.2 T at 77 K for the B35-FeSn transforms to a doublet with a quadrupole splitting close to 1 mm/s on reaching the Néel temperature of 370 K (Djéga-Mariadassou et al. 1966). Figure 6.20 shows the temperature-dependent average hyperfine field of the FeSn (B35) alloys (Häggström et al. 1975b). The average hyperfine field has been obtained from the spectrum fitted with six sextets below 340 K and three above up to 361 K, which subsequently collapsed to a doublet at 370 K. The average B_{hf} at 180, 295 and 359 K are 14.0, 11.0 and 5.6 T, respectively. The six low-temperature sextets reach a mean saturation B_{hf} in three groups of 17.0, 15.5 and 13.9 T. The structure is not a simple antiferromagnet but has a canted spin structure at the basal plane.

Mössbauer spectrometry of Fe_3Sn_2 at various temperatures has shown a spin direction close to the c-axis in the temperature range 220 K to the T_c of 657 K (Le Caër et al. 1978). At lower temperatures, the spin direction is perpendicular to the

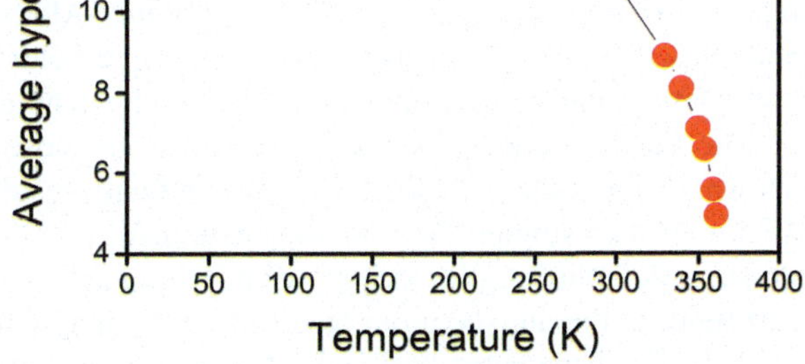

Fig. 6.20 The average hyperfine field with temperature of FeSn with the B35 structure. Data from (Häggström et al. 1975b)

c-axis. The mean hyperfine field at 4.2 K is 21.7 T, which decreases to 16.3 T in an applied field of 5.6 T, suggesting the negative B_{hf}. The room temperature mean hyperfine field and isomer shift values are 20.0 T and 0.33 mm/s, respectively.

$FeSn_2$ is a collinear antiferromagnet above 93 K up to its Néel temperature of 378 K, whereas, at low temperatures, it exhibits two different non-collinear antiferromagnetic spin structures canted along and away from the (001) basal plane (Venturini et al. 1987). At 100 K, the B_{hf} of $FeSn_2$ is 15.2 T, and in an applied field of 3.2 T, it could be fitted with three sextets of effective fields 18.3, 15.6 and 12.5 T (Le Caër et al. 1986). The B_{hf} of the second sextet makes an angle of 78° with the gamma ray direction while the other two hyperfine fields are parallel.

6.16 FeAl Alloys

FeAl alloys below 54 at.% Al exhibit disordered bcc, the ordered DO_3 type Fe_3Al and B2 type FeAl structures. Initial studies have shown broad spectra with an outer line splitting narrowing down with increasing nn Al (Flinn and Ruby 1961). In the region above 30 at.% Al, the FeAl shows a broad single line that narrows down close to the equiatomic composition (Stearns 1964). The isomer shift becomes more positive in the disordered region. The disordered FeAl has 8 to 4 nn Fe, whereas the ordered alloy has down to 2 nn Fe, and therefore, the hyperfine field for the ordered alloy is lower. The hyperfine fields of ordered Fe_3Al at 300 K are 22.9 and 29.9 T, and the extrapolated values at 0 K are 24.6 and 33.6 T, corresponding to A and D sites, respectively (Ono and Ishikawa 1962).

The magnetic properties of disordered FeAl up to 60 at.% Al suggest that the T_c close to that of Fe with lower Al concentrations decreases drastically above 16 at.% Al and reaches below 200 K for > 50 at.% Al (Yeluskov et al. 1992). The average hyperfine field from measurements below 77 K shows a linear decrease up to 16 at.% Al with a slope of – 0.17 T/at.% Al. The low-temperature data also indicate that the average isomer shift increases with Al reaching 0.24 mm/s at 60 at.% Al. The B_{hf} drastically reduces beyond 3 nn Al, which vanishes as the Al near neighbor reaches beyond eight. Disordered bcc FeAl could be obtained up to 47.5 at.% Al and the room temperature Mössbauer spectra with increasing Al composition show a broadening of the spectra, which becomes a single line at 47.5 at.% Al (Perez-Alcazar and da Silva 1987). The average hyperfine field decreases to 19.0 T as the Al concentration reaches 40 at.%, represented in Fig. 6.21 (Perez-Alcazar and da Silva 1987; Lesoille and Gielen 1970). The distribution of hyperfine fields depends on the number of Fe and Al nn in the Fe environment, related to the synthesis conditions.

The FeAl ordered alloys with compositions close to the equiatomic region are shown to have three distinct Fe environments resulting in hyperfine parameters arising from (i) singlet corresponding to undistorted B2, (ii) temperature-independent quadrupole splitting due to the vacancy-induced local distortion and (iii) 8 nn Fe surrounding the antisite atom resulting in the magnetic hyperfine splitting (Bogner et al. 1998). The center shift at 4.2 K is close to 0.18 mm/s for the Fe near vacancy

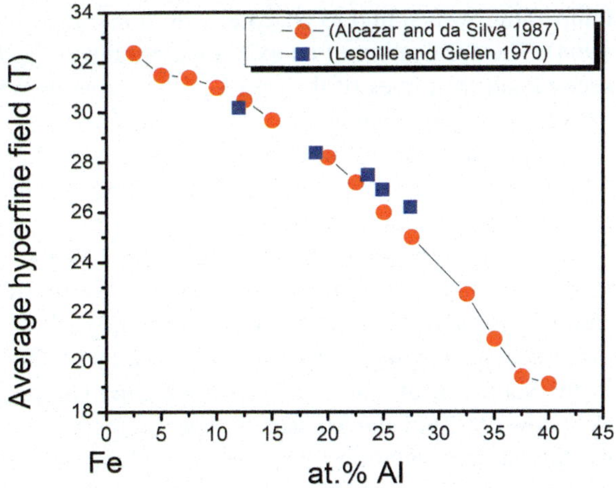

Fig. 6.21 Room temperature average hyperfine field of disordered FeAl alloys with composition

and 0.14 mm/s for the Fe antisite atom as well as corner atoms around Fe antisite. In an applied magnetic field of 13.5 T at 4.2 K, the corner atoms around the antisite indicate an internal field close to − 3 T, whereas the Fe antisite atom shows a higher value close to − 6 T. Low-temperature Mössbauer experiments in an applied field of 7 T have revealed canted antiferromagnetic spin arrangement in $FeAl_2$ (Kaptás et al. 2006).

6.17 FeGa Alloys

FeGa alloys exhibit disordered bcc B2 structure (α) up to 20 at.% Ga beyond which it orders up to 30 at.% Ga forming DO_3 structure (α″). The fcc β-Fe_3Ga exists with the $L1_2$ structure in the composition range of 25–30 at.% Ga. The ordered bcc Fe_3Ga show A- and D- site hyperfine fields of 20.5 and 31.2 T, respectively, at room temperature (Kawamiya et al. 1972). The extrapolated values at 0 K are 23.0 and 36.0 T. The fcc β phase hyperfine fields are 27.4 and 26.9 T at 0 K and RT, respectively. The average hyperfine fields from the two sites of disordered alloys are 30.2 and 32.5 T for 8.3 at.% Ga that decreases to 28.5 and 32.4 T for 20.5 at.% Ga (Dunlap et al. 2006).

6.18 FeAu Alloys

Initial studies on FeAu up to 11.5 at.% Fe concluded ferromagnetic nature of the alloys with Curie temperature as low as 7 K for 0.85 at.% Fe that increased to 35 K for 11.5 at.% Fe (Borg et al. 1963). In another study with Fe_5Au_{95} alloy, the transition temperature was found to be 23 K with a B_{hf} of 26.0 T at 4.2 K (Craig and Steyert 1964). The application of an axial external field of 3.08 T resulted in non-vanishing second and fifth lines of the sextet without line broadening or shift, suggesting the antiferromagnetic nature.

A rapid increase in Curie temperature of FeAu beyond 16 at.% Fe and B_{hf} below 33.0 T at 5 K for < 24.7 at.% Fe was reported (Gonser et al. 1965). RKKY-type exchange interaction is suggested at lower Fe concentrations for the FeAu alloys. However, low-temperature Mössbauer effect experiments with Fe up to 12.8 at.% suggested short-range magnetic-ordering due to strong nn magnetic interaction, resulting in a considerable departure from the Brillouin curve (Rideout 1969). It was discussed that the quadrupole splitting arises with the first Fe nn, and the distribution in the values results in increased linewidth. The point charge model-based fitting (Window 1972) of the Mössbauer spectra recorded at 4.2 and 300 K concluded (i) random spin orientation for one Fe neighbor, (ii) spin orientation along <111> for more than one near neighbor, (iii) randomly oriented along <111> for higher concentration from the clusters and (iv) long-range ferromagnetism at 16 at.% Fe. The transition to ferromagnetism is not abrupt, and B_{hf} varies gradually with composition. On the other hand, the composition-dependent (with Fe at.%) transition temperature (T_t) plot exhibits an abrupt variation near 12–16 at.% Fe, as shown in Fig. 6.22. This deviation with composition has also been observed from the perturbed angular correlation spectrometry (Cauteren and Hermans 1987).

The Fe-Au alloys with 17–19 at.% Fe exhibited ferromagnetic ordering at 165–170 K and a spin glass like transition below 45–60 K (Lauer and Keune 1982; Campbell et al. 1983). The field-cooled measurements at 4.2 K under a longitudinal applied field of 2.0 T showed the $\Delta m = 0$ lines (Lauer and Keune 1982). The intensity of the $\Delta m = 0$ line appeared at 45 K and increased with reducing temperature. This suggested the rotation of frozen Fe spins, forming an average angle away from the applied field direction at low temperatures. The 4.2 K spectrum also showed a

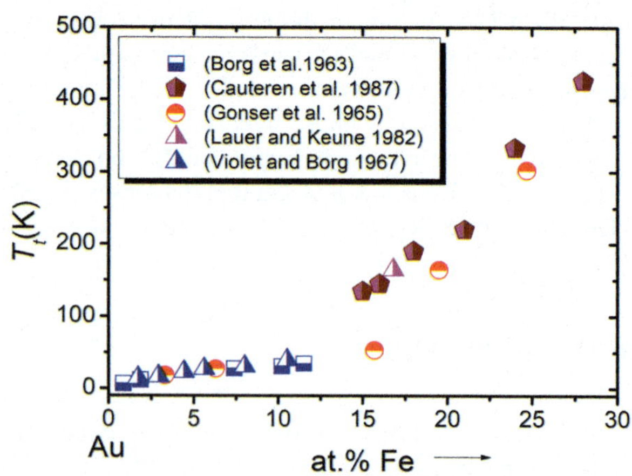

Fig. 6.22 The composition-dependent transition temperature in FeAu reported from Mössbauer spectrometry and perturbed angular correlation spectrometry

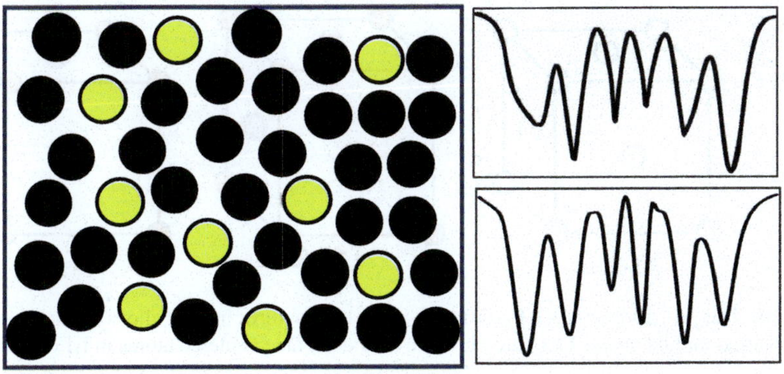

Fig. 6.23 The random arrangement of Fe atoms with Fe-rich and Au neighbors produces various hyperfine field distribution patterns below the transition temperature

distribution of hyperfine fields with peaks at 27.0 and 32.5 T attributed to Au and Fe-rich neighbor clusters, respectively.

FeAu in the composition range 10.5–33 at.% Fe has shown the presence of two sextet components with B_{hf} in the range 26.9–33.4 T at 4.2 K (Violet and Borg 1983). These two sextets are suggested due to two different phases, a solid solution and a Fe-rich environment, equivalent to α-Fe at higher concentrations. The two sextets in $Au_{82}Fe_{18}$ are reinterpreted as due to 3 nn Fe and 5 or 6 nn Fe environments rather than from two phases (Beck 1983). The nucleation of α-Fe and the clustering of Fe increases with its concentration in FeAu alloys (5–20 at.% Fe) (Whittle and Campbell 1985). Although long-range order (LRO) was reported initially at lower Fe concentrations, it was later shown that LRO prevails only above 20 at.% Fe (Cauteren et al. 1987). The random distribution of Fe atoms and their near neighbor environment results in hyperfine distributions and various spectral patterns below the transition temperatures, as represented in Fig. 6.23. The Mössbauer spectrometry of the FeAu system yielded various interpretations among different research groups (Monod and Campbell 1984; Violet and Borg 1984) while it has been accepted that the sample preparation history plays a significant role in the short-range order and precipitation of Fe in FeAu (Fratzl et al. 1991).

6.19 FePt Alloys

FePt alloys exhibit a disordered fcc structure (A1) and an ordered face-centered tetragonal (fct), $L1_0$ structure. The Fe- and Pt- rich phases that are Fe_3Pt and $FePt_3$, respectively, show an fcc $L1_2$ structure. The A1, $L1_0$ and $L1_2$ structures and their corresponding atomic positions are shown in Fig. 6.24. In the disordered fcc structure, the Fe and Pt atoms occupy random positions, whereas the face-centered positions are occupied by one and the corners by the other in $L1_2$ structure. In the $L1_0$ ordered

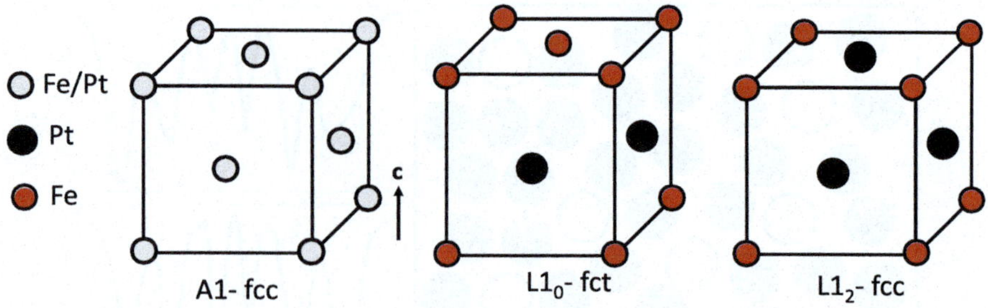

Fig. 6.24 The A1 (disordered), $L1_0$ (ordered) and $L1_2$ structures in FePt alloys. Four atoms in the face-centered middle layer of $L1_0$ are Pt, while all the six face-centered atoms in $L1_2$ are Pt

structure, the top and basal planes are occupied by one type of atom and the fcc lattice is elongated along the *c*-axis, resulting in fct structure.

FePt alloys with very low concentrations of Fe itself show long-range ordering where the T_c increases drastically from 5.4 to 104 K as the Fe is increased from 0.9 to 6.3 at.% (Crangle and Scott 1965). Mössbauer studies with Pt-rich FePt alloys of < 15 at.% Fe have shown an increase in T_c with Fe reaching a maximum close to 200 K (Segnan 1967). Consequently, the spectra of FePt with 3, 6.9 and 13.1 at.% Fe show sextets at 4 K, which broadens with reduced intensity due to distribution in B_{hf}, transforming to a broad single line above 32, 112 and 198 K, respectively. The spectra of Fe-rich $Fe_{72}Pt_{28}$ alloy show sextets with B_{hf} of 34.2 and 36.3 T at 5 K and T_c of 505 and 371 K for the ordered and disordered states, respectively (Sumiyama et al. 1976). The temperature-dependent average hyperfine field rapidly falls for the disordered state rather than the ordered state. In the invar composition range of 28–32 at.% Pt, nearly 80–100 K increase in the T_c could be obtained with increasing Pt, whereas there is no drastic variation in the B_{hf} at 4.2 K (Sumiyama et al. 1978). On the other hand, the disordered alloys show reduced T_c, higher B_{hf} and larger widths at 4.2 K.

Mössbauer studies have been carried out on the disordered and ordered fcc Fe_3Pt with the composition $Fe_{72}Pt_{28}$ at various temperatures in a longitudinal applied field of 5 T (Gonser et al. 1979). The disordered Fe-rich Fe_3Pt exhibits broad distribution in the hyperfine field compared with the ordered alloy. The RT and 13 K spectra for the disordered alloy are fitted with six sextets with Fe nn coordination from 6 to 11. The B_{hf} ranges from 23.5 to 31.2 T at RT, which increases to 32.7–40.7 T at 13 K. The application of a longitudinal field of 5 T on the sample at 13 K resulted in the complete disappearance of the $\Delta m = 0$ lines for both the ordered and disordered alloy, along with the reduction of the effective hyperfine field. Although the average hyperfine field for both states is similar, the ordered alloy could be distinguished from its narrow hyperfine field distribution compared to the disordered alloy that extends to high temperatures. The sextet transforms to a single line spectrum at 368 K for the disordered alloy corresponding to its Curie temperature, whereas it is higher by 150 K for the ordered alloy.

Apart from the studies with Fe-rich alloy, the equiatomic disordered FePt shows a single sextet with a broad distribution of B_{hf}, and an additional sextet with higher B_{hf} appears at RT for the partially ordered alloy (Tsurin et al. 1976). The room temperature spectrum for the ordered equiatomic FePt indicates B_{hf} of 27.1 T and a quadrupole shift of 0.29 mm/s due to tetragonal distortion (Goto et al. 1990). Remarkably, a higher B_{hf} component of 35.0–36.0 T with a relative intensity of 15–20% was obtained for the 36–40 at.% Pt ordered alloy suggesting the role of Pt neighbor in enhancing the B_{hf}. Depending on the degree of short-range order, the disordered alloys could exhibit distribution in hyperfine fields of 31.0–31.8 T (Goto et al. 1994).

Chemically synthesized FePt is usually found to be in the superparamagnetic fcc phase. The Mössbauer spectrum of equiatomic FePt nanoparticles of diameter 4 nm showed relaxation doublet at 220 K, whereas a large average B_{hf} component of 46.4 T was obtained at 10 K (Stahl et al. 2002) from the shell-interface. The Brownian motion of the non-interacting particles results in a significant drop in the count rate above 250 K due to reduced f-factor.

SiO_2 coated fcc $Fe_{51}Pt_{49}$ was annealed at 1173 K in a reducing atmosphere for 1 h to obtain $L1_0$ phase and made into an oriented polystyrene composite magnet in an applied field of 9 T (Tamada et al. 2008). The Mössbauer spectra at RT were recorded in the transverse and longitudinal direction of the oriented magnet without any external applied fields. Figure 6.25 shows the RT Mössbauer spectra obtained by applying γ-ray in (a) perpendicular and (b) parallel orientations. The strong orientation of the FePt magnet could be seen from the large relative intensity of the second and fifth lines of the sextet for the perpendicular geometry compared to the parallel case. The FePt could be fitted with core (site 1) and surface (site 2) sextets from $L1_0$ phase and doublet (site 3) from the fcc phase. The core and surface sextets show B_{hf} of 27.3 and 23.9 T, respectively, in both parallel and perpendicular orientations. The isomer shifts and quadrupole splitting values of both the sextets are similar (0.3 mm/s). The isomer shift and quadrupole splitting values of the doublet are 0.35 and 0.85 mm/s, respectively. The relative intensity of the core spin is 15%, and the orientation angles are 82° in the perpendicular and 22° in the parallel geometries. The orientation angles (angle between the magnetic moment easy axis and γ-ray) are similar for the core and surface spins. The orientation angle of 22° is less than the value of 54.7° required for random orientation. Almost 20% of the relative intensity is assigned to the superparamagnetic doublet component, implying an incomplete transformation to fct FePt even at an annealing temperature of 1173 K.

High-pressure Mössbauer studies on disordered $Fe_{72}Pt_{28}$ alloy show decreasing T_c and B_{hf} with pressure. The transformation from the high spin to low spin state occurs at a critical pressure of 4.2 GPa that enhances to 7.1 GPa at 4.2 K (Abd-Elmeguid and Micklitz 1989).

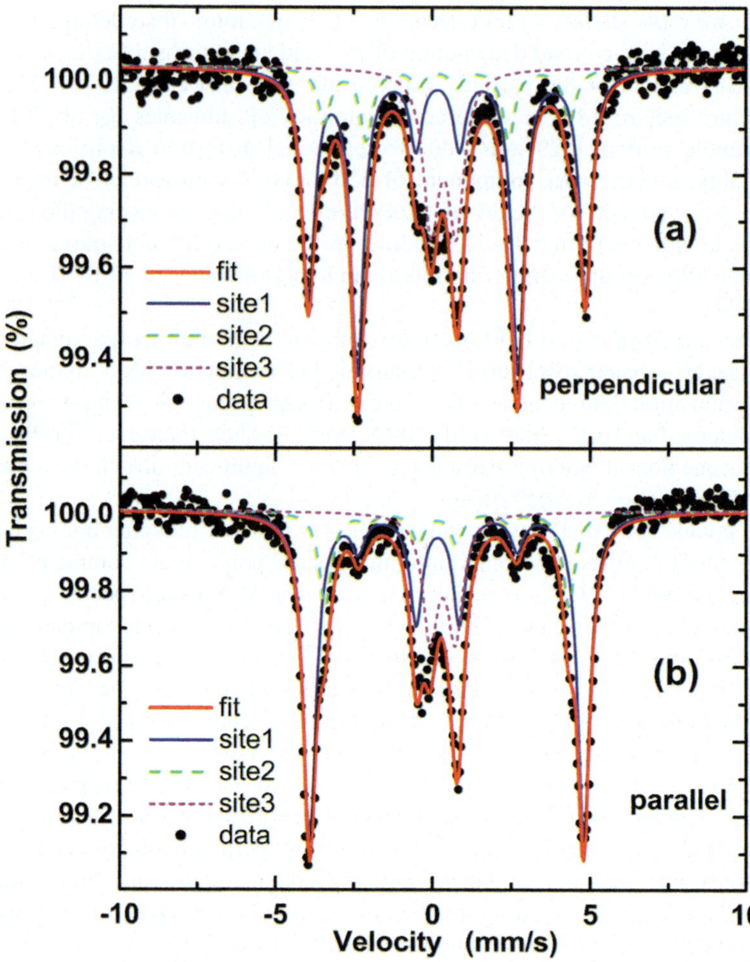

Fig. 6.25 Mössbauer spectra of the $L1_0$ FePt nanomagnet in the absence of external field with gamma ray in the **a** perpendicular and **b** parallel direction of the orientation. The site 3 doublet in the figure arises from fcc FePt. Reprinted with permission (Tamada et al. 2008), © The American Physical Society

6.20 FePd Alloys

Fe impurity in Pd exhibits a significant magnetic moment, and Mössbauer spectrometry has shown the Brillouin curve with $J = 13/2$ and a saturation hyperfine field of − 29.5 T (Craig et al. 1962). ^{57}Co isotope diluted in CoPd alloys with 3–100 at.% Co from 4 K to RT indicates almost constant B_{hf} in the range of − 31.0 to − 31.2 T, suggesting insignificant variation of 3d spin with Co in Pd (Nagle et al. 1962). However, the B_{hf} of − 29.5 T for Fe diluted in Pd reaches a value of − 33.5 T up to 50 at.% Fe in $Fe_{1-x}Pd_x$ (x = 0 to 0.5) alloys (Craig et al. 1965a, b). This suggests long-range conduction electron polarization and the contribution of the large electronic density of states of Pd. The average electron density of states is modified according to the concentration of Fe, giving a linear variation before reaching a constant B_{hf}.

FePd alloys exhibit ordered ($L1_0$), disordered and Pd-rich $FePd_3$ phases similar to FePt. Low-temperature Mössbauer effect experiments down to 4.2 K on a composition close to disordered Fe_3Pd_{97} have indicated saturation hyperfine field of − 31.4 T, $J = 1$ and T_c of 90 K (Craig et al. 1965a, b). The B_{hf} is proportional to the magnetization from the experiments either in the presence or absence of an applied field, except for discrepancies at high temperatures. In an applied field of 2.0 T, the low-temperature B_{hf} is − 29.4 T as the external field adds to the negative field of − 31.4 T, whereas above 90 K, the positive field obscures the T_c.

The magnitudes of the hyperfine field for the equiatomic ordered FePd alloy are 26.8 and 28.1 T at RT and 4.2 K, respectively, whereas the disordered alloy exhibits a higher value of 32.0 T at RT (Longworth 1968). On the other hand, the ordered $FePd_3$ alloy with 27 at.% Fe presents B_{hf} of 28.6 and 33.4 T at RT and 4.2 K, respectively. This is presumed due to the fct and fcc structures of the FePd and $FePd_3$ compositions, respectively. The T_c for the ordered FePd and $FePd_3$ are obtained as 728 and 500 K, respectively. The disordered alloys show line broadening and fluctuation in magnetization near the transition temperature arising from the short-range interactions. The RT B_{hf} of the disordered $Fe_{27}Pd_{73}$ alloy is 27.8 T, which decreases to 26.7 T for the $Fe_{22}Pd_{78}$ composition.

The ferromagnetic ground state of the ordered $FePd_3$ transforms to a non-magnetic state at high pressure beyond 12 GPa, as observed from the significant decrease in the amplitude of the quantum beats from nuclear forward scattering experiments (Winterrose et al. 2009). Although this may be either due to the high spin to low spin transformation or the decrease of T_c below RT, with Fe moments being retained, the study suggested the latter.

Fe-rich compositions of FePd alloys exhibit B_{hf} higher than Fe in bulk and thin film forms. Figure 6.26 shows the hyperfine field of disordered bcc and fcc $Fe_{1-x}Pd_x$ alloys reported in the literature (Klimars et al. 1985; Zhang et al. 1988; Kuang et al. 1988). It is observed that bcc FePd could be obtained for x < ~0.22 both in sputter deposited and bulk samples (Klimars et al. 1985; Zhang et al. 1988) and the B_{hf} near this transition region show maximum values, noticed from Fig. 6.26. CEMS studies have shown that the bcc FePd has RT B_{hf} > 34.5 T above 64 at.% Fe in disordered thin films (Laggoun et al. 2014).

6.21 FeRh Alloys

FeRh alloys are ferromagnetic up to 50 at.% Rh and become antiferromagnetic above that. The ferromagnetic alloys are bcc, and they exist in the disordered (α) and ordered (α′) states (CsCl type), whereas the Rh-rich antiferromagnetic phase is fcc (γ). The bcc–fcc phase boundary may vary according to the synthesis conditions.

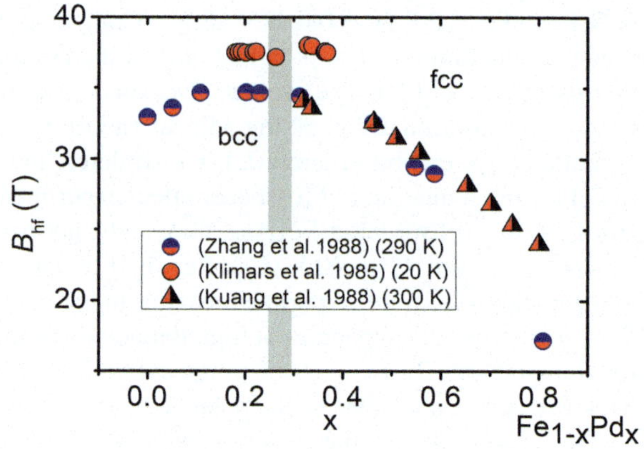

Fig. 6.26 The composition-dependent B_{hf} of bcc and fcc disordered FePd alloys

The Mössbauer studies have been carried out on the bcc FePd up to 50 at.% Rh for the ordered and disordered alloys (Shirane et al. 1963). In the ordered phase (20–50 at.% Rh), two distinct B_{hf}, one corresponding to 8 nn Fe and another with 0–5 nn Fe are observed. The B_{hf} of the Fe site with 8 nn Fe exhibits a maximum value close to 40.0 T, corresponding to the 25–30 at.% Rh. The second component shows hyperfine field that is below the disordered phase (~35.5 T) for the 25–30 at.% Rh composition. The reduction in B_{hf} is drastic, near 50 at.% Rh. The line broadening increases with composition as well as with disorder up to 25–30 at.% Rh and decreases thereafter. Despite higher B_{hf}, the 8 nn Fe may show a low magnetic moment. The temperature-dependent magnetic behavior of equiatomic bcc FeRh alloy exhibits antiferromagnetic to ferromagnetic transition at 338 K. The B_{hf} that decreases with temperature presents a discontinuity just above the transition temperature with a slightly higher value (> 25.0 T).

In another study, the fcc phase was already obtained with 37.5 at.% Rh in quenched FeRh alloy, resulting in a singlet at RT (Hao et al. 1971). The low-temperature studies on nearly equiatomic (49 at.% Fe) fcc FeRh have indicated broad sextet signifying a distribution of B_{hf} with an average value of 15.0 T at 4.2 K (Sumiyama et al. 1972). The fcc phase is antiferromagnetic with a Néel temperature close to 90 K. While the bcc FePd presents an antiferromagnetic to ferromagnetic phase transformation with temperature, it was also shown that the Fe-rich fcc alloy may transform to bcc, resulting in an enhanced B_{hf} with temperature (Rosenberg et al. 1998; Filoti et al. 1998). Slight variations in the composition may shift the transformation temperature and the Curie temperature.

6.22 Amorphous Alloys

The Mössbauer spectrum of amorphous alloys such as FeB metglass in ribbon form shows broad lines due to structural disorder. The spectrum can be described by distribution of hyperfine fields. The 300 K spectrum of the $Fe_{80}B_{20}$ metglass ribbon with the gamma ray perpendicular to the ribbon plane shows an average hyperfine field of 25.0 T and a FWHM distribution of 2.5 T, on fitting with a natural linewidth of 0.22 mm/s (Franke and Rosenberg 1978). The ratio of the intensity of the second line to the first line (I_2/I_1) decreases from 1.08 at 300 K to 0.39 at 75 K, indicating the orientation of the magnetization direction from in-plane toward perpendicular to the ribbon plane at low temperature. The I_2/I_1 at 75 K is 0.13 for the Mo-substituted $Fe_{78}Mo_2B_{20}$ amorphous alloy with a slightly larger average hyperfine field of 26.1 T. Heat treatment of the alloy beyond 700 K leads to the crystallization of the Fe_3B, Fe_2B and FeB phases with hyperfine fields 25.8, 21.6 and 9.2 T, respectively.

The room temperature hyperfine fields of crystalline Fe_2B and FeB are reported to be 24.2 and 11.8 T, respectively (Shinjo et al. 1964). The large difference in the hyperfine field of Fe_2B and FeB arises from the boron neighbors that are 4 and 6, respectively. In the case of Fe_3B amorphous alloy, the hyperfine field distribution appears from three different B environments, giving an average hyperfine field of 26.8 T at 80 K (Vincze et al. 1980). The variation in the hyperfine fields at 4.2 K with the Fe and B near neighbors in FeB alloys obtained from the literature are represented in Fig. 6.27 (Barault and Greneche 1995). The average B_{hf} is in the order of B_{hf} ($Fe_3B > Fe_2B >$ FeB), with the most prominent B neighbor in FeB resulting in the lowest value.

The thermal stresses resulting from the rapid quenching of the melt, induce spin texture as shown in $Fe_{40}Ni_{38}Mo_4B_{18}$ amorphous metglass ribbons from Mössbauer experiments (Greneche et al. 1982; Greneche and Varret 1982). The inhomogeneities

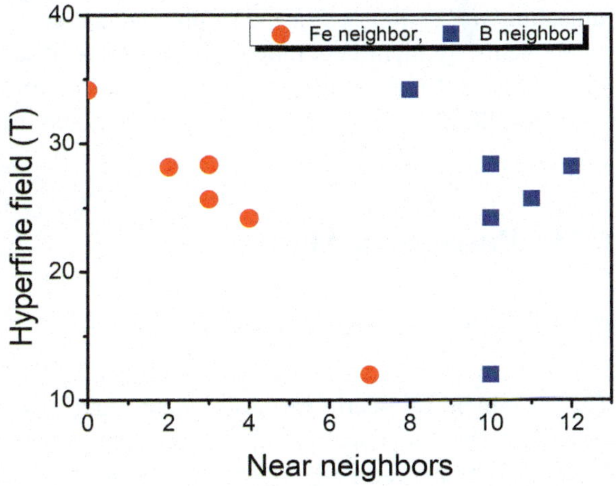

Fig. 6.27 The hyperfine field variation at 4.2 K with Fe and B near neighbors in FeB alloys. Data from (Barault and Greneche 1995)

in the metglass ribbons may result in a Curie temperature distribution of 2–3 K, obtained from Mössbauer thermal scanning experiments (Varret et al. 1988).

The FINEMET ($Fe_{73.5}Cu_1Nb_3Si_{13.5}B_9$) alloys are shown to have an amorphous component with an average hyperfine field of 20.7 T, while the annealed sample shows an amorphous component with a reduced average hyperfine field of 16.9 T and nearly six sextet components of larger hyperfine fields from the Fe_3Si phase (Rixecker et al. 1992, 1993). Mössbauer studies on nanocrystalline FINEMET alloy with the composition $Fe_{66}Cr_8CuNb_3Si_{13}B_9$ have shown that the amorphous matrix has a Curie temperature of 403 K while the crystalline α-FeSi phase in the amorphous matrix exhibits a much higher T_c (Jedryka et al. 1995). It was also shown that the crystalline, amorphous and interfacial components could be identified from the Mössbauer investigation of $Fe_{89}Zr_7B_4$ alloy in the temperature range of 4.2–550 K and in an applied magnetic field (Slawska-Waniewska and Greneche 1997). The choice of the fitting procedure helps in the identification of hyperfine field distribution from the amorphous matrix that are 14.5–8.8 T, while the interface exhibits 27.4–28.9 T, and the crystalline phase has a narrow hyperfine field of 32.6 T for the FeCuZrB alloy (Miglierini and Greneche 1997a, 1997b). The spectrum recorded at 373 K for the annealed FeCuZrB alloy is composed of amorphous, interface and crystalline components. The hyperfine field component and the resulting distribution are narrow for the crystalline part, whose origin could be found by employing an appropriate fitting method for Si free alloys, while broad and lowest values are encountered for the amorphous matrix.

A superparamagnetic component of nearly 5% relative fraction was reported in FINEMET alloys prepared through mechanical alloying due to the contribution from smaller particles (Manivel Raja et al. 2000). The RT Mössbauer spectra of the melt-spun FINEMET alloy with the composition $Fe_{73.5}Cu_1Nb_3Si_{13.5}B_9$ in the as-prepared amorphous form and those annealed at various temperatures are shown in Fig. 6.28 (Ponpandian et al. 2003). The average hyperfine field of the FINEMET alloy at RT is 21.0 T, which increases to 21.9 T due to the crystallization of Fe_3Si phase on annealing at 723 K/1 h. For the sample annealed at 923 K/1 h, six hyperfine field components are identified, ranging from 32.4 to 19.6 T corresponding to various Fe sites of the Fe_3Si phase along with Fe_2B. The evolution of other Fe-B phases occurs with annealing above 923 K, which suppresses the soft magnetic nature of the alloy.

6.23 Rare-Earth Alloys

6.23.1 Rare-Earth Binary Alloys

The prominent rare-earth transition metal binary alloys are the cubic Laves phase type RFe_2, rhombohedral RFe_3, hexagonal or rhombohedral R_2Fe_{17} and cubic R_6Fe_{23} where R is a rare-earth alloy that includes Sc, Y and La to Lu. Compounds such as

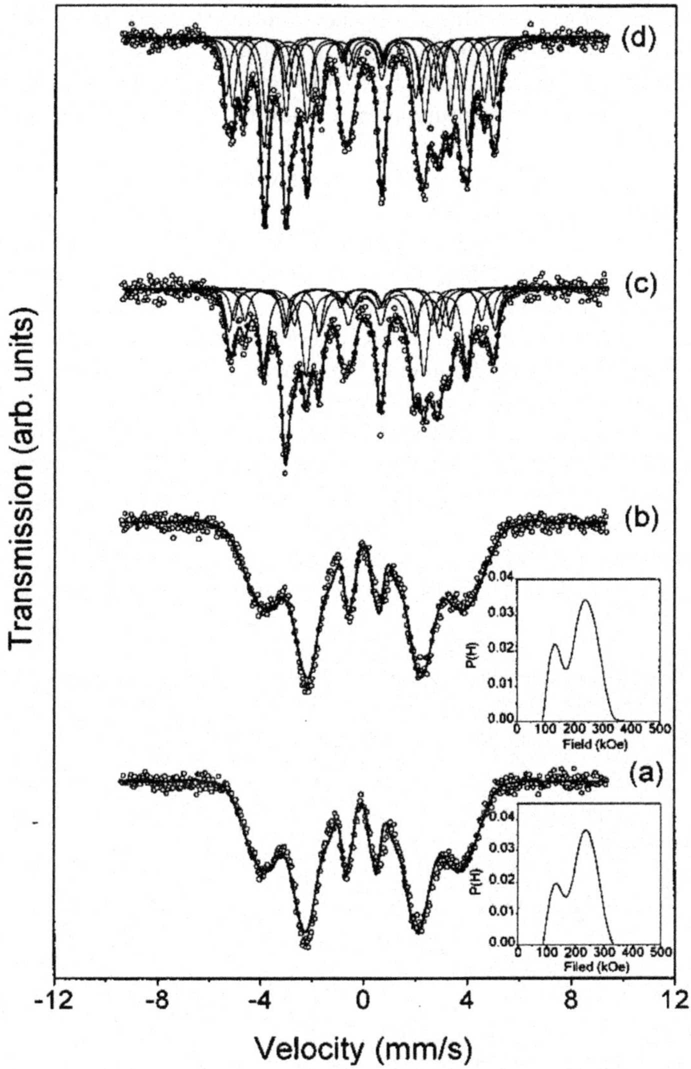

Fig. 6.28 The RT Mössbauer spectra of $Fe_{73.5}Cu_1Nb_3Si_{13.5}B_9$, **a** amorphous alloy, annealed at **b** 723, **c** 823 and **d** 923 K. Reprinted with permission (Ponpandian et al. 2003), ©AIP Publishing

$SmCo_5$ and Sm_2Co_{17} of the space group *P6/mmm* and $R\overline{3}m$, respectively, are well-known permanent magnetic materials with large magnetocrystalline anisotropy. The number of non-equivalent crystallographic Fe sites ranges from 1 for the cubic RFe_2, 4 for the cubic R_6Fe_{23} and 3–4 for the other phases.

The hyperfine field of various rare-earth intermetallic compounds of the type RFe_2 are listed in Table 6.3. B_{hf} in the range 22.0–23.0 T has been observed in most of the cubic Laves phase compounds at 77 K (Wertheim and Wernick 1962;

Wertheim et al. 1964). The Mössbauer spectrum of the cubic phase of $HfFe_2$ has been fitted with two sextet components at RT of hyperfine fields 18.4 and 17.6 T with the respective intensity ratio of 3:1 arising from the non-equivalent Fe sites while exhibiting negative quadrupole interaction (− 0.41 mm/s) for both the sites (Livi et al. 1976).

The B_{hf} of $CeFe_2$ is 15.6 T at 80 K, much lower than the other RFe_2 compounds owing to the lower magnetic moment and Curie temperature due to the transfer of 4f electron from Ce to Fe 3d band (Atzmony and Dariel 1974). The easy axis of magnetization of $CeFe_2$ is along [100] up to 150 K, beyond which the angle between the easy direction and [100], θ, increases to 30°. $GdFe_2$ shows a large hyperfine field and Curie temperature among the RFe_2 compounds, where the Gd-Fe exchange interaction outweighs the Fe–Fe interaction. Indeed, $GdFe_2$ has the largest T_c among various rare-earth intermetallic compounds.

The R_2Fe_{17} compounds exhibit a hexagonal phase (Th_2Ni_{17} type, $P6_3/mmc$) with six non-equivalent sites, while the rhombohedral phase (Th_2Zn_{17} type, $R\overline{3}m$) has five, of which four are Fe sites for both the phases. The hexagonal phase has the Fe sites (6g, 12j, 12k, 4f) represented in Wyckoff notation, while the rhombohedral phase has the Fe sites (6c, 9d, 18f, 18h). Mössbauer studies have shown that Tm_2Fe_{17} has c-axis anisotropy below 72 K while others in the group are of the basal plane (Gubbens et al.

Table 6.3 Mössbauer hyperfine fields at various temperatures for few rare-earth compounds of the type RFe_2 and their Curie temperatures

Compound	T	B_{hf} (T)	T_c (K)	References
$SmFe_2$	78 K	22.0	674	Wertheim and Wernick (1962)
$DyFe_2$	78 K 298 K	22.8 21.6	663	
$HoFe_2$	78 K 298 K	22.6 18.8	608	
$ErFe_2$	78 K 298 K	23.2 18.9	473*	
$TmFe_2$	78 K	23.1	613	
$TmFe_2$	78 K 298 K 473 K	21.6 ($\theta = 70.5°$) 19.1 ($\theta = 70.5°$) 14.5 ($\theta = 70.5°$)		Wertheim et al. (1964)
$CeFe_2$	80 K	15.6	230	Atzmony and Dariel (1974)
$GdFe_2$	80 K 300 K	22.6 21.7	785	
$LuFe_2$	80 K 300 K	21.5 18.5	610	
$HfFe_2$ Cubic	293 K	18.4 17.6		Livi et al. (1976)
YFe_2	0 K	20.7 ($\theta = 0°$) 21.3 ($\theta = 70.5°$)	540	Nishihara et al. (1982)

* T_c reported from other studies is larger by 100 K

1974). The hyperfine fields of individual sites vary among various reports according to the fitting procedure and the synthesis method, often fitted with more than two sextets for each site, whereas certain general trends are observed. The 6c site, also known as dumb bell site, has the largest isomer shift and hyperfine field among all the other sites. The order of the hyperfine field for the rhombohedral phase is B_{hf} (6c > 9d > 18f > 18h), while for the hexagonal phase, it is B_{hf} (4f > 6g > 12j > 12k). The weighted average hyperfine fields of the R_2Fe_{17} compounds at 85 K are shown to have values in the order B_{hf} (Th_2Fe_{17} < Pr_2Fe_{17} < Nd_2Fe_{17} < Sm_2Fe_{17} < Pr_2Fe_{17}) (Long et al. 1994). The hyperfine fields at various temperatures for a few of the R_2Fe_{17} compounds are presented in Table 6.4. The relative intensities for the 6c:9d:18f:18h sites for the R_2Fe_{17} are assigned 6:9:18:18 while the magnetically inequivalent 9d, 18f and 18h sites further give sextets of relative intensities 6:(6:3):(12:6):(12:6) (Long et al. 1992; 1994). Nitrogen and carbon interstitials in R_2Fe_{17} increase the Curie temperature beyond 673 K (Coey and Sun 1990). The weighted average hyperfine field of Nd_2Fe_{17} increases from 29.2 to 33.4 T for $Nd_2Fe_{17}N_{2.6}$ (Long et al. 1992).

The hyperfine fields at 15 K of the various Fe sites in Sm_2Fe_{17} have shown the largest B_{hf} of 36.0 T for the 6c site and 27.7 T for the 18h site (Hu and Coey 1991). The temperature-independent quadrupole splitting is of the order Δ (18f > 6c > 9d > 18h) in the range 0.4 to – 0.4 mm/s. The weighted average hyperfine field is 30.1 T at 15 K, which decreases to 22.1 T at 290 K. The magnetization direction is along the *b*-axis in the entire temperature range of 15–290 K.

The exchange integral as a function of the distance between the Fe–Fe pair in the Sm_2Fe_{17} and $Sm_2Fe_{17}N_y$ nitride system has been obtained from the hyperfine fields at various temperatures (Li and Morrish 1997). The $\mathcal{J}_{\text{Fe–Fe}}$ is evaluated using the expressions for i^{th} and j^{th} atom as

$$B^i_{\text{hf}} = B^i_{\text{hf}}(0)B_S(\alpha) \tag{6.8}$$

Table 6.4 Hyperfine fields from various sites for a few of the R_2Fe_{17} alloys

Compound	T (K)	B_{hf} (T)				$\overline{B}_{hf}$ (T)	References
		6c	9d	18f	18 h		
Nd_2Fe_{17}	78	35.0	30.0 27.4	28.3 31.4	27.1 26.6	29.1	Long et al. (1992)
	295	18.6	17.5 15.0	15.1 15.3	15.1 14.1	15.7	
Sm_2Fe_{17}	77	34.5	29.3	25.8	25.3	27.3	Chen et al. (1993)
Y_2Fe_{17}	77	32.1	29.7	26.8	26.2	27.7	
Pr_2Fe_{17}	85	34.3	28.2 30.6	29.8 26.4	26.6 26.3	28.6	Long et al. (1994)
	225	25.9	21.2 22.7	22.2 19.8	19.9 19.7	21.4	

$$\alpha = \frac{S_{\mathrm{Fe}}}{k_{\mathrm{B}}T}\left[\sum \frac{2\mathcal{J}^{ij}_{\mathrm{Fe-Fe}} Z^{ij}_{\mathrm{Fe-Fe}} \beta H^{j}_{\mathrm{hf}}}{g_{\mathrm{Fe}}}\right] - \mathcal{J}_{\mathrm{Fe-R}} Z^{i}_{\mathrm{R-Fe}} \gamma \mu_{\mathrm{R}}(T) \tag{6.9}$$

$$\mu_{\mathrm{R}}(T) = \mu_{\mathrm{R}}(0) B_S\left(\frac{S_{\mathrm{R}}}{k_{\mathrm{B}}T}\mathcal{J}_{\mathrm{R-Fe}} \sum_{j} \frac{Z^{j}_{\mathrm{R-Fe}} \mu^{j}_{\mathrm{Fe}}(T)}{g_{\mathrm{Fe}}}\right) \tag{6.10}$$

where $B_S(\alpha)$ is the Brillouin function given in terms of the exchange integral ($\mathcal{J}$) for the Fe–Fe pair as in (6.9). The temperature-dependent rare-earth magnetic moment $\mu_{\mathrm{R}}(T)$ can be obtained from the Brillouin function described as per (6.10) considering the $\mathcal{J}_{\text{R-Fe}}$ obtained from the Curie temperature. Z is the number of near neighbors, β is a constant, γ is related to the g-factor of rare-earth (R) as $\gamma = -\frac{g_R - 1}{g_R}$ and S is the spin value with 1 and 5/2 for Fe and Sm, respectively. The results suggest that the exchange integral is negative when the distance between Fe–Fe pair is less than 2.45 Å.

6.23.2 $Nd_2Fe_{14}B$ Alloy

One of the technologically important permanent magnetic materials is $Nd_2Fe_{14}B$, with the space group ($P4_2/mnm$) that has the most energy product (64 MGOe) as of now, with a theoretical possibility of 120 MGOe. The unit cell has 68 atoms from the four formula units, and the tetragonal structure (a = 8.80 Å and c = 12.20 Å), with various Nd, Fe and B sites are shown in Fig. 6.29 (Herbst et al. 1984). The T_c of $R_2Fe_{14}B$ lies in the range 420–660 K, with $Nd_2Fe_{14}B$ showing a value of 585 K. The spin reorientation temperature of $Nd_2Fe_{14}B$ is 135 K, where the magnetization direction is canted away from the c-axis towards the [110] direction, reaching an angle ~ 30° at 4 K (Herbst 1991).

Site assignment of sextets in $Nd_2Fe_{14}B$ is complicated since it has six Fe sites that are (4c, 4e, $8j_1$, $8j_2$, $16k_1$, $16k_2$), and the hyperfine parameters overlap with each other. The room temperature spectrum of $Nd_2Fe_{14}B$, along with the most suitable fitting of the sextets assigned to the Fe sites and their corresponding hyperfine parameters, are shown in Fig. 6.30 and Table 6.5, respectively (Onodera et al. 1987). The largest hyperfine field among all the sites is exhibited by the $8j_2$ site, whereas the 4e site has the lowest value due to two boron atoms at a distance of 2.096 Å and two Nd atoms. B_{hf} of $16k_1 < 16k_2$ since the $16k_1$ site has a non-magnetic boron, two Nd atoms and nine Fe atoms, while the $16k_2$ site has one excess Fe along with the absence of boron. The $B_{\mathrm{hf}}(8j_1 < 8j_2)$ since $8j_2$ has 12 Fe that effectively screens the Nd neighbor than $8j_1$. The isomer shifts are δ ($16k_1 > 16k_2$) and δ ($8j_1 < 8j_2$).

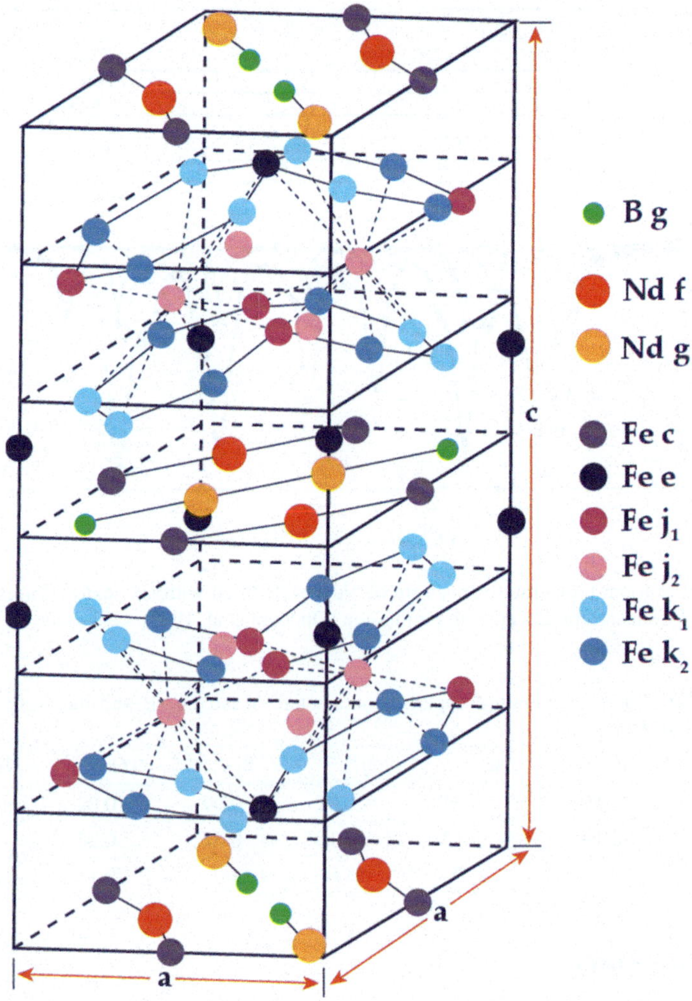

Fig. 6.29 The unit cell of $Nd_2Fe_{14}B$ with the Nd, Fe and B sites in the tetragonal structure (not to scale). Reproduced with permission (Herbst et al. 1984), © The American Physical Society

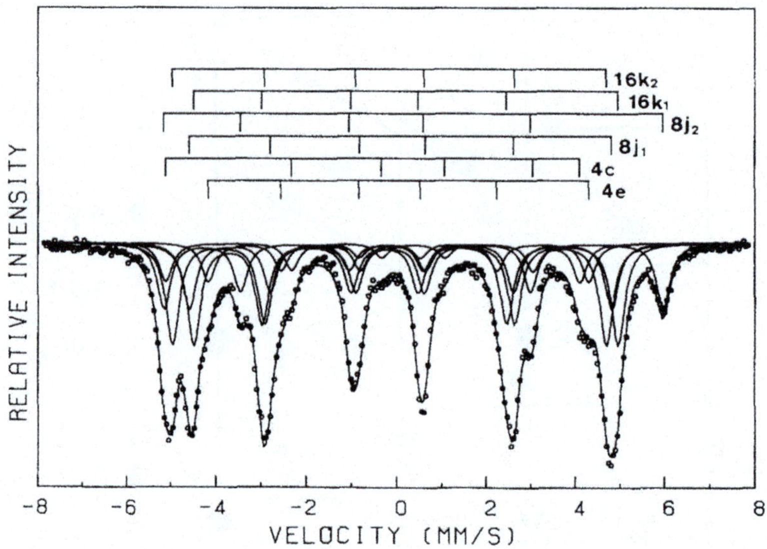

Fig. 6.30 The room temperature spectrum of $Nd_2Fe_{14}B$ fitted with six sextets corresponding to the various Fe sites. Reproduced with permission (Onodera et al. 1987), ©Elsevier

Table 6.5 The room temperature hyperfine parameters of $Nd_2Fe_{14}B$ and their site assignment (Onodera et al. 1987)

Site	4c	4e	$8j_1$	$8j_2$	$16k_1$	$16k_2$
δ (mm/s)	– 0.07	– 0.04	– 0.01	0.09	– 0.02	– 0.14
Δ (mm/s)	– 0.44	0.12	0.10	0.32	0.25	– 0.01
B_{hf} (T)	28.5	25.9	28.8	34.0	29.1	29.7

6.24 Summary

The Mössbauer spectrometry of various phases of Fe and Fe-based alloys is presented. The temperature and composition-dependent hyperfine parameters of the Fe-based metals and their alloys reported in the literature are discussed.

References

M.M. Abd-Elmeguid, H. Micklitz, Physica B **161**, 17 (1989)
M.M. Abd-Elmeguid, U. Hobuss, H. Micklitz, B. Huck, J. Hesse, Phys. Rev. B **35**, 4796 (1987)
U. Atzmony, M.P. Dariel, Phys. Rev. B **10**, 2060 (1974)
P. Bak, M.H. Jensen, J. Phys. C: Solid State Phys. **13**, L881 (1980)
G. Barault, J.-M. Greneche, Solid State Commun. **96**, 155 (1995)

A. Barla, H. Wilhelm, M.K. Forthaus, C. Strohm, R. Rüffer, M. Schmidt, K. Koepernik, U.K. Rößler, M.M. Abd-Elmeguid, Phys. Rev. Lett. **114**, 016803 (2015)
P.A. Beck, Phys. Rev. B **28**, 5336 (1983)
D. Bessas, I. Sergueev, K. Glazyrin, C. Strohm, I. Kupenko, D.G. Merkel, G.J. Long, F. Grandjean, A.I. Chumako, R. Rüffer, Phys. Rev. B **101**, 035112 (2020)
S.C. Bhargava, P.K. Iyengar, Phys. Status Solidi **38**, 601 (1976)
J. Bogner, W. Steiner, M. Reissner, P. Mohn, P. Blaha, K. Schwarz, R. Krachler, H. Ipser, B. Sepiol, Phys. Rev. B **58**, 14922 (1998)
R.J. Borg, R. Booth, C.E. Violet, Phys. Rev. Lett. **11**, 464 (1963)
L. Brossard, G.A. Fatseas, J.L. Dormann, P. Lecocq, J. Appl. Phys. **42**, 1306 (1971)
S.J. Campbell, P.E. Clark, T.J. Hicks, J. Phys. F: Metal Phys. **6**, 249 (1976)
I.A. Campbell, S. Senoussi, F. Varret, J. Teillet, A. Hamzic, Phys. Rev. Lett. **50**, 1615 (1983)
X. Chen, D.H. Ryan, Z. Altounian, L.X. Liao, J. Appl. Phys. **73**, 6038 (1993)
C.L. Chien, S.H. Liou, D. Kofalt, W. Yu, T. Egami, T.R. McGuire, Phys. Rev. B **33**, 3247 (1986)
A.M. Clogston, B.T. Matthias, M. Peter, H.J. William, E. Corenzwit, R.C. Sherwood, Phys. Rev. **125**, 541 (1962)
J.M.D. Coey, H. Sun, J. Magn. Magn. Mater. **87**, L251 (1990)
G. Cort, R.D. Taylor, J.O. Willis, J. Appl. Phys. **53**, 2064 (1982)
P.P. Craig, D.E. Nagle, W.A. Steyert, R.D. Taylor, Phys. Rev. Lett. **9**, 12 (1962)
P.P. Craig, W.A. Steyert, Phys. Rev. Lett. **13**, 802 (1964)
P.P. Craig, B. Mozer, R. Segan, Phys. Rev. Lett. **14**, 895 (1965a)
P.P. Craig, R.C. Perisho, R. Segan, W.A. Steyert, Phys. Rev. **138**, A1460 (1965b)
J. Crangle, W.R. Scott, J. Appl. Phys. **36**, 921 (1965)
B.R. Cuenya, W. Keune, D. Li, S.D. Bader, Phys. Rev. B **71**, 064409 (2005)
B. deMayo, D.W. Forester, S. Spooner, J. Appl. Phys. **41**, 1319 (1970)
B. deMayo, Phys. Rev. B **24**, 6503 (1981)
J. Desimoni, F.H. Sanchez, M.B.F. van Raap, H. Bearnas, C. Clerc, X.W. Lin, Phys. Rev. B **51**, 86 (1995)
J.K. Desmarais, W. Bi, J. Zhao, M.H. Hu, E. Alp, J.S. Tse, J. Chem. Phys. **154**, 214104 (2021)
J.W. Drijver, F. van der Woude, S. Radelaar, Phys. Rev. B **16**, 985 (1977a)
J.W. Drijver, F. van der Woude, S. Radelaar, Phys. Rev. B **16**, 993 (1977b)
J.W. Drijver, K. De Groot, F. van Der Woude, J. Phys. Colloques **35**, C6-465 (1974)
S.M. Dubiel, J. Zukrowski, J. Magn. Magn. Mater. **23**, 214 (1981)
R.A. Dunlap, J.D. McGraw, S.P. Farrell, J. Magn. Magn. Mater. **305**, 315 (2006)
D. Eliezer, S. Nadiv, M. Ron, Appl. Phys. Lett. **26**, 340 (1975)
R.D. Ellerbrock, A. Fuest, A. Schatz, W. Keune, R.A. Brand, Phys. Rev. Lett. **74**, 3053 (1995)
T. Ericsson, W. Karner, L. Häggström, K. Chandra, Phys. Scr. **23**, 1118 (1981)
M. Fanciulli, C. Rosenbland, G. Weyer, A. Svane, N.E. Christensen, Phys. Rev. Lett. **75**, 1642 (1995)
G. Filoti, V. Kuncsea, E. Navarro, A. Hernando, M. Rosenberg, J. Alloys Compd. **278**, 60 (1998)
P.A. Flinn, S.L. Ruby, Phys. Rev. **124**, 34 (1961)
J.B. Forsyth, C.E. Johnson, P.J. Brown, Philos. Mag. **10**, 713 (1964).
J.E. Frąckowiak, Phys. Status Solidi **87**, 109 (1985)
H. Franke, M. Rosenberg, J. Magn. Magn. Mater. **7**, 168 (1978)
P. Fratzl, F. Langmayr, Y. Yoshida, Phys. Rev. B **44**, 4192 (1991)
B. Fultz, Phys. Rev. B **44**, 9805 (1991)
A.G. Gavriliuk, V.V. Struzhkin, S N. Aksenov, A.A. Mirnonovich, I.A. Troyan, A.G. Ivanova, I.S. Lyubutin, JETP Lett. **117**, 126 (2023).
E. Germagnoli, C. Lamborizio, S. Mora, I. Ortalli, II Nuovo Cimento **42B**, 314 (1966)
U. Gonser, C.J. Meechan, A.H. Muir, H. Wiedersich, J. Appl. Phys. **34**, 2372 (1963)
U. Gonser, R.W. Grant, C.J. Meechan, A.H. Muir Jr., H. Wiedersich, J. Appl. Phys. **36**, 2124 (1965)
U. Gonser, S. Nasu, W. Kappes, J. Magn. Magn. Mater. **10**, 244 (1979)
T. Goto, H. Utsugi, K. Watanabe, Hyperfine Interact. **54**, 539 (1990)

T. Goto, H. Utsugi, K. Watanabe, J. Alloys Compd. **2014**, 173 (1994)
R.G. Grant, D.C. Cook, Hyperfine Interact. **94**, 2309 (1994)
J.-M. Greneche, F. Varret, J. Phys. C: Solid State Phys. **15**, 5333 (1982)
J.-M. Greneche, M. Henry, F. Varret, J. Magn. Magn. Mater. **26**, 153 (1982)
P.C.M. Gubbens, J.J. Van Loef, K.H.J. Buschow, J. Phys. Colloques **35**, C6-617 (1974)
D. Guenzburger, D.E. Ellis, Phys. Rev. B **52**, 13390 (1995)
L. Häggström, L. Granas, R. Wäppling, S. Devanarayananan, Phys. Scr. **7**, 125 (1973)
L. Häggström, T. Ericsson, R. Wappling, E. Karlsson, Phys. Scr. **11**, 55 (1975)
L. Häggström, T. Ericsson, R. Wäppling, K. Chandra, Phys. Scr. **11**, 47 (1975b)
R. Halbauer, U. Gonser, J. Mag. Magn. Mater. **35**, 55 (1983)
H.H. Hamdeh, B. Fultz, D.H. Pearson, Phys. Rev. B **39**, 11233 (1989)
H.H. Hamdeh, M.R. Al-Hilali, N.S. Dixon, L.S. Fritz, Phys. Rev. B **45**, 2201 (1992)
S.S. Hanna, J. Heberle, C. Littlejohn, G.J. Perlow, R.S. Preston, D.H. Vincent, Phys. Rev. Lett. **4**, 177 (1960)
C.C. Hao, P. Duwez, C.C. Tsuei, J. Appl. Phys. **42**, 4282 (1971)
J.F. Herbst, J.J. Croat, F.E. Pinkerton, W.B. Yelon, Phys. Rev. B **29**, 4176 (1984)
J.F. Herbst, Rev. Mod. Phys. **63**, 819 (1991)
J. Hesse, Hyperfine Interact. **47**, 357 (1989)
B.P. Hu, J.M.D. Coey, J. Less Common Met. **171**, 33 (1991)
Y. Ishikawa, Y. Endoh, J. Phys. Soc. Jpn. **23**, 205 (1967)
E. Jedryka, N. Randrianantoandro, J.-M. Greneche, A. Slawska-Waniewska, H.K. Lachowicz, J. Magn. Magn. Mater. **140–144**, 451 (1995)
C.E. Johnson, M.S. Rideout, T.E. Cranshaw, Proc. Phys. Soc. **81**, 1079 (1963)
K. Kanematsu, T. Ohoyama, J. Phys. Soc. Jpn. **20**, 236 (1965)
S.G. Kang, H. Onodera, H. Yamamoto, H. Watanabe, J. Phys. Soc. Jpn. **36**, 975 (1974)
J. Kapoor, J. Andres, F. Mezei, Y. Li, C. Polaczyk, D. Riegel, W.D. Brewer, E. Beck, S.B. Legoas, S. Frota-Pessôa, Phys. Rev. Lett. **77**, 2806 (1996)
D. Kaptás, E. Sváb, Z. Somogyvári, G. André, L.F. Kiss, J. Balogh, L. Bujdosó, T. Kemény, I. Vincze, Phys. Rev. B **73**, 012401 (2006)
P. Karipoth, A. Thirumurugan, S. Velaga, J.-M. Greneche, R.J. Joseyphus, J. Appl. Phys. **120**, 123906 (2016)
N. Kawamiya, K. Adachi, Y. Nakamura, J. Phys. Soc. Jpn. **33**, 1318 (1972)
W. Keune, R. Halbauer, U. Gonser, J. Lauer, D. Williamson, J. Appl. Phys. **48**, 2976 (1977)
C.W. Kimball, W.C. Phillips, M.V. Nevitt, R.S. Preston, Phys. Rev. **146**, 375 (1966)
C. Kimball, W.D. Gerber, A. Arrot, J. Appl. Phys. **34**, 1046 (1963)
T.A. Kitchens, W.A. Steyert, R.D. Taylor, Phys. Rev. **138**, A467 (1965)
S. Klimars, J. Hesse, B. Huck, J. Magn. Magn. Mater. **51**, 183 (1985)
D. Kodama, K. Shinoda, K. Sato, Y. Konno, R.J. Joseyphus, K. Motomiya, H. Takahashi, T. Matsumoto, Y. Sato, K. Tohji, B. Jeyadevan, Adv. Mater. **18**, 3154 (2006)
J.C. Krause, J. Schaf, M.I. da Costa Jr., C. Paduani, Phys. Rev. B. **61**, 6196 (2000)
J.P. Kuang, M. Matsui, K. Adachi, J. Phys. Colloques **49**, C8-135 (1988)
A. Laggoun, A. Guittoum, S. Bahamida, M. Boudissa, A. Fnidiki, Eur. Phys. J. Appl. Phys. **68**, 20301 (2014)
A. Laggoun, A. Hauet, J. Teillet, Hyperfine Interact. **54**, 825 (1990)
L. Larsen, H.R. Poulsen, N.O.R. Poulsen, L. Vistisen, J.M. Knudsen, Phys. Rev. Lett. **48**, 1054 (1982)
J. Lauer, W. Keune, Phys. Rev. Lett. **48**, 1850 (1982)
G. Le Caër, B. Malaman, B. Roques, J. Phys. F: Metal Phys. **8**, 323 (1978)
G. Le Caër, B. Malaman, G. Venturini, H.G. Wagner, U. Gonser, Hyperfine Interact. **28**, 631 (1986)
M.R. Lesoille, P.M. Gielen, Phys. Status Solidi **37**, 127 (1970)
D. Li, M. Freitag, J. Pearson, Z.Q. Qiu, S.D. Bader, Phys. Rev. Lett. **72**, 3112 (1994)
Z.W. Li, A.H. Morrish, Phys. Rev. B **55**, 3670 (1997)
F.P. Livi, J.D. Rogers, P.J. Viccaro, Phys. Status Solidi A **37**, 133 (1976)

G.J. Long, O.A. Pringle, F. Grandjean, K.H.J. Buschow, J. Appl. Phys. **72**, 4845 (1992)
G.J. Long, S. Mishra, O.A. Pringle, F. Grandjean, K.H.J. Buschow, J. Appl. Phys. **75**, 5994 (1994)
G. Longworth, Phys. Rev. **172**, 572 (1968)
Ph. Mangin, G. Marchal, J. Appl. Phys. **49**, 1709 (1978)
M. Manivel Raja, K. Chattopadhyay, B. Majumdar, A. Narayanasamy, J. Alloys Compd. **297**, 199 (2000)
C. Djéga-Mariadassou, P. Lecocq, G. Trumpy, J. Träff, P. Østergaard, Nuovo Cimento B **46**, 35 (1966)
S. M. Dubiel, Crit. Rev. Solid State Mater. Sci. (2024)
M. Miglierini, J.-M. Greneche, J. Phys. Condens. Matter **9**, 2303 (1997a)
M. Miglierini, J.-M. Greneche, J. Phys. Condens. Matter **9**, 2321 (1997b)
P. Monod, I.A. Campbell, Phys. Rev. Lett. **52**, 2096 (1984)
A. Monza, A. Meffre, F. Baudelet, J.P. Rueff, M. d'Astuto, P. Munsch, S. Huotari, S. Lachaize, B. Chaudret, A. Shukla, Phys. Rev. Lett. **106**, 247201 (2011)
J.A. Moyzis Jr., H.G. Drickamer, Phys. Rev. **171**, 389 (1968)
P. Munayco, J. Larrea J., Y.T. Xing, H. Micklitz, E.M. Baggio-Saitovitch, Phys. Rev. B **74**, 014423 (2006)
D.E. Nagle, P.P. Craig, P. Barrett, D.R.F. Cochran, C.E. Olsen, R.D. Taylor, Phys. Rev. **125**, 490 (1962)
Y. Nakamura, M. Shiga, N. Shikanzono, J. Phys. Soc. Jpn. **19**, 1177 (1964)
A. Narayanasamy, S. Ramasamy, T. Nagarajan, Solid State Commun. **41**, 769 (1982)
A. Narayanasamy, T. Nagarajan, P. Muthukumarasamy, T.S. Radhakrishnan, J. Phys. F: Metal Phys. **9**, 2261 (1979)
S. Nasu, T. Sasaki, T. Kawakami, T. Tsutsui, S. Endo, J. Phys. Condens. Matter **14**, 11167 (2002)
G.S. Neumann, L. Stixrude, R.E. Cohen, Proc. Natl. Acad. Sci. U.S.A. **101**, 33 (2004)
M. Nicol, G. Jura, Science **141**, 1035 (1963)
Y. Nishihara, T. Katayama, S. Ogawa, J. Phys. Soc. Jpn. **51**, 2487 (1982)
H. Ohno, M. Mekata, J. Phys. Soc. Jpn. **31**, 102 (1971)
K. Ono, Y. Ishikawa, A. Ito, J. Phys. Soc. Jpn. **17**, 1747 (1962)
H. Onodera, A. Fujita, H. Yamamoto, M. Sagawa, S. Hirosawa, J. Magn. Magn. Mater. **68**, 6 (1987)
C. Paduani, E.G. da Silva, G.A. Perez-Alcazar, Hyperfine Interact. **73**, 233 (1992)
G. Paduani, E.G. da Silva, G.A.P. Alcazar, M. McElfresh, J. Appl. Phys. **70**, 7524 (1991)
G.A. Perez Alcazar, E.G. da Silva, J. Phys. F: Metal Phys. **17**, 2323 (1987)
Yu.I. Petrov, E.A. Shafranovsky, Y.F. Krupyanskii, S.V. Essine, J. Appl. Phys. **91**, 352 (2002)
D.N. Pipkorn, C.K. Edge, P. Debrunner, G. de Pasquali, H.G. Drickamer, H. Franenfelder, Phys. Rev. **135**, A1604 (1964)
N. Ponpandian, A. Narayanasamy, K. Chattopadhyay, M. Manivel Raja, K. Ganesan, C.N. Chinnasamy, B. Jeyadevan, J. Appl. Phys. **93**, 6182 (2003)
R.S. Preston, S.S. Hanna, J. Heberle, Phys. Rev. **128**, 2207 (1962)
R.S. Preston, D.J. Lam, M.V. Nevitt, D.O. Van Ostenburg, C.W. Kimball, Phys. Rev. **149**, 440 (1966)
N. Randrianantoandro, F. Gaffet, J. Mira, J.M. Greneche, Solid State Commun. **111**, 323 (1999)
G.N. Rao, Hyperfine Interact. **24–26**, 1119 (1985)
M.S. Rideout, J. Phys. C: Solid State Phys. **2**, 1258 (1969)
G. Rixecker, P. Schaaf, U. Gonser, J. Phys. Condens. Matter **4**, 10295 (1992)
G. Rixecker, P. Schaaf, U. Gonser, Phys. Status Solidi A **139**, 309 (1993)
L.D. Roberts, J.O. Thomson, Phys. Rev. **129**, 664 (1963)
M. Rosenberg, V. Kuncser, O. Crisan, A. Hernando, E. Navarro, G. Filoti, J. Magn. Magn. Mater. **177–181**, 135 (1998)
P.J. Schurer, N.J.G. Hall, A.H. Morrish, Phys. Rev. B **18**, 4860 (1978)
R. Segnan, Phys. Rev. **160**, 404 (1967)
M. Shiga, Y. Nakamura, J. Phys. F: Metal Phys. **8**, 177 (1978)
T. Shinjo, Y. Nakamura, J. Phys. Soc. Jpn. **18**, 797 (1963)
T. Shinjo, F. Itoh, H. Takaki, Y. Nakamura, N. Shikazono, J. Phys. Soc. Jpn. **19**, 1252 (1964)

G. Shirane, C.W. Chen, P.A. Flinn, R. Nathans, Phys. Rev. **131**, 183 (1963)
D.A. Shirley, G.A. Westenbarger, Phys. Rev. **138**, A170 (1965)
A. Slawska-Waniewska, J.-M. Greneche, Phys. Rev. B **56**, R8491 (1997)
B. Stahl, N.S. Gajbhiye, G. Wilde, D. Kramer, J. Ellrich, M. Ghafari, H. Hahn, H. Gleiter, J. Weißmüller, R. Würschum, P. Schlossmacher, Adv. Mater. **14**, 24 (2002)
M.B. Stearns, J. Appl. Phys. **35**, 1095 (1964)
P. Steiner, V.V. Zdrojewski, D. Gumprecht, S. Hüfner, Phys. Rev. Lett. **31**, 355 (1973)
Y. Sumitomo, T. Moriya, H. Ino, F.E. Fujita, J. Phys. Soc. Jpn. **35**, 461 (1973)
K. Sumiyama, M. Shiga, Y. Nakamura, Phys. Status Solidi A **13**, K75 (1972)
K. Sumiyama, M. Shiga, K. Tachi, Y. Nakamura, J. Phys. Soc. Jpn. **40**, 1002 (1976)
K. Sumiyama, M. Shiga, Y. Kobayashi, Y. Nishi, Y. Nakamura, J. Phys. Soc. Jpn. **8**, 1281 (1978)
K. Sumiyama, T. Yoshitake, Y. Nakamura, Trans. Japan Inst. Metals **26**, 217 (1985)
K. Swieca, Y. Kondo, F. Pobell, Phys. Rev. B **56**, 6066 (1997)
T. Takabatake, H. Mazaki, T. Shinjo, Phys. Rev. B **21**, 2706 (1980)
Y. Tamada, S. Yamamoto, S. Nasu, T. Ono, Phys. Rev. B **78**, 214428 (2008)
J.L. Tholence, R. Tournier, Phys. Rev. Lett. **25**, 867 (1970)
S. Tomiyoshi, H. Yamamoto, H. Watanabe, J. Phys. Soc. Jpn. **21**, 709 (1966)
S. Tomiyoshi, H. Yamamoto, H. Watanabe, J. Phys. Soc. Jpn. **30**, 1605 (1971)
G. Trumpy, E. Both, C. Djéga-Mariadassou, P. Lecoeq, Phys. Rev. B **2**, 3477 (1970)
V.A. Tsurin, A.E. Ermakov, Y.G. Lebedev, B.N. Filippov, Phys. Status Solidi A **33**, 325 (1976)
J. Van Cauteren, L. Hermans, M. Rots, Phys. Rev. B **35**, 1939 (1987)
F. Varret, T. Besnard, J.M. Greneche, J. Magn. Magn. Mater. **73**, 157 (1988)
G. Venturini, B. Malaman, G. Le Caër, D. Fruchart, Phys. Rev. B **35**, 7038 (1987)
I. Vincze, I.A. Campbell, A.J. Meyer, Solid State Commun. **15**, 1495 (1974)
I. Vincze, T. Kemeny, S. Arajs, Phys. Rev. B **21**, 937 (1980)
C.E. Violet, R.J. Borg, Phys. Rev. **162**, 608 (1967)
C.E. Violet, R.J. Borg, Phys. Rev. Lett. **51**, 1073 (1983)
C.E. Violet, R.J. Borg, Phys. Rev. Lett. **52**, 2098 (1984)
J.T. Wang, D.S. Wang, Y. Kawazoe, Appl. Phys. Lett. **88**, 132513 (2006)
R. Wäppling, L. Häggström, E. Karlsson, Phys. Scr. **2**, 233 (1970)
R. Wäppling, L. Häggström, Phys. Lett. **28A**, 173 (1968)
G.K. Wertheim, J.H. Wernick, Phys. Rev. **125**, 1937 (1962)
G.K. Wertheim, V. Jaccarino, J.H. Wernick, Phys. Rev. **135**, A151 (1964)
G.L. Whittle, S.J. Campbell, J. Phys. F: Metal Phys. **15**, 693 (1985)
J.M. Williams, J.B. Dunlop, B. Leontic, Mater. Sci. Eng. **23**, 307 (1976)
D.L. Williamson, S. Bukshpan, R. Ingalls, Phys. Rev. B **6**, 4194 (1972)
B. Window, J. Phys. C: Metal Phys. **3**, S323 (1970)
B. Window, Phys. Rev. B **6**, 2013 (1972)
M.L. Winterrose, M.S. Lucas, A.F. Yue, I. Halevy, L. Mauger, J.A. Munoz, J. Hu, M. Lerche, B. Fultz, Phys. Rev. Lett. **102**, 237202 (2009)
H.P. Wit, L. Niesen, H. De Waard, Hyperfine Interact. **5**, 233 (1978)
H. Yamamoto, J. Phys. Soc. Jpn. **20**, 2166 (1965)
H. Yamamoto, J. Phys. Soc. Jpn. **21**, 1058 (1966)
H. Yamauchi, H. Watanabe, Y. Suzuki, H. Saito, J. Phys. Soc. Jpn. **36**, 971 (1974)
E.P. Yelsukov, E.V. Voronina, V.A. Barinov, J. Magn. Magn. Mater. **115**, 271 (1992)
S.L. Zhang, K. Sumiyama, Y. Nakamura, J. Magn. Magn. Mater. **73**, 58 (1988)
Y. Zhou, D. Wang, Y. Kawazoe, Phys. Rev. B **59**, 8387 (1999)

Chapter 7
Mössbauer Spectrometry of Spinel Ferrites

7.1 Spinel Ferrites

Spinel ferrites are ferrimagnetic materials with two sublattices, A and B, occupied by metal ions in the spinel structure and the magnetic behavior is manifested through superexchange interaction. They have the general formula $AO{\cdot}B_2O_3$, where the A-site (tetrahedral site) cation is surrounded by four oxygen anions, and the B-site [octahedral site] cation is surrounded by six oxygen anions. The ferrites can be prepared by numerous chemical and physical methods leading to different morphologies (single crystal, crystalline powders, pellets, thin films, nanoparticles) opening up the possibility of tayloring their physical properties for numerous applications. Their magnetic behavior which is manifested by a dominant superexchange interaction are highly dependent on the nature and distribution of cations in their respective sites, as well as the morphology and size of nanocrystallites, all of which related to the synthesis process and experimental conditions. As a result, different magnetic behaviors can be observed such as ferrimagnetic, antiferromagnetic, non-collinear, Yafet-Kittel and spin glass. They behave like soft and hard magnets, and most of them have high electrical resistivity, low electrical losses, and very good chemical stability. Soft ferrites with their low coercivity can act as conductors of magnetic fields, and are used in the electronics industry for high-frequency inductors, transformers, antennas and in various microwave components. Some hard ferrites can be used to produce permanent magnets for applications such as loudspeakers, refrigerator and electric micromotors.

There are eight formula units in a unit cell of spinel ferrite, as shown in Fig. 7.1, with each formula unit having 7 ions, constituting 56 ions per unit cell. Of the 64 A-sites, only 8 are occupied, and among the 32 available B-sites, 16 sites are occupied, making a total of 24 cation sites with the remaining 32 oxygen ions. The cation sites are in the interstitial positions of the fcc cubic spinel structure ($Fd\overline{3}m$). The spinel ferrites are classified into normal, inverse and mixed spinels according to the cation distribution, as shown in Table 7.1. The normal spinel has 8 divalent metal ions (M)

R. J. Joseyphus and J.-M. Greneche, *Fundamentals of ^{57}Fe Mössbauer Spectrometry*,
https://doi.org/10.1007/978-981-99-8653-8_7

in the A-site and all the ferric ions in the B-site, while inverse spinel has 8 divalent metal ions in the B-site and an equal number of ferric ions in both the A- and B-sites. Mixed spinel has divalent metal ions in both A- and B- sites along with the ferric ions.

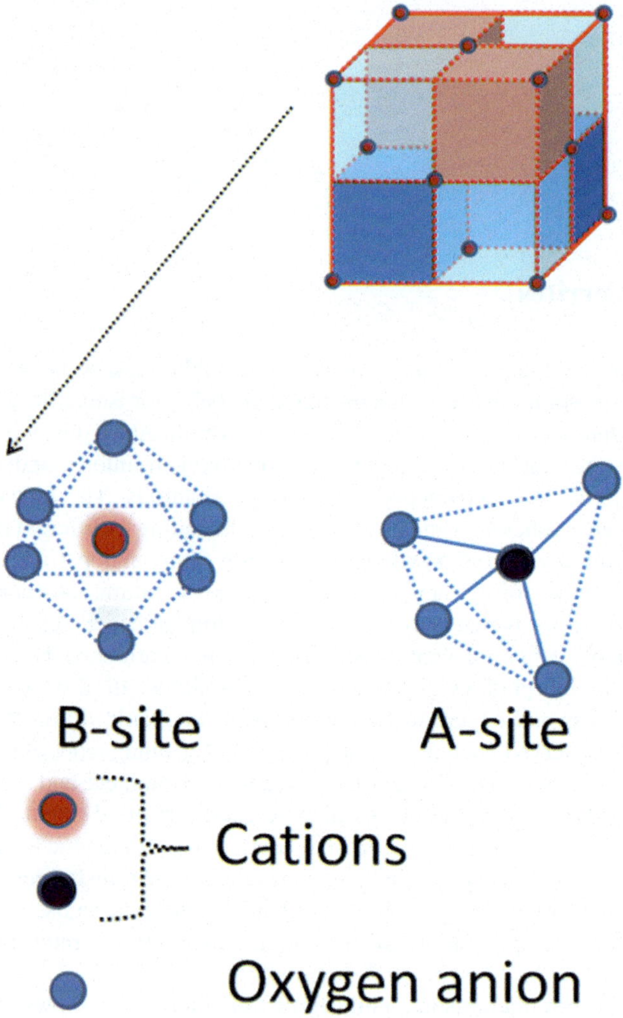

Fig. 7.1 Unit cell of a spinel ferrite has eight formula units with 64 A-sites (tetrahedral) and 32 B-sites [octahedral]. The A-site cation has four oxygen ion neighbors, whereas the B-site cation has six

Table 7.1 The cation distribution in A- and B- sites according to the types of spinel ferrite

Site	Normal spinel	Inverse spinel	Mixed spinel
A-site divalent ion	8 M^{2+}	–	$(1 - x)M^{2+}$
A-site trivalent ion	–	$8Fe^{3+}$	xFe^{3+}
B-site divalent ion	–	$8M^{2+}$	yM^{2+}
B-site trivalent ion	$16Fe^{3+}$	$8Fe^{3+}$	$(1 - y)Fe^{3+}$
Total cations	24	24	24

7.2 Magnetite (Fe_3O_4)

Magnetite, Fe_3O_4, is a mixed valence Fe oxide which has concentrated great attention since a long time because this famous natural mineral is probably the oldest magnetic material known. Bulk magnetite has the inverse spinel structure where the Fe^{3+} ions occupy the tetrahedral site, and the octahedral site is occupied by both Fe^{2+} and Fe^{3+} ions, represented as $\left(Fe^{3+}\right)[Fe^{3+}Fe^{2+}]O_4^{2-}$. The Mössbauer spectrum of Fe_3O_4 with an applied field at various temperatures revealed the ferrimagnetic behavior.

The Mössbauer spectra of Fe_3O_4 from room temperature to Curie temperature are shown in Fig. 7.2 (van der Woude et al. 1968). The A- and B- site hyperfine fields could be easily resolved at room temperature (295 K) (Fig. 7.2a). With increasing temperature, the hyperfine field decreases (Fig. 7.2b), and the A- and B- site hyperfine fields are distinct even up to 825 K (Fig. 7.2c). A paramagnetic singlet is observed at a temperature above 859 K (Fig. 7.2 (d)). Usually, the hyperfine field for A-site (50.8 T) is higher than B-site, whereas a range of B-site hyperfine fields are reported (45–48.9 T) (Van Loef 1966; Häggström et al. 1978).

The temperature-dependent A- and B- site hyperfine fields and isomer shift values for polycrystalline Fe_3O_4 from 130 K are shown in Fig. 7.3. (De Grave et al. 1993). The isomer shift obtained with respect to Fe at room temperature is distinct for the A- and the B- sites. The smallest value associated with the least intense magnetic component is clearly attributed to the Fe^{3+} species while the other corresponds neither to Fe^{3+} nor Fe^{2+}. In fact, the perfectly intermediate value, can be attributed to the presence of an average valence value of 2.5+ per Fe ion in B sites, i.e. $Fe^{2.5+}$ species, that is consistent with the presence of delocalized electron. This exchange electron between ferrous and ferric sites is always observed above the Verwey temperature ($T_V \approx 120$ K). Below T_V, magnetite undergoes a sharp first-order metal–insulator transition associated with a decrease of the conductivity estimated at about two orders of magnitude and resulting from charge ordering mechanism of Fe^{2+} ions on the B sublattice with the formation of charged (001) planes alternately occupied by ferric and ferrous ions. In addition, this description is consistent with the $Fe^{2+}Fe^{3+}{}_2O_4$ formula of stoichiometric magnetite. The linewidths are found to vary from 0.25 to 0.40 mm/s, increasing with temperature to reach 0.8 mm/s. The decrease in isomer shift values and the resolved A- and B- site hyperfine fields with temperature could be

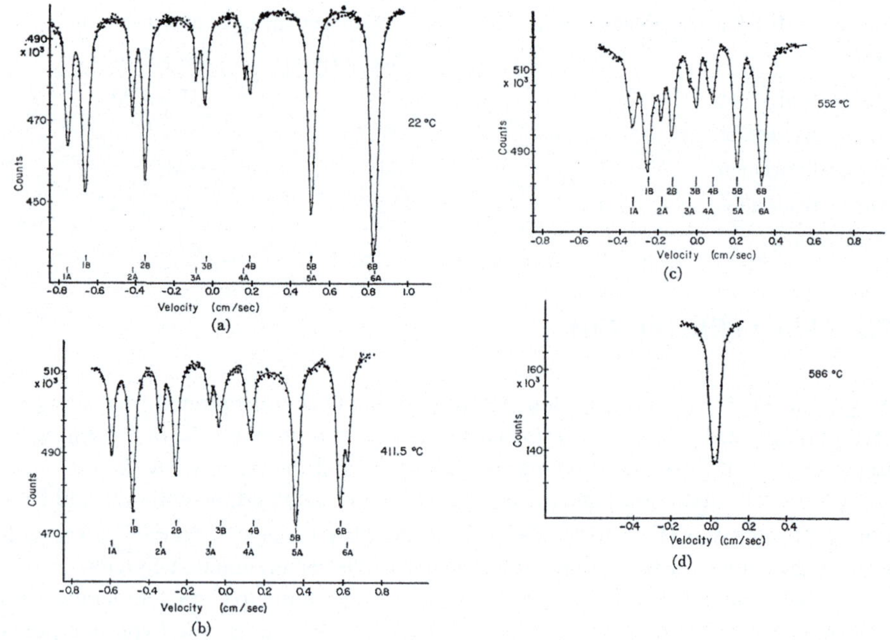

Fig. 7.2 Mössbauer spectra of Fe_3O_4 at **a** 295, **b** 685, **c** 825 and **d** 859 K. Reprinted with permission (van der Woude et al. 1968), © The American Physical Society

observed for polycrystalline Fe_3O_4. The A- and B- site intensities are 34–35 and 64–65%, respectively, in the entire temperature range. The quadruple shift continuously decreases from 0.2 to −0.2 mm/s with temperature, which may arise from local distortions. For the B-site, the net spin could be obtained using the model of non-localized electrons strongly coupled to the localized electrons. The fitting gives a net spin (S) from local spin (S_1) + electron spin (1/2), $S = S_1 + 1/2 = 2.9$, Curie temperature, $T_c = 875$ K and B-site saturation field of $B_{hf}(0) = 48.0$ T. In an external magnetic field, the A-site hyperfine field adds up while that of B-site gets subtracted. The values of the Mössbauer parameters of Fe_3O_4 form the basis for comparing the changes with other metal ion substitutions.

The Mössbauer spectra of stoichiometric and non-stoichiometric magnetite at 300 K in the absence of a magnetic field and in a transverse magnetic field of 0.84 T (Daniels and Rosencwaig 1969) are shown in Fig. 7.4. On application of an external transverse field ($B_\perp$) of 0.84 T to magnetite, the magnitude of hyperfine field of A-site increases whereas that of B-site decreases, resulting in the resolved sixth line of the sextet, as observed from Fig. 7.4b. The Mössbauer parameters of the stoichiometric magnetite at room temperature with and without an external applied field (Daniels and Rosencwaig 1969) and that of natural magnetite (Doriguetto et al. 2003) are presented in Table 7.2. The A-site hyperfine field has a larger magnitude than the B-site. Usually, the isomer shift of the A-site is lower than the B-site in

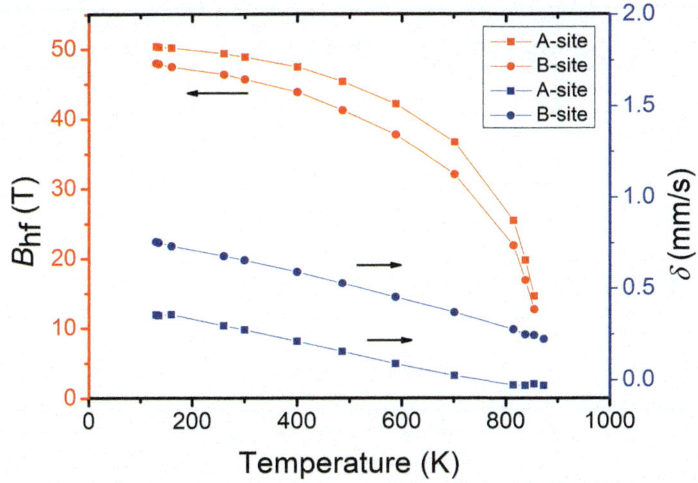

Fig. 7.3 The temperature-dependent hyperfine field and isomer shift (relative to Fe at RT) values of polycrystalline Fe_3O_4. Data from (De Grave et al. 1993)

spinel ferrites due to its lower oxygen coordination, and the quadrupole shift of both sites is negligible due to the cubic environment. The hyperfine parameters are close to each other for the stoichiometric and non-stoichiometric magnetite. However, the ratio of intensity between the A- and B- site is 0.58 for the stoichiometric magnetite, 0.98 for the non-stoichiometric and 0.51 for the natural magnetite. It was concluded that most of the tetrahedral sites are occupied by ferric ions, while vacancies are predominant in the octahedral site of the non-stoichiometric magnetite. The A-site hyperfine parameters arise from the Fe^{3+} ions while it is difficult to distinguish the Fe^{2+} and Fe^{3+} in the B-site due to fast electron exchange resulting in broadening (Kündig and Hargrove 1969).

Below the Verwey temperature, magnetite transforms from cubic to orthorhombic accompanied by a sharp increase in resistivity. Also, the A- and B- site hyperfine fields together result in almost five sextets (two Fe^{2+} and two Fe^{3+} at B-site) (Hargrove and Kündig 1970). At 82 K, in transverse and longitudinal fields on Fe_3O_4 single crystals cut along [100] direction, the isomer shift and quadrupole shift values could be as large as 1.2 and −2.62 mm/s, respectively, for Fe^{2+} in B-site (Hargrove and Kündig 1970). Low-temperature measurements conducted by field cooling through T_V exhibit narrow linewidths. The tetrahedral site cation environment is closer to cubic than the octahedral site.

The Mössbauer spectrum at 4.2 K for the Fe_3O_4 single crystal cut perpendicular to the [111] direction shows narrow linewidths without the requirement of field cooling and well-resolved A-site hyperfine field at various external applied fields, as displayed in Fig. 7.5 (Berry et al. 1998). The Mössbauer parameters without and with an applied field of 13 T on the single crystal Fe_3O_4 at 4.2 K are listed in Table 7.3. A single hyperfine field sextet corresponding to A-site and four B-site

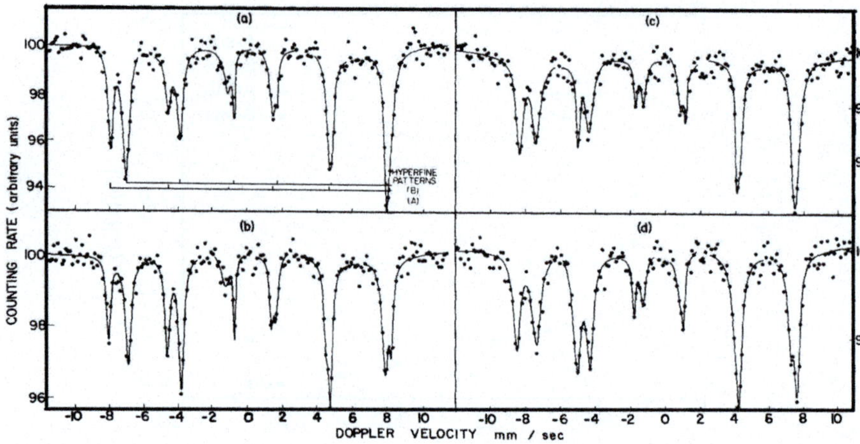

Fig. 7.4 Mössbauer spectra of stoichiometric magnetite (Fe_3O_4) at 300 K **a** without and **b** with a transverse magnetic field of 0.84 T, and non-stoichiometric magnetite **c** without and **d** with a transverse magnetic field of 0.84 T. The outermost lines are better resolved in an applied field. Reprinted with permission (Daniels and Rosencwaig 1969), © Elsevier

Table 7.2 The Mössbauer parameters for stoichiometric Fe_3O_4 at 300 K without and with an applied field

Sample details	Site assignment	B_{hf}/ B_{eff}(T)	Isomer shift (mm/s)	References
Fe_3O_4 at 300 K without external field	A-site	49.8	+0.4 (SS)	Daniels and Rosencwaig (1969)
	B-site	46.3	+0.8 (SS)	
Fe_3O_4 at 300 K with a transverse external field of 0.84 T	A-site	50.4	+0.4 (SS)	
	B-site	45.7	+0.7 (SS)	
Fe_3O_4 at 300 K without external field	A-site	48.9	+0.25	Doriguetto et al. (2003)
	B-site	45.7	+0.65	

sextets are demonstrated from these experiments apart from the [111] spin alignment. The relative intensity ratio of A- and B- sites is 28:78, and the quadrupole shift value for Fe^{2+} is near zero with an external field. This established that the electric field gradient (V_{zz}) lies along the [001] direction. The notion of charge state ordering has been refuted, and the studies below T_V are still being pursued by various researchers (Garcia and Subias 2004).

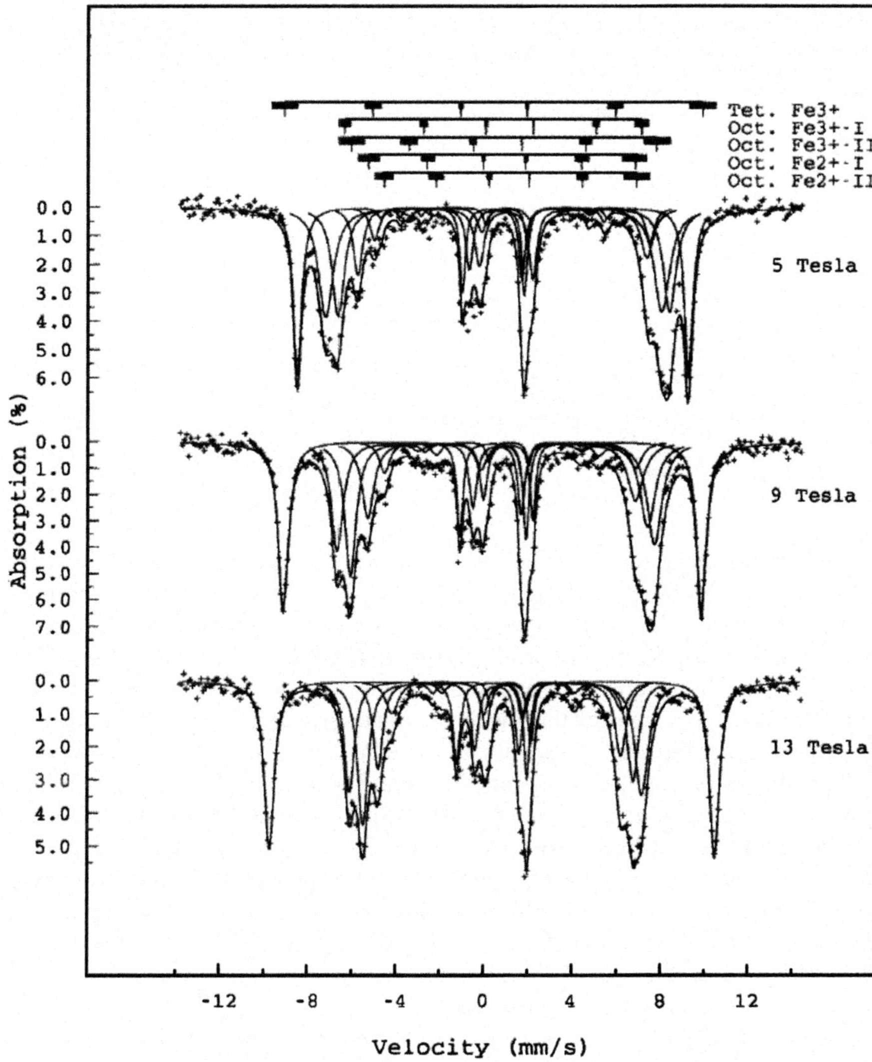

Fig. 7.5 Mössbauer spectra of Fe_3O_4 in applied fields of 5, 9 and 13 T at 4.2 K for a single crystal aligned along [111]. Reprinted with permission (Berry et al. 1998), © IOP Publishing Ltd

7.2.1 High Pressure Studies

Moderate pressure of up to 15 GPa on polycrystalline magnetite reduced the average hyperfine field of tetrahedral and octahedral sites (Halasa et al. 1974). The results suggest electron exchange between ferrous and ferric sites. The magnetite transforms into a monoclinic phase that coexists with the spinel phase at pressures beyond 25 GPa (Pasternak et al. 1994). Further studies with high pressures up to 140 GPa have

Table 7.3 The Mössbauer parameters for single crystal Fe_3O_4 obtained from the spectrum at 4.2 K in the absence of a field and in a 13 T external magnetic field (γ-ray directed along [111] direction of the crystal) (Berry et al. 1998)

Sample condition	Site information	B_{hf}/B_{eff} (T)	δ (mm/s)	Δ (mm/s)
0 T	A-Fe^{3+}	50.1	0.4	0.0
	B-Fe^{3+}(I)	52.2	0.5	0.0
	B-Fe^{3+}(II)	49.8	0.8	−0.3
	B-Fe^{2+} (I)	48.2	1.0	−0.4
	B-Fe^{2+} (II)	35.9	1.0	0.9
13 T	A-Fe^{3+}	62.8	0.4	0.0
	B-Fe^{3+}(I)	40.0	0.8	−0.4
	B-Fe^{3+}(II)	39.3	0.7	0.1
	B-Fe^{2+} (I)	34.2	0.8	−0.1
	B-Fe^{2+} (II)	32.4	1.1	−0.1

shown that complete transformation to the high-pressure monoclinic phase occurs beyond 60 GPa, as observed from the emergence of a doublet in the spectra shown in Fig. 7.6 (Xu et al. 2004). The sextet to doublet transformation temperature (spinel to monoclinic phase) decreases with pressure. At 55 GPa, where the non-magnetic doublet is present, the quadrupole splitting and isomer shift are 0.85 and 0.1 mm/s, respectively. It was suggested that the ferric species had undergone a Mott transition due to charge-transfer gap closure beyond 55 GPa.

Experiments on single-crystal magnetite were conducted to clarify structural or spin transition ambiguity with high pressure (Glazyrin et al. 2012). The Mössbauer experiments were conducted with an applied pressure up to 25 GPa using a 370 MBq ^{57}Co source in a Rh matrix. Comparing the experimental results from resistivity, single crystal X-ray diffraction and Mössbauer spectrometry, a high-spin to intermediate-spin transition and inverse to normal spinel transformation were ruled out. A gradual decrease in the A-site hyperfine field with pressure was attributed to the decrease in the effective spin magnetic moment contribution at the A-site. The A-site isomer shift value was nearly constant, whereas the B-site isomer shift decreased with pressure due to 3d electron delocalization, resulting in a larger s-electron density. The increase in Jahn–Teller distortion was also inferred from increased B-site quadrupole splitting values with high pressure. The studies suggested the increase in hybridization of majority spin electrons up to 15 GPa and a decrease in B-site hyperfine field resulting in lower magnetization of magnetite with further increase in pressure.

Synchrotron Mössbauer spectrometry experiments were performed on Fe_3O_4 single crystals at 10–290 K and pressure up to 40 GPa along with neutron diffraction (Kozlenko et al. 2019). The high-pressure studies showed an increase in Curie temperature of 20.5 K/GPa up to 4.5 GPa and suppression of T_V above 8 GPa. The structure changes from cubic to orthorhombic beyond 28 GPa and a ferrimagnetic T_c of 420 K was obtained, from the fit using the Brillouin function. A high-spin to low-spin crossover of Fe^{3+} is evident at pressures greater than 28 GPa at 290 K.

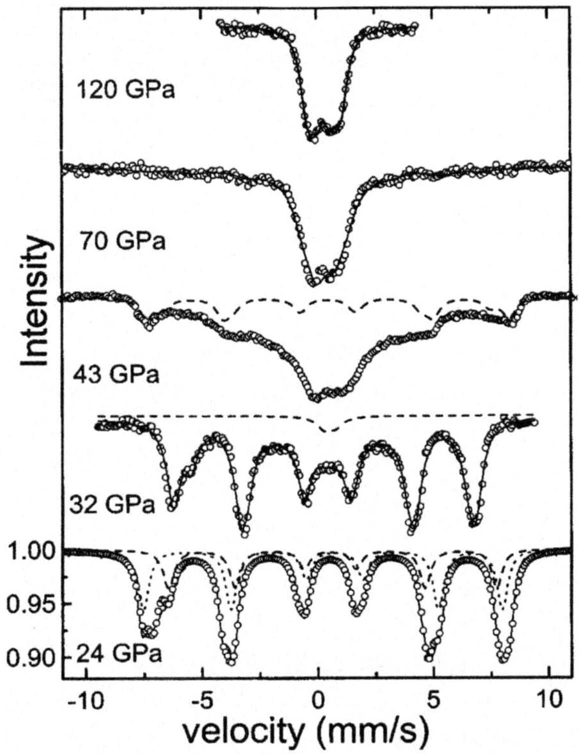

Fig. 7.6 The room temperature Mössbauer spectra of magnetite under various applied pressures. Reprinted with permission (Xu et al. 2004), © The American Physical Society

The pressure-induced increase in hybridization strength between Fe-d orbital and O-p orbital leads to enhanced electronic delocalization (Kozelnko et al. 2019). Based on the pressure-dependent magnetic properties and from the literature, the phase diagram of Fe_3O_4 has been arrived at, as shown in Fig. 7.7. The Verwey phase is present at low pressures and low temperatures, whereas the orthorhombic phase with lower T_c occurs at high pressures. The inverse spinel to normal spinel transformation has not been observed in high-pressure studies.

At pressures beyond 70 GPa, the orthorhombic *Bbmm* space group transforms to the subgroup *Pmma*, which is a second-order transition and does not give any volume change (Greenberg et al. 2017). However, the observed decrease in volume beyond 70 GPa and metallic behavior at 90 GPa accompanied by the loss of magnetic interaction in the octahedral site suggest site-selective Mott transition. The Mott transition was not complete even up to 120 GPa due to the persistence of high spin ferric components.

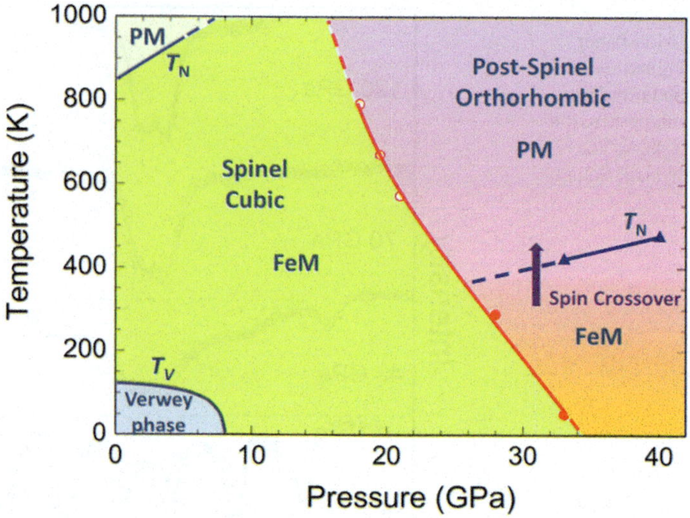

Fig. 7.7 The structural, magnetic and electronic phase diagram of magnetite (FeM and PM are ferrimagnetic and paramagnetic states, respectively). Reproduced from (Kozlenko et al. 2019)

7.3 Maghemite (γ-Fe_2O_3)

Maghemite has the fcc inverse spinel structure similar to magnetite; therefore, it is difficult to identify through X-ray diffraction. The structure has only Fe^{3+} cations with a net magnetic moment of 2.5 μ_B per formula unit and a Curie temperature as high as 950 K. The hyperfine field for A- and B- sites could not be resolved easily as the B_{hf} is close to 50 T with almost similar isomer shift values for both the sites (Bauminger et al. 1961). The A- and B- site B_{hf} could be resolved using an external field of 1.7 T (Armstrong et al. 1966). The ratio of the relative intensity of the A- and B- sites is close to 0.6 implying entire vacancies in the B-site. The reduction in saturation magnetization due to surface spin effects was examined in 5–7 nm γ-Fe_2O_3 (Coey 1971) through Mössbauer spectrometry at 5 K. The non-zero second and fifth line of the sextets in an applied field of 5 T is a signature of non-collinear spin arrangement in the A- and B- sites, as shown in Fig. 7.8.

Ultrafine 6.5 nm γ-Fe_2O_3 is superparamagnetic as observed from the Mössbauer spectrum recorded at 296 K with and without applied field (Fig. 7.9) (Coey and Khalafalla 1972). Figure 7.9a shows a broad absorption peak, signifying a relaxation time of the order of 10^{-9} s. This broad absorption further resolves into sextets by applying an external field of 5 T parallel to the γ-ray direction (Fig. 7.9b). Nearly 34% of cations have their spins canted at angles between 0 and 90°.

Mössbauer spectra of rod-shaped γ-Fe_2O_3 particles at 295 and 4.2 K in the absence of external field and with external transverse and longitudinal fields were studied at 315 and 4.2 K (Pollard 1988). A transverse field of 1.65 T could resolve the B_{hf} components even at a higher temperature of 315 K in contrast to the unresolved A-

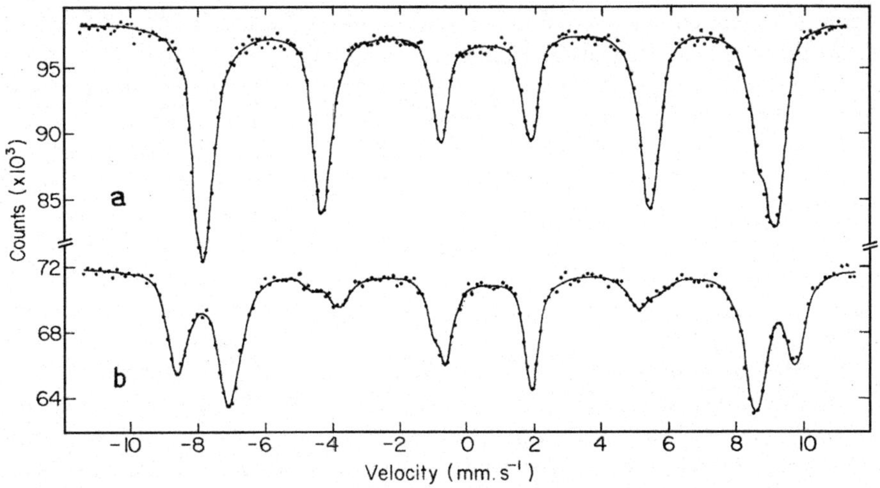

Fig. 7.8 Mössbauer spectra of ultrafine γ-Fe_2O_3 at 5 K **a** without and **b** with an applied field of 5 T. The non-vanishing second and fifth lines with an applied field indicate a non-collinear spin arrangement while the A- and B- site hyperfine fields are well resolved. Reprinted with permission (Coey 1971), © The American Physical Society

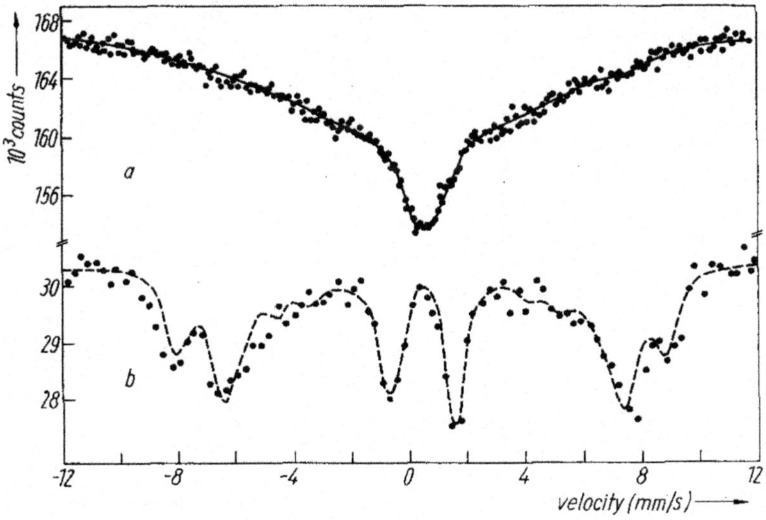

Fig. 7.9 Mössbauer spectra of 6.5 nm γ-Fe_2O_3 at 296 K **a** without external field and **b** with an external field of 5 T. Reprinted with permission (Coey and Khalafalla 1972), © John Wiley & Sons, Inc

and B- sites without an external magnetic field. The difference in B_{hf} of A- and B-sites was as large as 10 T at 4.2 K in a longitudinal field of 6.1 T. However, it was suggested that an external field above 6 T was required to rule out hematite (α-Fe_2O_3) impurity, often misinterpreted as spin canting (Pollard 1990).

The spin canting has been studied using NMR and in-field Mössbauer spectrometry that established that both the octahedral and tetrahedral sites are canted (Daou et al. 2010). Mössbauer spectra of the nanosized and commercial maghemite under external fields, as shown in Fig. 7.10, result in a canting angle of 22–23° for the A-site and 15–18° for the B-site. The measurements from nanosized and commercial powders also suggest that the canting arises due to the disordered structure and surface effects.

The Mössbauer parameters of γ-Fe_2O_3 from various studies are listed in Table 7.4. Generally, the A- and B- site hyperfine fields are not well resolved without an external applied field. The close hyperfine fields around 50 T for both the sites split into A-site field with a larger B_{hf} in an external field. Conversely, the B-site field decreases below 50 T with an external magnetic field. The quadrupole shifts are close to zero for both

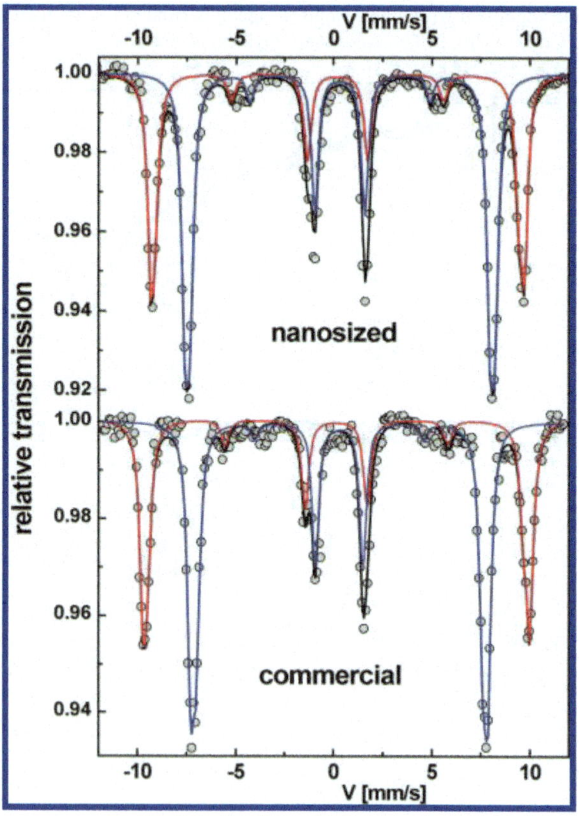

Fig. 7.10 Mössbauer spectra of the nanosized and commercial maghemite under external fields of 6 and 8 T, respectively, recorded at 11 K. Reprinted with permission (Daou et al. 2010), © American Chemical Society

Table 7.4 The Mössbauer parameters for γ-Fe_2O_3 at various temperatures and applied fields

Sample	Temperature	Applied field	B_{hf}/ B_{eff}(T)	δ (mm/s)	References
γ-Fe_2O_3	RT	0	48.8 (A) 49.9 [B]	0.04 (SS) 0.05 (SS)	Armstrong et al. (1966)
γ-Fe_2O_3	RT	0	50.0	0.32	de Bakker et al. (1990)
	80 K	0	51.7	0.40	
γ-Fe_2O_3 obtained from lepidocrocite and annealed	10 K	6.0 T	55.8 (A) 48.0 [B]	0.39 0.48	Da Costa et al. (1996)
γ-Fe_2O_3	450 K	0	47.6 (A) 46.5 [B]	0.15 0.27	
	300 K	0	50.1 (A) 50.0 [B]	0.24 0.36	
	80 K	0	51.9 (A) 52.9 [B]	0.35 0.47	
	4.2 K	6.0 T	57.7 (A) 47.3 [B]	0.33 0.45	
Nanosized γ-Fe_2O_3	11 K	6.0 T	58.1 (A) 48.0 [B]	0.39 0.52	Daou et al. (2010)
Commercial γ-Fe_2O_3	11 K	8.0 T	60.5 (A) 46.0 [B]	0.37 0.51	

sites. Most of the studies also report spin canting angles that may vary according to the synthesis conditions of the sample.

7.4 Zn Substituted Fe_3O_4

The inverse spinel Fe_3O_4 could turn to a normal spinel with Zn^{2+} substitution that prefers the tetrahedral site. Room temperature Mössbauer spectra of $Zn_xFe_{3-x}O_4$ (x = 0.2, 0.4, 0.6, 0.8, 1) have shown that the tetrahedral ferric ion hyperfine field is more prominent than octahedral ferrous and ferric ion fields which are indistinguishable and a quadrupole doublet appears when x = 0.6 (Dobson et al. 1970). The variations of A- and B- site hyperfine fields at 300 and 77 K with composition for $Zn_xFe_{3-x}O_4$ (x = 0,0.2, 0.4, 0.6, 0.8) are shown in Fig. 7.11 (Srivastava et al. 1976). The A-site isomer shift $\delta(A) \approx 0.4$ mm/s < $\delta(B) \approx 0.5$–0.8 mm/s, both at 300 and 77 K. Significant reduction in the hyperfine field is observed with more than half of Zn substitution. The variation in the hyperfine field with Zn substitution is due to A-B and B-B super-transferred hyperfine fields. The A- and B- site hyperfine fields are resolved at 300 K as well as at 77 K while it vanishes at RT for x = 0.8.

Low-temperature experiments at 4.2 K have shown that the A- and B- site ferric ion hyperfine fields are indistinguishable with Zn substitution in Fe_3O_4 (Dickof et al. 1980). Using in-field Mössbauer spectrometry, the magnetic properties are explained

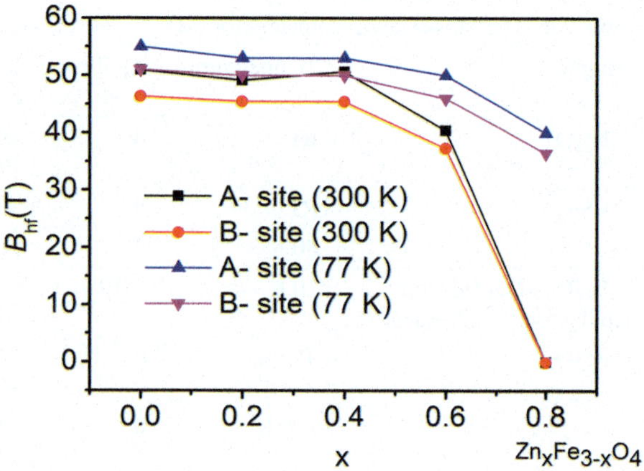

Fig. 7.11 The variation of A- and B- site hyperfine fields at 300 and 77 K with composition for $Zn_xFe_{3-x}O_4$ (x = 0, 0.2, 0.4, 0.6, 0.8). Data from (Srivastava et al. 1976)

based on a spin canting that is dominant with higher Zn substitution, especially beyond x = 0.4. Cd substitution show effects similar to Zn substitution, and the difference in isomer shift values between the A- and B- sites without substitution and with x = 0.3 is nearly 0.2 mm/s (Ok and Evans 1976).

7.5 Zinc Ferrite ($ZnFe_2O_4$)

Zinc ferrite is a normal spinel and shows interesting magnetic properties with nano-materials. The Zn^{2+} divalent ion occupies the A-site, completely replacing the ferrous ion and has zero net magnetic moment. The equal proportion of Fe^{3+} ions in the B-site cancels each other in an antiparallel arrangement, and therefore, the Zn-ferrite shows a non-magnetic nature with a T_N of 10 K. At room temperature, the Mössbauer spectrum of bulk zinc ferrite is characterized by a paramagnetic doublet (Mizoguchi and Tanaka 1963) and the quadrupole splitting is in the range 0.33–0.36 mm/s at various temperatures above RT, as shown in Table 7.5.

Ultrafine $ZnFe_2O_4$ particles in the size range of 10–20 nm prepared using the hydrothermal method have shown a doublet at RT and an antiferromagnetic sextet

Table 7.5 The Mössbauer parameters of $ZnFe_2O_4$ at various temperatures (Mizoguchi and Tanaka 1963)

Temperature (K)	Isomer shift[a] (mm/s)	Quadrupole splitting (mm/s)
299	0.45	0.36
471	0.36	0.34
629	0.26	0.34
783	0.22	0.33

[a] Relative to stainless steel

below its T_N of 13 K (Pannaparayil et al. 1990). The doublet at room temperature with a large quadrupole splitting of 1.18 mm/s suggested the ultrafine nature of the particles. At 4.2 K, the sextet was fitted with a hyperfine field of 52.0 T corresponding to Fe^{3+} on the B-site. The 4.2 K spectrum hardly exhibited splitting in an applied longitudinal field of 10 T, indicating the antiferromagnetic nature of $ZnFe_2O_4$. The ultrafine $ZnFe_2O_4$ obtained from the hydrothermal synthesis using Zn-α-Fe_2O_3 particles produces a relaxation spectrum at low temperatures, as shown in Fig. 7.12 (Kopcewicz et al. 2019). The hyperfine field of 47 T and an isomer shift of 0.45 mm/s were obtained from the spectrum recorded at 12 K.

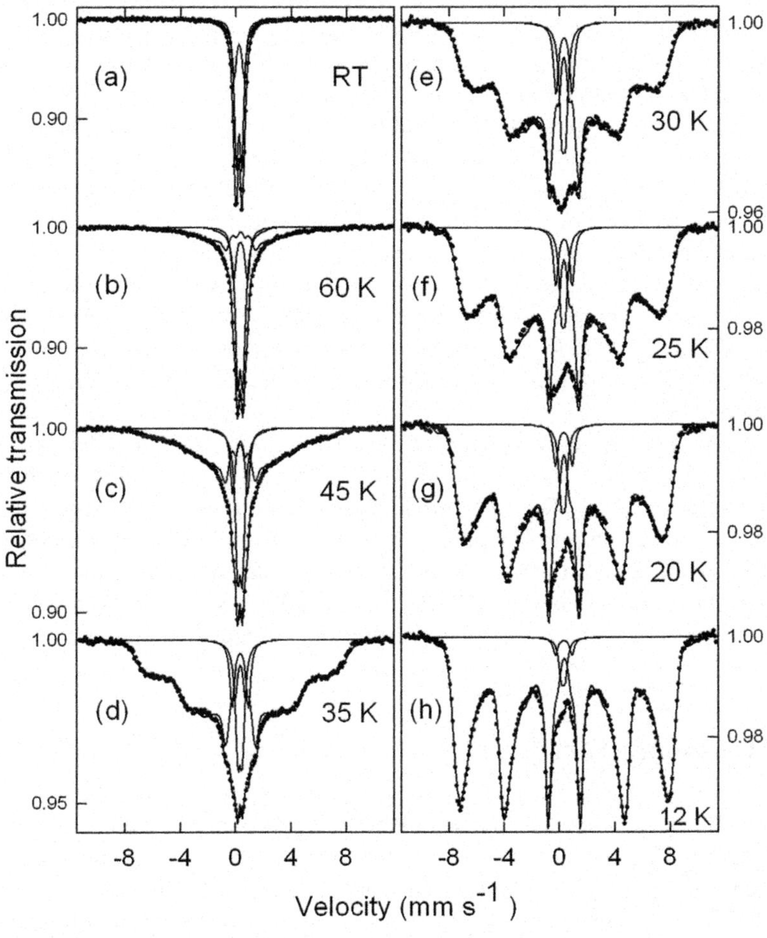

Fig. 7.12 Mössbauer spectra at various temperatures for the $ZnFe_2O_4$ nanoparticles obtained through hydrothermal synthesis. Reprinted with permission (Kopcewicz et al. 2019), © Wiley

In another experiment, zinc ferrite nanoparticles prepared through chemical synthesis showed a canted spin structure from low-temperature high-field Mössbauer effect measurements (Oliver et al. 1999). The A-site canting angle has been determined to be 140–160° antiparallel to the applied field, whereas that of the B-site occurred at 30–55° at the measurement temperature of 4.2 K. Long range order was suggested even beyond 25 K. However, the fast relaxation process hindered the exact determination of the inversion parameter.

Interestingly, the normal spinel structure with the antiferromagnetic sublattices could be overturned to ferrimagnetic alignment by mechanical milling (Chinnasamy et al. 2000). The average grain size of the $ZnFe_2O_4$ decreases with milling and reaches 11 nm for a milling duration of 20 h. The room temperature saturation magnetization of 10 emu/g for the 20 h milled sample dropped to paramagnetic values on annealing, suggesting the ferrimagnetic nature with higher milling duration. The ferrimagnetic alignment could be witnessed from the spectra shown in Fig. 7.13, where the application of a 6 T field resolves the A- and B- site hyperfine fields with unequal intensity for the 20 h milled sample. The average hyperfine field at 10 K in an external field of 6 T for the A-site is 56.4 T, while that of B-sites are 47.1 and 43.7 T, indicating two different Fe^{3+} environments. Canting angles larger than 35° are obtained for the A- and B- sites. Later, the ferrimagnetic ordering was confirmed from alternative techniques such as NMR, and it was reported that A-B interaction is ferrimagnetic, whereas B-B interaction is antiferromagnetic in nanocrystalline samples (Shim et al. 2006). Determining superparamagnetic transition in $ZnFe_2O_4$ requires careful analysis as they are masked by the paramagnetic transition and relaxation effects.

High-pressure studies with $ZnFe_2O_4$ showed two additional doublets and a paramagnetic ferric doublet due to orthorhombic structure at pressures beyond 25 GPa (Greenberg et al. 2017). The Néel temperature was 80 K at 47 GPa with B_{hf} of 40.3 and 46.4 T. A high spin to low spin transition has been inferred from the study on increasing the pressure further due to a nearly 4% decrease in volume and a 12% decrease in octahedral volume. Indeed, the T_N decreases with pressure, and almost half of the ferric ions remain in the high spin state even up to a pressure of 110 GPa.

7.6 Cadmium Ferrite

Cadmium ferrite is a normal spinel similar to zinc ferrite and exhibits room temperature quadrupole splitting and isomer shift of 0.8 and 0.2–0.4 mm/s, respectively, with negative quadrupole coupling constant (Hudson and Whitfield 1967; Evans et al. 1971). The normal spinel could be converted to mixed spinel on mechanical milling (Chinnasamy et al. 2001). Figure 7.14 shows the Mössbauer spectra of the mechanically milled $CdFe_2O_4$ at (a) 77 K for the 7 and 20 h milled samples and (b) 20 h at 10 K in an applied field of 6 T. The spectra at 77 K show a magnetic component of average hyperfine fields 24–26 T with significant relaxation effects along with the quadrupole doublet. The in-field spectrum for the 20 h milled sample

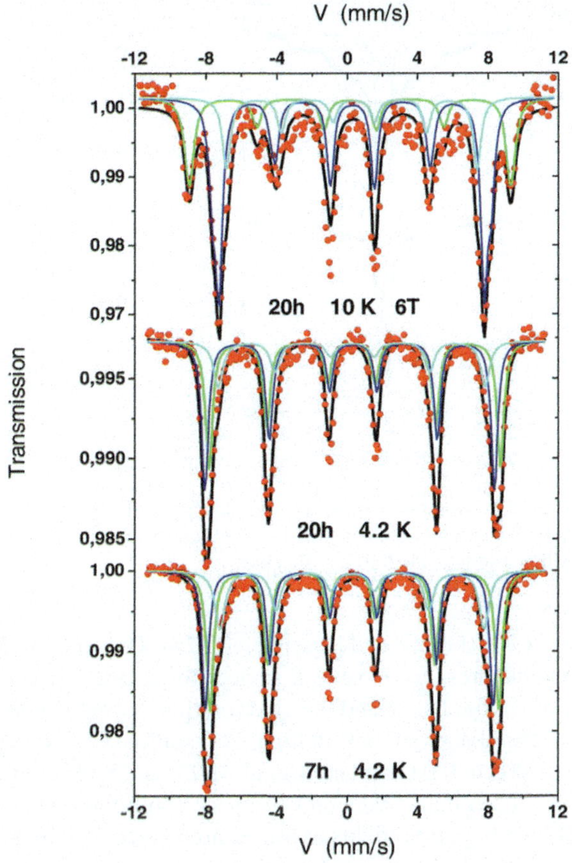

Fig. 7.13 Zero-field Mössbauer spectra of $ZnFe_2O_4$ milled for 7 and 20 h at 4.2 K and the 20 h milled sample in an applied field of 6 T at 10 K. Reprinted with permission (Chinnasamy et al. 2000), © IOP Publishing

(Fig. 7.14b) shows resolved A- and B- site ferric ion hyperfine fields with more than one component for each site. The effective hyperfine fields for the A- and B- sites have relative intensities of 18–35 and 65–82%, respectively, signifying the deviation from normal spinel structure. The average canting angle is estimated to be close to 50° for the A- and B- sites.

7.7 Co and Ni Substituted Fe_3O_4

$Co_xFe_{3-x}O_4$ with x up to 2.76 have shown that the replacement of Fe^{3+} by Co^{3+} ion weakens the A-B superexchange interaction, reducing the Curie temperature. When $x \geq 28$, the compound becomes paramagnetic (Murray and Linnett 1976) due to the transformation towards an environment corresponding to Co_3O_4. $Co_xFe_{3-x}O_4$ with x up to 0.7 have shown five B-site sextets and hyperfine field values ranging from 43.0

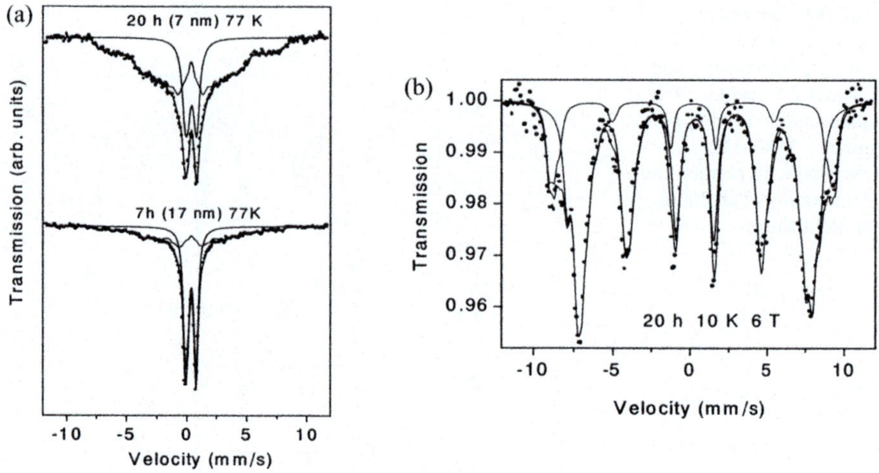

Fig. 7.14 Mössbauer spectra of the mechanically milled $CdFe_2O_4$ at **a** 77 K for the 7 and 20 h milled samples and **b** 20 h milled sample at 10 K in an applied field of 6 T. Reprinted with permission (Chinnasamy et al. 2001), © Elsevier

to 51.7 T (Franke and Rosenberg 1977). The composition-dependent hyperfine fields and the presence of several sextets could be explained by the random distribution of Co^{2+} in the B-site and the electron transfer between Fe^{2+} and Fe^{3+} ions.

Mössbauer studies on highly crystalline $Co_xFe_{3-x}O_4$ in the composition range x = 0.005 to 0.04 (De Grave et al. 1993) and with x = 0.1 to 0.9 (Persoons et al. 1993) have been undertaken and compared with the non-substituted samples. The A- and B- site hyperfine fields at 300 K are nearly 48.9 and 45.8 T, respectively, up to x = 0.04 in the studies involving lower concentrations of substituent. Co ion substitutes into the octahedral site, and Verwey transition temperature shifts to 70 K for the composition up to x = 0.2. A slight but systematic decrease in A-site isomer shift with substitution is present. The Curie temperature decreases with substitution and reaches a value of 816 K when x = 0.8. The dipolar field contribution at 0 K is – 0.5 T. Two distinct B-site hyperfine field components are obtained in the composition range x = 0.1 to 0.4 below 650 K, attributed to substantial B-site octahedral occupation by Co^{2+} in one and none on the other.

$Ni_xFe_{3-x}O_4$ with x up to 1 revealed a decrease in the A-site hyperfine fields in the range 49.6–50.0 T and a ferric ion isomer shift of 0.55 mm/s (Linnet and Rahman 1972). The mixed Fe^{2+} and Fe^{3+} in the B-site gave isomer shifts close to 1 mm/s. The B-site could be fitted with two to three sextets with B_{hf} from 46.2 to 53.9 T according to the composition that determines the site occupancy. The cation distribution is thus given by $(Fe^{3+})[Ni_x^{2+}Fe^{3+}Fe_{1\text{-}x}^{2+}]O_4^{2-}$.

7.8 Cobalt Ferrite ($CoFe_2O_4$)

Cobalt ferrite is an inverse spinel with Co^{2+} in the B-site exhibiting a T_c of 860 K. Cobalt ferrite may also deviate from the inverse spinel type where Co^{2+} is pushed from the B-site to the A-site with an energy estimated as 0.19 eV (Sawatzky et al. 1969). The distribution of Co and Fe nearest A-site neighbors causes many B-site sextets, giving an A-site hyperfine field of 49.0 T and a B-site average hyperfine field of 48.4 T. At 0 K, the $\overline{B}_{hf}$ for A- and B- sites are 51.1 and 54.1 T, respectively, where the spread in the B-site field can be as large as 7.0 T (van der Woude and Sawatzky 1971). The hyperfine field contribution is mainly from the covalency effects caused by overlap distortion of B-site cations s orbitals with that of the ligand.

Cobalt ferrites of various particle sizes in the 8–100 nm range were prepared using chemical synthesis methods, and the critical single domain size was determined to be 40 nm (Chinnasamy et al. 2003). Since the magnetocrystalline anisotropy constant of cobalt ferrite is large compared to most of the spinel ferrites, the critical superparamagnetic size is less than 10 nm and therefore, superparamagnetic fraction is not expected for larger particles. Usually, smaller particles coexist with larger particles in chemically synthesized samples, reducing the coercivity. Mössbauer spectrometry has been used to determine the superparamagnetic fraction in chemically synthesized samples, as shown in Fig. 7.15. The weighted average hyperfine field for the sample with an average particle size of 40 nm is 48.0 T. This reduces to 45.0 T for the 20 nm particles due to surface spin effects. The cobalt ferrite particles in the critical single domain regime show a large coercivity of 0.47 T at room temperature due to the size selected synthesis process. In the case of $CoFe_2O_4$ nanoparticles (mean diameter about 6 nm), prepared by a thermal decomposition technique (Peddis et al. 2011), the in-field Mössbauer spectrum showed a complex hyperfine structure, unlike other cobalt ferrite nanoparticles, requiring a particular fitting model. In fact, the only physically reasonable model capable of describing the hyperfine structure was based on effective magnetic field distributions derived from canting angle distributions correlated with hyperfine field distributions, for the A- and B- sites. The cationic degree of inversion estimated from Mössbauer analysis of the in-field spectrum is consistent with that estimated from neutron diffraction measurements, indicating non-collinear magnetic structures in the A- and B- sublattices. This study has led to a better understanding of the correlation between spin canting and cation distribution, probably with inhomogeneous cation distributions in these nanoparticles, even with such a small diameter.

The polyol process was used to obtain cobalt ferrite nanoparticles of various sizes from 6.1 to 2.4 nm through hydrolysis (Artus et al. 2011). The coercivity for all the particles was in the range 1.3–1.5 T at 5 K. The ratio of ferric ion population between A- and B- sites was 0.63, which deviates from the ratio of 1.00, usually observed for inverse spinel structure. The polyol process has been found to produce spinel ferrites that are different from other synthesis methods in terms of cation distribution, reduced surface spin effects and high crystallinity. Figure 7.16 shows the Mössbauer spectra at 77 and 12 K in an applied field of 8 T for the cobalt ferrite

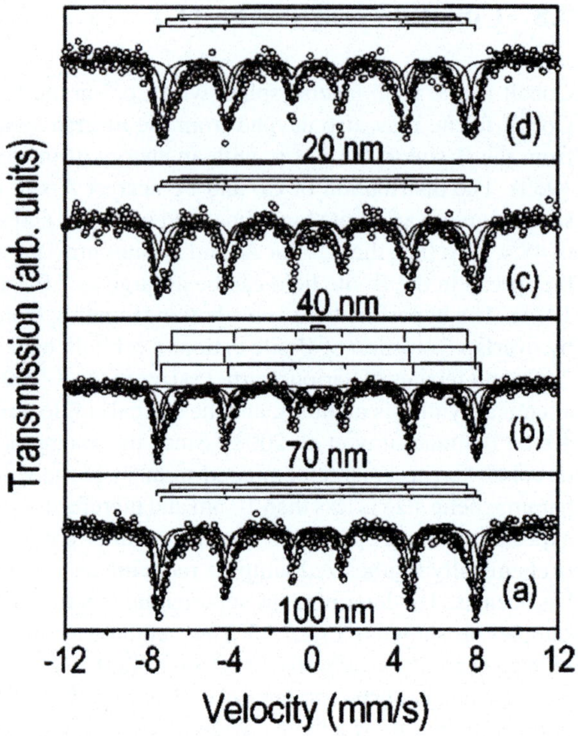

Fig. 7.15 Room temperature Mössbauer spectra of nearly monodispersed cobalt ferrite nanoparticles of various sizes obtained through aqueous methods. Reprinted with permission (Chinnasamy et al. 2003), © AIP Publishing

nanoparticles with an average particle size of 5.5 nm obtained through hydrolysis in a polyol medium. The spectrum at 77 K is not well resolved but could be fitted with two sextets corresponding to A- and B- sites exhibiting hyperfine fields of 50.6 and 53.6 T, respectively. These sextets are well resolved with an applied field of 8 T at 12 K with relative areas of 39 and 61 corresponding to A- and B- site ferric ions, respectively, implying the deviation from the ideal inverse spinel structure.

7.9 Nickel Ferrite

Nickel ferrite is an inverse spinel with a ferrimagnetic T_c close to 868 K. The A- and B- site hyperfine fields could be resolved from room to low temperature for bulk $NiFe_2O_4$. The A- and B- site hyperfine fields of nickel ferrite at 4.2 K are 50.6 and 54.8 T, respectively with a collinear Néel-type ferrimagnetic arrangement (Chappert and Frankel 1967). However, a non-collinear spin arrangement has been established in nanoparticles from the non-vanishing $\Delta m = 0$ lines in an applied magnetic field parallel to the γ-ray (Morrish and Haneda 1981).

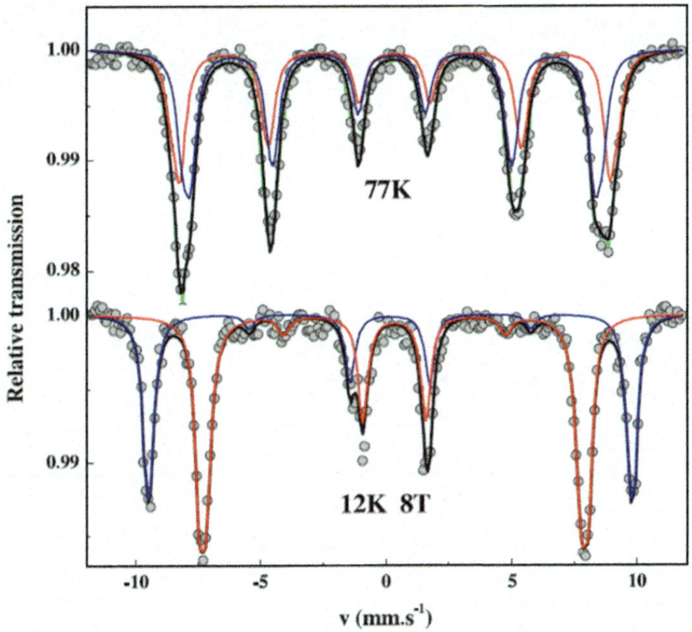

Fig. 7.16 Mössbauer spectra for the 5.5 nm cobalt ferrite, obtained through hydrolysis in polyol medium, at 77 and 12 K in an applied field of 8 T. Reprinted with permission (Artus et al. 2011), © IOP Publishing

Mechanical milling was found to transform the nickel ferrite into a mixed spinel structure. In-field Mössbauer spectrometry in an applied field of 5 T at 4.2 K was used to establish the spin canting and mixed spinel structure of mechanically milled $NiFe_2O_4$ (Chinnasamy et al. 2001). The A- and B- site intensity ratio for 60 nm nickel ferrite has been determined to be 0.9 for the 1 h milled sample, whereas it decreases to 0.5 with 30 h milling.

Nickel ferrite nanoparticles with an average size close to 4 nm synthesized using forced hydrolysis in polyol medium exhibited superparamagnetic doublet at room temperature (Chkoundali et al. 2004). Figure 7.17 shows the room-temperature and low-temperature spectra with and without an applied magnetic field for the $NiFe_2O_4$ obtained through polyol hydrolysis. A broad doublet at room temperature is obtained due to superparamagnetic relaxation that transforms into resolvable sextets at low temperatures. At 4.2 K, without an applied magnetic field, the spectrum could be fitted with two sextets corresponding to A- and B- sites. The two sextets are well resolved in an applied field of 6 T at 10 K without the second and fifth-line intensities. The intensity ratio of A- and B- site sextets is 0.75, suggesting deviation from inverse spinel structure. Thus, it is shown that the deviation from inverse spinel structure occurs not only with mechanical milling but also through chemical methods.

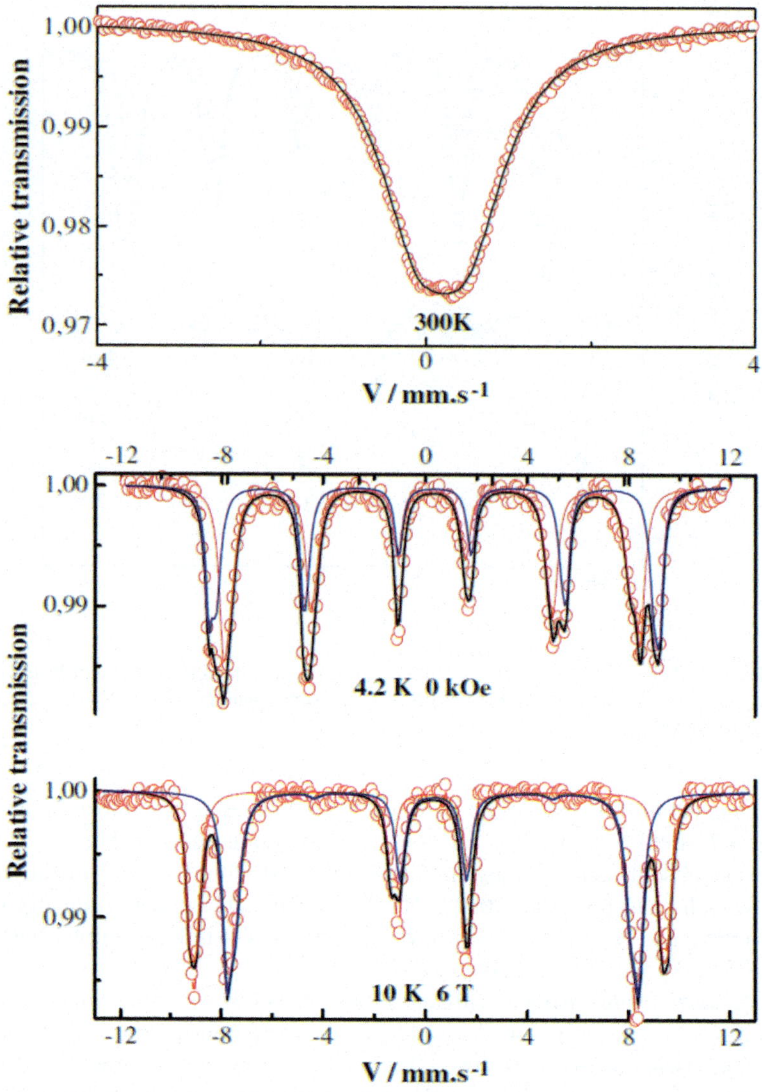

Fig. 7.17 Room temperature and low-temperature in-field Mössbauer spectra of $NiFe_2O_4$ nanoparticles obtained through forced hydrolysis in polyol medium. Reprinted with permission (Chkoundali et al. 2004), © IOP Publishing

7.10 Manganese Ferrite

Mn ferrite is a nearly normal spinel with Mn^{2+} occupying a fraction of about 0.8 in the tetrahedral site (Hasting and Corliss 1956). The Curie temperature of 12 nm nanoparticles was enhanced to 620 K compared to a bulk value of 573 K that was initially suggested due to finite-size effects (Kulkarni et al. 1994). The zero-field Mössbauer spectrum could not confirm the cation distribution due to unresolved A- and B- site hyperfine fields and relaxation effects. $MnFe_2O_4$ synthesized through the coprecipitation method with an average particle size of 11.5 nm showed a cation distribution obtained from Mössbauer analysis, as $(Mn_{0.33}Fe_{0.67})[Mn_{0.67}Fe_{1.33}]O_4$ that could be transformed to $(Mn_{0.6}Fe_{0.4})[Mn_{0.4}Fe_{1.6}]O_4$ on annealing in an inert atmosphere (Chen et al. 1996). The Mn^{2+} occupying the tetrahedral site is less than 50% for the samples prepared through coprecipitation. Consequently, the ferrimagnetic Curie temperature of nanoparticles was higher than the bulk value by nearly 50 K and showed increasing values with decreasing particle size.

Mn ferrite nanoparticles of average sizes 4 and 50 nm prepared through a modified coprecipitation method showed higher Curie temperature for the large particles (Yang et al. 2009). Figure 7.18 shows the Mössbauer spectra at 300 K (left) and at 15 K in an applied field of 8 T (right) for the 4 and 50 nm Mn ferrite nanoparticles, obtained through a modified coprecipitation method. The similarity in the spectra at room temperature could be noticed for the 4 nm (top left) and 50 nm (bottom left) particles, both having unresolved ferrimagnetic and superparamagnetic components from larger and smaller particles, respectively. At low temperatures in an applied magnetic field, the A- and B- sites are well resolved, showing nearly 63% ferric ions in A-site with 4 nm particles that decreased to 54% on increasing the particle size to 50 nm. These studies have shown that the cation distribution influences the Curie temperature, which could depend on the synthesis method.

7.11 Magnesium Ferrite

Magnesium ferrite shows equally distributed Mg^{2+} and Fe^{3+} in the A- and B- sites. Polycrystalline sintered Mg ferrite prepared through slow cooling exhibits well-resolved A- and B- site sextets that show decreasing hyperfine field and isomer shift(Pd) values as depicted in Fig. 7.19 (De Grave et al. 1979). The room temperature hyperfine fields of the A- and B- sites are 46.4 and 49.6 T, respectively. The isomer shift values are negative at higher temperatures, and the sextets transform to doublet at 673 K, corresponding to the paramagnetic state. The extrapolated hyperfine fields for the A- and B- sites at 0 K are 50.5 and 54.0 T, respectively. High-field spectra show three B-site sextets due to supertransferred hyperfine interaction. Mg ferrite subjected to high pressure undergoes a high spin to low spin transformation with pressure similar to Zn ferrite.

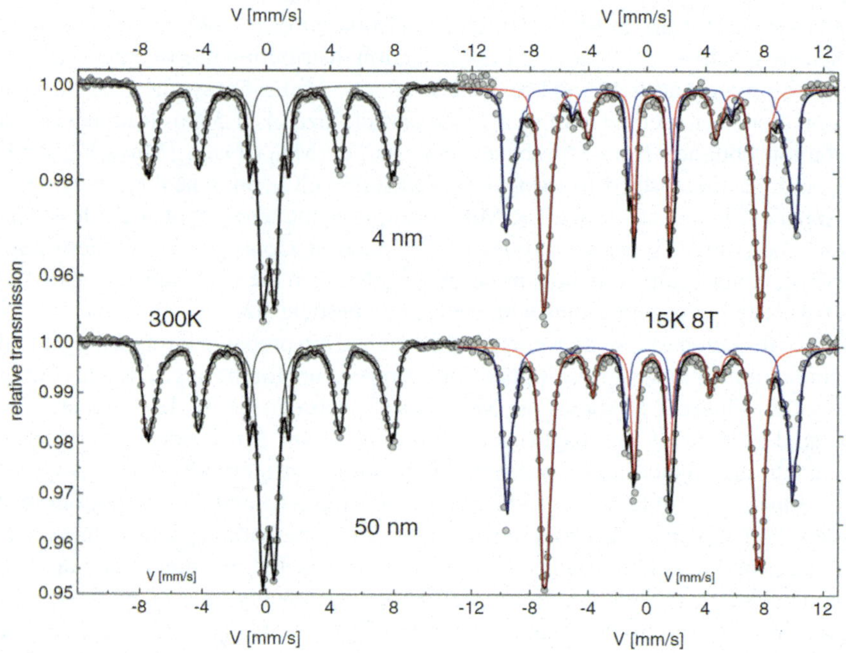

Fig. 7.18 Mössbauer spectra of 4 and 50 nm Mn ferrite at 300 K (left) and at 15 K in an applied magnetic field of 8 T (right). Reprinted with permission (Yang et al. 2009), © IOP Publishing

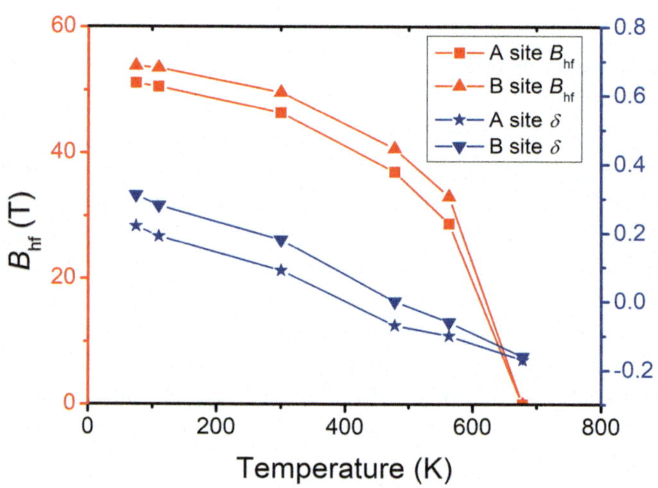

Fig. 7.19 The temperature-dependent hyperfine field (B_{hf}) and isomer shift (δ)(Pd) for the A- and B- sites of polycrystalline $MgFe_2O_4$. Data from (De Grave et al. 1979)

7.12 Copper Ferrite

Copper ferrite exhibits a cubic inverse spinel structure and a low-temperature tetragonal structure. The cubic and tetragonal Cu ferrites were obtained in quenched and slow-cooled samples, respectively (Evans and Hafner 1968). Cubic Cu ferrite show a B_{hf} of 47.5 T and a δ of +0.28 mm/s at room temperature for the sample prepared using the ceramic method. With applied magnetic fields of more than 3.2 T, the spectrum could be fitted with two sextets of hyperfine fields 49.6 and 45.7 T. The spectrum of tetragonal $CuFe_2O_4$ single crystals could be fitted with two sextets of B_{hf} 48.1 and 50.5 T and their corresponding δ values are +0.27 and +0.37 mm/s, respectively. The octahedral B-site is distorted, resulting in a negative quadrupole interaction parameter. The ferrimagnetic T_c of the tetragonal phase is close to 800 K and may coincide with the cubic phase. The temperature-dependent hyperfine field may not lead to a conclusion about the type of phase since the tetragonal to cubic transformation occurs beyond 663 K.

Tetragonal bulk $CuFe_2O_4$ was mechanically milled for 60 min, and the average grain size was reduced to 15 nm (Prabhu et al. 2011). The in-field Mössbauer spectra at 10 K with 8 T applied field for the bulk and the mechanically milled sample did not show a significant change in the cation distribution as the A- to B- site intensity ratio was 0.8 in both the samples. The A-site spin canting increased with grain size reduction. It was shown that the reduction of grain size alters the tetragonal to cubic phase transformation temperature from 643 K corresponding to the bulk to 563 K for the 15 nm sample.

7.13 Substituted Ferrites

Mössbauer spectrometry on several di-ionic substituted ferrites have been undertaken at various temperatures and applied fields. The transformation from ferrimagnetic sextets to paramagnetic doublet has been encountered in most studies with divalent Zn substitution at higher concentrations. Non-magnetic ions with Zn^{2+} show paramagnetism in the whole composition range. The presence of relaxation effect requires the experiments to be conducted at low temperatures, although they could be avoided in a few cases. The findings from Mössbauer spectrometry for a few substituted ferrites are listed in Table 7.6. The nanoscale magnetic characteristics of the substituted ferrites could be inferred from their parent ferrites as they depend on the magnetic anisotropy and particle size. The particle size-dependent superparamagnetic blocking temperatures obtained from observing superparamagnetic doublet using Mössbauer spectrometry for a few ferrites are presented in Table 7.7. The superparamagnetic doublet coexists with the sextet along with relaxation broadening due to the distribution in particles, and therefore, more than 50% of superparamagnetic doublet intensity is considered to estimate the blocking temperature (Winkler 1984).

Table 7.6 The Mössbauer studies on Zn and Cr substituted ferrites at various temperatures and applied fields, and the findings reported in the literature

Ferrite	Experimental conditions (*B, T*)	Findings	References
$Co_{1-x}Zn_xFe_2O_4$	0 to 8.0 T, 4 K	$(Zn_xFe_{1-x-y})[Co_{1-x-y}Fe_{1+x+y}]O_4$ Inverse spinel, B-site canting depending on Zn concentration, Néel temperature for x = 0.4 and 0.6 are 513 and 322 K, respectively	Petitt and Forester (1971)
$(ZnO)_x(NiO)_{1-x}Fe_2O_3$	1.35 T, 300 K, 77 K	Zn substitution does not affect A- and B- sites δ. Yafet-Kittel (YK) type spin canting increases with Zn. Distribution of relaxation rates suggesting disorder	Daniels and Rosencwaig (1970)
$Ni_{1-x}Zn_xFe_2O_4$	5 T, 7 K	$(Zn_xFe_{1-x})[Ni_{1-x}Fe_{1+x}]O_4$ Decrease in B-site B_{hf} due to supertransferred hyperfine fields. YK-type spin arrangement. Significant Fe^{3+} spin canting for x < 0.5	Leung et al. (1973)
$Ni_{1-x}Zn_xFe_2O_4$	5 and 9 T, 4.2 K	$(Zn_xFe_{1-x})[Ni_{1-x}Fe_{1+x}]O_4$ No A-site canting, B-site canting increases with Zn	Clark and Morrish (1973)
$Ni_{0.25}Zn_{0.75}Fe_2O_4$	4.2–77 K, 8 T	Increase in linewidth with *T*. Blocking temperature ~ 250 K. $T_{YK} << T_N$ non-collinear to antiparallel spin. Local spin canting. Two B-site B_{hf} with canting in only one. $B_{hf}(A) \approx B_{hf}(B) = 51.5$ T at 4.2 K. $B_{eff}(A) = 59.5$ T, $B_{eff}(B_1) = 44.0$ T, $B_{eff}(B_2) = 44.8$ T at 4.2 K and 8 T. $T_N > 350$ K	Bhargava and Zeman (1980)
$Mn_{1-x}Zn_xFe_2O_4$	Various *T*	B_{hf} decreases with Zn neighbors (6 nn model used). T_c decreases with Zn (> 500 K for x = 0 to < 300 K for x > 0.65)	Cser et al. (1968)

(continued)

Table 7.6 (continued)

Ferrite	Experimental conditions (B, T)	Findings	References
$Mn_{1-x}Zn_xFe_2O_4$	5 and 9 T, 4.2 K	For x = 0 up to 8% Fe^{3+} in A-site. Collinear up to 0.5. Largest B_{hf} at x = 0.5. At least 99% Fe^{3+} in B-site for x > 0.5 with canted spins. No Mn^{2+} in B-site with Zn substitution. Canting degree depends on Zn^{2+} in A-site. Very small change in B_{hf} with x. With 6 nn Zn^{2+}, B-site Fe ion spin is reversed. Ferrimagnetic even with x = 0.95	Morrish and Clark (1975)
$Cu_{1-x}Zn_xFe_2O_4$	RT	$(Zn_xFe_{1-x})[Cu_{1-x}Fe_{1+x}]O_4$ Paramagnetic at 300 K for x ≥ 0.6; B_{hf}(B) = 52.0 to 44.0 T and B_{hf}(A) = 48.0 to 45.0 T from x = 0 to 0.4	Patil and Kulkarni (1979)
$Li_{0.5-x/2}Zn_xFe_{2.5-x/2}O_4$	295–973 K	B_{hf}(A) = 42.2 T, B_{hf}(B) = 40.4 T for x = 0; one of the B-site is paramagnetic for x = 0.2; paramagnetic for x > 0.4. Inverse to normal spinel with increasing x	Young and Smit (1971)
$Li_{0.5-x/2}Zn_xFe_{2.5-x/2}O_4$	4.2–300 K, 7 T	Spin canting for x > 0.6, relaxation effects and flipped Fe^{3+}(B). For x = 0.7, B_{eff}(A) = 50.8 T, $B_{eff}(B_{1\text{-}4})$ = 43.1–52.3 T	Rosenberg et al. (1982)
$Li_{(1-x)/2}Zn_xFe_{(5-x)/2}O_4$ (x up to 0.5) single crystals	300 K	B_{hf}(A) = 50.1 T, B_{hf}(B) = 51.5 T, I_A = 37% for x = 0; B_{hf}(A) = 49.2 T, B_{hf}(B) = 49.8–45.9 T, I_A = 28% for x = 0.2; B_{hf}(A) = 48.5 T, B_{hf}(B) = 49.7–43.2 T, I_A = 22% for x = 0.31; B_{hf}(A) = 46.9 T, B_{hf}(B) = 48.1–41.1 T, I_A = 21% for x = 0.42; B_{hf}(A) = 44.5 T, B_{hf}(B) = 45.8–38.7 T, I_A = 18% for x = 0.5; Average reduction in B_{hf}(B) = 2.2 T per Zn^{2+}	Xu (2003)
$Mg_{1-x}Zn_xFe_2O_4$ (x ≤ 0.6)	77 K	$(Zn_xFe_{1-x})[Mg_{1-x}Fe_{1+x}]O_4$ B_{hf}(A) ≈ 50.0 T, B_{hf}(B) ≈ 53.0 T, I_A/I_B = 0.76 for x = 0.1; B_{hf}(A) ≈ 45.5 T, B_{hf}(B) ≈ 48.5 T, I_A/I_B = 0.26 for x = 0.5; B_{hf} (A,B) decreases with x. Relaxation effects with x > 0.5	Kulkarni and Joshi (1985)

(continued)

Table 7.6 (continued)

Ferrite	Experimental conditions (B, T)	Findings	References
$Mg_{1-x}Zn_xFe_2O_4$ ($x \geq 0.8$)	4.2 K, 6 T	$B_{eff}(A) = 57.5$ T, $B_{eff}(B) = 46$ T for x = 0.8; $B_{eff}(A) = 56$ T, $B_{eff}(B) = 47$ T for x = 0.85; $B_{eff}(A) = 53$ T, $B_{eff}(B) = 45.5$ T for x = 0.9; $B_{eff}(A) = 58.5$ T, $B_{eff}(B) = 45$ T for x = 0.95; B-site canting angle 40-47°	Teillet et al. (1991)
$Cd_{1-x}Zn_xFe_2O_4$	300 K	All are paramagnetic. Quadrupole splitting increases with x(>0.8 mm/s)	Chakrabarti et al. (2007)
$ZnCr_xFe_{2-x}O_4$ x = 0.8, 1.2	295–823 K	Both are paramagnetic. QS = 0.38 mm/s for x = 0.8; QS = 0.63 mm/s for x = 1.2 both at RT; QS independent of T; change in oxygen parameter increases QS with x; trigonal field at B-site	Muthukumarasamy et al. (1980)
$Zn_{1+x}Sn_xFe_{2-2x}O_4$	77, 295 K	All are paramagnetic; QS increases from 0.35 mm/s for x = 0 to 0.47 mm/s for x = 0.9. QS(T) independent	Baijal et al. (1984)
$CoCr_xFe_{2-x}O_4$	300 K	$B_{hf}(A) \approx B_{hf}(B) \approx 50.0$ T for x = 0; $B_{hf}(A) \approx 40.0$ T, $B_{hf}(B) \approx$ 31.0 T for x = 0.8	Krieble et al. (2006)

A—A-site, B—B-site, B_{hf}—hyperfine field, I—intensity, QS—quadrupole splitting, B_{eff}—effective hyperfine fields at the given T and applied field, B

Cobalt ferrite with a mean diameter of 7.6 nm has a mean blocking temperature of 270 K, corresponding to an effective anisotropy constant of 1.76×10^5 J/m^3 (Hamdeh et al. 2005). The blocking temperature of cobalt ferrite with an average particle size of 8.5 nm is 420 K, where a doublet is obtained (Rondinone et al. 2000). The effective anisotropy constant increases from 14×10^5 J/m^3 for the cobalt ferrite of average size 3 nm to 22.7×10^5 J/m^3 for the 2 nm particles while their blocking temperatures estimated from Mössbauer studies are 160 and 100 K, respectively (Moumen et al. 1996). Since most of the other ferrites have a lower magnetocrystalline anisotropy constant than cobalt ferrite, the nanoparticles of comparable sizes are superparamagnetic at room temperature.

Table 7.7 The superparamagnetic blocking temperatures (T_B) of various spinel ferrite nanoparticles obtained from Mössbauer spectrometry

Ferrite	Parameters	Particle characteristics	T_B (K)	References
Fe_3O_4	$K_{eff} = 6 \times 10^3$ J/m^3; τ_o = 1 ns	100 nm, ellipsoid	292	McNab et al. (1968)
Fe_3O_4	$K_{eff} = 3.7 \times 10^4$ J/m^3	8 nm particles	130	Winkler (1984)
γ-Fe_2O_3	Critical single domain diameter, $D_{sd} = 100$ nm	6.5 nm, agglomerates	295	Coey and Khalafalla (1972)
$MnFe_2O_4$	$K_{eff} = 8 \times 10^4$ J/m^3	10 nm particles	295	Rechenberg and Tourinho (1991)
$CuFe_2O_4$	Mechanically milled	7 nm grain size	296	Goya and Rechenberg (1998)
$MgFe_2O_4$	Relaxation spectrum	6 nm particles	95	Chen et al. (1999)
$NiFe_2O_4$	Relaxation spectrum	5 nm particles obtained by annealing at 573 K	80	Albuquerque et al. (2001)
$CoFe_2O_4$	$K_{eff} = 1.7 \times 10^5$ J/m^3; $\tau \approx \tau_o$; $\tau_o = 5$ ns	7.6 nm, spherical	270	Hamdeh et al. (2005)
$CoFe_2O_4$	$K_{eff} = 22.7 \times 10^5$ J/m^3	2 nm particles	100	Moumen et al. (1996)
	$K_{eff} = 14 \times 10^5$ J/m^3	3 nm particles	160	Moumen et al. (1996)
$CoFe_2O_4$	$K_{eff} = 2.23 \times 10^5$ J/m^3	8.5 nm particles	420	Rondinone et al. (2000)

K_{eff}—effective magnetic anisotropy

7.14 Hercynite ($FeAl_2O_4$)

$FeAl_2O_4$ known as hercynite, was initially reported to be a normal spinel, and later, Mössbauer spectrometry established that it could exhibit an inverted spinel structure based on the synthesis method. Accordingly, only one quadrupole doublet of Δ = 1.6 mm/s and δ = 1.08 mm/s was obtained at 295 K (Rossiter 1965) for the sample sintered at 1323 K. The quadrupole splitting increased to 2.81 mm/s with a δ of 1.28 mm/s at 77 K. The temperature-dependent quadrupole splitting and the parameters suggested ferrous ions in the tetrahedral site. Subsequently, ferrous ions in both A- and B- sites indicated the inverted spinel for the sample prepared at a sintering temperature greater than 1323 K (Yagnik and Mathur 1968).

Spin glass like behavior was reported in $FeAl_2O_4$ from the low-temperature Mössbauer spectrometry, which showed shape-varying cooling and heating transitions in the temperature range of 13.5 and 16.5 K, respectively (Dormann et al. 1990). Later, studies reported a spin glass like ground state where the Fe^{2+} in the tetrahedral crystal field is Jahn–Teller active and suppressed at low temperatures, forming spin-orbital liquid (Tristan et al. 2005). $FeAl_2O_4$ has the following magnetic parameters: spin, $S = 2$; Curie–Weiss temperature, $\Theta_C = -130$ K; spin glass transition temperature, T_f of 12 K; effective paramagnetic moment, 5.32 μ_B; and frustration parameter (Θ_C/T_f) = 11, bringing them under the class of frustrated systems.

7.15 Coulsonite (FeV_2O_4)

FeV_2O_4 or coulsonite is a normal spinel in which the Fe^{2+} occupies the tetrahedral site. Similar to most other ferrites, depending on the synthesis method, it exhibits partial inverse cation distribution (Gupta and Mathur 1975), leading to $\left(Fe^{2+}_{1-x}V^{3+}_{x}\right)\left[Fe^{2+}_{x}Fe^{3+}_{y}V^{3+}_{2-x-y}\right]O^{2-}_{4}$. The cubic structure undergoes a series of structural transformations with reducing temperature from tetragonal (~140 K) to orthorhombic (~110 K) and again to tetragonal (~70 K). The Mössbauer spectrum of coulsonite shows a singlet at room temperature and transforms to a doublet at 250 K due to the Jahn–Teller effect. As it transforms into the tetragonal structure, the quadrupole splitting increases linearly from 1–2 mm/s as the tetragonal distortion increases (Tanaka et al. 1966). The B_{hf} develops with a reduction in temperature around 110 K corresponding to the ferrimagnetic transition, reaches a maximum, and decreases further, reaching a constant value as it approaches lower temperature (Boutron and Imbert 1968).

Low-temperature Mössbauer experiments with applied external fields on FeV_2O_4 single crystals have shown collinear Fe^{2+} and V^{3+} ferrimagnetic spin arrangement along the *a*-axis for the orthorhombic phase (Nakamura et al. 2018). For the low-temperature tetragonal structure, both the spins incline from the *c*-axis to form a canted spin structure. The A- and B- site B_{hf} at 18 K are 8.3 and 45.8 T, respectively. An interplay between the Jahn–Teller and spin–orbit stabilization effects determines the structure and magnetic characteristics. When vanadium ion is substituted in magnetite, the well-resolved hyperfine field sextets at room temperature broaden as it becomes Fe_2VO_4 with the cation distribution $\left(Fe^{2+}\right)\left[FeV^{3+}\right]O^{2-}_{4}$ and becomes paramagnetic at 473 K (Abe et al. 1974).

7.16 Chromite ($FeCr_2O_4$)

Similar to coulsonite, chromite, with the formula $FeCr_2O_4$, is cubic and undergoes a tetragonal transformation below 135 K and becomes orthorhombic below 70 K. The room temperature singlet becomes a doublet as the temperature decreases, indicating Jahn–Teller distortions and exhibits ferrimagnetic ordering at 70–90 K (Tanaka et al. 1966) followed by a conical spiral magnetic ordering below ~ 35 K.

The Mössbauer spectra at high pressure up to 93 GPa at room temperature and low temperatures are shown in Fig. 7.20 (Xu et al. 2017a, b). The singlet at ambient

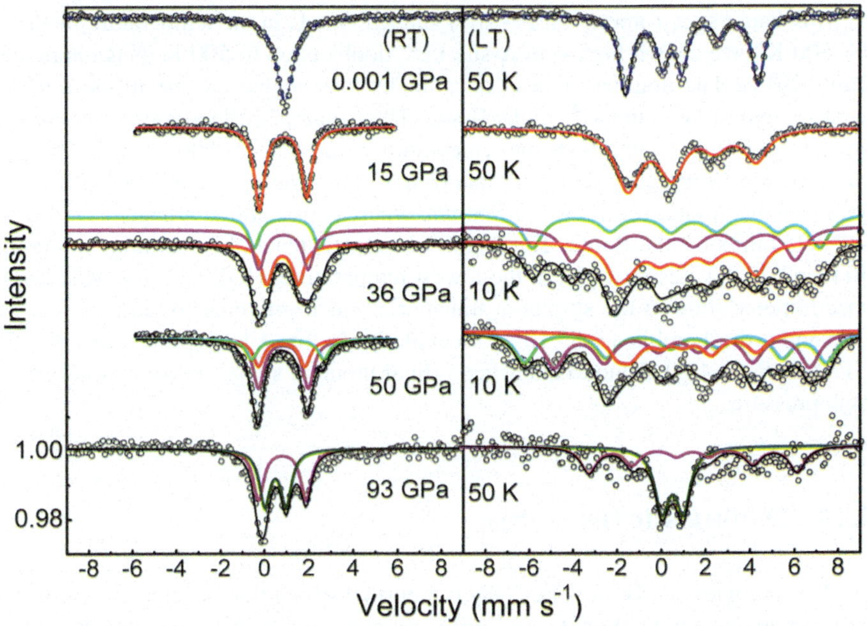

Fig. 7.20 The Mössbauer spectra of the $FeCr_2O_4$ subjected to a pressure up to 93 GPa at room temperature (RT) and low temperatures (LT). Reprinted with permission (Xu et al. 2017a, b), © The American Physical Society

temperature pressure transforms to a doublet at a pressure of even 2 GPa, although the structure is cubic due to Jahn–Teller distortion. The cubic to tetragonal transformation occurs at room temperature at a pressure of ~ 12 GPa. Beyond ~ 30 GPa, the Mössbauer spectrum at 10 K shows B_{hf} of 34–42 T indicative of a quenched orbital moment. Tetrahedral sites with partial orbital quenching and Fe–Cr ion cation inversion associated with unit cell volume reduction are observed in the pressure range 30–60 GPa. Beyond ~ 60 GPa, an octahedral low-spin doublet coexists with the tetrahedral high-spin doublet. The low spin state is stabilized to nearly 50% of Fe sites with pressures up to 93 GPa.

7.17 Ulvite (Fe_2TiO_4)

Ulvite has the cation distribution $(Fe^{2+})[Fe^{2+}Ti^{4+}]O^{2-}{}_4$, where ferrous ion occupies both the A- and B- sites. It transforms to a tetragonal phase below 163 K and an antiferromagnetic transition below 140 K. High-field Mössbauer spectrometry at various temperatures performed on the cubic phase proved the A- and B- site ferrous ion quadrupole interactions as positive and negative, respectively (Vanleerberghe and Vandenberghe 1985).

Mössbauer spectrometry on Fe_2TiO_4 single crystals in the temperature range of 16–500 K have shown high spin A-site Fe^{2+} doublets up to 500 K (Nakamura and Fuwa 2014). The doublet, instead of a singlet was suggested due to the random arrangement of Fe^{2+} and Ti^{4+} in the B-site. The isomer shift increases with temperature from 0.9 to 1.2 mm/s as the temperature reduces from 500 to 16 K for both sites. The quadrupole coupling constant ($e^2qQ/2$) increases with decreasing temperature for the A-site, reaching ~ 3.5 mm/s, whereas it acquires a negative value of ~ −0.25 mm/s for the B-site. The B_{hf}(A) and B_{hf}(B) reach the maximum values of 40.0 and 22.0 T, respectively, at the lowest temperature. High-pressure Mössbauer spectrometry showed the structural transformation from cubic to tetragonal at ~8 GPa and orthorhombic at ~55 GPa (Xu et al. 2017a, b). A low spin doublet at ~ 55 GPa and B_{hf} ~ 42.0 T due to quenched orbital moment suggested spin crossover at high pressure.

7.18 Daubréelite ($FeCr_2S_4$)

Daubréelite with the formula $FeCr_2S_4$ is a normal spinel that is cubic and magnetically ordered at 180 K. With decreasing temperature, the B_{hf} becomes 20.6 T at 77 K, reaches a maximum of ~ 21.0 T at 70 K and drops again to 18.6 T at 22 K (Eibschutz et al. 1967). The decrease in B_{hf} with reducing temperature is attributed to the easy direction change along $< 100 >$, identified from the appearance of quadrupole splitting. It was also suggested that a temperature-dependent lattice distortion associated with the magnetostrictive effect exists below 172 K, and the dynamic Jahn–Teller (J-T) effect was ruled out (Hoy and Singh 1968).

From the low-temperature Mössbauer experiments down to 2 K, an abrupt change from positive to negative electric field gradient was observed at 10 K (Spender and Morrish 1972). It was stated that tetrahedral distortion was a result of J-T stabilization of the ferrous ground state below 10 K and dynamic J-T stabilization above 10 K. Later, it was revealed that polycrystalline samples might be an orbital liquid that exhibits orbital ordering near 10 K and stoichiometric single crystals may show orbital glass behavior due to geometric frustration (Fichtl et al. 2005). The Cr^{3+} sublattice and Fe^{2+}-Cr^{3+} interactions are ferro and antiferromagnetic, respectively, whereas the interaction among ferrous ions in the sublattice is weak. The sulfur-coordinated Fe^{2+} is J-T active in contrast to the octahedrally coordinated Cr^{3+}. Hence J-T effect can trigger the orbital order transition in polycrystalline samples.

High-pressure Mössbauer spectrometry on polycrystalline $FeCr_2S_4$ recorded up to 20 GPa at room temperature and below the ferrimagnetic T_c of 90 K have shown phase transformation with pressure (Amiel et al. 2011). The ferrimagnetic spinel phase transforms to the high spin doublet when the pressure increases beyond 7 GPa and transforms to a Cr_3S_4 type phase at 11 GPa. This phase was characterized by a doublet with a lower isomer shift of 0.4 mm/s, quadrupole splitting of 0.3 mm/s, and the magnetic state unchanged down to 5 K. $GeFe_2O_4$ is another example of normal spinel with a T_N ~ 11 K and $V_{zz} < 0$ at 0 K (Boutron and Imbert 1968).

7.19 Summary

This chapter has presented the Mössbauer parameters of various spinel ferrites and their site occupancies. The high and low temperature, and external magnetic field facilities were utilized to carefully analyze the Mössbauer spectra, to estimate the occupancy of the Fe sites, the stoichiometry of the iron-based ferrite phase and the magnetic structure as a function of morphology of certain spinel ferrites. The magnetic properties as a function of synthesis methods, post-treatments, pressure and crystallite sizes of the spinel ferrites investigated through Mössbauer spectrometry are also discussed.

References

M. Abe, M. Kawachi, S. Nomura, J. Solid State Chem. **10**, 351 (1974)
A.S. Albuquerque, J.D. Ardissoon, W.A.A. Macedo, J.L. Lopez, R. Paniago, A.I.C. Persiano, J. Magn. Magn. Mater. **226–230**, 1379 (2001)
Y. Amiel, G.K. Rozenberg, N. Nissim, A. Milner, M.P. Pasternak, M. Hanfland, R.D. Taylor, Phys. Rev. B **84**, 224114 (2011)
R.J. Armstrong, A.H. Morrish, G.A. Sawatzky, Phys. Lett. **23**, 414 (1966)
M. Artus, L.B. Tahar, F. Herbst, L. Smiri, F. Villain, N. Yaacoub, J.-M. Greneche, S. Ammar, F. Fiévet, J. Phys. Condens. Matter **23**, 506001 (2011)
J.S. Baijal, C. Prakash, P. Krishan, K.K. Laroia, J. Phys. C: Solid State Phys. **17**, 5993 (1984)
R. Bauminger, S.G. Cohen, A. Marinov, S. Ofer, E. Segal, Phys. Rev. **122**, 1447 (1961)
F.J. Berry, S. Skinner, M.F. Thomas, J. Phys. Condens. Matter **10**, 215 (1998)
S.C. Bhargava, N. Zeman, Phys Rev B21:1717&1726 (1980)
F.H. Boutron, P. Imbert, J. Appl. Phys. **39**, 775 (1968)
M. Chakrabarti, D. Sanyal, A. Chakrabarti, J. Phys. Condens. Matter **19**, 236210 (2007)
J. Chappert, R.B. Frankel, Phys. Rev. Lett. **19**, 570 (1967)
C.P. Chen, C.M. Sorensen, K.J. Klabunde, G.C. Hadjipanayis, E. Devlin, A. Kostikas, Phys. Rev. B **64**, 9288 (1996)
Q. Chen, A.J. Rondinone, B.C. Chakoumakos, Z.J. Zhang, J. Magn. Magn. Mater. **194**, 1 (1999)
C.N. Chinnasamy, A. Narayanasamy, N. Ponpandian, K. Chattopadhyay, K. Shinoda, B. Jeyadevan, K. Tohji, K. Nakatsuka, T. Furubayashi, I. Nakatani, Phys. Rev. B **63**, 184108 (2001)
C.N. Chinnasamy, A. Naryanasamy, N. Ponpandian, K. Chattopadhyay, H. Guerault, J.-M. Greneche, J. Phys. Condens. Matter **12**, 7795 (2000)
C.N. Chinnasamy, B. Jeyadevan, K. Shinoda, K. Tohji, D.J. Djayaprawira, M. Takahashi, R. Justin Joseyphus, A. Narayanasamy, Appl. Phys. Lett. **83**, 2862 (2003)
C.N. Chinnasamy, A. Narayanasamy, N. Ponpandian, R. Justin Joseyphus, K. Chattopadhyay, K. Shinoda, B. Jeyadevan, K. Tohji, K. Nakatsuka, H. Guérault , J.-M. Greneche, Scripta Mater. **44**, 1411 (2001)
S. Chkoundali, S. Ammar, N. Jouini, F. Fiévet, P. Molinié, M. Danot, F. Villain, J.-M Greneche, J. Phys. Condens. Matter **16**, 4357 (2004)
P.E. Clark, A.H. Morrish, Phys. Status Solidi A **19**, 687 (1973)
J.M.D. Coey, D. Khalafalla, Phys. Status Solidi A **11**, 229 (1972)
J.M.D. Coey, Phys. Rev. Lett. **27**, 1140 (1971)
L. Cser, I. Dezsi, I. Gladkih, L. Keszthelyi, D. Kulgawczuk, N.A. Eissa, E. Sterk, Phys. Status Solidi **27**, 131 (1968)

G.M. Da Costa, C. Laurent, E. De Grave, R.E. Vandenberghe, in Mineral Spectroscopy, ed. by M.D. Dyar, C. McCammon, M.W. Schaefer (The Geochemical Society, 1996)
J.M. Daniels, A. Rosencwaig, Canadian. J. Phys. **48**, 381 (1970)
J.M. Daniels, A. Rosencwaig, J. Phys. Chem. Solids **30**, 1561 (1969)
P.M.A. de Bakker, E. De Grave, R.R. Vandenberghe, L.H. Bowen, Hyperfine Interact. **54**, 493 (1990)
E. De Grave, A. Govaert, D, Chambaere, G. Robbrecht, Physica B+C **96**, 103 (1979)
E. De Grave, R.M. Persoons, R.E. Vandenberghe, P.M.A. de Bakker, Phys. Rev. B **47**, 5881 (1993)
P.A. Dickof, P.J. Schurer, A.H. Morrish, Phys. Rev. B **22**, 115 (1980)
D.C. Dobson, J.W. Linnet, M.M. Rahman, J. Phys. Chem. Solids **31**, 2727 (1970)
A.C. Doriguetto, N.G. Fernandes, A.I.C. Persiano, E.N. Filho, J.-M. Greneche, J.D. Fabris, Phys. Chem. Miner. **30**, 249 (2003)
J.L. Dormann, M. Seqqat, D. Fiorani, M. Nogues, J.L. Soubeyroux, S.C. Bhargava, P. Renaudin, Hyperfine Interact. **54**, 503 (1990)
E. Eibschutz, S. Shtrikman, Y. Tenenbaum, Phys. Lett. **24A**, 563 (1967)
B.J. Evans, S.S. Hafner, H.P. Weber, J. Chem. Phys. **55**, 5282 (1971)
B.J. Evans, S.S. Hafner, J. Phys. Chem. Solids **29**, 1573 (1968)
R. Fichtl, V. Tsurkan, P. Lunkenheimer, J. Hemberger, V. Fritsch, A.A.K. von Nidda, E.W. Scheidt, A. Loidl, Phys. Rev. Lett. **94**, 027601 (2005)
H. Franke, M. Rosenberg, J. Magn. Magn. Mater. **4**, 186 (1977)
J. Garcia, G. Subias, J. Phys. Condens. Matter **16**, R145 (2004)
K. Glazyrin, C. McCammon, L. Dubrovinsky, M. Merlini, K. Schollenbruch, A. Woodland, M. Hanfland, Amer. Miner. **97**, 128 (2012)
G.F. Goya, H.R. Rechenberg, Nanostruct. Mater. **10**, 1001 (1998)
E. Greenberg, W.M. Xu, M. Nikolaevsky, E. Bykova, G. Garbarino, K. Glazyrin, D.G. Merkel, L. Dubrovinsky, M.P. Pasternak, G.K. Rozenberg, Phys. Rev. B **95**, 195150 (2017)
M.P. Gupta, H.B. Mathur, J. Phys. C: Solid State Phys. **8**, 370 (1975)
L. Häggström, H. Annersten, T. Ericsson, R. Wäppling, W. Karner, S. Bjarman, Hyperfine Interact. **5**, 201 (1978)
N.A. Halasa, G. DePasquali, H.G. Drickamer, Phys. Rev. B **10**, 154 (1974)
H.H. Hamdeh, W.M. Hikal, S.M. Taher, J.C. Ho, N.P. Thuy, O.K. Quy, N. Hanh, J. Appl. Phys. **97**, 064310 (2005)
R.S. Hargrove, W. Kündig, Solid Sate Commun. **8**, 303 (1970)
J.M. Hasting, L.M. Corliss, Phys. Rev. **104**, 328 (1956)
G.R. Hoy, K.P. Singh, Phys. Rev. **172**, 514 (1968)
A. Hudson, H.J. Whitfield, Molecular Phys. **12**, 165 (1967)
T.J. Daou, J.-M. Greneche, S.J. Lee, S. Lee, C. Lefevre, S. B. Colin, G. Pourroy, J. Phys. Chem. C **114**, 8794 (2010)
M. Kopcewicz, A. Grabias, I.K. Kudelska, W. Dobrowolski, Phys. Status Solidi B 1800223 (2019)
D.P. Kozlenko, L.S. Dubrovinsky, S.E. Kichanov, E.V. Lukin, V. Cerantola, A.I. Chumakov, B.N. Savenko, Sci. Rep. **9**, 4464 (2019)
K. Krieble, C.C.H. Lo, Y. Melikhov, J.E. Snyder, J. Appl. Phys. **99**, 08M912 (2006)
G.U. Kulkarni, K.R. Kannan, T. Arunarkavalli, C.N.R. Rao, Phys. Rev. B **49**, 724 (1994)
R.G. Kulkarni, H.H. Joshi, Solid State Commun. **53**, 1005 (1985)
W. Kündig, R.S. Hargrove, Solid State Commun. **7**, 223 (1969)
L.K. Leung, B.J. Evans, A.H. Morrish, Phys. Rev. B **8**, 29 (1973)
J.W. Linnet, M.M. Rahman, J. Phys. Chem. Solids **33**, 1465 (1972)
T.K. McNab, R.A. Fox, A.J.F. Boyle, J. Appl. Phys. **39**, 5703 (1968)
T. Mizoguchi, M. Tanaka, J. Phys. Soc. Jpn. **18**, 1301 (1963)
A.H. Morrish, K. Haneda, J. Appl. Phys. **52**, 2496 (1981)
A.H. Morrish, P.E. Clark, Phys. Rev. B **11**, 278 (1975)
N. Moumen, P. Bonville, M.P. Pileni, J. Phys. Chem. **100**, 14410 (1996)
P.J. Murray, J.W. Linnett, J. Phys. Chem. Solids **37**, 619 (1976)
P. Muthukumarasamy, T. Nagarajan, A. Narayanasamy, J. Phys. C: Solid State Phys. **13**, 3961 (1980)

S. Nakamura, A. Fuwa, Hyperfine Interact. **226**, 267 (2014)
S. Nakamura, Y. Kobayashi, S. Kitao, M. Seto, Physica B: Condens. Matter **536**, 620 (2018)
H.N. Ok, B.J. Evans, Phys. Rev. B **14**, 2956 (1976)
S.A. Oliver, H.H. Hamdeh, J.C. Ho, Phys. Rev. B **60**, 3400 (1999)
T. Pannaparayil, S. Komarneni, R. Marande, M. Zadarko, J. Appl. Phys. **67**, 5509 (1990)
M.P. Pasternak, S. Nasu, K. Wada, S. Endo, Phys. Rev. B **50**, 6446 (1994)
V.U. Patil, R.G. Kulkarni, Solid State Commun. **31**, 551 (1979)
D. Peddis, N. Yaacoub, M. Ferretti, A. Martinelli, G. Piccaluga, A. Musinu, C. Cannas, G. Navarra, J.-M. Greneche, D. Fiorani, J. Phys. Condens. Matter **23**, 426994 (2011)
R.M. Persoons, E. De Grave, P.M.A. de Bakker, R.E. Vandenberghe, Phys. Rev. B **47**, 5894 (1993)
G.A. Petitt, D.W. Forester, Phys. Rev. B **4**, 3912 (1971)
R.J. Pollard, Hyperfine Interact. **41**, 509 (1988)
R.J. Pollard, J. Phys. Condens. Matter **2**, 983 (1990)
D. Prabhu, A. Narayanasamy, K. Shinoda, B. Jeyadevan, J.-M. Greneche, K. Chattopadhyay, J. Appl. Phys. **109**, 013532 (2011)
H.R. Rechenberg, F.A. Tourinho, Hyperfine Interact. **67**, 627 (1991)
A.J. Rondinone, A.C.S. Samia, Z.J. Zhang, Appl. Phys. Lett. **76**, 3624 (2000)
M. Rosenberg, P. Deppe, S. Dey, U. Janssen, C.E. Patton, C.A. Edmondson, IEEE Trans. Magn. **18**, 1616 (1982)
M.J. Rossiter, J. Phys. Chem. Solids **26**, 775 (1965)
G.A. Sawatzky, F. van der Woude, A.H. Morrish, Phys. Rev. **187**, 747 (1969)
J.H. Shim, S. Lee, J.H. Park, S.J Han, Y.H. Jeong, Y.W. Chao, Phys. Rev. B 73, 064404 (2006).
M.R. Spender, A.H. Morrish, Solid Sate Commun. **11**, 1417 (1972)
C.M. Srivastava, S.N. Shringi, R.G. Srivastava, Phys. Rev. B **14**, 2041 (1976)
M. Tanaka, T. Tokor, Y. Aiyama, J. Phys. Soc. Jpn. **21**, 262 (1966)
J. Teillet, J.L. Dormann, A. Hauet, M. Nogues, F. Varret, M. Zemirli, J.-M. Greneche, Hyperfine Interact. **68**, 345 (1991)
N. Tristan, J. Hemberger, A. Krimmel, H.A. Krug von Nidda, V. Tsurkan, A. Loidl, Phys. Rev. **B 72**, 174404 (2005)
F. van der Woude, G.A. Sawatzky, A.H. Morrish, Phys. Rev. **167**, 533 (1968)
F. van der Woude, G.A. Sawatzky, Phys. Rev. B **9**, 3159 (1971)
J.J. Van Loef, Physica **32**, 2102 (1966)
R. Vanleerberghe, R.E. Vandenberghe, Hyperfine Interact. **23**, 75 (1985)
W. Winkler, Phys. Status Solidi A **84**, 193 (1984)
M.X. Xu, G.R. Hearne, S. Layek, D. Levy, J.P. Itie, M.P. Pasternak, G.K. Rozenberg, E. Greenberg, Phys. Rev. B **95**, 045110 (2017a)
W.M. Xu, G.R. Hearne, S. Layek, D. Levy, J.P. Itie, M.P. Pasternak, G.K. Rozenbery, E. Greenberg, Phys. Rev. B **96**, 045108 (2017b)
W.M. Xu, G.Y. Machavariani, G.K. Rozenberg, M.P. Pasternak, Phys. Rev. B **70**, 174106 (2004)
Z.C. Xu, J. Appl. Phys. **93**, 4746 (2003)
C.M. Yagnik, H.B. Mathur, J. Phys. C: Solid State Phys. **1**, 469 (1968)
A. Yang, C.N. Chinnasamy, J.-M. Greneche, Y. Chen, S.D. Yoon, Z. Chen, K. Hsu, Z. Cai, K. Ziemer, C. Vittoria, V.G. Harris, Nanotechnology **20**, 185704 (2009)
J.W. Young, J. Smit, J. Appl. Phys. **42**, 2344 (1971)

Chapter 8
Mössbauer Spectrometry of Garnets, Hexaferrites and Orthoferrites

8.1 Ferrimagnetic Garnets

Rare-earth iron garnets (RIG) have the structural formula $R_3Fe_5O_{12}$, where R is a rare-earth ion. The cubic garnet structure has the 24d and 16a sites occupied by Fe^{3+} ions and 24c sites occupied by R^{3+} ions. The various sites of magnetic garnets are shown in Fig. 8.1. The (d), [a] and {c} sites are the tetrahedral, octahedral and dodecahedral sites, respectively, and the brackets represent their respective site notations. The (d) and [a] site ferric ion moments are antiparallel to each other and strongly coupled. The {c} site moments are weakly coupled antiferromagnetically to the resultant (d)−[a] magnetic moments. Therefore, the temperature-dependent magnetic properties are determined by the type of rare-earth ion, giving rise to variations in the compensation temperature (T_{comp}) and almost the same Curie temperature (T_c) of 564 K due to the ferric moments, as shown in Fig. 8.2. Three types of patterns in the variation of the magnetic moment per formula unit (f.u.) could be noticed with temperature (Néel et al. 1964): (i) drastic decrease in magnetic moment arising from the rare-earth moments which increase beyond T_{comp}, (ii) gradual decrease in the magnetic moment up to T_c; and (iii) gradual increase in magnetic moment followed by decrease up to T_c. Since the Y^{3+} ion is non-magnetic, the temperature-dependent magnetic behavior of the yttrium iron garnet is solely determined by the Fe^{3+} ions in the tetrahedral and octahedral sites as in Fig. 8.2 (ii). However, small variations in the transition temperatures (T_c and T_{comp}) are observed between magnetic and Mössbauer studies and among different samples of the same RIG.

Rare-earth ions such as Tb, Dy, Gd, Ho and Er exhibit a large magnetic moment close to 0 K antiparallel to the much smaller ferric magnetic moment. At 0 K, the magnetic moment of the RIG is in the order of TbIG > DyIG > GdIG > HoIG > ErIG > SmIG > LuIG > YIG > LaIG > EuIG > YbIG (Geller et al. 1963, 1965). The temperature-dependent sublattice magnetic moments and the net magnetic moment are schematically shown in Fig. 8.3. With temperature, the net magnetic moment

R. J. Joseyphus and J.-M. Greneche, *Fundamentals of ^{57}Fe Mössbauer Spectrometry*,
https://doi.org/10.1007/978-981-99-8653-8_8

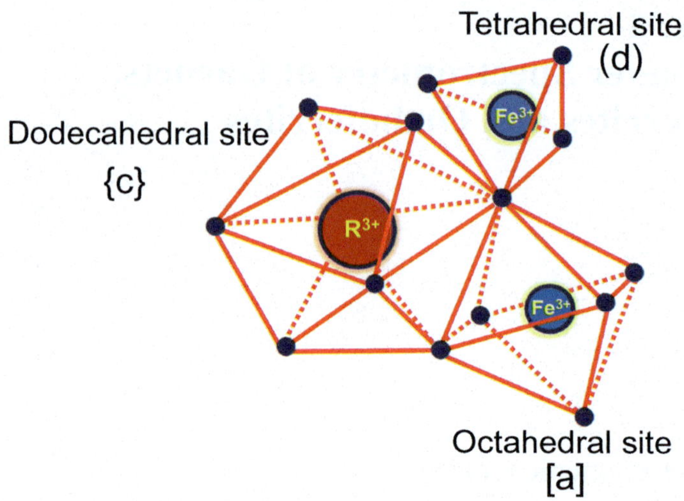

Fig. 8.1 The tetrahedral (d), octahedral [a] and dodecahedral {c} sites of RIG

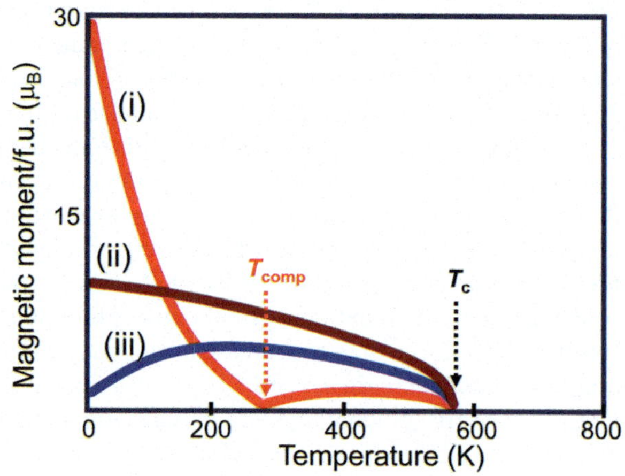

Fig. 8.2 The temperature-dependent magnetic moment of various RIG follows (i) a decreasing trend to reach a minimum at T_{comp}, increases and drops again on approaching the T_c, (ii) continously decreases with temperature and (iii) reaches a maximum and then decreases again

becomes zero when the rare-earth and the ferric ion moments are exactly antiparallel, giving antiferromagnetic-like arrangement at T_{comp}. At higher temperatures, the magnetic moment of the rare-earth ion becomes negligible, and the magnetic moment of the ferric site drops rapidly as it reaches the Curie temperature. Although the net magnetic moment may be negative for $T > T_{comp}$, bulk magnetization increases beyond T_{comp} and vanishes at T_c.

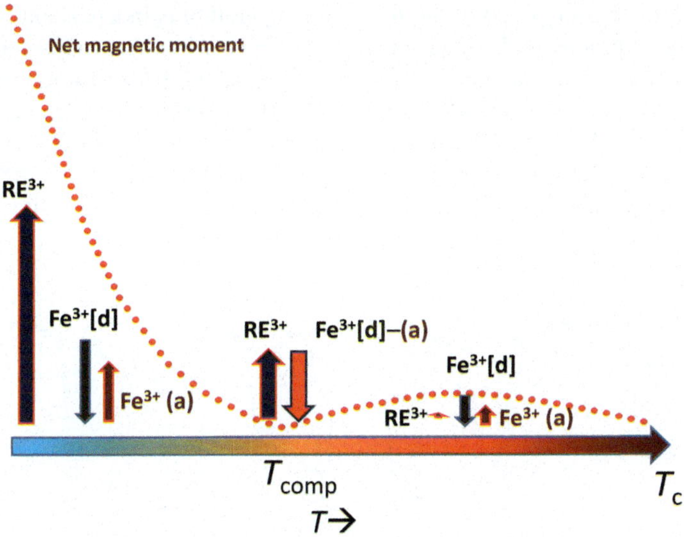

Fig. 8.3 The variation in the magnetic moment for various sublattices with temperature in RIG. The experimental magnetic moment shows the largest value at low temperatures due to the large rare-earth (RE) magnetic moment and becomes zero at the compensation temperature (T_{comp}). The RIG's becomes paramagnetic at the T_c; determined only by ferric ion moments

8.2 Yttrium Iron Garnet ($Y_3Fe_5O_{12}$)

The room temperature Mössbauer spectrum of yttrium iron garnet (YIG) was initially recorded on a single crystal with a 0.50 T applied field along the [111] and [100] direction (Alff and Wertheim 1961). The direction of magnetization was chosen such that the angle (θ) between the applied field and the z-axis of electric field gradient are minimum. The tetrahedral d-site hyperfine field B_{hf}(d) of 29.0 T and octahedral a-site B_{hf}(a) of 46.7 T are obtained along the [100] magnetization direction. Along the [111] direction, the B_{hf}(d) and B_{hf}(a) are 39.4 and 48.1 T, respectively. The average hyperfine fields, $\overline{B}_{hf}$ are 39.2 and 47.4 T, while the quadrupole interaction energies are 75 and 90 neV for the tetrahedral and octahedral sites, respectively. The B_{hf}(d) and B_{hf}(a) of polycrystalline YIG are 39.0 and 48.5 T, respectively, at 300 K (Bauminger et al. 1961). The low-temperature measurements at 85 K results in $B_{hf}(d) = 46.0$ T and $B_{hf}(a) = 53.5$ T.

The Mössbauer spectra obtained in the range 300–775 K have shown a linear decrease in relative intensity of the tetrahedral (I_d) and octahedral (I_a) sites with temperature, preserving the intensity ratio of 3:2 (I_d:I_a) (Sawatzky et al. 1969). The YIG transforms from the ferrimagnetic to the paramagnetic state with two superimposed tetrahedral and octahedral site doublets at 560 K. The temperature-dependent B_{hf} for the tetrahedral site was fitted with four sextets and the octahedral site with two, reasoned as due to rhombohedral ($R\overline{3}$ symmetry) distortion (Vandormael et al.

2001). The decreasing trend of the B_{hf} with temperature is independent of the rare-earth ion substitutions as obtained with Dy^{3+} and Eu^{3+} while they follow the Brillouin function with $S = 5/2$. The typical Mössbauer spectra of YIG at various temperatures below room temperature are shown in Fig. 8.4. (Long et al. 2016). The B_{hf}(d) value is in the range of 55 T at 7.5 K and decreases to 49 T at 295 K. The B_{hf}(a) values are 47 and 39.6 T at 7.5 and 295 K, respectively. The angle θ between the EFG and the B_{hf} is distributed around 54.7° in YIG for the d-site while those for a-site are around 70.5 and 0°.

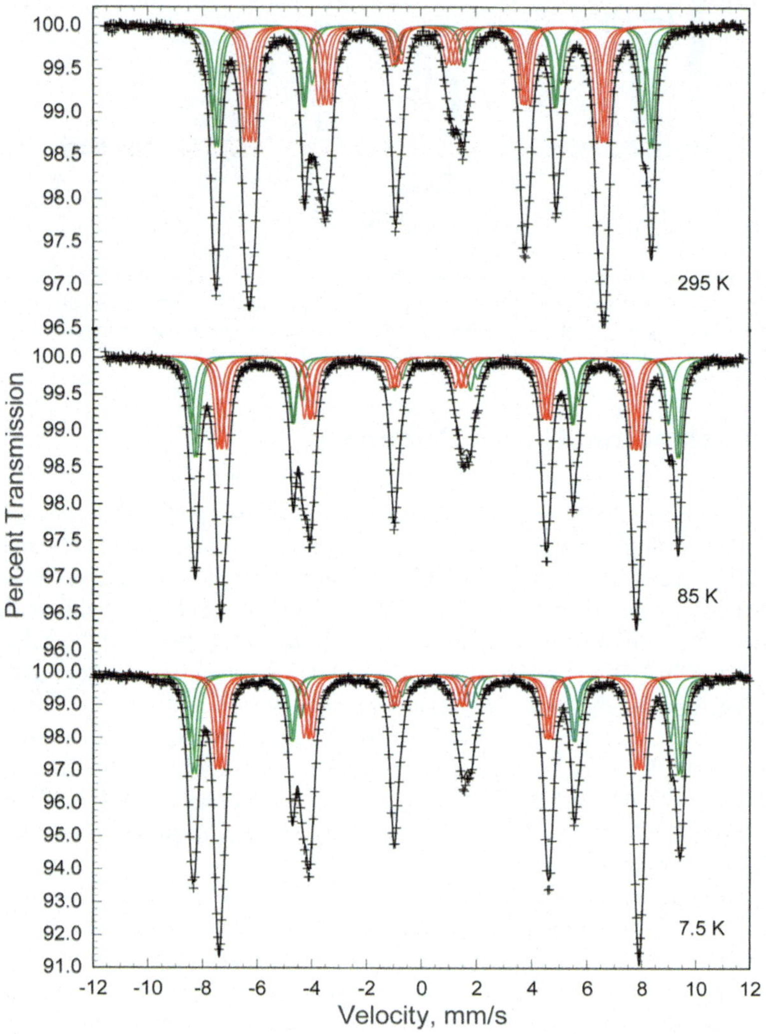

Fig. 8.4 The Mössbauer spectra of YIG at various temperatures. Reprinted with permission (Long et al. 2016), © American Chemical Society

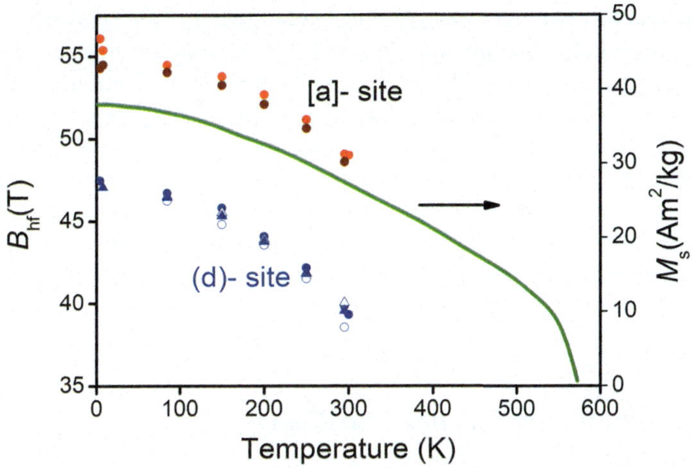

Fig. 8.5 The M_s and B_{hf} with temperature for the tetrahedral (d) and octahedral [a] sites of YIG. Data from (Vandormael et al. 2001; Greneche et al. 1988; Anderson 1964; Long et al. 2016)

Figure 8.5 shows the plot of [a] and (d) site hyperfine fields and saturation magnetization (M_s) data of the YIG (Vandormael et al. 2001; Greneche et al. 1988; Anderson 1964; Long et al. 2016). The M_s values are 38 and 27.4 A·m^2/kg at 0 and 292 K, respectively (Anderson 1964). Molecular field theory predicts 76 and 114 A·m^2/kg for the octahedral and tetrahedral sites, respectively at 0 K, which gives a net magnetization of 38 A·m^2/kg that is in excellent agreement with the experimental value.

It was shown that Al, Ga, In and Sc prefers tetrahedral site, and 25% Sc in the d-site reduces the Curie temperature to 450 K from the bulk YIG value of 533 K, as observed from the disappearance of the hyperfine field (Coey 1972). The preference of 25% Ga in the octahedral site reduces the T_c to 467 K. Charge compensated substitutions such as Co^{2+} and Ge^{4+} in YIG drastically reduce the T_c even below 100 K according to the substitution level resulting in paramagnetic doublets at room temperature (Geller et al. 1983).

The amorphous YIG showed a paramagnetic doublet at 77 K and RT quadrupole splitting (QS) of 1.12 mm/s (Gyorgy et al. 1979). The paramagnetic doublet transformed to a sextet at ~30 K and at 4.2 K a distribution in hyperfine fields with $\overline{B}_{hf}$ of 45.0 T and ferric ion average isomer shift of 0.45 mm/s was obtained (Eibschütz and Lines 1982). Indeed, the transformation temperature to a sextet from the paramagnetic doublet varies according to the synthesis method that influences the extent of amorphization. Polycrystalline YIG subjected to neutron irradiation induces amorphization that reduces the $\overline{B}_{hf}$ by 2.0 T at 4.2 K, and a broad T_c of 70–90 K was obtained (Greneche et al. 1988).

The polycrystalline YIG also transforms to an irreversible amorphous YIG phase subjected to a pressure beyond 48 GPa (Lyubutin et al. 2005). Amorphous

YIG could also be obtained using chemical synthesis, and the heat-treated samples showed nanocrystalline nature with $\overline{B}_{hf}$ at 4.2 K slightly lower than the bulk YIG due to tetrahedral distortion (Vaqueiro et al. 1997). Substitution of $0.3Si^{4+}$ per formula unit in YIG induces spin reorientation from [111] to [100] above 80 K (Geller et al. 1980a, b). The spin reorientation was observed from the splitting of the tetrahedral relative intensity. Unlike spinel ferrites, cation inversion is not possible in garnets, and the YIG transforms to orthoferrites and other oxides on mechanical milling (Joseyphus et al. 2006). YIG single crystal film is used in synchrotron-based Mössbauer spectrometry as a nuclear Bragg diffractor along the {200} plane that is electronically forbidden (Gerdau et al. 1985).

8.3 Samarium Iron Garnet ($Sm_3Fe_5O_{12}$)

Samarium iron garnet (SmIG) shows higher magnetic moment than YIG below room temperature (Geller et al. 1963). Further, at low temperatures, the magnetic moment follows $L - S$ behavior, and at room temperature, it is $L + S$ due to the admixture of closely spaced multiplets. The cross over temperature is difficult to determine due to weak magnetic moment and is widely reported in the range 195–330 K. The tetrahedral and octahedral hyperfine fields at 295 K are 40.8 and 49.6 T, respectively (van Loef 1968). The cross-over temperature greater than room temperature resulted in declaring [110] as the easy direction in the initial reports.

Mössbauer spectra were recorded for polycrstalline SmIG in the temperature range 5–298 K (Geller and Balestrino 1980a, b). The easy magnetization direction above 60 K was found to be along [111], whereas it was close to [110] below 50 K, and another one below 16.5 K was suggested to be along [100]. Later it was assigned to ⟨740⟩ direction from measurements on a single crystal (Balestrino et al. 1984). In the entire temperature range, SmIG undergoes multiple spin reorientations from ⟨111⟩ in the temperature range 60–560 K, along ⟨110⟩ in the 18–60 K range, and at very low temperatures below 20 K, the direction is unclear as other reports also suggest ⟨210⟩ (Guillot et al. 1991).

8.4 Gadolinium Iron Garnet ($Gd_3Fe_5O_{12}$)

The large magnetic moment with reducing temperature for the Gd^{3+} results in a compensation temperature at 290 K where the net Fe^{3+} and Gd^{3+} magnetic moments of equal magnitude are antiparallel to each other for GdIG. Mössbauer spectra recorded in the temperature range from 4.2 to 714 K have shown the easy direction as [111] without any spin reorientation and a T_c of 563 K (Cook and Li 1986). Above T_c, the quadrupole splitting of the tetrahedral site was found to be negative with a value of – 0.43 mm/s, whereas only the magnitude of 0.89 mm/s could be determined for the octahedral site. Assuming the [111] as easy axis, the tetrahedral to octahedral intensity ratio (I_d): [I_a] could be further distributed as (I_d):[I_{a2}:I_{a1}] where the a-site is subdivided giving the values (6): [3:1]. For the [100] easy axis, the d-site intensity ratio is further subdivided into the form (I_{d1}:I_{d2}): [I_a], resulting in the ratio as (2:4): [4].

Experiments in the temperature range from 80 to 310 K confirmed [111] as the easy axis of magnetization without spin canting (Siedel et al. 1992). The B_{hf}(d), B_{hf}(a_2), B_{hf}(a_1) values are 40.9, 50.0 and 49.2 T, respectively, at the compensation temperature of 286 K. It further increased to 47.4, 54.5 and 53.6 T at 80 K. The extrapolated saturation hyperfine field values are 47.6, 54.6 and 53.7 T, respectively.

Mechanical milling was attempted on GdIG to observe any change in cation distribution or spin canting with size reduction. Mössbauer experiments from room temperature to 16 K showed decomposition of GdIG to products such as $GdFeO_3$ with milling (Joseyphus et al. 2006). Figure 8.6 shows the room temperature Mössbauer spectra of the (a) bulk and (b) 10 h milled GdIG. The RT spectrum shows B_{hf}(d) and B_{hf}(a) of 40.2 and 49.2 T, respectively, corresponding to polycrystalline GdIG. The mechanically milled samples exhibit broad doublet originating from the size-reduced phases that are superparamagnetic. The magnetization studies suggested a decrease in the compensation temperature from 290 K for the bulk to 268 K for the 10 h milled GdIG due to the weakening of the bonds with milling.

8.5 Hyperfine Parameters of RIG

^{57}Fe Mössbauer parameters of various RIG reported in the literature are listed in Table 8.1. The B_{hf} for the tetrahedral and octahedral sites at a particular temperature are close to each other irrespective of the rare-earth ions. The tetrahedral site hyperfine field and isomer shift are smaller than the octahedral site. The slight differences between the B_{hf} are attributed to orientation and site distortions. The hyperfine fields of DyIG and EuIG at various temperatures were studied using ^{161}Dy and ^{151}Eu Mössbauer spectrometry (Cohen et al. 1964). ^{161}Dy Mössbauer effect in DyIG at 20, 77 and 300 K have shown hyperfine fields of 485, 330 and 72 T, respectively. The Eu site exhibits hyperfine fields of 67, 57, 33.5 T at 20, 81 and 300 K, respectively. The DyIG thus shows large magnetization at low temperatures but drops rapidly with temperature, thereby resulting in a compensation temperature compared with EuIG.

8.6 TmIG

The temperature-dependent hyperfine field of thulium nuclei was calculated using a mean-field model with two Tm sites due to strong anisotropy (Meyer et al. 1994). The temperature-dependent hyperfine fields for the ferric ion [a] and (d) sites and that for {c_1} and {c_2} sites of Tm are shown in Fig. 8.7. The magnetization orientation along

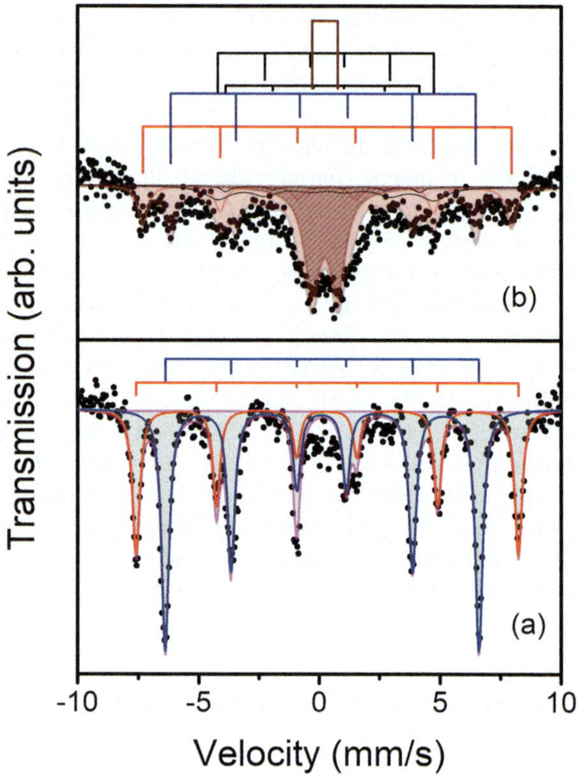

Fig. 8.6 The Mössbauer spectra at 300 K for the **a** bulk and **b** 10 h milled GdIG. Reprinted with permission (Joseyphus et al. 2006), © Elsevier

the crystal axis resulted in two equally populated chemically equivalent sites that were verified experimentally using ^{169}Tm Mössbauer spectrometry (Cohen 1963). The agreement between the calculated and the experimental values for the Tm sites can be seen in Fig. 8.7. Moreover, the magnetic contribution from the two ferric ion sites is dominant beyond room temperature while the hyperfine field of Tm diminishes as observed from the theoretical curves of TmIG.

8.7 Hexaferrites

Hexagonal ferrites are classified into various types, such as M, W, Z and Y that resemble the magnetoplumbite structure. The formula and abbreviation for various hexaferrite and their magnetic moment arrangement in the sites [], () and { } corresponding to octahedral, tetrahedral and trigonal bipyramid sites, respectively, are shown in Table 8.2. The general formula for M-type hexaferrite is $MFe_{12}O_{19}$ where M is a divalent ion of Ba, Sr, Pb and Ca, also represented as BaM, SrM, PbM and CaM,

Table 8.1 ^{57}Fe Mössbauer parameters for the RIG at various temperatures

RIG	T	δ (mm/s)	$\overline{B}_{hf}$ (T)	References
EuIG ($T_c = 566$ K)	4.2 K	0.21	47.5 (d)	Nowik et al. (1967), Crecelius et al. (1969), Litterst et al. (1979)
		0.46	54.5 [a]	
DyIG ($T_c = 563$ K)	4.2 K	0.20	48.0 (d)	
		0.42	55.1 [a]	
SmIG	295 K	–	40.7 (d)	van Loef (1968)
		–	49.6 [a]	
EuIG	85 K	0.28	46.8 (d)	Vandormael et al. (2001)
		0.51	54.4 [a]	
	295 K	0.18	40.2 (d)	
		0.40	49.4 [a]	
DyIG	85 K	0.3(SS)	46.0 (d)	Bauminger et al. (1961)
		0.6(SS)	54.0 [a]	
	300 K	0.5(SS)	39.5 (d)	
		0.6(SS)	48.5 [a]	
DyIG	295 K	0.19	39.4 (d)	Sankaranarayanan and Gajbhiye (1994)
		0.44	48.8 [a]	
ErIG	295 K	0.19 0.35	39.2 (d) 48.9 [a]	
TmIG ($T_c = 550$ K)	90 K	–	47.0 (d)	Meyer et al. (1994)
			55.0 [a]	
	300 K		39.0 (d)	
			48.0 [a]	
HoIG	RT	0.16	40.7 (d)	Lataifeh et al. (1999)
		0.38	50.1 [a]	
	77 K	0.25	47.6 (d)	
		0.52	55.3 [a]	
TbIG ($T_c = 560$ K; $T_{comp} = 260$ K)	13 K	0.15	48.1 (d)	Hong et al. (2004)
		0.38	55.3 [a]	
		0.48	53.2 [a]	
	260 K	0.05	41.5 (d)	
		0.28	50.4 [a]	
	295 K	0.04	40.2 (d)	
		0.26	49.4 [a]	
	500 K	−0.07	27.4 (d)	
		0.14	35.6 [a]	
GdIG	300 K	0.22 0.43	40.2 (d) 49.0 [a]	Joseyphus et al. (2006)

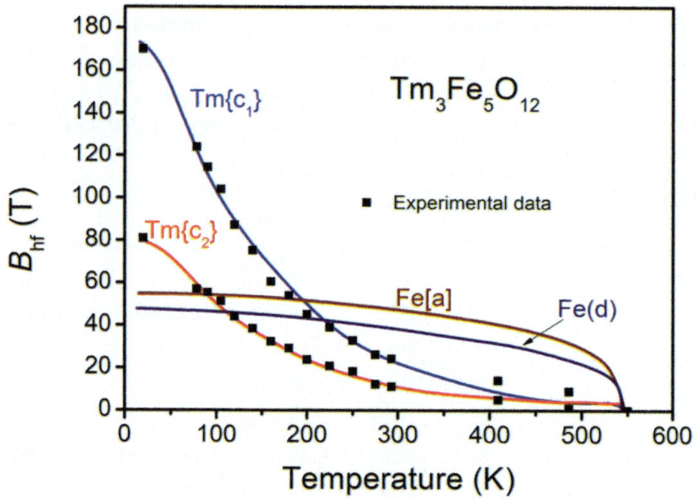

Fig. 8.7 The theoretical temperature-dependent hyperfine fields for the Tm and Fe sites in TmIG. The experimental B_{hf} data from the ^{169}Tm sites are also plotted. Data from (Meyer et al. 1994; Cohen 1963)

respectively. The ferric ion occupies tetrahedral, octahedral and trigonal bipyramid sites.

The total magnetic moment in BaM is determined by

$$m(\mathrm{T}) = m\langle 6[12\mathrm{k}]-2[4\mathrm{f}_2]+1[2\mathrm{a}] - 2(4\mathrm{f}_1) + 1\{2\mathrm{b}\}\rangle$$

Table 8.2 Various types of hexagonal ferrites and their magnetic arrangements

Hexaferrite type	Formula	Abbreviation	$m\{\}$	$m[]$	$m()$	
M	$MFe_{12}O_{19}$ (M = Ba, Sr, Pb,..)	BaM SrM PbM	1↑	7↑2↓	2↓	
W	$BaMe_2Fe_{16}O_{27}$ (Me = Co, Zn,..)	Me_2W	1↑	11↑2↓	4↓	
Y	$Ba_2Me_2Fe_{12}O_{22}$ (Me = Co, Zn,..)	Me_2Y	–	8↑2↓	4↓	
Z	$Ba_3Me_2Fe_{24}O_{41}$ (Me = Co, Zn,..)	Me_2Z	1↑	15↑4↓	6↓	
X	$Ba_2Me_2Fe_{28}O_{46}$ (Me = Co, Zn, …)	Me_2X	2↑	10↑4↓	2↓	
U	$Ba_4Me_2Fe_{36}O_{60}$ (Me = Co, Zn, …)	Me_2U	2↑	22↑6↓	8↓	

where the brackets in the Wykoff notations represent the various sites. The magnitude of the magnetic moment is further given outside the site notations, where the 12k site gives the largest magnetic moment. Since the magnetic moment is contributed only by the ferric ions (5 μ_B) as Ba^{2+}, Sr^{2+} are non-magnetic, the net magnetic moment at 0 K is given as

$$m(0) = 6 \times 5\,\mu_B - 2 \times 5\,\mu_B + 5\,\mu_B - 2 \times 5\,\mu_B + 5\,\mu_B = 20\,\mu_B$$

.

The structure of SrM, an M type hexaferrite is shown in Fig. 8.8. The hyperfine field components for the BaM were initially fitted with four sextets that followed a decreasing trend with temperature, vanishing at the Curie temperature of 730 K similar to the magnetization data (van Loef and Franssen 1963). The bi-pyramidal (2b) site has quadrupole splitting greater than 1 mm/s that was found to increase with decreasing temperature (Kreber et al. 1975). The overlapping hyperfine fields and the close isomer shift values make the site assignment difficult for a few. Later on, it was shown that the room temperature hyperfine field is of the order of $B_{hf}\{2b\} < B_{hf}[12k] < B_{hf}(4f_1) < B_{hf}(2a) < B_{hf}[4f_2]$ for both BaM and SrM (Evans et al. 1987). However, the relative isomer shift values are uncertain, with values ranging from 0.3–0.4 mm/s.

The site assignment is now clear with five hyperfine sextets having relative intensities 6:2:2:1:1 corresponding to $[12k]:[4f_2]:(4f_1):(2a):\{2b\}$ and the sites could be identified from the isomer shift and quadrupole splitting values (Šepelák et al. 2014). The room temperature Mössbauer spectrum of polycrystalline BaM is shown in Fig. 8.9 and the corresponding hyperfine parameters are listed in Table 8.3. The linewidth of the sextets is 0.26 mm/s, obtained using the ^{57}Co/Rh source. The octahedrally coordinated sites [$4f_2$, 12k, 2a] are characterized by larger isomer shift (> 0.19 mm/s). The bi-pyramidal and tetrahedral sites have a lower isomer shift than the octahedral sites, and the bi-pyramidal site could be distinguished from its large quadrupole splitting (> 1 mm/s) and lowest B_{hf}. The [$4f_2$] site has the largest B_{hf} and δ values.

The comparison of the hyperfine parameters among BaM, SrM and PbM single crystals along parallel and perpendicular orientations have shown that the relative intensity of the {2b} subspectrum and its quadrupole splitting is least in PbM (Thompson et al. 1993). The {2b} displacement parallel to the *c*-axis is found to be larger in PbM than SrM and BaM which determines their magnetic properties.

Mössbauer spectrometry was undertaken in the $BaZn_{2-x}Co_xFe_{16}O_{27}$ ($x \leq 2$) single crystals, and the spin arrangement in Co_2W (x = 2) ferrite was shown to be along *c*-axis at temperatures above 503 K, but reaches a cone of angle 70° at 463 K down to 77 K (Paoluzi et al. 1988). The quadrupole splitting of the Co_2W is close to − 0.2 mm/s, and drastic variation up to + 0.5 mm/s occurs at the spin reorientation temperature of 503 K. Further, the T_c and the spin orientation temperature increase with x, suggesting the role of Zn in decreasing the transition temperatures similar to previous studies (Albanese et al. 1986).

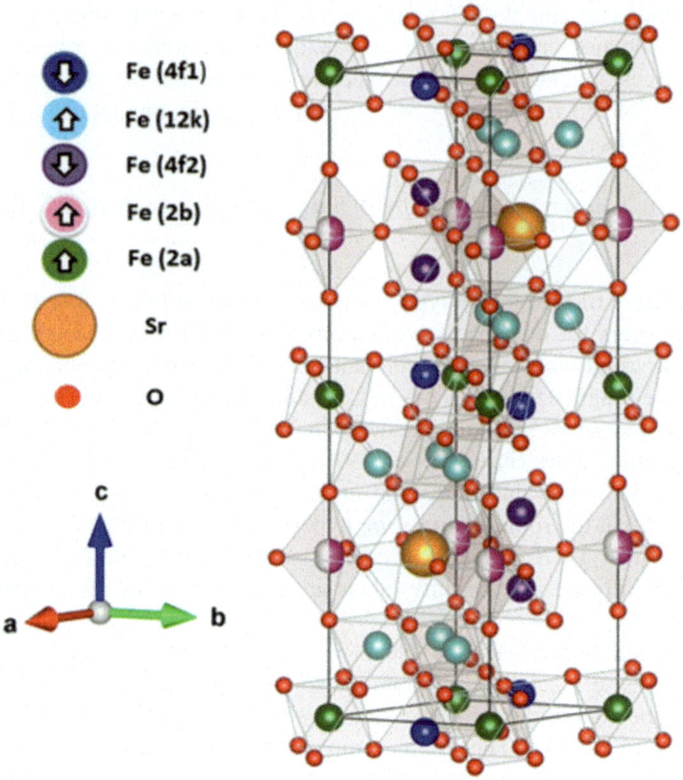

Fig. 8.8 The structure of an M-type hexaferrite (SrM) (Soria et al. 2019)

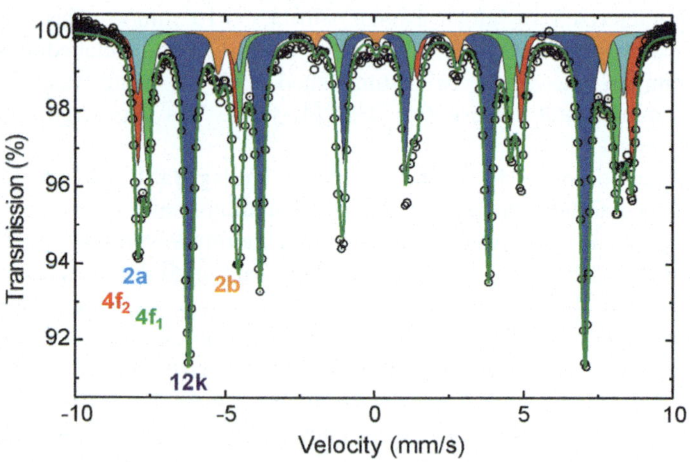

Fig. 8.9 The room temperature Mössbauer spectrum of $BaFe_{12}O_{19}$. (Šepelák et al. 2014)

Table 8.3 The room temperature hyperfine parameters for the polycrystalline BaM (Šepelák et al. 2014)

Site	δ (mm/s)	QS (mm/s)	B_{hf} (T)
{2b}	0.14	1.08	40.0
[12k]	0.22	0.21	41.1
($4f_1$)	0.14	0.11	48.7
(2a)	0.20	0.02	50.4
[$4f_2$]	0.25	0.10	51.2

Polycrystalline $Ba_3Co_{2-x}Zn_xFe_{24}O_{41}$ (x = 0, 0.5, 1.5) could be fitted with only six distinguishable sextets from among ten metal ion sites ($4f_{IV\uparrow}$, $4f_{IV\downarrow}$, $12k_{VI\uparrow}$, $4f_{VI\uparrow}$, $4e_{IV\downarrow}$, $12k_{VI\uparrow}$, $2d_{V\uparrow}$, $2a_{VI\uparrow}$, $4f_{VI\downarrow}$, $4e_{VI\downarrow}$) (Lim et al. 2014a, b). It was shown that Zn ion prefers tetrahedral sites. Mössbauer studies with $x \leq 2$ have shown the conical to a planar spin reorientation in the Co_2Z (x = 2) ferrite around 250 K (Lim et al. 2015). From the application of a magnetic field parallel to the γ-ray, it was shown that the canting angle decreases from 24° for x = 0 to 8° for x = 2 at 5 T. The spectrum at 4.2 K was not well resolved and composed of ten sextets, whereas the application of magnetic field resolved into two major sextet components, as shown in Fig. 8.10. These were assigned to A- and B- sublattice components with small and large intensities, respectively. The average hyperfine field was 55.0 T at 4.2 K, which decreased to less than 30.0 T at 600 K for Co_2Z.

The Me_2Y type hexaferrites have close-lying hyperfine fields of nearly the same relative intensities and hence fitted with only one sextet component. The average B_{hf} of Mg_2Y and Co_2Y are 48.4 and 50.8 T, respectively, at 78 K (Albanese et al. 1975). The room temperature spectrum for the Mg_2Y with a single B_{hf} of 41.5 T could be resolved into two sextets with an applied field of 2.5 T that showed effective hyperfine fields of 45.0 and 39.0 T. The Mg^{2+} prefers octahedral coordination with almost $0.7Mg^{2+}$ in $6c_{VI\downarrow}$ site and another 1.3 in spin-up sites. Co_2Y showed effective B_{hf} of 49.0 and 42.5 T with an applied field of 2.5 T at RT. The Co^{2+} prefers the $6c_{VI\downarrow}$ site occupied with $0.9Co^{2+}$ and another 1.1 is distributed among $3a_{VI}$, $18h_{VI}$ and $3b_{VI}$. The Curie temperatures are 553 and 613 K for the Mg_2Y and Co_2Y, respectively. The decrease in B_{hf}, canting angle and T_c with Zn has been reported (Lim et al. 2014a, b).

8.8 Orthoferrites

The perovskite structure with the general formula ABX_3 exists in cubic, orthorhombic and tetragonal forms. Orthoferrites have the orthorhombically distorted perovskite structure with the general formula $RFeO_3$ (R = rare-earth). The ABX_3 structure representing the rare-earth, Fe and oxygen ions positions is shown in Fig. 8.11. The site representation is {R}[Fe]O_3, where {} and [] are the dodecahedral and octahedral sites, respectively. Each unit cell has four R^{3+} and Fe^{3+} ions. The Fe^{3+} ions are antiferromagnetically aligned through a strong superexchange bond of 180°,

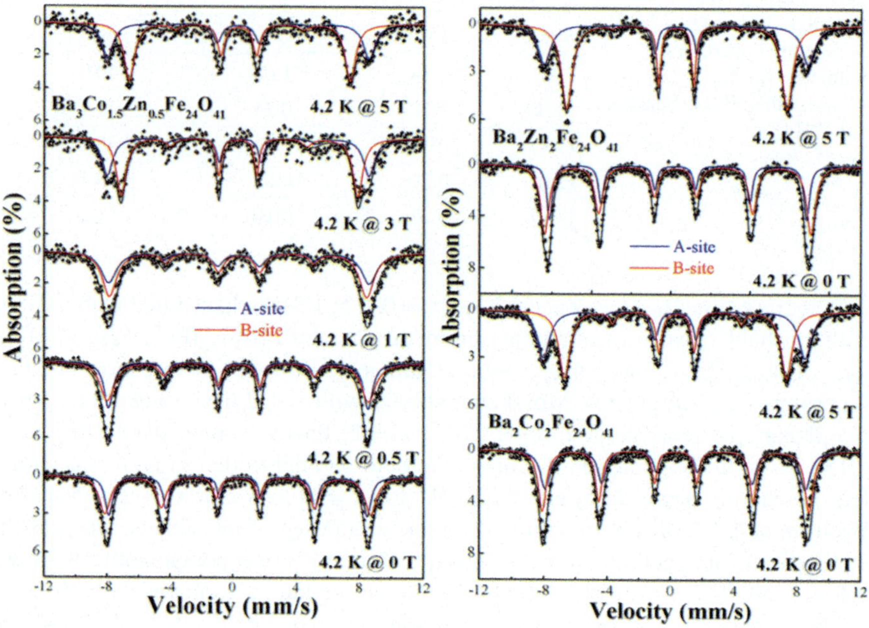

Fig. 8.10 Mössbauer spectra of $Ba_3Co_{2-x}Zn_xFe_{24}O_{41}$ with x = 0.5 (left), x = 0 and 2 (right) at 4.2 K with various applied fields up to 5 T. Reprinted with permission (Lim et al. 2015), © AIP Publishing

resulting in T_N above 650 K. Ferrimagnetism in orthoferrites arises from spin canting. As the temperature is reduced, it may exhibit a compensation point where the rare-earth ions of antiparallel spins cancel the ferric ion magnetic moments and spin reorientation occurs, where the magnetic orientation changes from the *c*- to *a*- axis.

Figure 8.12 shows the hyperfine fields at RT and 0 K along with the Néel temperature of various orthoferrites that show a decreasing trend from $LaFeO_3$ to $LuFeO_3$ (Eibschütz et al. 1964; Eibschütz et al. 1967). The B_{hf} at 0 K decreases from 56.4 to 54.6 T with the atomic number of rare-earth ion in $RFeO_3$. This trend is also seen in the room temperature B_{hf} (51.9–48.9 T) and the T_N (738–622 K) due to decreasing ionic radius (1.22–0.99 Å) consecutively decreasing the Fe–Fe distance. Since the rare-earth ionic radius in $YFeO_3$ (1.06 Å) is between $DyFeO_3$ (1.07 Å) and $HoFeO_3$ (1.05 Å), the B_{hf} and T_N values are to be placed between them.

The B_{hf} and T_N of orthoferrites increase with medium pressure up to 14 GPa, whereas the isomer shift decreases (Halasa et al. 1974). This implies the delocalization of d electron density, thereby resulting in an increase in the 3s electron density at the nucleus. The increase in T_N with pressure is consistent with the increase in exchange integral due to decreasing interatomic distance. The quadrupole splitting also becomes more negative with pressure due to the distortions. The pressure dependence of the $\overline{B}_{hf}$, T_N and δ for $LaFeO_3$ is shown in Fig. 8.13. The trend in the

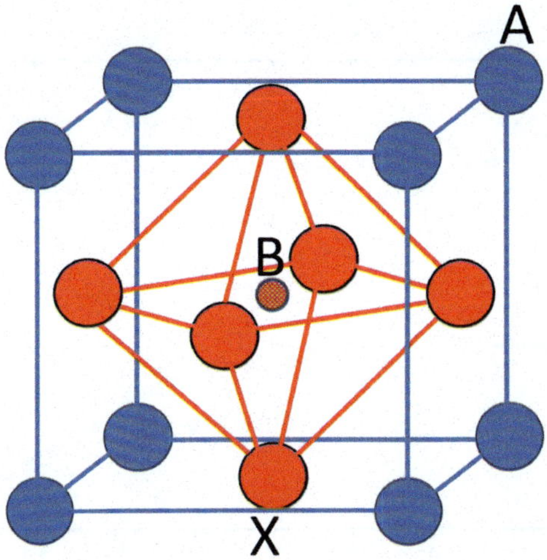

Fig. 8.11 The crystal structure of perovskite where A, B and X are the site positions of rare-earth, ferrric and oxygen ions, respectively

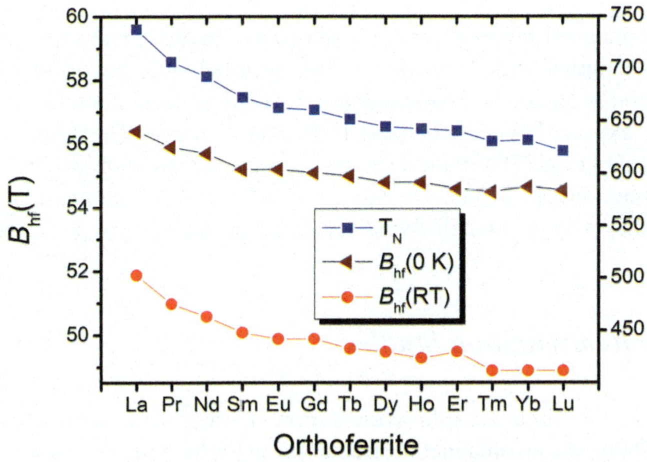

Fig. 8.12 The ^{57}Fe hyperfine fields and the Néel temperature (T_N) of various rare-earth orthoferrites (Eibschütz et al. 1964, 1967)

hyperfine parameters and Néel temperature with pressure is similar to other orthoferrites. At high pressure beyond 30 GPa, a doublet component arises that is complete beyond 45 GPa (Hearne et al. 1995), later identified as a tetragonal phase.

High-pressure studies at low temperatures were undertaken on $LaFeO_3$ and $PrFeO_3$ up to 170 GPa (Xu et al. 2001). The orthorhombic perovskite phase transforms to a tetragonal high-pressure phase beyond 30 GPa at ambient temperature.

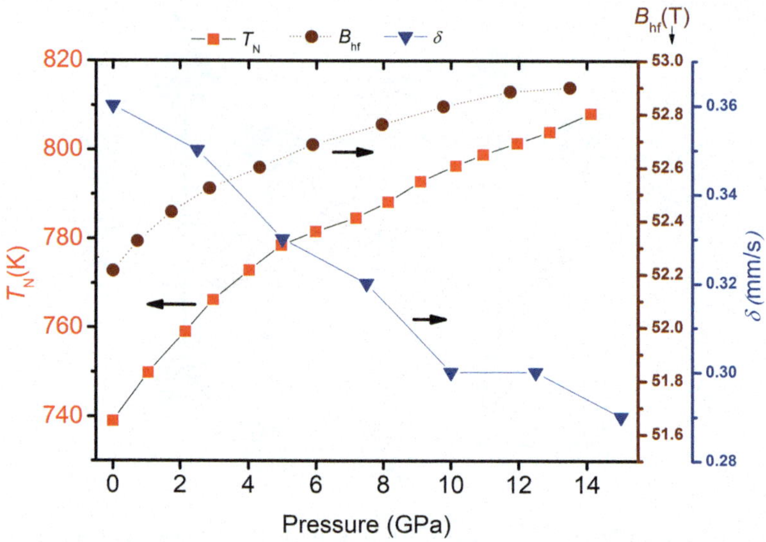

Fig. 8.13 The variation of $\overline{B}_{hf}$, T_N and δ with pressure for $LaFeO_3$. Data from (Halasa et al. 1974)

Also, pressure-induced high spin to low spin transition occurs at this intermediate pressure with complete transformation to low spin behavior beyond 65 GPa. The onset of metallic behavior and paramagnetic relaxation occurs beyond 120 GPa in $LaFeO_3$ and is expected to occur beyond 170 GPa in $PaFeO_3$. The relaxation time τ at 300 K for $LaFeO_3$ at 170 GPa is 1 ps, which increases to 48 ns at 8 K. Although these transformations with pressure are expected to be the common features of orthoferrites, they may occur at different values according to the type of orthoferrite.

8.8.1 Spin Reorientation Studies

Temperature and field-induced spin reorientation in orthoferrites are also extensively investigated. Pure antiferromagnetic orientation exhibits first-order spin reorientation, whereas canted antiferromagnets show gradual spin reorientation of second order. Bertaut notations are used to represent the spin configurations in rare-earth orthoferrites that can be of eight types (Γ_j, $j = 1$ to 8), among which the four major configurations are listed in Table 8.4 (Bertaut 1963; Yamaguchi 1974).

The base vectors {A} and {B} represent the ferric and rare-earth ion spins, respectively. The ferromagnetic spin alignment (F) and the three types of antiferromagnetic spin alignments (G, C, A) are represented in Fig. 8.14. The spin arrangement in the four-cornered lattice, arranged clockwise, can give the following vector sum combinations.

Table 8.4 The spin arrangement symmetry notations

Notation	Ferric ion base vector {A}	Rare-earth ion base vector {B}
Γ_4	G_x, A_y, F_z	F_z
Γ_3	C_x, F_y, A_z	C_x, F_y
Γ_2	F_x, C_y, G_z	F_x, C_y
Γ_1	A_x, G_y, C_z	C_z

The crystallographic *a*, *b* and *c* -axes are represented as *x*, *y*, *z*. G, C and A denote the types of antiferromagnetic spin arrangement as shown in Fig. 8.14 while F is for ferromagnetic spins

$F = S_1 + S_2 + S_3 + S_4$ all the spins are parallel
$G = S_1 - S_2 + S_3 - S_4$ alternate spins are antiparallel
$C = S_1 + S_2 - S_3 - S_4$ left and right spin pairs are antiparallel
$A = S_1 - S_2 - S_3 + S_4$ upper and lower spin pairs are antiparallel.

The spin configurations could be assigned *x*, *y* and *z*. Thus, F_x represents ferromagnetic spin alignment along the *a*-axis and so on. The configurations ($\Gamma_j, j = 5$ to 8) involve sign changes and can be constructed from the missing {B} base vectors.

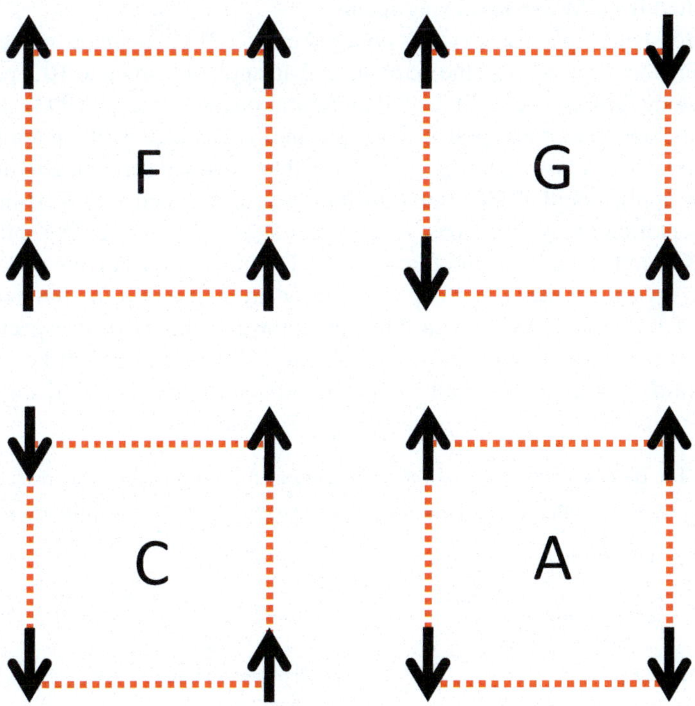

Fig. 8.14 The Bertaut ferromagnetic (F) and antiferromagnetic (G, C, A) spin configurations

For example, Γ_8 has the {B} as A_x, G_y compared with Γ_1; Γ_7 has G_z compared with Γ_2; Γ_6 has A_z compared with Γ_3; and Γ_5 has G_x, A_y compared with Γ_4.

According to the Bertaut notation Γ_4, the rare-earth ions are ferromagnetically aligned along *c* (F_z). Similar spin configuration for the ferric ions can be understood in terms of axes *a*, *b* and *c* as G_x, A_y and F_z, respectively. The spin reorientation in $YbFeO_3$ can be better represented using Bertaut notation as given in Table 8.5 (Davidson et al. 1975). There are three regions where the temperature limits are assumed from experiments (i) a high-temperature region where the configuration is Γ_4, (ii) an intermediate region (Γ_{42}) and (iii) low-temperature region (Γ_2). In the Γ_4 configuration, the ferromagnetic contribution comes from slightly canted ferric and rare-earth ions along the *c*-axis. Similarly, the Γ_2 configuration at low temperatures gives ferromagnetic contribution from ferric and rare-earth ions along the *a*-axis. The two pairs of rare-earth ions spin lie along the *ab* plane. In the intermediate temperature range, we have the intermediate spin configuration Γ_{42} giving an angle γ from the *c*-axis and 2ϕ between the Yb ion spin pairs.

Rare-earth Yb^{3+} orientation in the Yb-orthorferrite was studied using ^{170}Yb Mössbauer spectrometry, where it was found that the angle ϕ in which the Yb spins make with the *a*-axis is found to be close to 27° at 4.2 K, for the Γ_2 configurations (Davidson et al. 1975). The Fe spins are canted at an angle of 1.7° from the *c*- towards the *a*-axis. The low temperature Γ_2 and $\Gamma_2 \rightarrow \Gamma_4$ spin configurations with temperature were verified from Mössbauer experiments.

^{57}Fe Mössbauer spectrometry was undertaken in a $YbFeO_3$ single crystal at 4.2 K by exposing the gamma ray to the *bc* plane and an applied field up to 10 T perpendicular to it along the easy *c*-axis in the Γ_4 configuration (Brown et al. 1993). The single sextet in the absence of an applied field split into two sextets with an applied field, and their relative intensities along with $\Delta m = 0$ lines varied according to the magnitude of the applied field. A spin flop transition was established with field-induced Fe spin reorientation along the *c*-axis, resulting in $\Gamma_2 \rightarrow \Gamma_4$ at 4.2 K. The spin rotation angle was found to be 7° to the *a*-axis at 10 T. Similar experiments at 4.2 K with the single crystal *ab* plane and exposing the γ-ray first along the *a*-axis and next along the *b*-axis showed that the Fe spins rotate towards the *a*-axis irrespective of the direction of the applied fields above 5 T. The reorientation angles of Fe spins from the *c*- towards *a*- axis are 43° and 73° for the first and second case, respectively, at

Table 8.5 The Bertaut representation of the spin reorientations in $YbFeO_3$ (Davidson et al. 1975)

Temperature (K)	Bertaut spin notation	Yb spins	Fe spin alignment
(T_2) $7.8 \le T \le 630$ (T_N)	Γ_4	All along *c*-axis, F_z	Along *a*, G type (G_x)
(T_1) $6.6 \le T \le 7.8$ (T_2)	Γ_{42}	F_x, C_y, F_z	*ac* plane
$1.3 \le T \le 6.6$ (T_1)	Γ_2	In *ab* plane with one pair along *a*, F_x	Along *c*, G_z

10 T. The reorientation along the *a*-axis was suggested due to the Fe-Yb spin decoupling, resulting in the unusual high-temperature configuration. Apart from $YbFeO_3$, $TmFeO_3$, $NdFeO_3$, $SmFeO_3$ and $ErFeO_3$ also undergoes spin reorientation from $\Gamma_2 \rightarrow \Gamma_{42} \rightarrow \Gamma_4$ with temperature whereas Γ_4 persists down to low temperature for La-, Lu- and Y- orthoferrites.

An illustration of the spin reorientation results with $ErFeO_3$ single crystal is presented in Fig. 8.15. The ferric antiferromagnetic (AFM) spin directions are along *a*-axis at 95 K (Γ_4) that rotates to *c* at 85 K and along *b* at 4.2 K (Grant and Geller 1969; Gorodetsky et al. 1969; Belakhovsky et al. 1976). The gamma ray direction along the spin direction results in the sextet line intensity ratios given by 3:0:1 while it is 3:4:1 along the perpendicular direction. The spin directions could be reversed, and the intensity ratios are influenced by the strength of the external magnetic field. The critical applied field along the *a*-axis for complete orientation of the $\Gamma_4 \rightarrow \Gamma_2$ in $ErFeO_3$ at 270 and 130 K are 10 and 6.5 T, respectively (Prelorendjo et al. 1982). The high temperature Γ_4 configuration at 130 K, in the absence of applied field, gradually transformed to Γ_2 with increasing applied field along the $ErFeO_3$ single crystal *a*-axis and the γ-ray along the *c*-axis. The single sextet splits apart with the application of an external field, and the $\Delta m = 0$ line gradually reduces with the applied field. In the absence of an external field, the $\Gamma_4 \rightarrow \Gamma_2$ orientation occurs at 95 K. At 78 K, an applied field of 10 T along the *c*-axis is required for the complete reorientation of $\Gamma_2 \rightarrow \Gamma_4$. Reaching low temperature from 4.2 to 2.2 K, $\Gamma_2 \rightarrow \Gamma_{12}$ configuration occurs that could be reverted with an applied field nearly 0.5 T along the *c*-axis. The study showed that the low-temperature spin orientation is determined by the DM interactions between Er-Fe ions.

Table 8.6 shows the spin–orientation parameters of Gd-, Tb- and Dy- orthoferrites at various temperatures and applied fields obtained from ^{57}Fe Mössbauer spectrometry. The orientation of $GdFeO_3$ spins along the *c*-axis requires a larger field of 7.6 T at 293 K, whereas it is only 0.7 T at 2 K, suggesting the preference towards the *c*-axis at low temperatures (Durbin et al. 1977). For the $TbFeO_3$, the preferred orientation is along *a*-axis down to 6 K. In contrast, the preferred orientation becomes the *c*-axis on reducing the temperature up to 4.2 K, which reorients to the *a*-axis on further reducing the temperature (Nikolov et al. 1994). This suggests a spin-reorientation configuration of $\Gamma_4 \rightarrow \Gamma_2 \rightarrow \Gamma_4$ with reducing temperature in the absence of applied field.

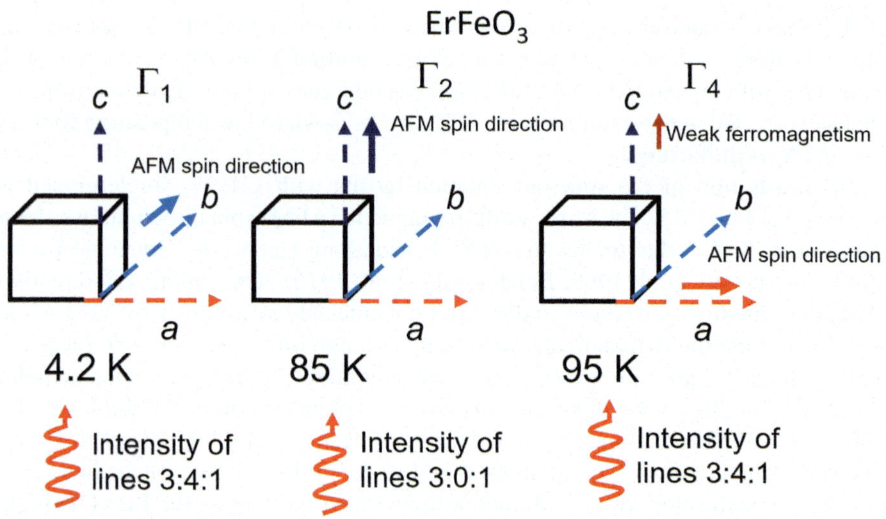

Fig. 8.15 The various spin configurations of single crystal $ErFeO_3$ at 4.2, 85 and 95 K. The sextet line intensity ratios are 3:0:1 when the gamma ray is parallel to the antiferromagnetic ferric spin direction and 3:4:1 with perpendicular orientation

Table 8.6 The influence of temperature and applied field on the spin-orientation direction of various single crystal orthoferrites

Orthoferrite	T_N (K)	Applied field	Orientation direction	Temperature (K)	Critical field (T)	References
$GdFeO_3$	661	*a*-axis	*a*-axis to *c*-axis	2 4.2 77 293	0.7 1.2 8.1 7.6	Durbin et al. (1977)
$TbFeO_3$	681	–	*a*-axis *c*-axis *a*-axis	Upto 6 6–4.2 < 4.2	–	Nikolov et al. (1994)
$DyFeO_3$	650	*a*-axis	*a*-axis to *c*-axis	77 130 270	4.5 6.5 9.3	Prelorendjo et al. (1980)
			b-axis to *c*-axis	4.2	1.6	
		b-axis	*b*-axis to *a*-axis	4.2	0.75	
		c-axis	*b*-axis to *a*-axis	4.2	0.7	

The $DyFeO_3$ showed a reduced critical field for the $\Gamma_4 \rightarrow \Gamma_2$ transitions as the temperature is lowered when the applied field is along the *a*-axis and γ-ray parallel to the *c*-axis (Prelorendjo et al. 1980). The extrapolated saturation field is 11 T for complete orientation above RT. The $\Gamma_1 \rightarrow \Gamma_4$ transition requires a lower critical field of 0.75 T along the *b*-axis at 4.2 K than the $\Gamma_1 \rightarrow \Gamma_2$(1.6 T). Applying γ-ray parallel to the *b*-axis, the angle between the Fe^{3+} spin and the crystal *b*-axis was obtained with temperature in the absence of an applied field. The angle change gave a Morin transition temperature (T_M) of 35 K for $DyFeO_3$. Below T_M, at the measurement temperature up to 4.2 K, the preferred configuration is Γ_1. For $HoFeO_3$, the critical field for spin flop at 270, 140 and 77 K are 8.1, 4.4 and 1.2 T, respectively, which are lower than $ErFeO_3$ (Johnson et al. 1980).

8.9 Rare-Earth Free Pervoskites

8.9.1 SrFeO₃

$SrFeO_x$ is composed of Fe^{4+} and Fe^{3+}, and Mössbauer studies on $SrFeO_{2.9}$ have shown a single absorption line at room temperature with positive isomer shift with reference to SS (Shirane and Cox 1962). Although the sample was composed of Fe^{4+} and Fe^{3+} from chemical analysis, the positive isomer shift compared with ferrous and ferric ions was suggested due to electron exchange between Fe^{4+}-Fe^{3+} charge states or impurities.

Cubic perovskite $SrFeO_3$ with Fe^{4+} showed an isomer shift of – 0.17 mm/s at RT with reference to Cu, which is still less negative than those predicted for pure $3d^4$ configurations (Gallagher et al. 1964). This is due to the enhanced covalency owing to the polarization of the oxygen ions with reduced ionic radius, resulting in the increased screening of the s electrons. The Néel temperature is 134 K, while the B_{hf} at 4 and 78 K are 33.1 and 29.0 T, respectively. The oxygen-deficient single phase $SrFeO_{2.9}$ is composed of Fe^{3+} charge state in a tetragonally distorted lattice. The consideration of quadrupole splitting resulted in RT isomer shifts of – 0.2 mm/s and + 0.26 mm/s corresponding to the ferric ions in different environments. The T_N is 80 K for this composition, and the 4 K spectrum showed two B_{hf} corresponding to 29.1 and 42.1 T, QS of + 0.29 mm/s and δ of – 0.26 mm/s and + 0.08 mm/s (relative to source in Cu at RT). The hyperfine parameters were assigned to originate from Fe^{3+} in an oxygen-deficient environment, whereas it resembled Fe^{4+}. More oxygen deficient $SrFeO_{2.5}$ resulted in orthorhombic brownmillerite structure with T_c greater than 600 K. High-pressure studies have shown the increase of T_N to RT under a pressure of 19 GPa and high spin to low spin configuration with pressure (Takano et al. 1991).

8.9.2 *BaFeO$_3$*

Stoichiometric $BaFeO_3$ was difficult to synthesize, and hence Mössbauer studies were carried out in oxygen-deficient $BaFeO_x$ having Fe^{3+} ion in the orthorhombic brownmillerite structure (Gallagher et al. 1965). The RT isomer shift was more negative than $SrFeO_3$ due to the larger ionic radius of Ba ion, resulting in reduced contribution from ligand to 3d orbital of Fe ion. The B_{hf} at 4 K for the $BaFeO_{9.5}$ are 25.5 and 50.1 T, and the corresponding δ(Cu) are − 0.35 and + 0.41 mm/s, respectively. The isomer shifts in $BaFeO_{9.5}$ were assigned to originate from close to Fe^{4+} and Fe^{3+} in an octahedral environment. Consequently, the more negative and lowest B_{hf} originates from the Fe^{4+}, and the barium ferrate has a T_N of 164 K.

Stochiometric $BaFeO_3$ with the cubic perovskite structure could also be obtained, as reported later on (Hayashi et al. 2011). The B_{hf} at 77 and 4 K are 22.1 and 29.1 T, respectively. Interestingly, the compound showed ferromagnetic behavior that reduced the B_{hf} with an applied field of 0.5 T parallel to the γ-ray and the disappearance of the $\Delta m = 0$ lines. The ferromagnetic transition temperature was 111 K. It was shown that $SrFeO_3$ and $BaFeO_3$ exhibit field-induced ferromagnetism due to their helical spin arrangement (Hayashi et al. 2013).

8.9.3 *BiFeO$_3$*

$BiFeO_3$ is an important G-type canted antiferromagnetic material that shows ferroelectricity below 1110 K. Mössbauer spectrometry of rhombohedrally distorted $BiFeO_3$ perovskite showed δ(SS) of 0.47 mm/s at 677 K, typical for Fe^{3+} state with a small covalency (Bhide and Multani 1965). The antiferromagnetic Néel temperature was determined as 643 K. The $BiFeO_3$ also exhibited a quadrupole splitting of 0.54 mm/s at 677 K. The sextet with B_{hf} of 49.3 T and quadrupole splitting of 0.37 mm/s at RT was found to originate from the ferric site (Biran et al. 1971). The temperature-dependent B_{hf} was obtained with a Brillouin curve of $S = 3/2$, B_{hf} of 55.5 T at 0 K and T_N of 640 K. The Mössbauer spectrum of $BiFeO_3$ in the temperature range 80–850 K indicated line broadening of the sextet below T_N with $\Gamma = 0.4$ mm/s, attributed to distribution in quadrupole splitting from small local asymmetries (Blaauw and van der Woude 1973). Later, the broadened lines were fitted with two sextet components from two non-equivalent octahedral ferric sites with B_{hf} of 50.1 and 49.8 T and quadrupole splitting of 0.4 and − 0.09 mm/s, respectively, at RT (De Sitter et al. 1976).

The solid solution of $BiFeO_3$ with other compounds, such as $BaTiO_3$ was attempted for enhanced multiferroic properties (Park et al. 2010). The room temperature Mössbauer spectra of $(BiFeO_3)_x$-$(BaTiO_3)_{1-x}$ solid solution with x = 1 to 0.5 are shown in Fig. 8.16. The bulk $BiFeO_3$ (x = 1) was fitted with two sextets of B_{hf} 49.8 and 49.4 T, quadrupole splitting of 0.34 and − 0.10 mm/s and δ of 0.38 and 0.39 mm/s, respectively. With increasing $BaTiO_3$, the B_{hf} decreases to 25.4 T at x =

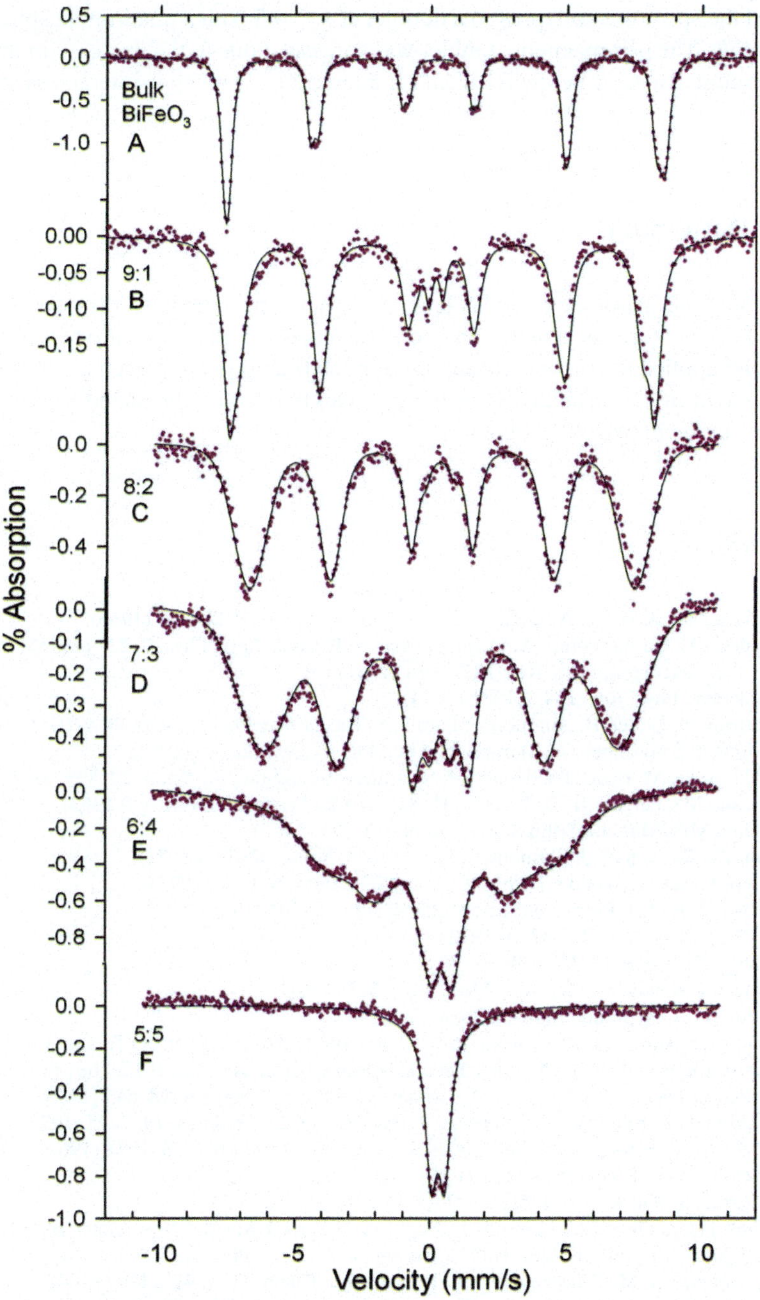

Fig. 8.16 The room temperature Mössbauer spectra of $(BiFeO_3)_x$-$(BaTiO_3)_{1-x}$ solid solution with decreasing x in the range $x = 1$ (A) to 0.5 (F) in an interval of 0.1. Reprinted with permission (Park et al. 2010), © The American Physical Society

0.6 and the spectrum collapses to a doublet at x = 0.5 with a quadrupole splitting of 0.50 mm/s. The paramagnetic doublet was suggested due to the decrease in the Néel temperature to a value below RT from the altered superexchange interaction strength by Ti^{4+}.

8.10 Summary

The hyperfine parameters of magnetic garnets, hexaferrites, orthoferrites and perovskites are discussed in this chapter. The magnetic spin orientation with temperature and applied magnetic field reported in the literature are presented. The substitution effect of cations and site occupancy can be further elucidated by noting the changes in the hyperfine parameters.

References

G. Albanese, E. Calabrese, A. Deriu, F. Licci, Hyperfine Interact. **28**, 487 (1986)
G. Albanese, M. Carbucicchio, A. Deriu, G. Asti, S. Rinaldi, Appl. Phys. **7**, 227 (1975)
C. Alff, G.K. Wertheim, Phys. Rev. **122**, 1414 (1961)
E.E. Anderson, Phys. Rev. **134**, A1581 (1964)
G. Balestrino, S. Geller, B. Antonini, P. Paroli, A. Tucciarone, Phys. Rev. B **30**, 4010 (1984)
R. Bauminger, S.G. Cohen, A. Marinov, S. Ofer, Phys. Rev. **122**, 743 (1961)
M. Belakhovsky, M. Boge, J. Chappert, J. Sivardiere, Solid State Commun. **20**, 473 (1976)
E.F. Bertaut, Magnetism III, G. T. Rado, H. Suhl (ed.), (Academic Press, NY, 1963)
V.G. Bhide, M.S. Multani, Solid State Commun. **3**, 271 (1965)
A. Biran, P.A. Montano, U. Shimony, J. Phys. Chem. Solids **32**, 327 (1971)
C. Blaauw, F. van der Woude, J. Phys. C: Solid State Phys. **6**, 1422 (1973)
S.R. Brown, I. Hall, J. Phys. Condens. Matter **5**, L99 & 4207 (1993)
J.M.D. Coey, Phys. Rev. B **6**, 3240 (1972)
R.L. Cohen, Phys. Lett. **5**, 177 (1963)
S.G. Cohen, I. Nowik, S. Ofer, Rev. Mod. Phys. **378** (1964).
D.C. Cook, J. Li, Hyperfine Interact. **28**, 495 (1986)
G. Crecelius, S. Hufner, D. Quitmann, Proc. Conf. Appl. Mössbauer Effect (Tihany, 1969) 507.
G.R. Davidson, B.D. Dunlap, M. Eibschütz, L.G. van Uitert, Phys. Rev. B **12**, 1681 (1975)
J. De Sitter, C. Dauwe, E. De Grave, A. Govaert, Solid State Commun. **18**, 645 (1976)
G.W. Durbin, C.E. Johnson, M.F. Thomas, J. Phys. C: Solid State Phys. **10**, 1975 (1977)
M. Eibschütz, G. Gorodetsky, S. Shtrikman, D. Treves, J. Appl. Phys. **35**, 1071 (1964)
M. Eibschütz, M.R. Lines, Phys. Rev. B **26**, 2288 (1982)
M. Eibschütz, S. Shtrikman, D. Treves, Phys. Rev. **156**, 562 (1967)
B.J. Evans, F. Grandjean, A.P. Lilot, R.H. Vogel, A. Gerard, J. Magn. Magn. Mater. **67**, 123 (1987)
P.K. Gallagher, J.B. MacChesney, D.N.E. Buchanan, J. Chem. Phys. **43**, 516 (1965)
P.K. Gallagher, J.B. MacChesney, D.N.E. Buchanan, J. Chem. Phys. **41**, 2429 (1964)
S. Geller, G. Balestrino, A.K. Ray, A. Tucciarone, Phys. Rev. B **27**, 326 (1983)
S. Geller, G. Balestrino, Phys. Rev. B **21**, 4055 (1980a)
S. Geller, G. Balestrino, Solid State Commun. **33**, 315 (1980b)
S. Geller, H.J. Williams, R.C. Sherwood, J.P. Remeika, G.P. Espinosa, Phys. Rev. **131**, 1080 (1963)
S. Geller, J.P. Remeika, R.C. Sherwood, H.J. Williams, G.P. Espinosa, Phys. Rev. **137**, 1034 (1965)

E. Gerdau, R. Ruffer, H. Winkler, W. Tolksdorf, C.P. Klages, Phys. Rev. Lett. **54**, 835 (1985)
G. Gorodetsky, L.M. Levinson, S. Shtrikman, D. Treves, B.M. Wanklyn, Phys. Rev. **187**, 637 (1969)
R.W. Grant, S. Geller, Solid State Commun. **7**, 1291 (1969)
J.-M. Greneche, H. Pascard, J.R. Regnard, Solid State Commun. **65**, 713 (1988)
M. Guillot, H. Le Gall, J.M. Desvignes, M. Artinian, J. Appl. Phys. **70**, 6401 (1991)
E.M. Gyorgy, K. Nassau, M. Eibschütz, J.V. Waszczak, C.A. Wang, J.C. Shelton, J. Appl. Phys. **50**, 2883 (1979)
N.A. Halasa, G. DePasquali, H.G. Drickamer, Phys. Rev. B **10**, 154 (1974)
N. Hayashi, T. Yamamoto, A. Kitada, A. Matsuo, K. Kindo, J. Hester, H. Kageyama, M. Takano, J. Phys. Soc. Jpn. **82**, 113702 (2013)
N. Hayashi, T. Yamamoto, H. Kageyama, M. Nishi, Y. Watanabe, T. Kawakami, Y. Matsushita, A. Fujimori, M. Takano, Angew. Chem. Int. Ed. **50**, 12547 (2011)
G.R. Hearne, M.P. Pasternak, R.D. Taylor, P. Lacorre, Phys. Rev. B **51**, 11495 (1995)
Y.H. Hong, J.S. Kum, I.B. Shim, C.S. Kim, IEEE Trans. Magn. **40**, 2808 (2004)
C.E. Johnson, L.A. Prelorendjo, M.F. Thomas, J. Magn. Magn. Mater. **15–18**, 557 (1980)
R.J. Joseyphus, A. Narayanasamy, A.K. Nigam, R. Krishnan, J. Magn. Magn. Mater. **296**, 57 (2006)
E. Kreber, U. Gonser, A. Trautwein, J. Phys. Chem. Solids **36**, 263 (1975)
M.S. Lataifeh, A.F.D. Lehlooh, S. Mahmood, Hyperfine Interact. **122**, 253 (1999)
J.T. Lim, C.S. Kim, J. Appl. Phys. **115**, 17A516 (2014a)
J.T. Lim, C.S. Kim, J. Appl. Phys. **117**, 17B743 (2015)
J.T. Lim, C.S. Kim, J. Appl. Phys. **115**, 17D706 (2014b)
F.J. Litterst, J. Tejada, G.M. Kalvius, J. Appl. Phys. **50**, 7636 (1979)
G.J. Long, F. Grandjean, X. Guo, A. Navrotsky, R.K. Kukkadapu, Inorg. Chem. **55**, 3413 (2016)
I.S. Lyubutin, A.G. Gavriliuk, I.A. Trojan, R.A. Sadykov, JETP Lett. **82**, 797 (2005)
W. Meyer, W. Sturhahn, E. Gerdau, Hyperfine Interact. **94**, 2059 (1994)
L, Néel R. Pauthenet, B. Dreyfus. Chapter VII The Rare-Earth Garnets. *Progress in Low Temperature Physics* (4, Elsevier, 1964) pp. 344–383
O. Nikolov, I. Hall, S.N. Barilo, S.A. Guretskii, J. Phys. Condens. Matter **6**, 3793 (1994)
I. Nowik, S. Ofer, Phys. Rev. **153**, 409 (1967)
A. Paoluzi, F. Licci, O. Moze, G. Turilli, A. Deriu, G. Albanese, E. Calabrese, J. Appl. Phys. **63**, 5074 (1988)
T.J. Park, G.C. Papaefthymiou, A.J. Viescas, Y. Lee, H. Zhou, S.S Wong, Phys. Rev. B **82**, 024431 (2010)
L.A. Prelorendjo, C.E. Johnson, M.F. Thomas, B.M. Wanklyn, J. Phys. C: Solid State Phys. **15**, 3199 (1982)
L.A. Prelorendjo, C.E. Johnson, M.F. Thomas, B.M. Wanklyn, J. Phys. C: Solid State Phys. **13**, 2567 (1980)
V.K. Sankaranarayanan, N.S. Gajbhiye, J. Mater. Sci. **29**, 762 (1994)
G.A. Sawatzky, F. van der Woude, A.H. Morrish, Phys. Rev. **183**, 383 (1969)
A. Seidel, L. Häggström, D. Rodic, Hyperfine Interact. **73**, 265 (1992)
V. Šepelák, M. Myndyk, R. Witte, J. Röder, D. Menzel, R.H. Schuster, H. Hahn, P. Heitjans, K.D. Becker, Faraday Discuss. **170**, 121 (2014)
G. Shirane, D.E. Cox, Phys. Rev. **125**, 1158 (1962)
G.D. Soria, P. Jenus, J.F. Marco, A. Mandziak, M. Sanchez-Arenillas, F. Moutinho, J.E. Prieto, P. Prieto, J. Cerdá, C. Tejera-Centeno, S. Gallego, M. Foerster, L. Aballe, M. Valvidares, H.B. Vasili, E. Pereiro, A. Quesada, J. de la Figuera, Sci. Rep. **9**, 11777 (2019)
M. Takano, S. Nasu, T. Abe, K. Yamamoto, S. Endo, Y. Takeda, J.B. Goodenough, Phys. Rev. Lett. **67**, 3267 (1991)
G.K. Thompson, B.J. Evans, J. Appl. Phys. **73**, 6297 (1993)
J.J. van Loef, J. Appl. Phys. **39**, 1258 (1968)
J.J. van Loef, P.J.M. Franssen, Phys. Lett. **7**, 225 (1963)

D. Vandormael, F. Grandjean, D. Hautot, G.J. Long, J. Phys. Condens. Matter **13**, 1759 (2001)
P. Vaqueiro, M.A.L. Quintela, J. Rivas, J.-M. Greneche, J. Magn. Magn. Mater. **169**, 56 (1997)
W.M. Xu, O. Naaman, G.K. Rozenberg, M.P. Pasternak, R.D. Taylor, Phys. Rev. B **64**, 094411 (2001)
T. Yamaguchi, J. Phys. Chem. Solids **35**, 479 (1974)

Chapter 9
Mössbauer Spectrometry of Antiferromagnetic Materials

9.1 Hematite (α-Fe_2O_3)

Bulk hematite (α-Fe_2O_3) with the corundum structure is known to be an antiferromagnet exhibiting weakly ferromagnetic behavior at high temperatures up to the Néel temperature, T_N. The corundum structure belongs to the trigonal crystal system with the $R\overline{3}c$ space group. The unit cell is often described in hexagonal or rhombohedral Bravais lattices. The isomer shift and quadrupole interaction were first demonstrated in α-Fe_2O_3 (Kistner and Sunyar 1960) and it was often used for Mössbauer spectrometer calibration in the past. The temperature-dependent internal hyperfine field was initially fitted with the $S = 5/2$ curve according to Weiss's theory (Ono and Ito 1962), resulting in a paramagnetic transition close to 963 K (Freier et al. 1962). The T_N was later fine-tuned and reported to be 955 K from Mössbauer measurements (Gilad et al. 1963).

Mössbauer experiments with ^{57}Fe enriched hematite showed a sextet at 300 K where the antiferromagnetic axis is perpendicular to c-axis, and at 100 K where both are parallel (Ono and Ito 1962). However, the temperature-dependent quadrupole shift changes from negative to positive values above the Morin transition temperature (T_M) of 260 K. The T_M is associated with the spin reorientation from the parallel [111] c-axis to the perpendicular axis of the unit cell, with the two ferric sublattices oriented at an angle to each other. Consequently, weakly ferromagnetic behavior, also known as parasitic ferromagnetism, occurs above T_M due to canted antiferromagnetic structure caused by Dzyaloshinskii–Moriya (DM) interaction.

The saturation hyperfine field was obtained as 54.4 T for the α-Fe_2O_3 powders, and the temperature-dependent reduced hyperfine field showed a discontinuity at the T_M (van der Woude 1966; Tobler et al. 1981). The quadrupole interaction energy also undergoes drastic variation from 9.6 neV below T_M to 5 neV above. Inhomogeneities in hematite samples would broaden the T_M transition region and decrease the effective hyperfine field. The decrease in the effective hyperfine field by 0.8 T above T_M in hematite can be evaluated from the various field contributions. The

R. J. Joseyphus and J.-M. Greneche, *Fundamentals of ^{57}Fe Mössbauer Spectrometry*,
https://doi.org/10.1007/978-981-99-8653-8_9

contact interaction, $B_c \approx -0.50$ T could not change significantly by the spin direction whereas contributions by other terms are difficult to separate and therefore, it can be conveniently taken as the combined influence of $B_L + B_D$ resulting in the B_{eff} variation (van der Woude 1966; Ruskov et al. 1976). The Brillouin curve with $S =$ 5/2 may deviate in hematite due to other contributions such as surface spin effects (Shinjo et al. 1983) and weak ferromagnetism. Al substitution in α-Fe_2O_3 as solid solution decreases the T_M and T_N (Srivastava and Sharma 1972).

The change in spin direction from the *c*-axis to the (111) plane with an applied field induces spin flop even below T_M with an applied field. This has been demonstrated using a hematite single crystal that showed the appearance of the $\Delta m = 0$ intensity lines of the sextet with an external magnetic field which was otherwise feeble with zero-fields at 80 K (Blum et al. 1965). The spin direction changes from *c*-axis to the (111) plane when the external field is increased to 6.6 T at 80 K. For the polycrystalline samples, the spin flop was observed from the broadening of the outer lines with an external field. External field applied perpendicular to the *c*-axis shifted the T_M to 246 K with increasing applied field up to 0.9 T (Simkin and Bernheim 1967). The abrupt spin flop from the *c*-axis to (111) plane shifting the T_M to 258 K was obtained with a higher field of 2 T (Cinader et al. 1967). The application of a transverse field of 10 T ($B \perp \gamma$, *c*-axis) on hematite single crystal induces spin reorientation from the *c*-axis to an angle closer to the (111) plane at 245 K, as observed from the transition line intensity ratios of 3:3.6:1 (Pankhurst et al. 1986). The $\Delta m = 0$ lines were not completely aligned as observed from the value of 3.6. The orientation angle between *c*-axis and applied field direction (θ_B) gradually changed away from the *c*-axis with applied magnetic field at 230 K reaching 14°, 51° and 62° at 3, 8 and 10 T, respectively. The orientation requires less energy with temperature resulting in a higher θ_B of 74° at 245 K with an applied field of 10 T.

The Mössbauer spectra of polycrystalline α-Fe_2O_3 at 4.2 K in longitudinal applied magnetic fields up to 9 T is shown in Fig. 9.1 (Pankhurst and Pollard 1990). According to the angle between the incident γ-ray and the effective field, the intensities of the lines are given as $3\left(1+\cos^2\theta\right) : 4\sin^2\theta : \left(1+\cos^2\theta\right)$. The isotropic distribution of θ would yield intensity ratios 3:2:1 in the absence of an external magnetic field. A longitudinal applied field results in the broadening of the outermost lines due to an increase in the effective field. The spin flop field is predicted as 6.2 T and therefore the fit of the spectrum is poor with an external field of 9 T. The isomer shift, quadrupole splitting, hyperfine field and anisotropy field are estimated as 0.49 mm/s, 0.40 mm/s, 53.8 T and 0.02 T, respectively.

The Mössbauer spectrum of hematite fine particles exhibited a superparamagnetic doublet at 300 K, and the T_M was absent in contrast to the bulk (Nakamura 1964). Superparamagnetic behavior of α-Fe_2O_3 with size reduction was achieved with particles in alumina or silica matrix (Constabaris et al. 1965). The hyperfine field of the 18 nm hematite nanoparticles reduces to 52.7 T compared with the bulk value of 54.2 T at 80 K (Kundig et al. 1966). The RT spectrum showed an increasing superparamagnetic fraction with a decreasing size from 18 nm and completely superparamagnetic when the particle size is less than 10 nm.

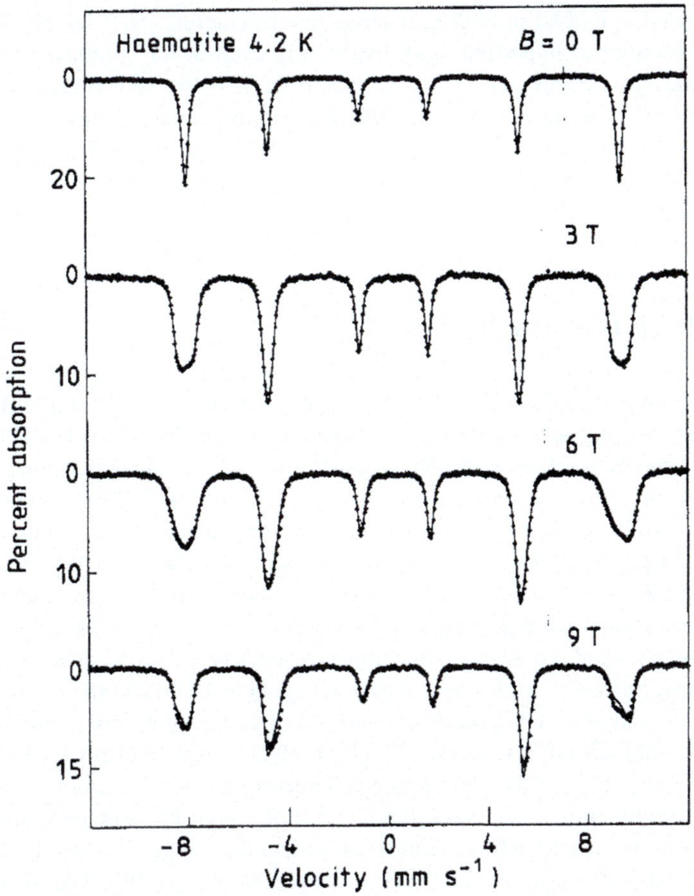

Fig. 9.1 The Mössbauer spectra of hematite polycrystals at 4.2 K show a single line sextet in zero applied field (top), and with increasing longitudinal magnetic field, the intensity of the outermost line is reduced and broadened. Reprinted with permission (Pankhurst and Pollard 1990), © IOP Publishing

The relaxation time of the surface ions in hematite nanoparticles in the size range 4–7 nm is much smaller than the 50 nm particles or bulk (van der Kraan 1973). The spectrum at 5 K for the hematite nanoparticles of average size 16 nm showed B_{hf} = 53.1 T, $\delta = 0.49$ mm/s and a quadrupole shift of − 0.10 mm/s (Bødker et al. 2000). The temperature-dependent studies indicated a superparamagnetic blocking temperature (T_B) of 143 K without any signature of T_M up to 5 K. The persistence of a weakly magnetic nature was also noticed down to the lowest temperature, as the saturation magnetization of 0.4 A·m^2/kg at 6 K was close to the RT value of the bulk single crystal. Indeed, the T_M is size dependent, decreasing with size reduction and suppressed below 20 nm due to the domination of surface spins (Ozdemir et al. 2008). The non-interacting hematite nanoparticles of 8 nm, coated with a non-magnetic layer, showed a doublet at 200 K, whereas the interacting particles of similar size exhibited broadened hyperfine field distribution even at RT (Frandsen and Mørup 2005). Interparticle interaction was found to deviate the spin direction from the (001) plane to an angle of 15° out of the plane in the nanoparticles due to changes in magnetic anisotropy, based on low-temperature measurements. Agglomerated particles may lead to the suppression of superparamagnetic relaxation depending on the interaction strength.

9.1.1 High Pressure Studies

The ambient pressure B_{hf} of 51.4 T for the α-Fe_2O_3 decreases by 1% up to 10 GPa and remains almost the same up to 20 GPa, whereas the quadrupole splitting decreases from + 0.4 mm/s to − 0.4 mm/s with pressure up to 20 GPa (Vaughan and Drickamer 1967). The positive to negative sign change occurs at 3 GPa. The variation in the quadrupole splitting was associated with the strong dependence of the crystalline electric field gradient on the oxygen parameter with pressure. Another reason for the transition to the negative quadrupole splitting with pressure was speculated to be due to T_M. Further increase in pressure up to 60 GPa showed a decrease in B_{hf}, δ and Δ close to 49.5 T, 0.07 mm/s and − 1.1 mm/s, respectively (Syono et al. 1984).

The appearance of a high-pressure phase begins at 48 GPa with coexisting sextet and doublet, and at 60 GPa, the B_{hf}, δ and Δ of the high-pressure phase are 32.4 T, 0.40 mm/s and 1.15 mm/s, respectively. The doublet shows δ and Δ values of 0.07 and 0.89 mm/s, respectively. These hyperfine characteristics for the sextet and doublet at 60 GPa resemble those of the low spin phase. Later high-pressure experiments have shown the first-order transformation from a corundum-type structure to a distorted Rh_2O_3(II)-type structure beyond 50 GPa (Pasternak et al. 1999). The high-pressure phase was non-magnetic and observed to be of pressure-induced insulator to metallic behavior due to the breakdown of the electronic d-d correlation leading to a Mott transition. Indeed, the maghemite (γ-Fe_2O_3) irreversibly transforms to hematite at 30 GPa (Kawakami et al. 2003).

9.2 Fe-Oxy-Hydroxides

The Fe-hydroxides and oxy-hydroxides we discuss in this section, are mostly ferric with few mixed valent compounds such as green rust and ferrous hydroxide that may show antiferromagnetic or ferrimagnetic nature (Murad 1996). The various Fe-hydroxides and their hyperfine parameters are listed in Table 9.1. Bulk magnetic properties are exhibited only by highly crystalline samples in these classes of materials. The Fe-oxy-hydroxides that form through various processes involving Fe-hydroxides transform into hematite on thermal treatment. The Mössbauer spectra of these compounds are often influenced by the presence of their counterpart compounds, superparamagnetic component, low crystallinity and variation in the Néel temperature.

9.2.1 Goethite (α-FeOOH)

Goethite (α-FeOOH) has the orthorhombic unit cell structure and belongs to the *Pbnm* space group. The magnetic spin arrangement is antiferromagnetic Fe^{3+} along the *c*-axis of the single crystal *Pbnm* structure, as depicted in Fig. 9.2. Conversely, it can be represented along the *b*-axis or inclined at an angle to the *b*-axis in *Pnma*. The T_N was reported over a wide range from 400 K to as low as 261 K according to the particle size, morphology and stoichiometry (Bocquet et al. 1992; Barrero et al. 2006a). Identifying superparamagnetic transition is difficult in goethite as it may be shifted according to the crystallinity. The T_N of defect-free bulk goethite is close to 400 K.

The temperature-dependent hyperfine field could be fitted with a Brillouin curve corresponding to $S = 5/2$ with a T_N of 400 K (Takada et al. 1964) and saturation $B_{hf}(0)$ of 50.7 T (de Grave and Vandenberghe 1986). Mössbauer spectra of single crystal natural goethite aligned with $\gamma \| b$ axis showed an intensity ratio of 3:4:1 indicating the magnetization vector along the *ac* plane (Forsyth et al. 1968). The magnetic spin direction is along the *c*-axis in the *Pbnm* space group, confirmed by neutron diffraction. The principal axis of EFG lies in the *ab* plane, making an angle of 40° with the *b*-axis. The RT spectrum was a single sextet with an effective hyperfine field close to 38.0 T and exhibited a ferric isomer shift of 0.35 mm/s. The quadrupole splitting above T_N was 0.6 mm/s, whereas a negative value of – 0.3 mm/s at low temperatures suggested magnetization direction change. At high temperature, goethite is unstable and transforms into hematite. Variations in the T_N and B_{hf} occur due to differences in the crystallinity of the samples (van der Woude and Dekker 1966).

Application of a longitudinal magnetic field at 4.2 K broadened the lines with non-vanishing second and fifth-line intensities due to the antiferromagnetic nature of the goethite (Pollard et al. 1992). The applied field, *B*, along the longitudinal direction, resulted in three sextets for the goethite, with broadened first and sixth line and less for the second and fifth line. The antiferromagnetic sextet with strong exchange and strong anisotropy broadens the first and last lines due to two additional contributions from $B_{hf} - B$ and $B_{hf} + B$ without a significant broadening of the second and fifth lines, as shown in Fig. 9.3.

9.2.1.1 Magnetic Anisotropy Constant

Goethite form rod-like elongated particles with ultrafine crystallite size that undergoes superparamagnetic relaxation resulting in line broadening, reduced B_{hf} and doublet at RT for the particles of lowest volume similar to hematite (Shinjo 1966).

Table 9.1 The hyperfine parameters for various Fe- oxy-hydroxides

Compound	Composition	T_N/T_c (K)	T(K)	δ (mm/s)	$\Delta/2\varepsilon$ (mm/s)	B_{hf} (T)	References
Goethite	α-FeOOH	T_N 400	300	0.37	− 0.37	38.1	Morris et al. (1985)
			4.2	0.49	− 0.24	50.3	Pollard et al. (1992)
Lepidocrocite	γ-FeOOH	T_N 77	294	0.37	0.53	–	Murad and Schwertmann (1984)
			4.2	0.34	0.02	45.6	
Akaganeite	β-FeOOH	$T_N \approx$ 290–299	295	0.39	0.95	–	Murad (1979]
				0.38	0.55		
			4.2	0.50	0.00	48.9 (43%) 47.9 (28%) 47.5 (29%)	Pollard et al. (1992)
				0.47	− 0.28		
				0.49	− 0.56		
Feroxyhyte	δ-FeOOH	T_c 455	300	~ 0.5	− 0.1	40.0–45.0	Dezsi et al. (1967)
			4.2	0.50	0.16	53.2 (22%) 51.6 (46%) 49.3 (32%)	Pollard et al. (1992)
				0.49	0.04		
				0.48	− 0.06		
Ferrihydrite (2L)	$Fe_5HO_8.4H_2O$, $Fe_{10}O_{14}(OH)_2$	$T_N \approx$ 350	4.2	0.48	0.00	50.6 (61%) 47.2 (39%)	
				0.43	− 0.04		
Ferrihydrite (6L)			4.2	0.50	− 0.06	50.5 (61%) 47.1 (39%)	
				0.42	− 0.08		
Schwertmannite	$Fe_{16}O_{16}(OH)_y(SO_4)_z.nH_2O$	T_N 75	4.2	0.49	− 0.41	45.6	Bigham et al. (1994)
Bernalite	$Fe(OH)_3$	T_N 427	300 81	0.38	− 0.004	41.5 55.7	McCammon et al. (1995)
				0.48	− 0.008		
Green rust	$Fe_x^{3+}Fe_y^{2+}(OH)_{3x+2y-z}(Cl^-)_z$	Variable	78	1.25	2.88	–	Genin et al. (1986)

(continued)

Table 9.1 (continued)

Compound	Composition	T_N/T_c (K)	T(K)	δ (mm/s)	$\Delta/2\varepsilon$ (mm/s)	B_{hf} (T)	References
				1.25	2.60		
				0.47	0.41		
	$Fe_x^{3+}Fe_y^{2+}(OH)_{3x+2y-z}(1/4SO_4^{2-})_z$		78	1.26	5.86	–	Olowe et al. (1989)
				1.16	5.23		
				0.47	0.77		
Ferrous hydroxide	$Fe(OH)_2$	T_N	300	1.05	2.98	–	Miyamoto (1976)
		34	4.2	1.48	3.06	20.0	

δ—isomer shift, 2ε—quadrupole shift, Δ—quadrupole splitting, B_{hf}—hyperfine field (Area fraction in %), T—measurement temperature, T_N—Antiferromagnetic Néel temperature, T_c—Curie temperature.

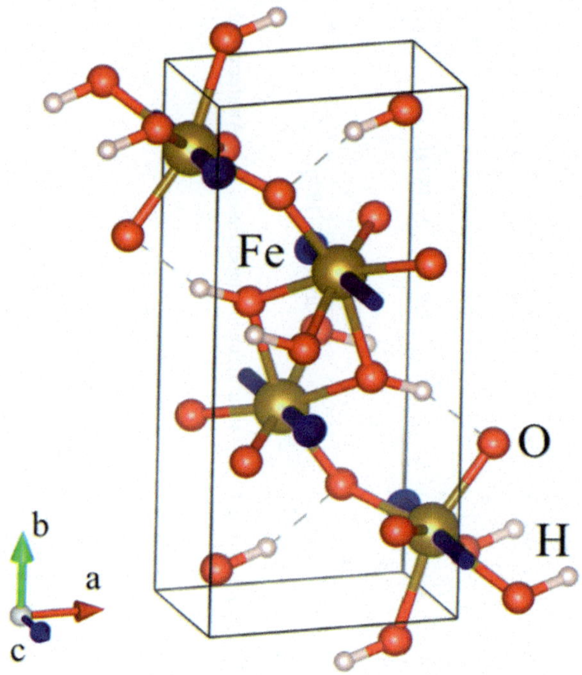

Fig. 9.2 Structure of goethite showing the Fe, H and O atom positions in the *Pbnm* space group (Oganessian et al. 2017)

The effective magnetic anisotropy constant (K) was estimated to be of the order 1 kJ/m^3 (with $\tau_o = 10^{-10}$ s) for the particles exhibiting T_B of 210 K. The estimation of T_B from the correlation time of fluctuation (τ_o) of $10^{-10} \sim 10^{-11}$ s, the experimental time scale (τ_m) of 2.2×10^{-8} s and $K = 45.7$ kJ/m^3 yielded T_B of the order 10^3 K casting doubt on the superparamagnetic relaxation of goethite ultrafine particles (Bocquet et al. 1992). This further requires the estimation of the K value as it leads to the difference in T_B and also in determining the antiferromagnetic or speromagnetic nature of most of the antiferromagnets (Pollard and Pankhurst 1992; Coey 1993; Pankhurst and Pollard 1993).

The significant K value, equivalent to hard magnetic materials, in antiferromagnets was explained by the collinear model using the anisotropy constant in the spin Hamiltonian given as (Pankhurst and Pollard 1990)

$$H_{eA} = -K \sum_i S_i^2 - K \sum_j S_j'^2 = -2KS^2 \tag{9.1}$$

where $K > 0$ is the effective anisotropy constant of the antiferromagnet that is equal for the identical sublattice spins S and S'. Solving for the energy density of uniaxial materials using the uniaxial anisotropy constant K_1 gives the following relationship

$$K_1 = 2NS^2K = 2NS^2K'\xi^2 \tag{9.2}$$

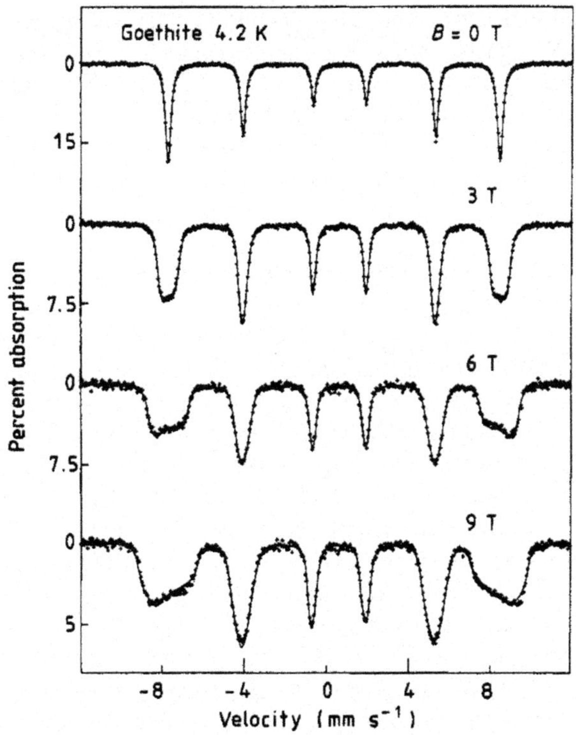

Fig. 9.3 The Mössbauer spectra of goethite at 4.2 K under various longitudinal applied magnetic field, *B*. The broadening of the first and sixth line intensities with increasing field (top to bottom) could be noticed. Reprinted with permission (Pankhurst and Pollard 1990), © IOP Publishing

where *N* is the number of sublattice spins per unit volume and $\xi = S'/S$ is the ratio of spins given by $\xi = 1$ for antiferromagnets and $\xi < 1$ for ferrimagnets. Consequently, the K_1 was as large as 250–400 kJ/m^3 and anisotropy field was 0.4 T at 4.2 K in goethite (Pankhurst and Pollard 1990, 1993). The above collinear spin model may be valid for antiferromagnets and not for speromagnets that are amorphous with randomly frozen moments.

Although calculations resulted in large K_1 values, experimentally it could be determined only from the spin flop transition in antiferromagnets. Antiferromagnets with strong exchange and large K_1 would require higher spin flop fields (B_{sf}) which could be obtained from the expression.

$$B_{sf} = \sqrt{2B_{ex}B_a - B_a^2} \tag{9.3}$$

where B_{ex} and B_a are the exchange and anisotropy fields, respectively. The spin flop experiments were conducted with a natural fibrous habit goethite ($Fe_{0.95}Al_{0.01}Si_{0.03}(OH)_{1.02}O_{0.99}$) that showed a T_N of 375 K (Coey et al.1995). The neutron diffraction results indicated the magnetic spin inclined at an angle of $\pm$ 13° to *c*-axis (*Pbnm* space group). The application of 6 T applied field at 20 K on the oriented fibrous powder parallel and perpendicular to the γ-ray resulted, respectively, in the diminished and increased second-fifth line intensities suggesting effective alignment of magnetization with the applied field. The spin flop was observed at an applied field beyond 20 T along the *b*-axis at 4.2 K. The lower limit of the K value obtained from the magnetization data was 8 kJ/m^3 at 4.2 K, while an upper limit gave a value of 30 kJ/m^3 at RT and 60 kJ/m^3 at 4.2 K. Collective superparamagnetic fluctuations of transverse magnetization has been proposed for the non-collinear structured goethite.

9.2.1.2 Superparamagnetism and Néel Temperature

It has been reported that the Néel temperature was reduced to 358 K with nanoparticles due to vacancy defects (Bocquet and Kennedy 1992). Mössbauer analysis by enriching (95.2%) a thin layer of ^{57}Fe on the surface of goethite prepared using pure ^{56}Fe isotope revealed the separation of the bulk and surface contribution (Yamamoto et al.1994). The bulk α-FeOOH showed the characteristic feature of antiferromagnetic nature with a single sextet, whereas the ^{57}Fe (surface)(α-FeOOH) showed contributions from the bulk and surface at 200 K. The temperature-dependent studies showed a broadening of the sextets collapsing to a doublet at 400 K corresponding to T_N. The reduced hyperfine field and temperature curves clearly deviated from the $S = 5/2$ Brillouin curve suggesting the role of surface effects in lowering the B_{hf} by more than 10 T at RT. Although the relaxation effect from surface spins could solve most of the problems in interpreting the spectrum of goethite, the reduced Néel temperature was a puzzle since that requires the alteration of the exchange interaction strength due to structural differences.

The structural variation with water content removal in goethite results in asymmetrically broadened sextets (Barrero et al. 2006a). The T_N of the highly crystalline goethite was 387 K, while that of the poorly crystalline non-stoichiometric sample was 310 K. The B_{hf} at 77 K in the poorly crystalline goethite was reduced from 47.0 to 37.0–38.0 T with annealing at 380 K for 30 min due to the vacancies left with water removal. The spectrum measured at 77 K which was a symmetric sextet without annealing the sample was broadened and asymmetric after annealing, as shown in Fig. 9.4. The study suggested that the superparamagnetism or cluster ordering is influenced by the hydroxyl to surface area ratio with the latter one for higher ratio.

The magnetic characteristics of the goethite samples prepared using ball milling were investigated using various characterization techniques, including Mössbauer spectrometry (Madsen et al. 2009; Brok et al. 2014). The nanograin aggregates of average size 5 nm showed narrow sextet at 20 K with B_{hf}, δ and 2ε of 49.5 T, 0.49 mm/s and − 0.26 mm/s, respectively, corresponding to goethite (Brok et al. 2014). A relaxation spectrum was obtained with increasing temperature, followed by a pure doublet above 240 K. The application of a magnetic field of 6 T to the sample at 260 K broadened the spectrum, as shown in Fig. 9.5. The broadening of the spectrum, which resulted in a hyperfine field distribution up to 40.0 T was from relaxation effect due to superparamagnetism since paramagnetic spectrum shall not induce broadening of this level. Thus, the doublet at 240 K was not due to the T_N of the goethite. The study also determined the K value close to 10^5 J/m^3 from inelastic

Fig. 9.4 The Mössbauer spectra at 77 K for the poorly crystalline goethite before removal of water (top) and after removal of water (bottom). (Barrero et al. 2006a)

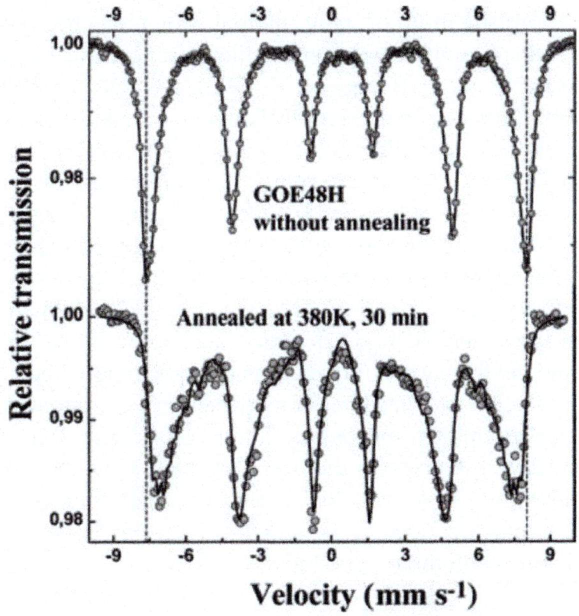

neutron scattering combined with Mössbauer spectrometry. It was clarified that the relaxation effects are in accordance with the super ferromagnetism model (Madsen et al. 2009).

Fig. 9.5 The 260 K Mössbauer spectra of 5 nm aggregated nanograins obtained by ball milling; without an applied field (top) and with a field of 6 T (bottom). Reprinted with permission (Brok et al. 2014), © IOP Publishing

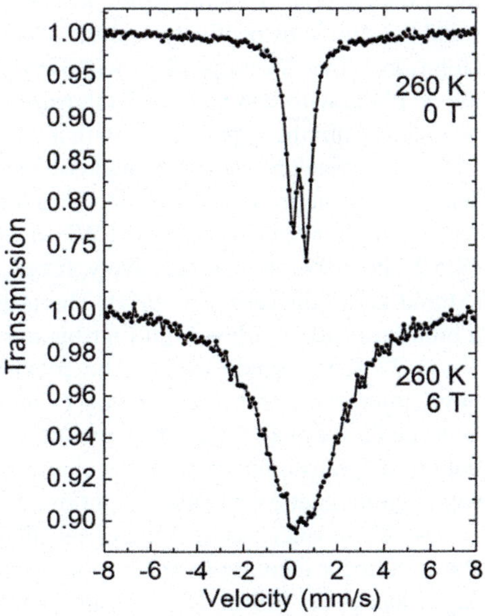

Investigation of pure natural stoichiometric goethite from Harz mountain, Germany and another nonstoichiometric goethite (α-$Fe_{0.92}[_{0.08}]O_{0.76}OH_{0.24}$, where [] is for vacancy) (Bocquet et al. 1992) revealed their T_N as 398 and 358 K, respectively, from muon spin relaxation (Pankhurst et al. 2012). The natural goethite showed a narrow sextet with a major fraction of goethite having a mean B_{hf}, δ and 2ε of 38.3 T, 0.34 mm/s and − 0.26 mm/s, respectively at RT. The nonstoichiometric goethite exhibited an asymmetric broadened spectrum with three sextets of B_{hf} 25.6, 31.5 and 35.3 T of almost equal area fractions corresponding to various Fe environments. The relaxation experiments with different time scales showed static behavior for measurement times shorter than 100 μs. The spin arrangement of Fe in the goethite structure was responsible for the dynamic behavior resulting in collective superparamagnetism and the formation of relaxation clusters in nonstoichiometric goethite.

Recent polarized neutron diffraction experiments have shown that goethite particles are antiferromagnetic, without surface spin effect, having an off-axis spin component with respect to the *c*-axis by more than 15°, different from the conventional structure (Brok et al. 2017). The differences in magnetic behavior are influenced by the spin direction of Fe in single-crystal, polycrystal, nanocrystal, stoichiometric and non-stoichiometric goethite.

9.2.2 *Lepidocrocite (γ-FeOOH)*

Lepidocrocite (γ-FeOOH) belongs to the *Cmcm* orthorhombic crystal system (previously *Amam* space group). It shows a ferric doublet above liquid nitrogen temperature (Takada et al. 1964; Rossiter and Hodson 1965). The RT Δ and δ are 0.6 mm/s and 0.3 mm/s, respectively, and the sextet at 4.2 K has an effective field of 46.0 T (Johnson 1969). Measurements with single crystal have shown the magnetization in the *ac* plane with γ||*b* and 3:4:1 intensity ratio. The collinear *c*-axis spin alignment was clear from the experiments with a 3.0 T applied field on the single crystal. The EFG is negative and makes an angle of 55° with the *c*-axis in the paramagnetic state. The poor crystallinity induces line broadening, and the 4.2 K sextet could be fitted with a single sextet of B_{hf} and FWHM of 45.6 T and 0.51 mm/s compared with 44.5 T and 1.09 mm/s, respectively, for the heat treated and as-synthesized samples (Murad and Schwertmann 1984). The presence of hyperfine field distribution and a broad transition below liquid nitrogen temperature resulted in the declaration of T_N as 77 K for lepidocrocite. The coexisting doublet and sextet at low temperatures, interpreted as corresponding to the superparamagnetic state, was ruled out later (de Grave et al. 1986). The poor crystallinity of lepidocrocite results in a distribution of T_N, as inferred from the studies with an applied magnetic field. However, superparamagnetic relaxation is not ruled out in lepidocrocite.

The Mössbauer studies on Al substituted γ-$Fe_{1-x}Al_xOOH$ (x = 0–0.14) have shown a linear increase in quadrupole splitting and FWHM with Al at RT and 80 K (de Grave et al. 1996). The measurements at 8 K revealed the decrease in B_{hf} by 0.6 T with Al up to x = 0.08, suggesting Al incorporation in the Fe site. The T_N

was shifted towards a lower temperature of 44 K from 50 K with Al up to x = 0.1. A distribution of hyperfine parameters with the exception of δ was observed for the lepidocrocite with or without Al (de Grave, 1986, 1996). This may be related to large inhomogeneities in the samples as defects or substitutions.

9.2.3 *Akaganeite (β-FeOOH)*

Akaganeite (β-FeOOH) has a monoclinic *I2/m* crystal structure where it is often occupied by a halide. The room temperature spectrum of akaganeite is a doublet with ferric isomer shift and quadruple splitting of 0.62–0.72 mm/s (Rossiter and Hodgson 1965; Johnston and Logan 1979). The measurements at 77 K revealed broadened sextets resulting in an average hyperfine field of 46.6 T, but with a wide variation of ± 4.9 T (Rossiter and Hodgson 1965), further refined to 47.5 T at 80 K (Dezsi et al. 1967). Akaganeite, after losing its water content, decomposes to hematite at 670 K. The hyperfine parameters are largely affected by the Cl and OH content in akaganeite. The low-temperature spectrum at 4 K could be fitted with sextets of B_{hf} in the range 47.3–48.6 T (Murad 1979). Satisfactory fits could not be obtained with one doublet at RT, and therefore two doublet fits were used, resulting in Δ values of 0.55 and 0.94 mm/s (Childs et al. 1980). The structural modification due to the halide was inferred as $(X,H_2O)_2Fe_8(O,OH)_{16}$ with $X = Cl^-$ or F^- where Cl content cannot exceed 4.6 wt.%. The maximum T_N of 299 K decreases with interstitial water content (Chambaere and de Grave 1984).

The fitting with four sextets at 77 K to account for the monoclinic structure and the Cl, based on the formula $Fe_8O_8(OH)_8.[2Cl].H_2.nH_2O$, resulted in B_{hf} values from 43.6 to 47.7 T with narrow linewidths and 2ε in the range of − 0.41 to − 0.12 mm/s (Barrero et al. 2006b). The [2Cl] represents the chloride in the tunnel structure where at most, two sites can be occupied, and one may be vacant. Low-temperature measurements with an applied magnetic field revealed a distribution in the hyperfine field due to canted spin arrangement from two sublattices. The ferric sublattice corresponds to two sites close to Cl and another two close to vacant sites. The speromagnetic structure was ruled out from the study because no 360° hyperfine field orientation was observed.

9.2.4 *Ferrihydrite ($Fe_5HO_8.4H_2O$)*

Ferrihydrite is usually represented by the chemical formula $Fe_5HO_8.4H_2O$ or $5Fe_2O_3.9H_2O$, has been shown to be the hexagonal $P6_3mc$ structure, better represented by the formula $Fe_{10}O_{14}(OH)_2$ (Michel et al. 2007). They are poorly crystalline 2–6 nm nanoparticles that show two broadened lines (2L) for smaller particles or more, usually six (6L) with larger particle size in the X-ray diffraction profile. The ideal structure has the FeO_4:FeO_6 polyhedra ratio 2:8, and the occupancy shall deviate with particle size. Ferrihydrite of the 2L form has a T_N close to 350 K, determined from neutron diffraction measurements (Seehra et al. 2000). The T_N obtained from Mössbauer measurements was often lower due to the superparamagnetic relaxation effects masking the Néel transition. Ferrihydrite is often associated with amorphous ferric gel $Fe(OH)_3.0.9H_2O$ that is superparamagnetic below 10 K (Coey and Readman 1973). The ferric gel shows a hyperfine field of 45.8 T at 4.2 K. The amorphous gel exhibits speromagnetic behavior due to the statistical distribution of spins at random directions, whereas it behaves as an antiferromagnet with better crystallinity. In contrast to amorphous ferric gel, ferrihydrites are identifiable from the 2L or 6L XRD lines. Bernalite ($Fe(OH)_3$), has a T_N higher than ferrihydrite.

The ferrihydrite has to be fitted with at least two doublets at RT with large quadrupole splitting of 0.78 and 0.86 mm/s for 2L, whereas that of 6L with 0.62 and 0.71 mm/s (Murad and Schwertmann 1980). The 2L ferrihydrite was fitted with two sextets at 4.2 K with $B_{hf} = 50.6$ T, $\delta = 0.48$ mm/s, $2\varepsilon = 0$ and $A = 61\%$ for the first sextet and $B_{hf} = 47.2$ T, $\delta = 0.43$ mm/s, $2\varepsilon = -0.04$ mm/s, $A = 39\%$, for the second sextet (Pollard et al. 1992). A very slight variation of hyperfine parameters for the 6L sextet could be observed, as shown in Table 9.1 corresponding to structural similarity. However, the sextets showed differences with an applied field of 9 T, possibly due to anisotropy variation.

The Mössbauer spectra of the 6L ferrihydrite obtained at various temperatures showed doublet at 300 K, relaxation spectrum corresponding to superparamagnetic state at intermediate temperature and clear sextet at 4.2 K (Guyodo et al. 2006). Although the average hyperfine field reduces drastically above 80 K, the superparamagnetic nature of the particles with an average size 5.4 nm was confirmed with an applied field of 5.5 T at 100 K that produced well-defined hyperfine splitting. The average hyperfine field (48.2–49.2 T at 4.2 K) reduces to 50% of the value at 105, 90 and 70 K for the particles of average sizes 5.4, 4.1 and 3 nm, respectively. Extensive studies on 2L, 5L and 6L ferrihydrite have shown doublets at 295 K, while the 5L and 6L ferrihydrites exhibit sextets at 20 K when the grain size is greater than 3 nm (Fig. 9.6) (Wang et al. 2016). The 6L ferrihydrite with an average grain size of 4 nm could be fitted with two sextets of hyperfine fields 45.8 and 39.5 T at 20 K arising from the size distribution-induced relaxation effects.

9.2.5 *Bernalite ($Fe(OH)_3$)*

Bernalite ($Fe(OH)_3$) has the corner-connected distorted perovskite-like structure with a vacant A-site cation, better represented by the *Pmmn* space group. The hyperfine parameters on natural bernalite determined from Mössbauer spectrometry showed $B_{hf} = 55.7$ T, $\delta = 0.48$ mm/s, $2\varepsilon = -0.01$ mm/s at 80.5 K and $B_{hf} = 34.4$ T, $\delta = 0.34$ mm/s, $2\varepsilon = -0.02$ mm/s at 350 K (McCammon et al. 1995). The natural sample also had goethite that could be distinguished from the hyperfine parameters at 80 K from the broad and less intense sextet with an increasing superparamagnetic component at 350 K, as shown in Fig. 9.7. The Weiss's molecular field model gives a

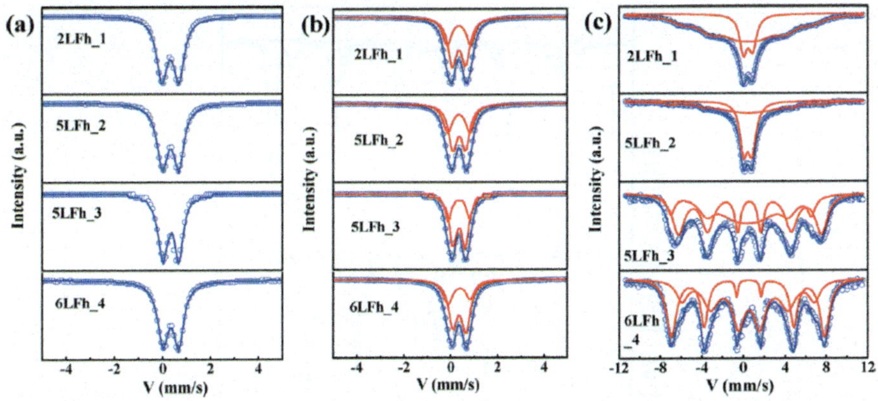

Fig. 9.6 The Mössbauer spectra of ferrihydrites: **a** fitted with single doublet at 295 K, **b** two doublets at 295 K and **c** ferrihydrites at 20 K. The first number indicates the XRD lines, and the last number is the grain size (Wang et al. 2016)

saturation hyperfine field of 55.7 T and T_N of 427 K. There may be a slight deviation from the collinear antiferromagnetic structure inferred from measurements with an applied magnetic field, resulting in a weak magnetic moment (McCammon et al. 1996).

9.2.6 Schwertmannite $(Fe_8O_8OH)_{8-2x}(SO_4)_x$ $(1 \leq x \leq 1.75)$

Schwertmannite, found in sulfate soils, is a poorly crystalline mesoporous material having the formula $(Fe_8O_8OH)_{8-2x}(SO_4)_x.nH_2O(1 \leq x \leq 1.75)$ with the close space group of *P4/m*. The asymmetric RT spectrum has to be fitted with two doublets of $\Delta = 0.64$ mm/s and slightly varying δ of 0.39 and 0.34 mm/s (Bigam et al. 1994). The spectrum at 4.2 K was fitted with a close-lying sextet pair of hyperfine fields 45.4–45.8 T and $\Delta = -0.37$ and -0.08 mm/s. The saturation hyperfine field is about 45.6 T and T_N is 75 K.

9.2.7 Feroxyhyte (δ-FeOOH)

Feroxyhyte (δ-FeOOH) with the hexagonal structure is known for its ferrimagnetic behavior among the Fe-oxy-hydroxides with a Curie temperature of 455 K. The magnetic behavior is reported to be as a result of ferrimagnetic domains from surface steps in the planar uncompensated antiferromagnetic layered structure (Koch et al. 1995). Well crystalline ferroxyhte may exhibit saturation magnetization as large as 14 $A{\cdot}m^2/kg$ at 4.2 K.

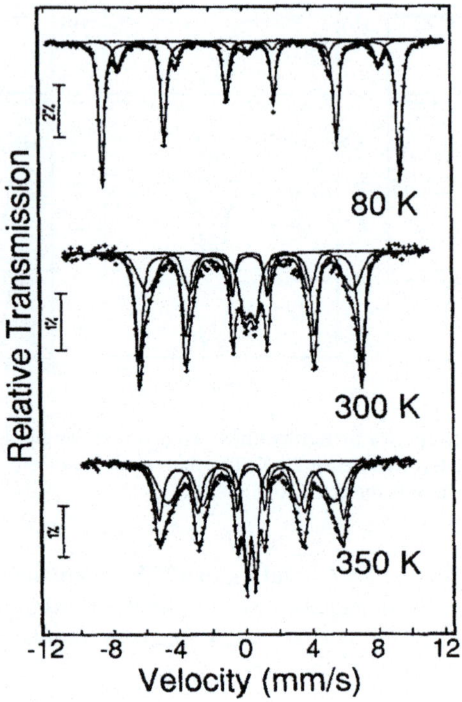

Fig. 9.7 The Mössbauer spectra of natural bernalite at various temperatures. The less intense broad sextet arises from goethite. Reprinted with permission (McCammon et al. 1995), © Springer Nature

The unresolved spectrum shows a sextet of B_{hf} = 53.3 T at 77 K with ferric δ (0.45 mm/s) (Rossiter and Hodgson 1965) whereas the resolved spectrum at 80 K has B_{hf} = 52.5 and 50.5 T (Dezsi et al. 1967). The hyperfine field of δ-FeOOH decreases with temperature, whereas the transformation to hematite begins at 643 K, even before its T_c, resulting in increasing B_{hf} at higher temperatures (Dezsi, et al. 1967). The hyperfine parameters beyond RT are influenced by the decomposition products. The feroxyhyte could be fitted with three sextets of B_{hf} = 53.2–49.3 T, δ = 0.5–0.48 and 2ε = 0.16 ~ − 0.06 at 4.2 K (Pollard et al. 1992) as shown in Table 9.1. The application of a 9.0 T longitudinal external field resulted in four sextet components with an effective hyperfine field in the range of 60.9 to 44.1 T at 4.2 K indicating ferrimagnetic alignment, unlike other FeOOH polymorphs that were fitted with sextets arising from antiferromagnetic alignment.

9.2.8 Green Rust

Three types of green rust are encountered with the general formula $Fe_4^{2+}Fe_2^{3+}(OH)_{12}X \cdot nH_2O$ where X = SO_4 and CO_3, n = 8 and 3, respectively, for the sulfate and carbonate green rust, while the chloride form is given as $Fe_3^{2+}Fe^{3+}(OH)_8Cl \cdot 1.5H_2O$ (Usman et al. 2018). The green rust brucite-like structure has ferrous and ferric ion interlayers and structural water known to the group of layered hydroxides. The sulfate group with the $P\overline{3}m1$ orthorhombic structure is classified as Green Rust I (GRI), while the carbonate and halide group with the $R\overline{3}m$ hexagonal structure are known to be Green Rust II (GRII). Green rust is an important corrosion product of iron and a soil component. They easily transform into other oxy-hydroxides and oxides with temperature and duration.

The chloride green rust GRI could be fitted with two ferrous ($\delta = 1.25$ mm/s) and one ferric ($\delta = 0.47$ mm/s) site with $\Delta = 2.88$, 2.60 and 0.41 mm/s, respectively at 78 K (Gancedo et al. 1976; Genin et al. 1986). The spectrum at 120 K for the non-oxidized carbonate green rust was fitted with two ferrous doublets and one ferric doublet assigned to three Fe sites, while the oxidized sample showed two ferric doublets and one ferrous doublet (Murad and Taylor 1984). The spectrum at 4.2 K showed a complex pattern with coexisting doublet and sextet of B_{hf} 48.0 T.

The Mössbauer spectrum of sulfate GRII was fitted with five doublets at 78 K corresponding to three ferrous and two ferric sites with Fe^{2+}:Fe^{3+} relative area 57:43, which could not be resolved beyond three doublets for the samples obtained from bacterial corrosion (Olowe et al. 1989). The ferrous doublets showed $\Delta = 2.72$–2.93 mm/s, $\delta = 1.12$–1.26 mm/s while the ferric sites showed $\Delta = 0.34$–0.53 mm/s and $\delta = 0.41$–0.43 mm/s for the GRII at 78 K. Sulphate GRII prepared using another method showed one ferrous and ferric site each with a quadrupole splitting of 2.63 and 0.50 mm/s, respectively at 293 K (Cuttler et al. 1990). Although the ferrous and ferric sextets could be identified at 4.2 K, definite intensity ratios could not be assigned.

Another form of green rust, such as the fougerite carbonate group layered double hydroxides with the formula $Fe^{2+}{}_4Fe^{3+}{}_2(OH)_{12}[CO_3]{\cdot}3H_2O$ is known as trebeurdenite and the oxidized form is known as Mössbauerite with the formula $Fe^{3+}{}_6O_4(OH)_8[CO_3]{\cdot}3H_2O$. The RT and 78 K doublets for the ferric Mössbauerite show Δ values 0.97 mm/s and close to 1.1 mm/s, respectively (Genin et al. 2014). The trebeurdenite is mixed valent and shows two doublets corresponding to the Fe^{2+} and Fe^{3+} with $\Delta = 2.9$ mm/s and 0.4–0.6 mm/s, respectively at 78 K.

9.2.9 Ferrous Hydroxide ($Fe(OH)_2$)

The ferrous hydroxide ($Fe(OH)_2$) is a hexagonal structured compound with ferrous ions in the octahedral site normal to the *c*-axis. The ferrous compound could easily oxidize to its mixed valent and ferric state oxides and oxy-hydroxides.

The quadrupole splitting at 300, 200, 90 and 4.2 K showed an increasing trend with decreasing temperature from 2.98 to 3.06 mm/s and ferrous δ from 1.05 to 1.48 mm/s (Miyamoto 1976). The B_{hf} at 4.2 K was 20.0 T. The T_N is 34 K while weak interaction between adjacent antiferromagnetic layers results in a ferrimagnetic state. The gradual oxidation of $Fe(OH)_2$ from ferrous to 70% ferric state with pressure and increase in T_N up to 63 K with pressure up to 40 GPa was observed (Pasternak et al. 2004). The irreversible ferric component showed $B_{hf} = 51.0$ T, $\Delta = 0.3$ mm/s and $\delta = 0.3$ mm/s at 60 K without alteration of the $Fe(OH)_2$ structure due to self-oxidation induced by altered O–H axial symmetry creating an effective dipole potential at Fe site. However, the phase change is imminent at very high pressure.

9.3 Cubic Perovskite Antiferromagnets

9.3.1 AFeF$_3$ (A = K, Rb)

The $AFeF_3$ group shows the cubic perovskite unit cell of the *Pm3m* space group at room temperature with the A = K, Rb occupying the corners, Fe at the body center and F at the face center positions. The $KFeF_3$ and $RbFeF_3$ undergo lattice distortion below the T_N close to 120 and 100 K, respectively, and exhibit G-type antiferromagnetic order. Stress-induced orientation effects are expected at low temperatures in these materials.

$RbFeF_3$ exhibits multiple temperature dependence such as: paramagnetic up to 100 K, antiferromagnetic up to 86 K, ferrimagnetic with further decrease in temperature due to inequivalent magnetic state of two Fe sites, and weakly ferromagnetic below 40 K (Wertheim et al. 1967). The cubic symmetry was evident from the singlet paramagnetic spectrum. $RbFeF_3$ undergoes a first-order transition from the antiferromagnetic tetragonal to orthorhombic distortion corresponding to its ferrimagnetic state. The antiferromagnetic state at 88 K showed a B_{hf} of 15.3 T and a quadrupole splitting of 0.15 mm/s for the powdered samples. The ferrimagnetic state was also clearly evident from the spectrum measured at 82 K, which showed two sextets of unequal intensity from the two Fe sites. The internal field for saturation was estimated as − 17.2 T, considering the core polarization contribution of the ferrous high spin state as − 41.2 T, orbital contribution of 24.0 T and neglecting dipole contribution that is less than 0.5 T. Below 39 K, all Fe sites are equivalent yielding an effective field of 25.2 T, quadrupole splitting of 2.7 mm/s and EFG axis at 40° to the direction of magnetization (Wertheim et al. 1967). However, Jahn-Teller-induced spin canting was considered as the ferromagnetic origin at low temperatures (Goodenough et al. 1970). The weak ferromagnetic structure at intermediate temperatures and antiferromagnetic structure below 20 K was also observed in Fe_2SiO_4 with the olivine structure (Kundig et al. 1967). At RT, Fe_2SiO_4 showed a symmetric doublet with $\Delta = 2.81$ mm/s and at low temperatures, the hyperfine parameters from two Fe sites were identified.

Experiments on stress-free $RbFeF_3$ powders in the intermediate temperature range showed clearly the two ferrous sites with B_{hf} 17.6 to 26.0 T for site I and 17.3 to 20.5 T for site II as the temperature is reduced from 81 to 42 K (Someya et al. 1981). The quadrupole splitting of site I varied from 1.77 to 2.82 mm/s with decreasing temperature while it underwent sign change from − 1.40 mm/s at 55 K to + 1.46 mm/s at 50 K for site II. This sign change was only associated with the change in the coordinate system as $|V_{zz}| > |V_{yy}|$ at 55 K whereas $|V_{yy}| > |V_{zz}|$ at 50 K. An external field of 0.1–0.4 T was applied on the single crystal along the [100] axis both under parallel and perpendicular orientation that showed a difference in intensities between the B_{hf} of the two sites. The experiments revealed the ferrimagnetic spin arrangement along the [100] direction. The stresses affecting the intensity ratio were also studied using stressed and stress-free single crystals.

The Mössbauer spectra of stress-free single crystal $RbFeF_3$ showed single sextet with B_{hf} from 25.0 to 23.8 T with increasing temperature from 4.2 to 35 K (Someya et al. 1981). The experiments with an applied field revealed that the easy axis direction is canted by 6° from the [011] direction leading to a spontaneous magnetization along $[0\bar{1}1]$. Thus, the canted spin arrangement was proved to be the origin of weak ferromagnetism.

$KFeF_3$, shows rhombohedral distortion as it becomes antiferromagnetic, and it is weakly ferromagnetic below 41 K along $[1\bar{1}0]$ due to spin canting (Haida and Kohn 1976). The spectrum above RT showed cubic nature with a ferrous isomer shift of 1.14 mm/s. In contrast, below 115 K, in the antiferromagnetic state, it exhibited relaxation spectra with positive EFG up to 95 K (Fatehally et al. 1967) that are usually absent in single crystals. Mössbauer analysis of single crystals at 77 K showed V_{zz} to be positive in the region of trigonal distortion with hyperfine field and magnetization parallel to <111> trigonal axis, establishing collinear antiferromagnetic alignment (Varret and Imbert 1973). The stressed single crystal showed a reduction of B_{hf} from 17.2 to 16.9 T and an increase in quadrupole splitting from 0.41 to 0.44 mm/s and orientation changes compared with stress-free crystals at 77 K. At the lowest temperature region, non-axial distortion begins at 36 K while the hyperfine parameters at 4.2 K are $B_{hf} = 17.8$ T, $\Delta = 2.0$ mm/s and $\delta = 1.26$ mm/s. The hyperfine field is oriented at an angle 20° away from the <111> axis of trigonal distortion. The transition temperatures from another separate study confirmed the T_N as 113 K, <111> parallel spin alignment, the absence of relaxation effects and weak ferromagnetism below 37 K (Davidson et al. 1973). The gradual increase in the quadrupole splitting from 0.01 mm/s at 112 K to 1.2 mm/s at 37 K was also observed. Further experiments in the weak ferromagnetic region showed that the spin canting is at an angle 5° to the [111] towards the $\left[1\bar{1}0\right]$ direction at 4.2 K and hyperfine parameters similar to $RbFeF_3$ (Ito and Morimoto 1975).

9.4 Low Dimensional Antiferromagnetic Compounds

Antiferromagnetic order can be contributed by Heisenberg-type materials that show 3D ordering, as we have discussed so far or lower dimensional ordering such as 1D and 2D. Low dimensional ordering arises if the spin alignments are different from the crystallographic sites and suppressed in certain directions resulting in directional-dependent magnetic behavior (Cortie et al. 2019). The interplanar exchange interaction results in planar antiferromagnetic nature exhibiting 2D order, known as quasi-2D order realizable in bulk materials. A_2FeF_4 (A = K, Rb) and $FePS_3$ are a few of the 2D materials that are of renewed interest for potential applications.

Although the 1D and 2D antiferromagnetic ordered Heisenberg system is not possible beyond absolute zero under ideal conditions, anisotropy due to spin–orbit coupling or dipole–dipole interaction leads to finite T_N. The quasi 2D systems are better described by the intra (J) and interplane (J') exchange coupling parameters that are related to the T_N using the semi-empirical inverse relationship between the logarithmic ratio of J'/J (Yasuda et al. 2005). In the 3D magnetic system, all spin orientations are equally favorable ($J_x = J_y = J_z \neq 0$; $S_x = S_y = S_z \neq 0$) whereas spin orientations are allowed only along the plane ($J_x = J_y \neq 0$, $J_z = 0$; $S_x = S_y \neq 0$, $S_z = 0$) in an ideal 2D X–Y system. The ideal 1D Ising system allows only one spin orientation where J_z, $S_z \neq 0$ whereas values in other coordinates are zero. Long-range order is not possible in ideal systems except for the 3D and 2D Ising systems. In quasi-dimensional systems, the exchange constant and spin values are experimental approximations. For example J_x,J_y >> J_z and S_x,S_y >> S_z in quasi 2D systems.

9.5 2D Planar Antiferromagnets

9.5.1 *A_2FeF_4 (A = K, Rb, Cs)*

Rb_2FeF_4 and K_2FeF_4 are Fe-based 2D planar ($S = 2$) Heisenberg antiferromagnets. The spin structure of K_2FeF_4 in the magnetically ordered state at 4.2 K representing three Fe layers in the magnetic unit cell, obtained from neutron diffraction (Thurlings et al. 1982), with the intermediate F and K atomic layers, is shown in Fig. 9.8. K_2FeF_4 has basal plane magnetization with T_N of 67 K and a narrow transition width of 3 K (Thurlings et al. 1978). The critical exponent was obtained as $\beta = 0.17$ from the fit of T dependent B_{hf} in the range of 17.5 to 38.0 T.

Cobalt substituted $K_2Fe_{1-x}Co_xF_4$ shows planar spin alignment perpendicular to the c-axis with lower x. Increasing the Co transforms it into a uniaxial type parallel to the c-axis, whereas it has an oblique antiferromagnetic (OAF) phase in the range $0.16 \leq x \leq 0.32$ (Fendler et al. 1984). The measurements at 6 K showed that the B_{hf} decreases with Co concentration and reaches a steady value close to 27.5 T at $x \approx 1$. The uniaxial magnetic order has a transition temperature of 40–50 K. The OAF behavior is also

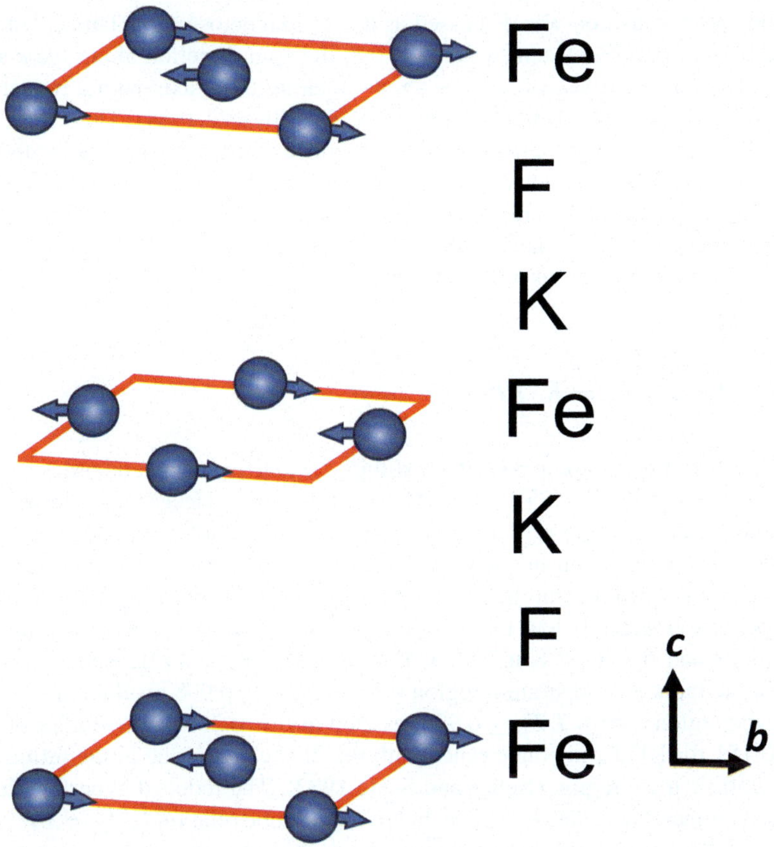

Fig. 9.8 The magnetic spin structure of ordered K_2FeF_4

found with the Ni-rich compound before the spin direction turns along the *c*-axis (Ito and Anma 1987). The T_N is in the range of 95–97 K when the Fe concentration is 0.006. $K_2Fe_{1-x}Zn_xF_4$ ($0 \leq x \leq 0.28$) single crystals showed paramagnetic ferrous doublet for $T > T_N = 23$ K when x = 0.26 and sextet for $T < 11$ K (Dikken et al. 1985). The neutron diffraction measurements showed an ordered region in the range 23 K < T < 11 K in contrast to the Mössbauer experiments due to the difference in the Mössbauer and neutron time scales of 10^{-8} and 10^{-11} s, respectively. A decrease in T_N was observed with decreasing Fe concentration reducing the hyperfine field. The substituted compound retains basal plane magnetic spin and exhibits 2D XY behavior below T_N.

The Rb_2FeF_4 with basal plane antiferromagnetic alignment similar to K_2FeF_4 has T_N between 50–60 K, and at 4.2 K, the B_{hf} becomes 36.4 T with other parameters being $\Delta = 2.77$ mm/s and $\delta = 1.27$ mm/s (Wertheim et al. 1968a, b). Measurements on the single crystal with the *c*-axis parallel to γ-ray showed larger intensity for the second and fifth line that suggested spins perpendicular to the *c*-axis (XY plane), and the broad transition was discussed as due to magnetostrictive lattice distortion. In the $Rb_2Fe_{1-x}Co_xF_4$ system, when x > 0.8, the spin direction along the *c*-axis in the antiferromagnetic region, makes an angle along the *c*-axis with a reduction of temperature, signifying a change from 2D to 3D (Hiraoka 1984). The T_N increases to 102 K with x = 0.98, whereas B_{hf} at 4.2 K reduces to 27.4 T compared with 36.4 T at x = 0. The Ba_2FeF_6 showed 2D antiferromagnetic structure with $T_N = 47.5$ K and temperature-dependent reduced B_{hf} fit in the range 20 K < T < 46 K yielded $\beta =$ 0.145 with normalization factor, $B = 1.05$ (Renaudin et al. 1983). The $B_{hf} = 21.5$ T, $\Delta = 2.43$ mm/s and ferrous $\delta = 1.41$ mm/s at 4.2 K.

9.5.2 $AFeF_4$ (A = K, Rb, Cs)

The $KFeF_4$ is a ferric compound that exhibits 2D antiferromagnetic order along the (001) plane with T_N = 137–141 K (Heger et al. 1971; Heeger and Geller 1972). Measurements on single crystals with γ-ray oriented along the *c*-axis have shown that the $\Delta m = 0$ lines vanishes at a low temperature far below T_N (Fig. 9.9). The room temperature spectrum exhibits a broad doublet and the antiferromagnetic hyperfine field pattern emerges below the Néel temperature. The RT $\Delta = 1.46$ mm/s, $\delta =$ 0.14 mm/s and B_{hf} at 4.2 K is 54.0 T. Fitting the temperature-dependent hyperfine field in the reduced temperature region $0.64 \leq T/T_N \leq 0.998$ resulted in $\beta = 0.182$ while the region $0.96 \leq T/T_N \leq 0.998$ resulted in $\beta = 0.209$. Later studies obtained $\beta = 0.151$ for $KFeF_4$ in concurrence with the 2D behavior due to the fitting in the asymptotic critical region (Keller and Savic 1983). The reduced hyperfine field vs. reduced temperature relation is given by the asymptotic relation $\frac{B_{hf}(T)}{B_{hf}(0)} = Bt^{\beta}$, where $t = 1 - T/T_N$, $t \rightarrow 0$ and the normalization factor $B < 1$ (0.92) in the range $3.8 \times 10^{-4} \leq t \leq 0.057$ for $KFeF_4$.

Mössbauer measurements on $RbFeF_4$ powders showed $T_N = 133$ K, $\beta = 0.245$ and at 4.2 K exhibited $B_{hf} = 53.0$ T and $\Delta = 1.6$ mm/s (Eibschütz et al. 1971). $CsFeF_4$ is another ferric layered antiferromagnet that has a T_N of 160 K (Eibschütz et al. 1972). The single crystals exhibit a large quadrupole splitting of 1.55 mm/s at 296 K and an isomer shift of 0.44 mm/s. The hyperfine parameters at 4.2 K are $B_{hf} = 54.0$ T, $\Delta = -1.54$ mm/s, $\delta = 0.54$ mm/s, $\eta = 0.3$ and $\theta = 5°$. The critical exponent, β for $CsFeF_4$ showed a comparatively higher value of 0.253–0.283, for $6 \times 10^{-3} \leq t \leq 0.056$ and B = 1.07–1.16, similar to $RbFeF_4$ that was suggested due to the crystallographic structure (Eibschütz et al. 1974a, b).

Mössbauer studies comparing $RbFeF_4$ and $KFeF_4$ single crystals showed T_N of 134–136 K for the two compounds, whereas the β values of 0.316 and 0.151 in the asymptotic region are indicative of 3D and 2D behavior, respectively (Keller and Savic 1983). The parameters extrapolated to 0 K are B_{hf} = 53.7 and 54.0 T, δ = 0.55 and 0.56 mm/s, $\Delta = -1.70$ and -1.47 mm/s for $RbFeF_4$ and $KFeF_4$, respectively. The isomer shifts indicate the ferric nature of the samples. The polar angle, θ between V_{zz} and B_{hf} that are close to the γ-ray tilt angle with the V_{zz} principal

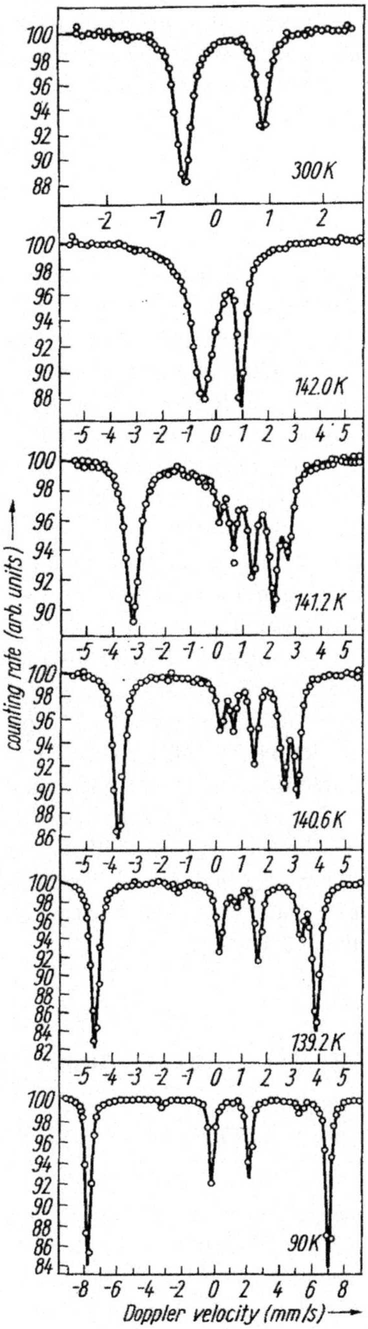

Fig. 9.9 Mössbauer spectra of $KFeF_4$ single crystals at various temperatures with c-axis parallel to gamma rays (Heger and Geller 1972)

axis, suggested magnetic moment alignment perpendicular to the layers for both the samples. 2D antiferromagnetic nature was also shown in $FeTa_2O_6$ that has T_N close to 10 K (Takano and Takada 1970; Eicher et al. 1986; Zawislak et al. 1995, 1997).

9.5.3 $FePS_3$

Iron thiophosphate ($FePS_3$) has the monoclinic structure, *C2/m* and is a quasi 2D layered Ising type antiferromagnet with T_N close to 120 K. The isomer shift value is close to 1.1 mm/s signifying the ferrous state and the quadrupole splitting is around 1.5–1.2 mm/s in the temperature range 296–195 K (Taylor et al. 1973). The Mössbauer spectrum above 120 K shows a single quadrupole doublet while the quadrupole splitting and isomer shift values increase with decreasing temperature reaching 1.67 and 1.04 mm/s, respectively, accompanied by a hyperfine field component of 9.7 T at 5 K (Jernberg et al. 1984). Intercalation of $FePS_3$ with an electron donor results in a charge transfer-induced change of magnetic properties. For the Li_xFePS_3 with Li intercalation, the hyperfine parameters for the two Fe sites (A and B) could be resolved at RT (Fatseas et al. 1987). The δ value for the A-site was 0.86 mm/s and almost identical to $FePS_3$ with x up to 1.45, while the B-site value was 0.43 mm/s due to substantial covalent bonding between Fe and S along with possible orbital distortion of core orbitals. The Δ also showed a reduced value of 0.48 mm/s for the B-site compared with 1.55 mm/s for the A-site and the relative intensity of the doublet reached 48% for $x = 1.45$. Many other intercalants have shown no change in antiferromagnetic or ferrimagnetic nature (Leaustic et al. 1996).

The selenide counterpart $FePSe_3$ of rhombohedral symmetry, *R3*, showed parameters similar to $FePS_3$ with $T_N = 123$ K and RT values of $\delta = 1.05$ mm/s and $\Delta = 1.45$ mm/s (Taylor et al. 1974). The principal axis of magnetization was along the hexagonal *c*-axis from measurements on a single crystal, and the B_{hf} at 77 K was obtained as 9.6 T.

9.6 1D Antiferromagnets

9.6.1 K_2FeF_5

Mössbauer spectra of the antiferromagnetic powder prepared from single crystal K_2FeF_5 1D antiferromagnet showed a T_N of 11.2 K and relaxation broadening, causing difficulty with fitting to obtain the β values (Gupta et al. 1977). The δ values of 0.42–0.53 mm/s in the temperature range 300–4.2 K indicated the ferric state of the compound. The inter-to-intrachain ratio of J'/J was obtained as 5×10^{-3} using $T_N/\Theta_C \approx 0.1$, where Θ_C is the Curie–Weiss constant suggesting the 1D character of the system. The Δ value was 0.7 mm/s in the whole temperature range, whereas the

$B_{hf}(0) = 41.0$ T. The low $B_{hf}(0)$ for the high spin ferric $S = 5/2$ state was believed to be due to the zero point spin reduction caused by the zero point energy of spin waves destroying long-range order. The EFG was determined to be negative with $\eta = 0.57$ in an applied field at 133 K in the paramagnetic region.

Further analysis of the K_2FeF_5 single crystal with an applied field parallel to the antiferromagnetic axis induced spin flop in a perpendicular direction towards the sample plane with a field of 3.7 T at 4.2 K (Gupta et al. 1978a, b). This could be observed from the appearance of $\Delta m = 0$ lines that increased in intensity beyond the first and sixth lines, whereas it vanished with a parallel applied field below 3.7 T. The applied field dependent low dimension characteristics could be noticed from the reduction of B_{hf} up to 34.2 T due to spin reduction at the critical field of 3.7 T that increased to 38.2 T when $B = 6.0$ T. Calculations also confirmed the J'/J of the order 10^{-3} and other spin wave characteristics of the 1D antiferromagnet. Rb_2FeF_5 also showed 1D character with $T_N = 9.3$ K, $B_{hf}(0) = 43.0$ T, $J'/J \approx 10^{-3}$ and zero-point reduction of 28% (Gupta et al. 1978a, b). The RT spectrum showed hyperfine parameters similar to K_2FeF_5 with negative EFG and ferric isomer shift.

The increase and decrease of T_N with applied field along the perpendicular and parallel easy axis, respectively, are the characteristics of quasi 1D antiferromagnets, also observed in K_2FeF_5 single crystal (Boersma et al. 1982). The difference in anisotropy along the *ac* plane is negligible, and the system showed uniaxial behavior. It was also demonstrated that the B_{hf} increases with applied field perpendicular to the spin direction even at absolute zero in quasi 1D K_2FeF_5 due to spin canting and reduced zero point energy with applied field (Cooper et al. 1982).

9.6.2 Polymorphous Compounds

The hydrated compound $K_2FeF_5.H_2O$ showed three-dimensional behavior below $T_N = 0.8$ K and its $B_{hf}(0) = 52.0$ T (Takacs and Reiff 1987). The hydrated β-$Rb_2FeF_5.H_2O$ belonging to the space group *Cmcm*, on the other hand, showed 1D antiferromagnetic ordering with $T_N = 9.7$ K, $\beta = 0.24$ ($B = 1.26$), $B_{hf}(0) = 39.6$ T, $\delta =$ 0.52–0.43 mm/s and $J'/J = 2.2 \times 10^{-3}$ (Fourquet et al. 1982). The $(NH_4)_2FeF_5.H_2O$ prepared through hydrothermal synthesis exhibited a paramagnetic nature up to 4.2 K with $\delta = 0.54$ mm/s and $\Delta = 0.41$ mm/s (Fourquet et al. 1988). The α-$(NH_4)_2FeF_5$, *Pbcn* space group that has cis-linked octahedra showed 1D behavior below $T_N =$ 7.5 K while its trans member β-$(NH_4)_2FeF_5$ showed two Fe sites of intensity ratio 8:2 and a transition temperature of 13 K (Calage et al. 1991). The B_{hf} at 4.2 K is 36.0 T for the former, while the latter showed 39.6 and 42.8 T with relative areas of 80 and 20%, respectively. The α-$Rb_2FeF_5.H_2O$ and α-$Cs_2FeF_5.H_2O$ showed 3D ordering at T_N of 4 and 2.7 K, respectively (Calage and Reiff 1994). However, they are expected to show a 1D feature at a much lower T than the measured temperature through Mössbauer spectroscopy.

9.7 Triangular Lattice Antiferromagnets

9.7.1 $AFeX_3$ (A = Rb, Cs; X = Br, Cl)

$RbFeBr_3$ belongs to the $AFeX_3$ (A = Rb, Cs; X = Br, Cl) class of materials with the space group $P6_3/mmc$, which is a quasi 1D antiferromagnet that has $(FeX_6)^{4-}$ triangular chains along the hexagonal *c*-axis (Kato et al. 1998). Successive crystallographic phase transformation to $P6_3cm$ occurs below 40 K, wherein the collinear ferrimagnetic or frustrated triangular configuration is determined by the relative exchange interaction between two ferrous sites (Nishiwaki et al. 2011). As the magnetic contribution is from the two A-sites and one B-site, collinear ferrimagnetism is dominant when the A-site exchange integral, J_{AA}, overtakes the AB exchange integral, J_{AB}, whereas triangular configuration arises when $J_{AA} < J_{AB}$.

$RbFeBr_3$ exhibits a 3D ordering T_N of 5.5 K and shows only quadrupole doublet up to 1.24 K. The Δ increases from 1.27 mm/s at RT to 1.74 mm/s at 1.54 K with corresponding δ of 1.04 to 1.20 mm/s, respectively, while the absence of B_{hf} at 1.54 K may arise from net cancellation of its individual components (Eibschütz et al. 1974a, b). $CsFeBr_3$ did not exhibit magnetic ordering up to 5.8 K, and the quadrupole splitting becomes more negative with much lower temperature reaching − 1.8 ~ − 1.9 mm/s for the ferrous singlet state (Bocquet 1988). The distorted octahedra are elongated along the *c*-axis of the crystal and produce a temperature-independent V_{zz} component with $\Delta = -1.03$ mm/s using two crystal field fit model.

9.7.1.1 Exchange Interaction Parameter

$RbFeBr_3$ shows interchain to intrachain exchange interaction, J'/J value ~ 0.04 and the interchain exchange parameter (J/k_B) is − 2.5 K, where the negative value signifies antiferromagnetic exchange (Lines and Eibschütz 1975). The $RbFeBr_3$ being a singlet state antiferromagnet, the $|J|_T$ value of 2.5 K is large to produce appreciable long-range order near absolute zero while $T_N = 5.5$ K can be explained by lower $|J'|_T$ of 0.1 K.

The T_N of linear chain Heisenberg antiferromagnets $FeOHSO_4$ and $FeOHCrO_4$ are found to be 111 and 104 K, respectively, from Mössbauer analysis (Townsend et al. 1985). Their respective intrachain exchange parameters are − 19 K and − 32 K. $((CH_3)_3NH)FeCl_3.2H_2O$, labeled as FeTAC, showed antiferromagnetic Ising-like chains along the *a*-axis with canting along the *b*-axis that is in contrast to ferromagnetic chains CoTAC and NiTAC (Landee et al. 1985). With a T_N of 3.2 K, the B_{hf} at 1.2 K is 20.1 T and fitting the data assuming $S = 1/2$ Ising chain yielded an interchain exchange parameter of − 13.3 K.

9.7.2 $AFeO_2$

The $AFeO_2$ class of quasi-2D triangular lattice antiferromagnets such as $CuFeO_2$, also known as delafossite, crystallizes in the hexagonal *P6₃/mmc* (also as rhombohedral $R\overline{3}m$) structure with octahedrally coordinated Fe. The RT doublet with $\delta = 0.4$ mm/s, indicative of ferric iron and $\Delta = 0.64$ mm/s while the sextet at 4.4 K has $B_{hf} = 52.0$ T for the polycrystalline sample (Muir and Wiedersich 1967). Measurements with a single crystal showed positive quadrupole interaction and magnetization direction along the *c*-axis below $T_N = 19$ K. The 2D antiferromagnetic arrangement from alternate ferromagnetic planes was proposed. Further measurements showed two magnetic transitions, one at 16 K with a partially disordered monoclinic (*C2/m*) magnetic unit cell and another at 11 K corresponding to the transition to the orthorhombic magnetic unit cell (Mekata et al. 1993). The study also presented the system as a triangular Ising spin configuration with successive magnetic transition and collinear magnetic structure contributed by the exchange interaction of the three spins as $4J_3 < 2J_2 \approx J_1 < 0$. Generally, the magnetic state changes from the paramagnetic to the partially disordered state at 14 K and a collinear four sublattice (4SL) phase along the *c*-axis (with spins ↑↑↓↓) at 11 K. The application of magnetic field (*B*) parallel to the *c*-axis at low temperature transforms the 4SL phase to multiple magnetic states such as ferroelectric incommensurate helical order in the applied field range $7\ \text{T} < B < 13$ T and collinear five sublattice (5SL) phase (↑↑↑↓↓) along *c*-axis in the range $13\ \text{T} < B < 20$ T (Lummen et al. 2010). Further increasing the field beyond 20 T results in the 3SL phase, canted 3SL (> 32 T) and canted high field phase (> 52 T).

The pressure-dependent magnetic phases were studied in the temperature range 4–300 K (Xu et al. 2004). The ambient pressure spectrum at 6 K presented a sextet of B_{hf} 51.0 T corresponding to the 4SL ground state that transformed completely to the 5SL phase at 19 GPa with a B_{hf} of 48.0 T. The 5SL phase underwent a relaxation spectrum and reached a T_N of 36 K at the pressure of 19 GPa. Further, an increase in pressure showed both ferric ($\delta \approx 0.4$ mm/s) and ferrous ($\delta \approx 0.62$ mm/s) states beyond 27 GPa, and both begin to transform to low spin phase ($\delta < 0.2$ mm/s, $Fe^{3+}(B_{hf}) <$ 10 T, $Fe^{2+}(B_{hf}) \rightarrow 0$) beyond 50 GPa, from the Mössbauer investigation up to 95 GPa at 5 K (Xu et al. 2016). A high magnetic ordering temperature of nearly 50 K was also observed with pressure.

The coexistence of hexagonal and rhombohedral polytypes in $AgFeO_2$ and the transformation of collinear sinusoidal spin density wave below 16 K has been reported (Siedliska et al. 2017). The Mössbauer spectra were shown to exhibit doublet above 16 K while the noncollinear magnetic structure occurs at temperatures below 8 K (Fig. 9.10). Al substitution induced non-collinear transformation from the collinear commensurate magnetic arrangement observed through asymmetric line intensity and relaxation broadening at 4.2 K without an applied field (Choi et al. 2008). The Mössbauer spectrum exhibits paramagnetic doublet at 18 K, sextet from collinear structure at 10 K and relaxation spectrum from incommensurate structure above 10 K for $CuFeO_2$. The B_{hf} of the four ferric sublattices are 48.5, 46.3, 42.3 and 38.7 T

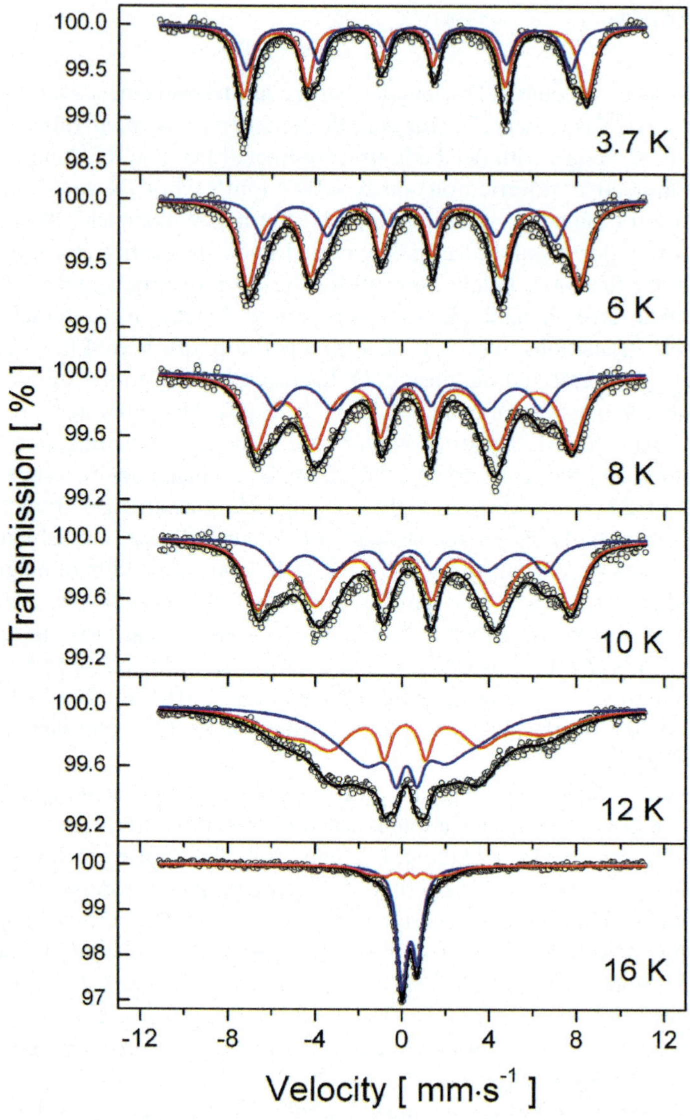

Fig. 9.10 Mössbauer spectra of $AgFeO_2$ powders at various temperatures (Siedliska et al. 2017)

at 12 K that show broadened linewidth for $CuFeO_2$ whereas Al substitution induces incommensurate broadened sextets at 4.2 K itself.

The complex non-collinear structure has been observed from the 4.2 K spectrum with Ga (0.035 in Fe) substituted $CuFeO_2$ single crystals (Nakamura et al. 2014). The magnetic spins are distributed in 3 or 12 directions in the (110) plane. The Δ and B_{hf} values at 4.2 K without substitution are 0.62 mm/s and 51.5 T, respectively, whereas these values are reduced for the three sextets in the Ga substituted sample. While the Δ value reduced uniformly to 0.53 mm/s for all the three sextets, the B_{hf} values are 51.3, 49.7 and 49.0 T from spin directions at an angle of 14°, 26° and 73° to the *c*-axis, respectively.

The Mössbauer measurements at 4.2 K in a longitudinal applied field of 5 T parallel to the *c*-axis of the $CuFe_2O_4$ crystal showed two collinear sextets ($\theta = 0$, 180°) of almost equal intensities in accordance to the 4SL (↑↑↓↓) phase (Nakamura et al. 2015). The 5SL phase was evident at an applied field of 14 T, resulting in collinear up spin and down spin sextets of effective fields 65.4 and 37.1 T, respectively, at 4.2 K. The intermediate field ranges showed an incommensurate phase for $CuFeO_2$. Ga substituted $CuFeO_2$ showed an incommensurate phase up to 14 T, beyond which it was similar to 5SL $CuFeO_2$.

9.8 Summary

The structure-dependent Mössbauer parameters for antiferromagnetic oxides are documented. The phase identification of antiferromagnetic materials from Mössbauer spectrometry is discussed. The database of hyperfine parameters of relevant antiferromagnetic materials is useful while analyzing the Mössbauer spectrum.

References

C.A. Barrero, J.D. Betancur, J.-M. Greneche, G.F. Goya, T.S. Berquo, Geophys. J. Int. **164**, 331 (2006a)

C.A. Barrero, K.E. Garcia, A.L. Morales, S. Kodjikian, J.-M. Greneche, J. Phys. Condens. Matter **18**, 6827 (2006b)

J.M. Bigam, L. Carlson, E. Murad, Miner. Magaz. **58**, 641 (1994)

N. Blum, A.J. Freeman, J.W. Shaner, L. Grodzins, J. Appl. Phys. **36**, 1169 (1965)

S. Bocquet, Phys. Rev. B **37**, 7840 (1988)

S. Bocquet, R.J. Pollard, J.D. Cashion, Phys. Rev. B **46**, 11657 (1992)

S. Bocquet, S.J. Kennedy, J. Magn. Magn. Mater. **109**, 260 (1992)

F. Bødker, M.K. Hansen, C.B. Koch, K. Lefmann, S. Mørup, Phys. Rev. B **61**, 6826 (2000)

F. Boersma, D.M. Cooper, W.J.M. de Jonge, D.P.E. Dickson, C.E. Johnson, A.M.C. Tinus, J. Phys. C: Solid State Phys. **15**, 4141 (1982)

E. Brok, C. Frandsen, D.E. Madsen, H. Jacobsen, J.O. Birk, K. Lefmann, J. Bendix, K.S. Pedersen, C.B. Boothroyd, A.A. Berhe, G.G. Simeoni, S. Mørup, J. Phys, D. Appl, Phys. **47**, 365003 (2014)

E. Brok, K. Lefmann, G.J. Nilsen, M. Kure, C. Frandsen, Phys. Rev. B **96**, 104426 (2017)

Y. Calage, M.C. Moron, J.L. Fourquet, F. Palacio, J. Magn. Magn. Mater. **98**, 79 (1991)

Y. Calage, W.M. Reiff, J. Solid State Chem. **111**, 294 (1994)

D. Chambaere, E. de Grave, J. Magn. Magn. Mater. **42**, 263 (1984)

C.W. Childs, B.A. Goodman, E. Paterson, F.W.D. Woodhams, Aust. J. Chem. **33**, 15 (1980)

D.H. Choi, I.B. Shim, C.S. Kim, J. Magn. Magn. Mater. **320**, e575 (2008)

G. Cinader, P.J. Flanders, S. Shtrikman, Phys. Rev. **162**, 419 (1967)

J.M.D. Coey, A. Barry, J.M. Brotto, H. Rakoto, S. Brennan, W.N. Mussel, A. Collomb, D. Fruchart, J. Phys. Condens. Matter **7**, 759 (1995)
J.M.D. Coey, J. Phys. Condens. Matter **5**, 7297 (1993)
J.M.D. Coey, P.W. Readman, Earth and Planet. Sci. Lett. **21**, 45 (1973)
G. Constabaris, R.H. Lindquist, W. Kündig, Appl. Phys. Lett. **7**, 59 (1965)
D.M. Cooper, G.P. Gupta, D.P.E. Dickson, C.E. Johnson, J. Phys. C: Solid State Phys. **15**, 3391 (1982)
D.L. Cortie, G.L. Causer, K.C. Rule, H. Fritzsche, W. Kreuzpaintner, F. Klose, Adv. Func. Mater. **30**, 1901414 (2019)
A.H. Cuttler, V. Man, T.E. Cranshaw, G. Longworth, Clay Miner. **25**, 289 (1990)
G.R. Davidson, M. Eibschütz, H.J. Guggenheim, Phys. Rev. B **8**, 1864 (1973)
E. de Grave, G.M. da Costa, L.H. Bowen, U. Schwertmann, R.E. Vandenberghe, Clays Clay Miner. **44**, 214 (1996)
E. de Grave, R.E. Vandenberghe, Hyperfine Interact. **28**, 643 (1986)
E. de Grave, R.M. Persoons, D.G. Chambaere, R.E. Vandenberghe, L.H. Bowen, Phys. Chem. Miner. **13**, 61 (1986)
I. Dezsi, L. Keszthelyi, D. Kulgawcsuk, B. Mclnar, N.A. Eissa, Phys. Status Solidi **22**, 617 (1967)
B.J. Dikken, A.F.M. Arts, H.W. de Wijn, W.A.H.M. Vlak, E. Frikkee, Phys. Rev. B **32**, 5600 (1985)
M. Eibschütz, G.R. Davidson, D.E. Cox, AIP Conf. Proc. **18**, 386 (1974a)
M. Eibschütz, G.R. Davidson, H.J. Guggenheim, Phys. Rev. B **9**, 3885 (1974b)
M. Eibschütz, H.J. Guggenheim, L. Holmes, J. Appl. Phys. **42**, 1485 (1971)
M. Eibschütz, H.J. Guggenheim, L. Holmes, J.L. Bernstein, Solid State Commun. **11**, 457 (1972)
S.M. Eicher, J.E. Greedan, K.J. Lushington, J. Solid State Chem. **62**, 220 (1986)
R. Fatehally, G.K. Shenoy, N.P. Sastry, R. Nagarajan, Phys. Lett. A **25**, 453 (1967)
G.A. Fatseas, M. Evain, G. Ouvrard, R. Brec, M.H. Whangbo, Phys. Rev. B **35**, 3082 (1987)
K. Fendler, G.V. Eynatten, Z. Phys, B. Conden, Matter **54**, 313 (1984)
J.B. Forsyth, I.G. Hedley, C.E. Johnson, J. Phys. C: Solid State Phys. **1**, 179 (1968)
J.L. Fourquet, F. Plet, Y. Calage, J. Solid State Chem. **74**, 34 (1988)
J.L. Fourquet, R. de Pape, J. Teillet, F. Varret, G.C. Papaefthymiou, J. Magn. Magn. Mater. **27**, 209 (1982)
C. Frandsen, S. Mørup, Phys. Rev. Lett. **94**, 027202 (2005)
S. Freier, M. Greenshpan, P. Hillman, H. Shechter, Phys. Lett. **2**, 191 (1962)
J.R. Gancedo, M.L. Martínez, J.M. Oton, J. Phys. Colloques **37**, C6–297 (1976)
J.M.R. Genin, S.J. Mills, A.G. Christy, O. Guerin, A.J. Herbillon, E. Kuzmann, G.O. Nguema, C. Ruby, C. Upadhyay, Miner. Magaz. **78**, 447 (2014)
J.M. Genin, P. Bauer, A.A. Olowe, D. Rezel, Hyperfine Interact. **29**, 1355 (1986)
P. Gilad, M. Greenshpan, P. Hillman, H. Shechter, Phys. Lett. **7**, 239 (1963)
J.B. Goodenough, N. Menyuk, K. Dwight, J.A. Kafalas, Phys. Rev. B **2**, 4640 (1970)
G.P. Gupta, D.P.E. Dickson, C.E. Johnson, B.M. Wanklyn, J. Phys. C: Solid State Phys. **10**, L459 (1977)
G.P. Gupta, D.P.E. Dickson, C.E. Johnson, B.M. Wanklyn, J. Phys. C: Solid State Phys. **11**, 3889 (1978a)
G.P. Gupta, D.P.E. Dickson, C.E. Johnson, J. Phys. C: Solid State Phys. **11**, 215 (1978b)
Y. Guyodo, S.K. Banerjee, R.L. Penn, D. Burleson, T.S. Berquo, T. Seda, P. Solheid, Phys. Earth Planet. Inter. **154**, 222 (2006)
M. Haida, K. Kohn, J. Phys. Soc. Jpn. **40**, 981 (1976)
G. Heger, R. Geller, D. Babel, Solid State Commun. **9**, 335 (1971)
G. Heger, R. Geller, Phys. Status Solidi B **53**, 227 (1972)
Y.S. Hiraoka, A. Ito, J. Phys. Soc. Jpn. **53**, 3957 (1984)
A. Ito, S. Morimoto, J. Phys. Soc. Jpn. **39**, 884 (1975)
A. Ito, T. Anma, Hyperfine Interact. **33**, 89 (1987)
P. Jernberg, S. Bjarman, R. Wappling, J. Magn. Magn. Mater. **46**, 178 (1984)
C.E. Johnson, J. Phys. C: Solid State Phys. **2**, 1996 (1969)

J.H. Johnston , N.E. Logan, J. Chem. Soc., Dalton Trans. **13** (1979)
T. Kato, T. Nishino, K. Morishita, H. Tanaka, K. Ito, J. Magn. Magn. Mater. **177–181**, 831 (1998)
T. Kawakami, S. Nasu, T. Tsutsui, T. Sasaki, T. Yamada, S. Endo, M. Takano, T. Katamoto, J. Phys. Soc. Jpn. **72**, 2640 (2003)
H. Keller, I.M. Savic, Phys. Rev. B **28**, 2638 (1983)
O.C. Kistner, A.W. Sunyar, Phys. Rev. Lett. **4**, 412 (1960)
C.B. Koch, C.A. Oxborrow, S. Mørup, M.B. Madsen, A.J. Quinn, J.M.D. Coey, Phys. Chem. Miner. **22**, 333 (1995)
W. Kundig, H. Bommel, G. Constabaris, R.H. Lindquist, Phys. Rev. **142**, 327 (1966)
W. Kundig, J.A. Cape, R.H. Lindquist, G. Constabaris, J. Appl. Phys. **38**, 947 (1967)
C.P. Landee, R.E. Greeney, W.M. Reiff, J.H. Zhang, J. Chalupa, M.A. Novtny, J. Appl. Phys. **57**, 3343 (1985)
A. Leaustic, J.P. Audiere, D. Cointereau, R. Clement, L. Lomas, F. Varret, H.C. Machado, Chem. Mater. **8**, 1954 (1996)
M.E. Lines, M. Eibschütz, Phys. Rev. B **11**, 4583 (1975)
T.T.A. Lummen, C. Strohm, H. Rakoto, P.H.M. van Loosdrecht, Phys. Rev. B **81**, 224420 (2010)
D.E. Madsen, L.C. Gontard, T. Kasama, R.E.D. Borkowski, C.B. Koch, M.F. Hansen, C. Frandsen, S. Mørup, J. Phys. Condens. Matter **21**, 016007 (2009)
C.A. McCammon, A. Pringe, H. Keppler, T. Sharp, Phys. Chem. Miner. **22**, 11 (1995)
C.A. McCammon, E. de Grave, A. Pring, J. Magn. Magn. Mater. **152**, 33 (1996)
M. Mekata, N. Yaguchi, T. Takagi, T. Sugino, S. Mitsuda, H. Yoshizawa, N. Hosoito, T. Shinjo, J. Phys. Soc. Jpn. **62**, 4474 (1993)
F.M. Michel, L. Ehm, S.M. Antao, P.L. Lee, P.J. Chupas, G. Liu, D.R. Strongin, M.A.A. Schoonen, B.L. Phillips, J.B. Parise, Science **316**, 1726 (2007)
H. Miyamoto, Mater. Res. Bull. **11**, 329 (1976)
R.V. Morris, H.V. Lauer Jr., C.A. Lawson, E.K. Gibson Jr., G.A. Nace, C.Stewart, J. Geophys. Res. **90**, 3126 (1985)
A.H. Muir, H. Wiedersich, J. Phys. Chem. Solids **28**, 65 (1967)
E. Murad, Clay Miner. **14**, 273 (1979)
E. Murad, Phys. Chem. Miner. **23**, 248 (1996)
E. Murad, R.M. Taylor, Clay Miner. **19**, 77 (1984)
E. Murad, U. Schwertmann, Miner. Magazine **48**, 507 (1984)
E. Murad, U. Schwertmann, Am. Mineral. **65**, 1044 (1980)
S. Nakamura, A. Fuwa, N. Terada, J. Phys. Soc. Jpn. **83**, 044701 (2014)
S. Nakamura, Y. Kobayashi, S. Kitao, M. Seto, A. Fuwa, N. Terada, J. Phys. Soc. Jpn. **84**, 024719 (2015)
T. Nakamura, T. Shinjo, Y. Endoh, N. Yamamoto, M. Shiga, Y. Nakamura, Phys. Lett. **12**, 178 (1964)
Y. Nishiwaki, A. Oosawa, K. Kakurai, K. Kaneko, M. Tokunaga, T. Kato, J. Phys. Soc. Jpn. **80**, 084711 (2011)
N.V.T. Oganessian, A.A. Guda, V.P. Sakhnenko, Sci. Rep. **7**, 16410 (2017)
A.A. Olowe, P. Bauer, J.M.R. Genin, J. Guezennec, Corrosion **45**, 229 (1989)
K. Ono, A. Ito, J. Phys. Soc. Jpn. **17**, 1012 (1962)
O. Ozdemir, D.J. Dunlop, T.S. Berquo, Geochem. Geophy. Geosyst. **9**, 1 (2008)
Q.A. Pankhurst, C.E. Johnson, M.F. Thomas, J. Phys. C: Solid Sate Phys. **19**, 7081 (1986)
Q.A. Pankhurst, L.F. Barquin, J.S. Lord, A. Amato, U. Zimmermann, Phys. Rev. B **85**, 174437 (2012)
Q.A. Pankhurst, R.J. Pollard, J. Phys. Condens. Matter **2**, 7329 (1990)
Q.A. Pankhurst, R.J. Pollard, J. Phys. Condens. Matter **5**, 7301 (1993)
M.P. Pasternak, A.P. Milner, G.K. Rozenberg, R.D. Taylor, R. Jeanloz, Phys. Rev. Lett. **92**, 085506 (2004)
M.P. Pasternak, G.K. Rozenberg, G.Y. Machavariani, O. Naaman, R.D. Taylor, R. Jeanloz, Phys. Rev. Lett. **23**, 4663 (1999)

R.J. Pollard, C.M. Cardile, D.G. Lewis, L.J. Brown, Clay Miner. **27**, 57 (1992)
R.J. Pollard, Q.A. Pankhurst, J. Phys. Condens. Matter **4**, L317 (1992)
J. Renaudin, J. Pannetier, S. Pelaud, A. Ducouret, F. Varret, G. Ferey, Solid State Commun. **47**, 445 (1983)
M.J. Rossiter, A.E.M. Hodgson, J. Inorg. Nucl. Chem. **27**, 63 (1965)
T. Ruskov, T. Tomov, S. Georgiev, Phys. Status Solidi A **37**, 295 (1976)
M.S. Seehra, V.S. Babu, A. Manivannan, J.W. Lynn, Phys. Rev. B **61**, 3513 (2000)
T. Shinjo, J. Phys. Soc. Jpn. **21**, 917 (1966)
T. Shinjo, M. Kiyama, N. Sugita, K. Watanabe, T. Takada, J. Magn. Magn. Mater. **35**, 133 (1983)
Siedliska K, T. Pikula, Z. Surowiec, D. Chocyk, E. Jartych, J. Alloys Compd. **690**, 182 (2017)
D.J. Simkin, R.A. Bernheim, Phys. Rev. **153**, 627 (1967)
Y. Someya, A. Ito, J. Phys. Soc. Jpn. **50**, 1891 (1981)
Y. Someya, A. Ito, S. Morimoto, J. Phys. Soc. Jpn. **50**, 1883 (1981)
J.K. Srivastava, R.P. Sharma, Phys. Status Solidi B **49**, 135 (1972)
Y. Syono, A. Ito, S. Morimoto, T. Suzuki, T. Yagi, S. Akimoto, Solid Sate Commun. **50**, 97 (1984)
L. Takacs, W.M. Reiff, J. Appl. Phys. **61**, 3667 (1987)
T. Takada, M. Kiyama, Y. Bando, T. Nakamura, M. Shiga, T. Shinjo, N. Yamamoto, Y. Endoh, H. Takaki, J. Phys. Soc. Jpn. **19**, 1744 (1964)
M. Takano, T. Takada, Mater. Res. Bull. **5**, 449 (1970)
B.E. Taylor, J. Steger, A. Wold, J. Solid State Chem. **7**, 461 (1973)
B. Taylor, J. Steger, A. Wold, E. Kostiner, Inorg. Chem. **13**, 2719 (1974)
M.P.H. Thurlings, E. Frikkee, H.W. de Wijn, Phys. Rev. B **25**, 4750 (1982)
M.P.H. Thurlings, E.M. Hendriks, H.W. de Wijn, Solid State Commun. **27**, 1289 (1978)
L. Tobler, W. Kundig, I. Savic, Hyperfine Interact. **10**, 1017 (1981)
M.G. Townsend, G. Longworth, J.E. Dutrizac, Phys. Rev. B **31**, 2966 (1985)
S. Usman, J.M. Byrne, A. Chaudhary, S. Orsetti, K. Hanna, C. Ruby, A. Kappler, S.B. Haderlein, Chem. Rev. **118**, 3251 (2018)
A.M. van der Kraan, Phys. Status Solidi A **18**, 215 (1973)
F. van der Woude, A.J. Dekker, Phys. Status Solidi **13**, 181 (1966)
F. van der Woude, Phys. Status Solidi **17**, 417 (1966)
F. Varret, P. Imbert, Phys. Status Solidi B **56**, 127 (1973)
R.W. Vaughan, H.G. Drickamer, J. Chem. Phys. **47**, 1530 (1967)
X. Wang, M. Zhu, L.K. Koopal, W. Li, W. Xu, F. Liu, J. Zhang, Q. Liu, X. Feng, D.L. Sparks, Environ. Sci. NANO **3**, 190 (2016)
G.K. Wertheim, H.J. Guggenheim, H.J. Levinstein, D.N.E. Buchanan, R.C. Sherwood, Phys. Rev. **173**, 614 (1968a)
G.K. Wertheim, H.J. Guggenheim, H.J. Williams, D.N.E. Buchanan, Phys. Rev. **158**, 446 (1967)
G.K. Wertheim, H.J. Guggenheim, H.J. Williams, D.N.E. Buchanan, J. Appl. Phys. **39**, 1253 (1968b)
W.M. Xu, G.R. Harne, M.P. Pasternak, Phys. Rev. B **94**, 035155 (2016)
W.M. Xu, M.P. Pasternak, R.D. Taylor, Phys. Rev. B **69**, 052401 (2004)
A. Yamamoto, T. Honmyo, M. Kiyama, T. Shinjo, J. Phys. Soc. Jpn. **63**, 176 (1994)
C. Yasuda, S. Todo, K. Hukushima, F. Alet, M. Keller, M. Troyer, H. Takayama, Phys. Rev. Lett. **94**, 217201 (2005)
L.I. Zawislak, G.L.F. Fraga, J.B.M. da Cunha, D. Schmitt, A.S. Carrico, C.A. dos Santos, J. Phys. Condens. Matter **9**, 2295 (1997)
L.I. Zawislak, J.B.M. da Cunha, A. Vasquez, C.A. dos Santos, Solid State Commun. **94**, 345 (1995)

Chapter 10
Mössbauer Spectrometry in Earth Sciences

10.1 Origin of Earth

The gravitational collapse of molecular clouds results in a protostar that develops into a star, and planets form from the surrounding dust grains in the star's accretion disc. Our Solar System was developed 4.5 Ga (1 Ga = 1 billion year) ago from the dust grains collected to form planetesimals, gradually growing into the planet Earth around the Sun. The leftover elements out of nucleosynthesis in the star-forming regions of molecular clouds were part of the accretion disc that later became planetary constituents. The collision and assemblage of the planetesimals generate heat in terrestrial planets, known as primordial heat, and the later cooling results in the condensation, crystallization and segregation of elements from minerals of Earth. The primary heat source in terrestrial planets is radiogenic heating through radioactive decay. The core of the Earth has 85% Fe, 5% Ni as its major constituents by weight, and other minor elements such as Si, S and $< 1\%$ O that are at a temperature (~6000 K), close to the Sun's surface temperature. The high pressure at the Earth's center increases the melting point of the Fe–Ni, resulting in a solid inner core and liquid outer core.

The elements found on the Earth are cosmic remnants formed from stellar nucleosynthesis. Hydrogen, followed by helium, are the most abundant elements in the universe, arising from the big bang. Elements up to Fe and Ni are formed through nuclear fusion, while neutron capture results in much heavier elements through stellar nucleosynthesis. The star-planet systems are born out of these remnants first through the condensation of elements with high melting points. Although refractory metals such as W, Re, Ta and Zr have high melting points, they are less abundant than the more stable Fe and Ni. The next condensates are metal oxides with a higher melting point, such as silicates that have undergone chemical modification. Zircon is found to be the oldest mineral on Earth, with an age of nearly 4.4 Ga. The elements that are part of the Earth might have undergone a lot of changes during its geological history, whereas the state of early minerals could be obtained from chondrites, which

R. J. Joseyphus and J.-M. Greneche, *Fundamentals of ^{57}Fe Mössbauer Spectrometry*,
https://doi.org/10.1007/978-981-99-8653-8_10

are meteorite remnants of the early Solar System. The study of minerals found in the Earth provides information about the history of the Earth and the Solar System. Considering the presence of numerous Sun-like stars in the universe and the discovery of innumerable exoplanets, it is expected that Earth-like habitable planets may be possible, although remote.

10.2 Geological History of Earth

The geological history of the Earth can be divided into four eons: Hadean, Archean, Proterozoic and Phanerozoic, marked by significant mineralization events as shown in Fig. 10.1. During the formation of the Earth, there were less than 250 minerals similar to that found in chondritic meteorites and increased to 500 at the end of Hadean Eon, while more than 4000 natural minerals were recognized after the great oxidation event (Hazen et al. 2008). Earth's mineral evolution is shaped by planetary accretion (>4.55 Ga), crust-mantle reworking (4.55–2.5 Ga) and bio-mediated mineralogy (>2.55 Ga). Apart from Fe–Ni, some of the Fe-bearing minerals that appeared during the early age are silicates such as ringwoodite, majorite, bridgmanite, pyroxene, orthoclase, plagioclase and wadsleyite that are found in Earth interiors. Subsequently, ankerite, siderite and iron ores, as banded iron formation (BIF), occurred before 2.5 Ga. After the formation of water, the great oxidation event, and the progression of plants and animals, it took nearly 4 billion years for the evolution of modern humans. The reducing atmosphere of the Earth was composed of H_2, CH_4 and NH_3 that transformed to CO_2, N_2 and H_2O due to volcanic activity, followed by lower CO_2 levels that replaced with O_2 (Kasting 1993).

10.3 Internal Structure of Earth

Understanding the minerals of the Earth requires us to know more about the structure of the Earth. Earth's interior is divided into layers known as the crust, upper mantle, lower mantle, outer core and inner core up to depths of 50, 410, 2890, 5150 and 6370 km, respectively, as shown in Fig. 10.2. The mass of the Earth is 5.97×10^{24} kg, and from the volume, the mean density is obtained as 5.5 g/cc. The mean surface temperature is 287 K, which increases drastically as we go deep into the Earth with temperatures of nearly 1800, 3500 and 5000 K as we reach the bottom layers of the upper mantle, lower mantle and outer core, respectively. The deepest drill hole (~12 km) could not reach even half the crustal depth for us to probe crustal minerals on site. The density and the minerals present inside the Earth could be estimated from the seismic primary and secondary (P- and S-) waves as they travel at different velocities according to the materials.

The variation in the density and pressure at various depths and the boundaries of various layers of the Earth are shown in Fig. 10.3. The 410–660 km region is marked

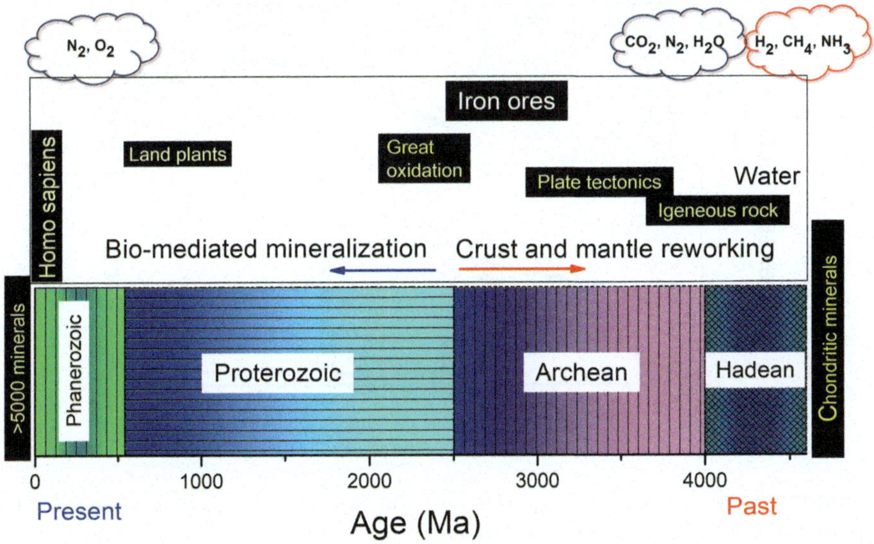

Fig. 10.1 The geological eons and the evolution of Earth minerals are influenced by various events (1 Ma = 1 million year). The early minerals are associated with a composition similar to chondrites. With less than 500 minerals up to the Hadean Eon, we now have more than 5000 natural and artificial minerals

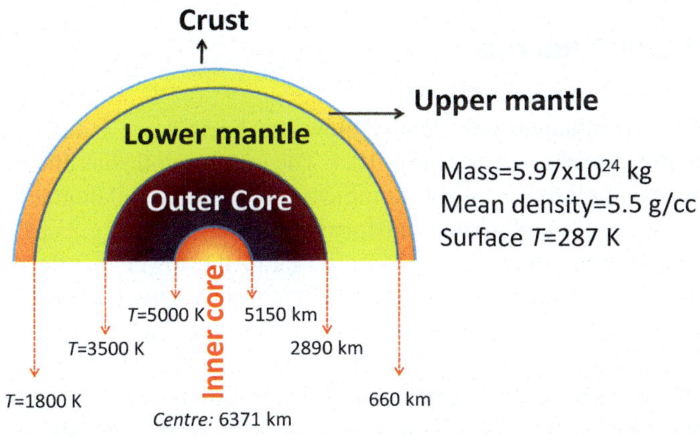

Fig. 10.2 The internal structure of the Earth is composed of various layers with extreme temperature and pressure towards the core. The internal properties of the Earth could be determined from seismology and mineral physics

by the transition zone. The region above the outer core has a 200 km thick D″ layer. The pressure, density and temperature increase with depth reaching a maximum close to 364 GPa, 13 g/cc and 6300 K, respectively, at the inner core (Anderson 2007).

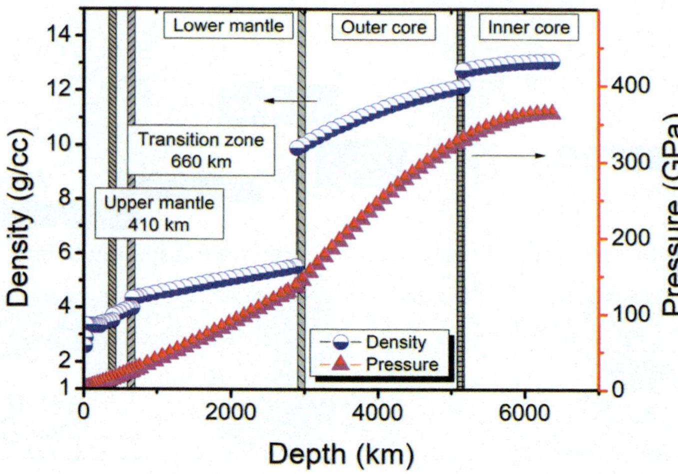

Fig. 10.3 Density and pressure variation of the Earth with depth. The upper mantle, 660 km transition zone, lower mantle, outer core and inner core are marked by density variation

The upper mantle of the Earth has olivine as the major mineral constituent, followed by pyroxene and garnet.

10.4 Mantle Minerals

The mineral constituents with depth below the Earth's crust range from silicate minerals to iron, as represented in Fig. 10.4. The upper mantle has olivine, garnet and pyroxene compounds that are orthorhombic orthopyroxene and monoclinic clinopryoxene. The pyroxenes and garnet transform to the majorite, (Mg, Fe, Ca)(Si, Al)O_3, in the transition zone that further decomposes to the lower mantle minerals. The transformation of olivine to perovskite and magnesiowustite (ferropericlase) occurs in the lower mantle. The outer liquid core and the inner solid core are made of Fe and other light elements. The minerals could be studied in the state as it is formed in the mantle as they could be preserved in natural diamonds. Also, high-pressure, high-temperature experiments on the samples synthesized at the laboratory could be used to predict the material properties and mantle conditions.

The phase transformation of the olivine as a function of depth in the transition zone of the upper mantle is represented in Fig. 10.5. The transition zone boundary has the orthorhombic olivine compound wadsleyite, (Mg, Fe)$_2$SiO$_4$, at a depth (D) of 410 km and pressure (P) of 13–14 GPa that transforms to the spinel ringwoodite, (Mg, Fe)$_2$SiO$_4$, ($P = 18$ GPa, $D = 510$ km) (Helffrich and Wood 2001). The ringwoodite further decomposes to the perovskite, (Mg, Fe)SiO$_3$, and magnesiowustite, (Mg,

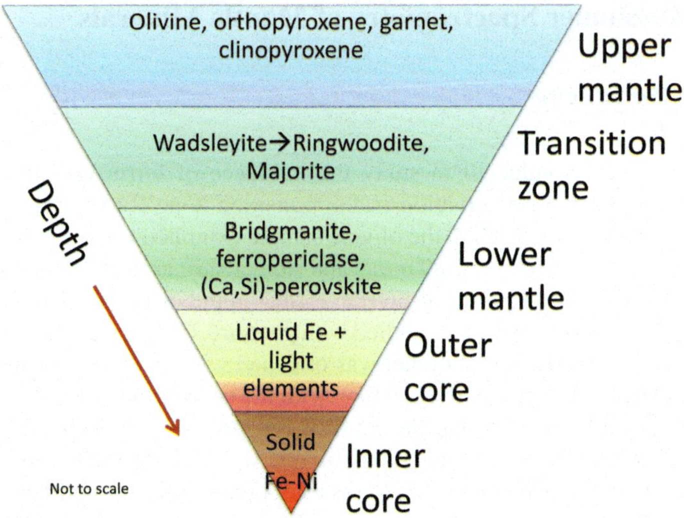

Fig. 10.4 The constituents of the Earth below the crust ranges from silicate minerals to iron, with minor light elements in liquid form at the outer core and solid form at the inner core

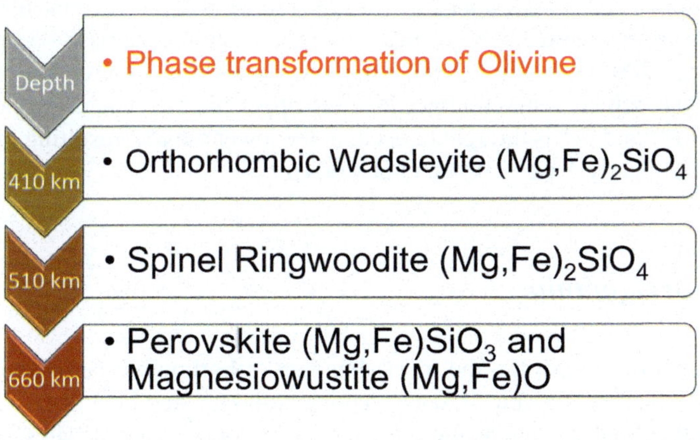

Fig. 10.5 The phase transformation of olivine with depth at the Earth's transition zone

Fe)O, at $P = 23$ GPa and $D = 660$ km, forming the lower transition zone boundary. Orthorhombic perovskite, (Mg, Fe) SiO_3, constitutes nearly 83 vol.% in the lower mantle. The transformation of pyroxene and garnet, $(Mg, Fe, Ca)_3(Si, Al)_2Si_3O_{12}$, is gradual with depth. Pyroxene transforms to garnet at $D = 350$–500 km, and at a depth of 580 km, separation of $CaSiO_3$ perovskite occurs along with the coexistence of garnet. The garnet phase dissolves into the (Mg, Fe) perovskite to form a high Al perovskite phase at 660–750 km depth, with the coexistence of $CaSiO_3$.

10.5 Mössbauer Spectrometry of Mantle Minerals

10.5.1 Wadsleyite

Experiments were conducted to study the influence of ferric ions in the olivine to wadsleyite phase transformation in the transition zone (Frost and McCammon 2009). At a pressure of 12 GPa, the olivine sample exhibited two Fe^{2+} doublets with quadrupole splitting 3.08 and 2.89 mm/s and isomer shift around 1.15 mm/s for both the doublets. The starting ratio of Fe/(Fe + Mg) in the olivine was varied, and the transformation to wadsleyite was studied with pressure. For the wadsleyite at a higher pressure of 15 GPa, the Fe^{3+} doublet was obtained with quadrupole splitting and δ of 0.94 and 0.28 mm/s, respectively, along with the two ferrous doublets. The Mössbauer experiments revealed that the ratio of Fe^{3+}/total-Fe in wadsleyite increased from 0 to 24 when Fe/(Fe + Mg) ratio in olivine was varied from 0.1 to 0.4, respectively. The results showed that ferric ion does not prefer olivine, while the minimum $Fe^{3+}/\Sigma Fe$ ratio in wadsleyite is suggested as 0.03. The ratio of the $Fe^{3+}/\Sigma Fe$ can be used to explain the depth variation of the 410 km transition zone boundary.

The oxygen fugacity of the upper mantle is high, whereas the olivine mineral does not have Fe^{3+} in it. The lower mantle compound bridgmanite has a higher concentration of Fe^{3+} ions than Fe^{2+}. Studies have also shown that the oxidation state of the upper mantle has remained almost constant for the past ~ 5 billion years (McCammon 2005). The $Fe^{3+}/\Sigma Fe$ in the lower mantle is suggested to have a ratio of at least 0.6, which is much larger than the upper mantle. The disproportionation of Fe^{2+} to Fe^{3+} and Fe^0 is thought to occur at the lower mantle, resulting in a metallic Fe content of nearly 1 wt.% (Frost et al. 2004).

10.5.2 Bridgmanite

In the Earth's lower mantle, the most abundant orthorhombic perovskite can exist as bridgmanite, (Mg, Fe)(Fe,Al,Si)O_3, having both Fe^{2+} and Fe^{3+} states. The Fe^{2+} is located in the A-site, whereas the Fe^{3+} can be in both A- and B- sites. While Fe^{3+} in the A-site remains in the high spin state with pressure, studies also suggested that the B-site ferric ion can transform to the low spin state. It was also shown that Fe^{2+} undergoes high spin to low spin transformation as one moves toward the lower mantle (McCammon et al. 2013).

High-pressure synchrotron Mössbauer spectrometry (SMS) experiments were conducted on bridgmanite ($Mg_{0.89}Fe_{0.12}Al_{0.11}Si_{0.89}O_3$) single crystals at 115 GPa and laser annealed at ~ 1800 K. The Mössbauer parameters deduced from SMS are listed in Table 10.1. (Lin et al. 2016). The experiments showed that the ferric ion to total Fe ratio remains the same before and after laser annealing. However, the high spin to low spin transformation of Fe^{3+} is not significantly observed under lower mantle conditions. Also, there is no migration of Fe^{3+} to the A-site. Under ambient

Table 10.1 The SMS hyperfine parameters of bridgmanite at ambient conditions and laser-annealed sample under a pressure of 115 GPa (Lin et al. 2016)

SMS at ambient conditions	SMS at 115 GPa and laser annealed
$\Delta = 0.91$ mm/s, $\delta = 0.45$ mm/s 80% Fe^{3+} in A-site	$\Delta = 0.84$ mm/s, $\delta < 0.6$ mm/s 80% Fe^{3+} in A-site
$\Delta = 1.96$ mm/s, $\delta = 1.14$ mm/s 20% Fe^{2+} in A-site	$\Delta = 1.81$ mm/s, $\delta > 1$ mm/s 14% Fe^{2+} in A-site
–	$\Delta = 3.21$ mm/s, $\delta > 1$ mm/s 6% Fe^{2+} in A-site

conditions, 20% Fe^{2+} is in the A-site, whereas under high pressure and temperature conditions, 6% undergoes a large quadrupole splitting of 3.21 mm/s due to lattice distortions. The charge balance in the lattice is achieved through the charge-coupled substitution ($Mg^{2+} + Si^{4+} = Al^{3+} + Fe^{3+}$) with Fe^{3+} in the A-site and Al^{3+} in the B-site accompanied by 11% Si^{4+} deficiency.

High-pressure SMS studies were carried out up to 85–130 GPa with Al-deficient $Mg_{0.94}Fe_{0.06}Al_{0.01}Si_{0.99}O_3$ and Al-rich $Mg_{0.89}Fe_{0.12}Al_{0.11}Si_{0.89}O_3$ bridgmanite samples (Mao et al. 2017). The quadrupole splitting of the Fe^{2+} doublets in Al deficient bridgmanite increased drastically from ~1.5 to > 3.6 mm/s between 13 and 32 GPa and increased further to 4 mm/s up to 85 GPa. Such drastic variation in Δ with pressure was not observed for the Al-rich bridgmanite up to 130 GPa. The synchrotron-based X-ray diffraction and SMS studies, along with other experimental signatures, show that A-site Fe^{2+} remains in the high spin state in the lower mantle. The high quadrupole splitting of 3.6–4.0 mm/s for the A-site Fe^{2+} is due to the changes in bond length, tilting angle of the octahedral and spontaneous shear strain without significant changes in the unit cell volume with pressure.

10.5.3 Silicate Phases

The olivine, $(Mg_xFe_{1-x})_2SiO_4$, has two non-equivalent octahedral sites M1 and M2. Figure 10.6 shows the Mössbauer spectra of synthetic olivine (top) and natural olivine (bottom) at 583 K (Shinno 1981). The synthetic olivine was subjected to annealing at a temperature of 1673 K. The isomer shift relative to metallic Fe of both the samples is 0.9 mm/s from the ferrous contribution while the ferric isomer shift is 0.04 mm/s for the natural olivine with a relative area of 3.4%. The ferrous component quadrupole splitting is 2–2.55 mm/s while that of ferric is 0.82 mm/s. The analysis suggested preference of ferrous ion in M1 site. Mössbauer studies were undertaken with γ-Fe_2SiO_4 spinel, Fe incorporated γ-$(Mg, Fe)_2SiO_4$ spinel, β-$(Mg, Fe)_2SiO_4$, $(Mg, Fe)SiO_3$ garnet and $(Mg, Fe)SiO_3$ clinopyroxene phases synthesized from olivine and pyroxene starting materials at high pressure and temperature (McCammon 1994; O'Beukk et al. 1993). The starting material was devoid of any ferric ions. The oxygen fugacity was controlled by buffering with excess SiO_2 and metallic Fe. The fitted

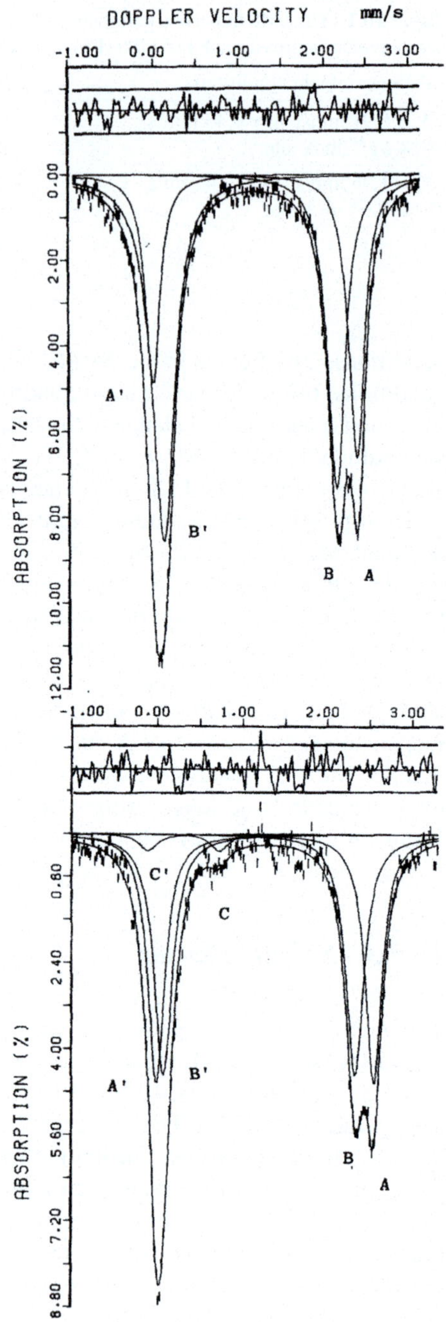

Fig. 10.6 The Mössbauer spectra of synthetic olivine (top) and natural olivine (bottom) at 583 K. The doublets A-A', B-B' are from ferrous ions while C–C' is ferric ion (Shinno 1981)

spectrum corresponds to the Fe^{3+} site. Metallic Fe could be seen in the spectrum of β and garnet phases. The upper mantle phase (Mg, Fe)SiO_3 clinopyroxene did not exhibit Fe^{3+}, whereas the $Fe^{3+}/\Sigma Fe$ was found to be 0.07 for the garnet phase. This suggested that the high-pressure phases were incorporated with Fe^{3+} even under lower oxygen fugacities.

10.5.4 Natural Mantle Minerals

Majorite and garnet inclusions in xenoliths, possibly originating from below 400 km depth, were separated, and Mössbauer spectrometry was undertaken under ambient conditions (Xu et al. 2017). The natural majorite was Al deficient compared with the garnet and synthetic majorite. The fitted spectrum exhibited doublet with Δ and δ values 3.57 and 1.28 mm/s, respectively, corresponding to Fe^{2+}. Another doublet with Δ and δ values of 1.28 and 0.39 mm/s, respectively, was assigned the Fe^{3+}. The $Fe^{3+}/\Sigma Fe$ was 0.81, and a high value is expected for the majorite synthesized under oxidizing conditions. The natural majorite was suggested to be from the deep upper mantle (~380 km), rapidly transported into a cold environment encapsulated by garnet overgrowth at a much shallow depth of 200 km. It was suggested that the higher Fe^{3+} may be due to a redox reaction involving carbonatitic magmas.

SMS of the Fe garnet phase preserved in diamond was undertaken in another study (Kisseva et al. 2018). The $Fe^{3+}/\Sigma Fe$ ratio determined from SMS suggested increasing value up to 0.3 with depth corresponding to the transition zone. This ratio is more than double for the garnets from shallow depths of ~200 km.

10.5.5 Ferropericlase

$Mg_{1-x}Fe_xO$ is a main mineral in the lower mantle and exists as a Mg-rich ferropericlase with a cubic rock salt structure. It has ferrous ions in the high spin state at ambient conditions. The Néel temperature decreases with Mg substitution from 190 K at ambient pressure for $Fe_{0.97}O$ to 25 K for $Mg_{0.8}Fe_{0.2}O$ (Speziale et al. 2005). The high spin to low spin transformation temperature decreases from 90 to 40 GPa without and with 80% Mg, respectively. As the lower mantle Mg-rich composition is at a much larger pressure, it should be in the low spin state.

Mössbauer studies on synthetic $Mg_{0.8}Fe_{0.2}O$ powders were performed at high pressure and compared with X-ray emission spectroscopy (XES) (Yu et al. 2006). The quadrupole splitting and isomer shift parameters of the doublet corresponding to high spin Fe^{2+} ($S = 2$) was found to transform to low spin Fe^{2+} state ($S = 0$) beyond 50 GPa. The signatures of low spin state are negligible quadrupole splitting and lower isomer shift. The complete transformation to the low spin state was reported beyond 100 GPa. While the high spin to low spin transition is well established, the analysis of XES in conjunction with Mössbauer spectrometry requires careful analysis as the transition pressure (~50–70 GPa) (Vanko et al. 2007) may be influenced by the experimental conditions. Indeed, theoretical studies suggest that the ferropericlase in the Earth's lower mantle exists in the low spin state (Leonov et al. 2017).

10.6 Crustal Minerals

The Earth's crust has plagioclase as the major constituent by volume followed by quartz, orthoclase, biotite, pyroxene, muscovite, amphibole, olivine, chlorite and other oxides, including Fe oxides, as represented in Fig. 10.7. High-density minerals such as olivine, pyroxine and amphibole are found deep inside the crust.

Among the various major crustal Fe-bearing rock minerals such as biotite, pyroxene, amphibole, olivine and chlorite, various other rock, clay minerals, and thousands of other minerals also exist. Olivine and pyroxene are also mantle minerals, as discussed in previous sections. Biotite and muscovite are few among the mica group of layered phyllosilicates.

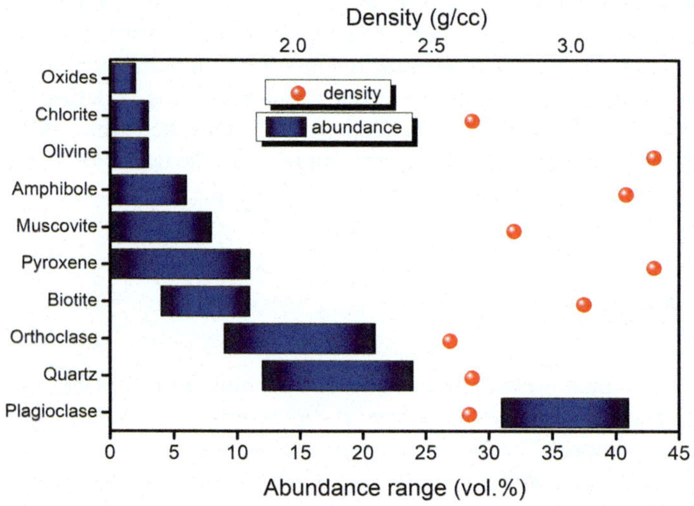

Fig. 10.7 The crustal minerals and their corresponding range of abundance and density. Plagioclase is the most abundant mineral in the Earth's crust in terms of volume. The average density of the crust is 2.7 g/cc

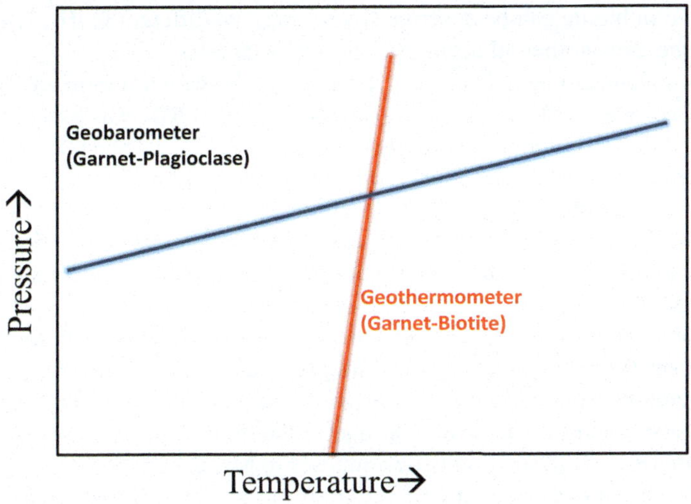

Fig. 10.8 The pressure–temperature (*P–T*) diagram showing geothermometer and geobarometer lines. The geothermometer translates left-right while it the up-down for the geobarometer. The intersection gives *P-T* condition

10.6.1 Geothermobarometer

The pressure–temperature (*P–T*) history of rocks is measured using geothermobarometry with garnet-biotite being one among the geothermometer. The geothermometer and geobarometer lines in the *P–T* diagram are shown in Fig. 10.8. The garnet-biotite geothermometer is based on the measured exchange of Fe^{2+} and Mg^{2+} ions among the minerals with temperature. The equilibrium constant (K_D), which relates to the ratio of Mg and Fe, is a function of temperature. For example, $K_D = 0.1$ and 0.2 fall close to 673 and 873 K, respectively, in the *P–T* diagram of the garnet-biotite geothermometer. The chemical reaction can be represented as.

$$Fe_3Al_2Si_3O_{12} + KMg_3AlSi_3O_{10}(OH)_2 \rightarrow Mg_3Al_2Si_3O_{12} + KFe_3AlSi_3O_{10}(OH)_2.$$

Almandine (garnet) + Philogopite (biotite) → Pyrope (garnet) + Annite (biotite).

The geobarometer, such as GBPQ, is based on the net transfer reactions among garnet, biotite, plagioclase and quartz.

10.6.2 Biotite

Biotite has a 2:1 layered mica group silicate structure where Si and Al occupy the tetrahedral site and elements such as Fe, Mg, Mn, Ti and Al are in the octahedral site. The Fe in biotite can be described as having two different O–H environments, one opposite and another adjacent, in distorted octahedra.

The room temperature Mössbauer spectrum of the natural biotite of the composition $K(Fe_{4.72}Mg_{0.48}Mn_{0.05}Ti_{0.46}Al_{0.02})Si_{5.63}Al_{2.37}O_{20}(OH)_4$ was assigned Fe^{2+}(I), Fe^{2+}(II) and Fe^{3+}(III) doublets with site populations 2.92, 1.05 and 0.75, respectively (Häggström et al. 1969). The isomer shift of Fe^{2+} was 1.13–1.11 mm/s, while that of Fe^{3+} was 0.55 mm/s. The quadrupole splitting value was 2.56–2.17 mm/s for the ferrous sites, whereas the ferric site showed a much smaller value of 0.55 mm/s. The values of isomer shift and quadrupole splitting of natural biotite were similar to the synthetic biotite except for their relative intensities.

The isomer shift increases according to the degree of oxidation of biotite, resulting in values from 0.3 to 0.6 mm/s (sodium nitroprusside standard) as the relative fraction of Fe^{3+} increases from 0.3 to 1.0, respectively (Bagin et al. 1980). This is due to the change in ionic nature with oxidation due to the elimination of O–H. At a higher temperature of 1370 K, the biotite decomposes to MgAl ferrispinel.

The Mössbauer data of natural biotite from various geographical areas suggested nearly 8% of Fe^{3+} among total Fe, irrespective of petrologic factors (Guidotti and Dyar 1991). Under reducing conditions, the ferric ion content was 4%, which coexists with graphite, whereas it could be as large as 46% of the total Fe under oxidizing conditions accompanied by hematite. The changes in Fe^{3+} content could influence the $Mg/(Mg + Fe_{total})$ and that of $Mg/(Mg + Fe^{2+})$ ratio, affecting the thermometry calculations. The appropriate rock type is vital for the study of garnet-biotite geothermometry.

10.6.3 Silicate Garnet

Silicate garnets have the formula $X_3Y_2Si_3O_{12}$, where X and Y are the eightfold (dodecahedral) and sixfold (octahedral) coordinated cations, respectively. The fourfold coordinated (tetrahedral) Si site could be occupied by other ions, such as Fe^{3+}. The Fe^{2+} occupies the 24c dodecahedral site, while the Fe^{3+} occupies the 16a octahedral and 24d tetrahedral sites. The isomer shift and quadrupole splitting values from the various sites of silicate garnets (Amthauer et al. 1976) are shown in Fig. 10.9. While Fe^{2+} avoids the octahedral site, it could occupy the tetrahedral site in Si garnets. It could be observed that the various sites have distinct isomer shift and quadrupole splitting values, with the dodecahedral site as the maximum, due to the exclusive oxygen coordination and valence states. The sign of V_{zz} of Fe^{2+} in the 24c site is negative, almost in the entire temperature range. The hyperfine parameters of the dodecahedral site are also not affected by the chemical composition. On the other hand, the electric field gradient of the octahedral Fe^{3+} is highly sensitive to electronic polarization and orbital overlap of nn anion.

Almandine and pyrope belong to the silicate garnet group $X_3Al_2Si_3O_{12}$, with X being Fe and Mg ions, respectively. Almandine is characterized by Fe^{2+} exclusively in the 24c site. Mössbauer spectra of natural and synthetic almandine showed similar

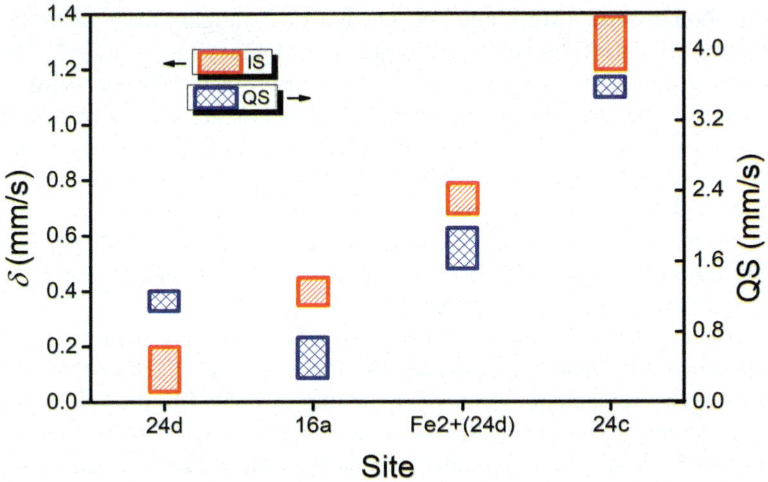

Fig. 10.9 The isomer shift and quadrupole splitting value ranges for the various sites of silicate garnets reported in the literature (Amthauer et al. 1976)

features with room temperature quadrupole splitting and isomer shift values close to 3.5 and 1.3 mm/s, respectively, and Néel temperature (T_N) below 10 K (Murad et al. 1987). In an almandine-pyrope solid solution, Fe or Mg ions occupy the dodecahedral site randomly. While the hyperfine parameters of the dodecahedral site are invariant irrespective of Fe^{2+} or Mg^{2+}, the T_N decreases with Mg^{2+} substitution (de Oliveira et al. 1989). Therefore, the T_N decreases from 10 K for almandine to < 2 K with 66% pyrope in almandine solid solution.

10.6.4 Pyroxene

Pyroxene minerals include clinopyroxenes and orthopyroxenes. The Fe bearing pyroxenes are aegerine, augite, esseneite, hedenbergite, jadeite, jervisite, omphacite, petundunnite and pigeonite, ferrosilite and hypersthene. According to the pressure and temperature conditions, the pyroxenes exist as monoclinic clinopyroxenes or orthorhombic orthopyroxene.

The distribution of Mg: Fe^{2+} according to Ca content finds its use in pyroxene thermometry. The Ca content of the pyroxene is accompanied by the changes in Mg and Fe^{2+} through exchange reaction as given below (Kretz 1982).

$$MgSiO_3 + CaFeSi_2O_6 = FeSiO_3 + CaMgSi_2O_6.$$

Enstatite + Ca rich pyroxene = Ferrosilite + Ca rich pyroxene

The orthopyroxene, $(Mg,Mn,Fe)_2Si_2O_6$, has Fe^{2+} ranging from 15 to 85% that shows ferrous doublets corresponding to M2 and M1 sites (Bancroft 1967). Orthopyroxene has infinite SiO_3 chains with octahedrally coordinated divalent cations in M1 and M2 sites. The M1 and the distorted M2 sites are surrounded by 1Si-6O and 1Si-4O ions, respectively. The inner and outer doublets of the orthopyroxene have quadrupole splitting values nearly 2.0 and 2.4 mm/s arising from M2 and M1 sites, respectively. The ferrous ions prefer the M2 site, which results in the increasing intensity of the inner doublet with Fe. The smaller Δ value for the M2 site is consistent with the Fe^{2+} in distorted coordination. The Δ values for both the sites reach similar values for lower Fe concentrations.

The orthopyroxene Fe end member, ferrosilite ($FeSiO_3$), exhibits the largest quadrupole splitting than the enstatites (Victor et al. 2001). The M1 site ferrous ion shows drastic temperature dependent Δ compared with M2 site Fe^{2+}. The strong preference for ferrous ions to the M2 site is due to the ligand field stabilization effects. The Δ of Fe^{2+} in M2 site is not affected much by the chemical composition. The orthopyroxene ferrosilite ($FeSiO_3$) shows the largest T_N of 45 K compared with 38 K for their clinopyroxene form and decreases with Fe content (Eeckhout et al. 2001).

Geothermometers utilizing garnet-clinopyroxene are based on the cation exchange between Mg and Fe^{2+}. However, Fe^{3+} is present in the rock system due to varying oxygen fugacities affecting the K_D in geothermometric estimations. The $Fe^{3+}/\Sigma Fe$ is found to be much more prominent in clinopyroxenes than the garnets (Sobolev et al. 1999). The equilibration temperature increases with $Fe^{3+}/\Sigma Fe$ in garnet, while that of clinopyroxene decreases. Here, the Mössbauer milliprobe analysis is more accurate than the electron microprobe analyses.

10.7 Clay Minerals

Clays are used in petroleum cracking, depolymerization of organic hydrocarbons, pottery, landfills to arrest contamination and building materials. Nontronite and vermiculite are smectite (swelling type) Fe-bearing minerals, whereas the non-swelling type includes berthierine, chlorite and illites (glauconite, celadonite).

Figure 10.10 shows the smectic type phyllosilicate clay minerals structure, which has an octahedral sheet that shares the oxygen atom with two tetrahedral sheets (2:1 layer), with the interlayer cations and water molecules sandwiched between the sheets. The tetrahedral and octahedral cations are represented as T and M, respectively. The unit cell of a 1:1 tetrahedral-octahedral layered type has six octahedral sites and four tetrahedral sites, while the 2:1 type has eight tetrahedral sites. If all the six octahedral sites are occupied, they are known as trioctahedral, while it is dioctahedral if four octahedral sites are occupied. The octahedral sheet unit cell has four cis-sites (M2) and two trans-sites (M1) according to the adjacent corner or opposite corner positions, respectively.

The 1:1 serpentine-kaolin group includes berthierine, lizardite, odinite, kaolinite and dickite without any net charge (Srodon 1999). The charge per half-unit cell (z)

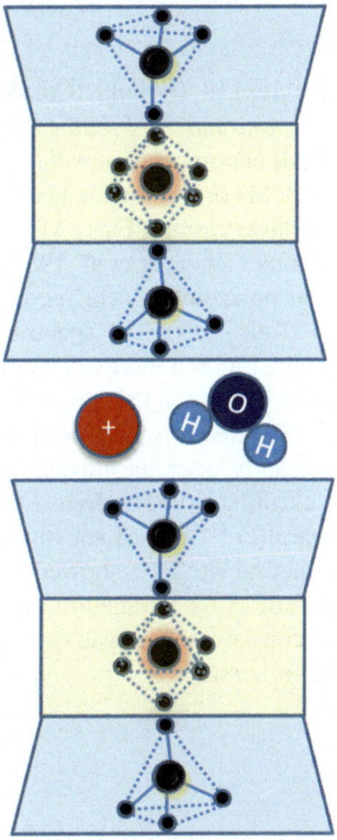

Fig. 10.10 The smectite structure comprises tetrahedral and octahedral layers with cations and water molecules sandwiched between them

of smectites is in the range 0.2–0.6. The 2:1 layer type vermiculite has a higher z of 0.2–0.9, similar to illites, while brittle mica species margarite has the largest z close to 2. The 2:1 type chlorites have variable z, with brucite/Fe cation interlayers, often represented as the 2:1:1 type. Chlorite/smectite(vermiculite) (C/S) mineral combinations called corrensite and illite/smectite have potential use in paleogeothermometry. The weathering process results in the following pathways: smectites→kaolinite; chlorite, mica→vermiculite. Smectites are transformed into illites and chlorites through weathering→early diagenesis→burial→low-temperature metamorphism. The smectite to illite conversion in shales is marked by the decrease in expandability of smectites with increasing temperature (increasing depth due to geothermal gradient).

10.7.1 Nontronite

Nontronite belongs to smectites and has the typical formula $(Ca, Na)_{0.3}Fe_2(Si, Al)_4O_{10}(OH)_2{\cdot}nH_2O$ where most of the ferric ion and Al occupies the dioctahedral sites, Si occupies the tetrahedral and alkali cations occupies the interlayer. In addition, Al and Fe^{3+} substitutions in tetrahedral, Mg or Fe^{2+} in octahedral sites result in a charge imbalance compensated by the interlayer cations. Early Mössbauer study on nontronite showed unresolved ferric ion doublet (Weaver et al. 1967).

Mössbauer spectra of synthetic and natural nontronites were recorded at room temperature and at low temperatures (Baron et al. 2017). The spectra of nontronites could be satisfactorily fitted with one tetrahedral and three octahedral site Fe^{3+} doublets. The δ values for the octahedral site Fe^{3+} were in the range 0.32–0.35 and 0.44–0.47 mm/s at 298 and 77 K, respectively. The deconvoluted doublets assuming various Si-Fe^{3+} environments did not significantly alter the covalency, observed from the close δ values. The Δ values increase from the 0.32–0.46 to 1.09–1.27 mm/s ranges due to increasing EFG from tetrahedral Fe^{3+} with Si substitutions around the ferric ion octahedral site cation. The tetrahedral site Fe^{3+} showed δ values ~0.2 and ~0.3 mm/s at 298 and 77 K, respectively. The Δ for tetrahedral Fe^{3+} was 0.59–0.64 mm/s. While the tetrahedral site showed consistent Δ, it was spread out for the octahedral site due to the three different environments.

10.7.2 Illites

Illites are dioctahedral minerals having the general formula $K_{0.8}(Al_{1.75}X^{2+}{}_{0.25})(Si_{3.5}Al_{0.5})O_{10}(OH)_2$ where X is Fe or Mg. It is also a layered 2:1 non-swelling type mica-like mineral. The MgFe-rich end member is known as glauconite, whereas the end member celadonite is MgAl-rich. Illites are increasingly studied for hydrocarbon exploration and paleothermometry (Pevear 1999). Glauconites are demonstrated for their potential use as geochronometers in calibrated dating of samples in the range 95–124 Ma (Fiet et al. 2006). Mössbauer studies suggested few natural glauconite samples as old as 642–1284 Ma (Zaitseva et al. 2018, 2019).

Illites may have the Fe^{3+} state alone or both Fe^{2+} and Fe^{3+} (Murad and Wagner 1994). Room temperature Mössbauer spectra of the illite samples obtained from various sites resulted in inner Fe^{3+} doublet with Δ in the range 0.6–0.8 mm/s and outer Fe^{2+} doublet with Δ in the range 2.6–3.0 mm/s exhibiting δ values 0.30–0.37 and 1.1–1.2 mm/s, respectively. The relative area of the Fe^{3+} showed more than 80% from most of the sites. The Fe-rich illites showed strong dioctahedral preference with less than 10% Fe^{2+} among the total iron ions, whereas that of Fe-poor illites was 20–44%.

10.7.3 Berthierine

Berthierine is a trioctahedral Fe-rich kaolinite-serpentine subgroup clay mineral with the formula $X_3(Si, Al)_2O_5(OH)_4$, where X = Fe, Mg and Al. It is a phyllosilicate made of 1:1 octahedral-tetrahedral sheets, also known to be among the green clay types. They occur as sea-water sediments around the mouth of rivers. The Fe in berthierine exists in the octahedral mixed valence states, whereas the Fe-rich end member greenalite, $Fe_3Si_2O_5(OH)_4$, has only octahedral Fe^{2+}.

Berthierine obtained from two different geographical sites showed Fe^{2+} and Fe^{3+} mixed valence states with δ values 1.11–1.22 and 0.22–0.35 mm/s, respectively, at RT (Mackenzie et al. 1984). The Δ values were 2.25–2.58 and 0.76–1.11 mm/s for the ferrous and ferric ions, respectively. Increasing the temperature beyond 473 K resulted in the air oxidization of berthierine resulting in the transformation of Fe^{2+} to Fe^{3+} and appearance of hematite beyond 973 K. Fe-rich berthierine has been found in Upper Cretaceous river sediments due to the prevalent reducing conditions (Toth and Fritz 1997).

Clearwater berthierine-rich samples (burial depths ~400–450 m) are fitted with three doublets, of which two are Fe^{2+} with a relative area of more than 70% (Hornibrook and Longstaffee 1996). The preservation of berthierine and reduction of Fe^{3+} to Fe^{2+} at a temperature less than 343 K during diagenesis is suggested during intense microbial activity and/or because of the strongly reducing environment created by the emplacement of hydrocarbons.

Figure 10.11 shows the Mössbauer spectra at (A) 300 K and (B) 77 K in an applied field of 0.04 T for the berthierine sample from the Sterling-Rock Falls area in Illinois (Blanc et al. 2014). The sample has the composition $(Al_{0.98}Fe^{3+}{}_{0.18}Fe^{2+}{}_{1.42}Mg_{0.16}X)(Si_{1.33}Al_{0.67})O_5(OH)_4$ where X = Li, Mn are in minor quantity. The room temperature spectrum could be fitted with ferrous (80%) and minor ferric doublets. The inner broad ferric component was inferred to be a superparamagnetic component. The 77 K spectrum with an applied field of 0.04 T was fitted with Fe^{2+} (δ = 1.25 mm/s, Δ = 2.81 mm/s, 78%) and Fe^{3+} (δ = 0.48 mm/s, Δ = 0.85 mm/s, 10%) doublets in octahedral coordination along with a sextet. The 12% sextet component with an isomer shift of 0.52 mm/s and B_{hf} of 46.0 T was assigned nano-goethite. The presence of a secondary phase, such as nano-goethite in a phyllosilicate, could be determined from Mössbauer spectrometry in correlation with other techniques.

10.7.4 Chlorites

The 2:1:1 type chlorites have four M2 and two M1 octahedral site positions. The 2:1 layer has the general formula $(Mg, Fe)_3Al_{1+x}Si_{3-x}O_{10}(OH)_2$, while the interlayer structure also has Fe cations. Chlorites are prevalent as tri-, di-tri- and di- octahedral types that show only two Fe^{2+} doublets, although four are possible in this mineral (Kallai and Rozenson 1981). The Δ value ranges from 2.3 to 2.7 mm/s. Although

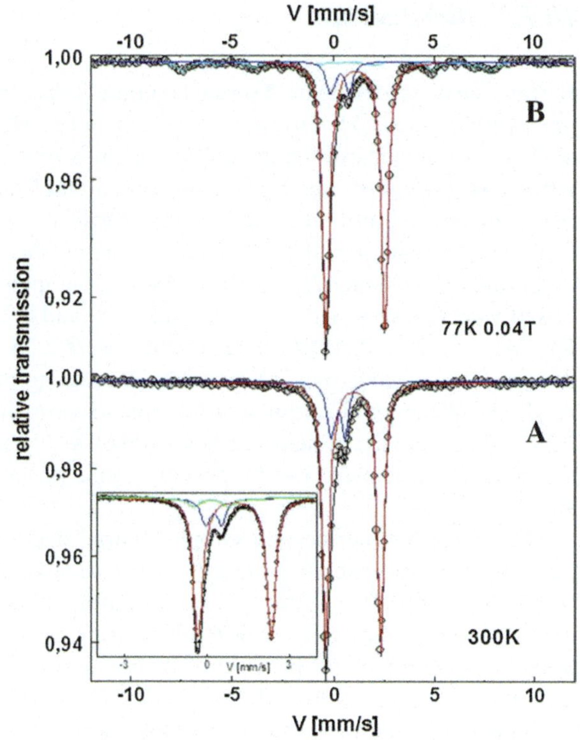

Fig. 10.11 Berthierine sample at 300 K (A) and 77 K (B) in an applied field of 0.04 T. The sextet is from goethite. The inset shows the ferric doublets. Reprinted with permission (Blanc et al. 2014), © Springer Nature

Fe^{3+} doublet is present in a few studies, their interpretation is difficult due to the low intensity and presence of impurities.

Mössbauer spectra of Fe-rich chlorite and ripidolite (green chlorite) were studied after heating the samples at 753 K for various durations in air and reducing atmosphere (Borggaard et al. 1982). The ripidolite showed a 90% relative area for Fe^{2+} with RT δ and Δ of 1.13 and 2.6 mm/s, respectively. Two Fe^{3+} doublets with Δ of 0.58 and 0.84 mm/s with δ values of 0.39 and 0.52 mm/s, respectively, were identified. The relative area of the Fe^{3+} increased with annealing duration (in air), reaching almost 98% for 1680 h. This could be reversed with annealing in reducing atmospheres such as H_2 and CO. The heat treatment at various atmospheres did not induce structural changes but changes in the cis- and trans- positions of O–H.

Natural clinochlorite could be fitted with two Fe^{2+} in M1 and M2 sites and almost 10% occupancy in interlayer Fe^{3+} trans- M4 site (Smyth et al. 1997). The M4 site Fe^{3+} is assigned to the doublet with δ and Δ of 0.33 and 0.8 mm/s, respectively.

10.8 Iron Ore Minerals

Common iron ore minerals are hematite, magnetite, goethite and maghemite, while other gangue minerals from which Fe is extracted are siderite, ankerite, pyrite, chlorite, ilmenite and chamosite. High-grade hematite iron ores contain > 60 wt.% Fe, which has other constituents such as Al_2O_3 and SiO_2 that occurs as banded iron formation (BIF) and granular iron formation (GIF) deposits (Lu 2021). The BIF and GIF are older iron ores > 2 Ga, while the low-grade rapitan iron formation (RIF) iron ores are younger (< 1 Ga).

The quality of iron ores could be ascertained by the Fe-minerals present in the ores and the impurities. The hyperfine parameters for magnetite and maghemite are discussed in the chapter on spinel ferrites. Other major Fe ore mineral constituents, such as hematite and goethite, are discussed further.

10.8.1 Hematite

The structure and properties of hematite exhibiting various hyperfine parameters are listed in Table 10.2. The hematite having the rhombohedral structure exhibits canted antiferromagnetic nature with a Néel temperature close to 955 K that becomes superparamagnetic with size reduction (Kündig et al. 1966). Consequently, the extrapolated B_{hf} at 0 K reduces to 53.3 T (Bødker et al. 2000) compared with the bulk value of 54.4 T, fitted with the Brillouin curve ($S = 5/2$) (van der Woude 1966). The nanoparticles show a range of hyperfine fields according to the particle size and measurement temperature. The 5 nm α-Fe_2O_3 could be fitted with two distinct hyperfine fields in the range 53.3–46.4 T at 4.2 K (Mansilla et al. 1999) whereas the 15–20 nm particles showed B_{hf} nearly 53.0 T in the temperature range 5–80 K. Thus, it is expected that pedogenic hematite having a distribution of large and ultrafine particles would exhibit a sextet of varying B_{hf} and relaxing doublet at room temperature.

Al_2O_3 is one of the impurities in iron ores. $Fe_2O_3{-}Al_2O_3$ solid solution showed a T_N from 955 to 859 K when the alumina content in hematite is increased from 0 to 10 wt. %, respectively (Srivastava and Sharma 1972). A linear decrease in B_{hf} from 51.6 to 50.3 T at RT was observed as the Al content in hematite increased to 15 mol. % (Fysh et al. 1982). In addition, poorly crystalline Al-hematite showed a linear decrease in B_{hf} at 4.2 K with Al content, whereas better crystalline counterparts showed almost stable B_{hf}. The quadrupole splitting at RT increases with Al substitution from 0.17 to 0.22 mm/s, while the δ values are invariably close to 0.36 mm/s. Hematite undergoes a spin-flop transition from perpendicular to parallel spin orientation to the *c*-axis below 263 K that is not observed in fine particles.

Ilmenite ($FeTiO_3$) and hematite form a solid solution in the rhombohedral structure, whose properties are shown in Table 10.2. There exists a ferrimagnetic cluster composition region that would show saturation magnetization nearly 25–30 $A{\cdot}m^2$/kg. The Néel temperature of hematite decreases linearly with increasing ilmenite,

Table 10.2 The structure and properties of hematite, one of the iron ore constituents, exhibit various hyperfine parameters under varying conditions

Mineral	Magnetic parameters	B_{hf}	References
Hematite α-Fe_2O_3 Corundum	Canted Antiferro $S = 5/2$ $T_N = 955$ K	51.5–51.8 T (RT) ~50.0 T (RT, 20 nm) ~54.0 T (80 K) ~53.0 T (5–80 K, 20 nm) 53.3–46.4 T (4.2 K, 5 nm) 54.4 T (0 K, bulk) 53.3 T (0 K, nano)	Kistner and Sunyar (1960) Ono and Ito (1962) van der Woude (1966) Kundig et al. (1966) Bødker et al. (2000) Mansilla et al. (1999)
Al-hematite α-(1-x)Fe_2O_3-xAl_2O_3 Rhombohedral solid solution	$T_N = 955$ to 859 K for 0 to 10 wt.% Al_2O_3	51.6 to 50.3 T (RT) (0 to 15 mol.% Al)	Srivastava and Sharma (1972) Fysh and Clark (1982)
Ilmenite-hematite α-(1-x)$FeTiO_3$.xFe_2O_3 Rhombohedral solid solution	$T_N = 55$ K for x = 0 Ferrimagnetic clusters for x = ~0.2–0.5 T_N increases with x	52.0 T (RT, x = 1) 44.5 T (RT, x = 0.5) 7.0 T (20 K, x = 0)	Shirane et al. (1962)

reaching 55 K (Shirane et al. 1962). In the (1-x)$FeTiO_3$.xFe_2O_3 composition close to x = 0.3–0.5, ferrimagnetic clusters result in a net magnetization. The ilmenite-rich sample showed a Fe^{2+} state ($\delta \approx 1.2$ mm/s, SS), while the isomer shift values indicated Fe^{3+} beyond x = 0.5. The RT spectrum could be resolved with two sextets for x = 0.7, resulting in hyperfine fields of 50.0 and 47.6 T and isomer shift (SNP standard) values of 0.52 and 0.8 mm/s, respectively (Warner et al. 1972). While the lower values indicated Fe^{3+}, the value of 0.8 mm/s suggested mixed ferric-ferrous states.

Magnetite has sextets arising from both ferrous and ferric states, whereas maghemite exhibits only a ferric state. In magnetite-hematite mixtures, Morin transition in hematite or Verwey transition in magnetite could not be used to identify the phases in natural samples due to variations in hyperfine parameters with structural parameters and particle size (Vandenberghe et al. 2000).

Figure 10.12 shows the 300 K Mössbauer spectra of Archean BIF, 2.9 Ga sample from Western Dharwar Craton, India and laterite soil from Minas Gerais, Brazil. The BIF sample was magnetic, and the magnetically separated sample was designated as MF from the rest of the sample (REST). The BIF showed an area fraction of 86% and 14% from hematite and goethite, respectively, while the Brazilian sample showed 100% hematite. The magnetically extracted sample exhibited 3% magnetite. The phases could be easily identified from the hyperfine fields, with hematite showing 51.5–51.7 T, goethite 37.8–38.0 T and magnetite 46.0–49.0 T. The Indian-REST sample had only hematite and goethite after the magnetic separation. The study

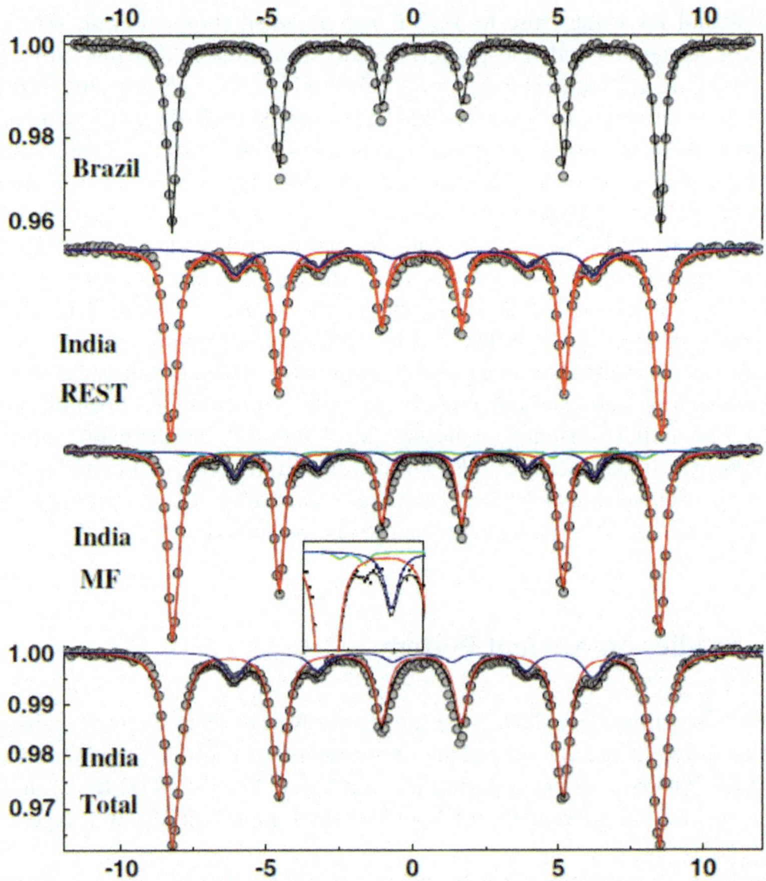

Fig. 10.12 RT Mössbauer spectra of Archean BIF, 2.9 Ga sample from Western Dharwar Craton, India and laterite soil from Minas Gerais, Brazil. MF- extracted magnetic fraction and REST- rest of the samples. The inset shows the magnified region of the inner sextet. Reprinted with permission (Orberger et al. 2014), © Springer Nature

proposed the transformation of magnetite and maghemite into hematite due to an earlier hypogene process.

10.8.2 Goethite

Goethite has an orthorhombic structure that exhibits antiferromagnetism with a T_N close to 400 K. Ferric sextet with a B_{hf} of 50.4 T in the temperature range

77–4.2 K and RT value close to 38.4 T was obtained (Forsyth et al. 1968). Stoichiometric variations and fine particles result in a broader distribution of hyperfine fields. Pure goethite exhibited a B_{hf} of 38.0 T at RT, and distribution down to 34.0 T due to structure and particle size deviations (Murad 1982). The temperature-dependent Mössbauer studies on well-crystalline natural goethite gave a saturation B_{hf} of 50.7 T and T_N of 400 K with a Brillouin fit using $S = 5/2$ (de Grave and Vandenberghe 1986). However, synthetic goethite exhibits a T_N that could be as low as 261 K according to the sample nature (Bocquet et al. 1992). Non-superparamagnetic defective goethite broadened the sextets, showing an average B_{hf} of 27.0 T at RT, 50.2 T at 4.2 K and a reduced T_N of 347 K. The doublet in the Mössbauer spectrum is influenced by collective superparamagnetic relaxation that may be misinterpreted as a paramagnetic state. Detailed literature leading to the conclusion of static and dynamic fluctuations in goethite has been discussed in Chapter.9 on antiferromagnetic materials. Mn^{3+} and Al^{3+} are the major substituents in goethite. The linear decrease in B_{hf} with Al substitution was similar to the hematite (Murad and Schwertmann 1983). Since the crystallinity also affects the hyperfine fields, due care has to be taken while analyzing the spectrum.

10.9 Studies on Ancient Pottery

Clay was used in pottery during the late Stone Age itself by various civilizations. The three different age system and their approximate transition periods are shown in Fig. 10.13. Pottery developed during the late Stone Age (~25,000 BCE) in various regions, followed by porcelain in China and steel utensils in the Iron Age (~1,500 BCE).

Mössbauer spectrometry has been extensively used in pottery to identify the pottery mineralogy and the prevalent firing conditions, as illustrated in Fig. 10.14. This could be seen from phase changes or only changes of valence states established from the Mössbauer data. The light-colored late medieval sherd from Surrey, UK showed Fe^{3+} doublet with δ and Δ values 0.1 and 0.94 mm/s, respectively (Cousins and Dharmawardena 1969). Another pottery showed Fe^{2+} and only 27% Fe^{3+} at the

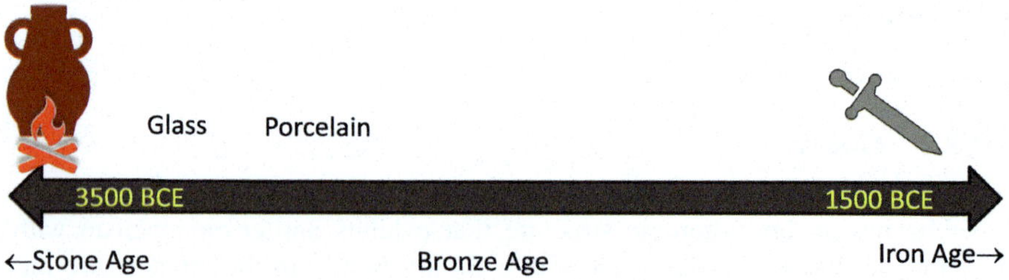

Fig. 10.13 Pottery dates back to the late Stone Age (~25,000 BCE), followed by glassware, porcelain and steel in the Iron Age (~1,500 BCE)

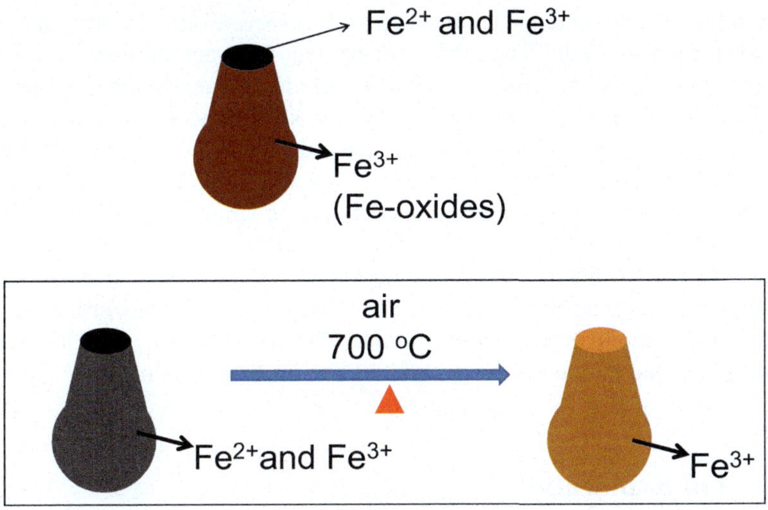

Fig. 10.14 The colored pottery (top) shows Fe^{3+} as in iron oxide-clay mixtures at the exterior, suggesting oxidizing conditions, whereas the interior black region shows mixed Fe valence states. The black pottery (bottom), when heated in air at a higher temperature, turns to a light color surface with only a change of state to Fe^{3+}

black interior, whereas complete transformation to Fe^{3+} was noted at the red surface, similar to oxidized clay. A lightly gritted blackware showed both ferrous and 27% ferric doublets that changed to Fe^{3+} with a buff color on heating in air at 973 K.

Ancient Greek pottery (600–800 BCE) showed Fe^{2+} and Fe^{3+} doublets from clay constituents in few samples, and Fe_2O_3/Fe_3O_4 with B_{hf} 48.0–51.0 T, for the samples obtained from other archaeological sites (Gangas et al. 1971). Unfired clay showed superparamagnetic doublet that transformed to sextets at 77 K, suggesting the presence of α-Fe_2O_3 and β-FeOOH (Kostikas et al. 1974). Fe-oxides could be ascertained from increased sextet component with firing temperature, whereas clay minerals shall still exhibit a doublet at RT. Indeed, most of the clay minerals have T_N below 10 K. Experimental results suggest superparamagnetic Fe-oxides in ancient pottery may arise due to aging effects associated with firing temperature (Gangas et al. 1976).

Mössbauer spectrometry was used to reconstruct the manufacture of ancient pottery. Hematite was found to be dominant when the firing temperature of clay exceeds 1073 K, whereas it decreases beyond 1373 K due to its incorporation in silicates (Wagner et al. 1998). The quadrupole splitting of Fe^{3+} doubles from 0.7 to 1.4 mm/s as the firing temperature increases from 573 to 973 K. Further, extreme temperature results in the destruction of the clay structure. However, closed heating with charcoal results in Fe^{2+} at 1073 K.

Mössbauer studies on sherds from China belonging to 4000 BCE old Majiabang culture showed the firing temperature to be less than 673 K (Hsia and Huang 2003). Color change of the pottery was achieved by firing in a reducing atmosphere first and later in an oxidizing atmosphere for a shorter duration, resulting in a hematite surface.

Jun porcelain belonging to the tenth century CE achieved various colors in the same material in the same kiln. The Fe^{2+} is more than 60% among total Fe, deducing the firing condition to be more than 1373 K in reducing atmosphere. Thousands of second century BCE Chinese warrior-horse life-size terracotta statues were found to be composed of grey and red ceramic materials. The grey ceramic materials had 40–60% Fe^{2+}, whereas the red was 95% Fe^{3+}. It was observed that Fe^{2+} at the surface exceeded the interior, implying firing in an oxidizing atmosphere and later on under reducing conditions.

Mössbauer spectrometry of medieval period stained glass showed that the color of the glass was determined by the ferrous and ferric contents (Longworth et al. 1982). Amber, blue and light green colored samples showed both Fe^{2+} and Fe^{3+}, whereas emerald green and purple samples exhibited Fe^{3+} in four coordination.

10.10 Oil Exploration

The use of Mössbauer spectrometry in oil exploration is related to the identification of Fe-bearing minerals associated with hydrocarbon generation. The Mössbauer spectrometry of minerals from Danish on-shore and off-shore bore wells were examined in one of the studies that revealed features as illustrated in Fig. 10.15. (Mørup et al. 1985).

The Cretaceous, Jurassic and Triassic sediments were found to occur at slightly different depths among several wells. Nearly 20–50% of the minerals in the samples are clay type showing Fe^{2+} and Fe^{3+} components throughout the depth up to ~km, apart from pyrite, carbonates and Fe-oxides. The off-shore wells showed the appearance of Fe-bearing carbonate, ankerite accompanied by diminishing siderite at depths below ~km, corresponding to the Jurassic sediments. It was suggested that the formation of ankerite, $CaMg_{0.5}Fe_{0.5}(CO_3)_2$, is due to the reaction of $CaCO_3$ at > 400 K with the Fe^{2+} and Mg^{2+} ions released by the transformation of smectite to illite and/or organic algals. The temperature gradient of the oil wells was 20–40 K/km, whereas the on-shore oil wells showed lower values. The lower burial depth of Jurassic sediments and temperatures < 373 K is suggested to be the reason behind the absence of ankerite, consequently resulting in dry wells. Mössbauer studies have provided information about the ratio of ankerite to siderite in this study. Siderite, $FeCO_3$ and ankerite have similar isomer shift value (1.2–1.3 mm/s) whereas the quadrupole splitting of siderite(1.8 mm/s) > ankerite (1.40–1.65 mm/s) at RT.

10.11 Mössbauer Spectrometry in Pedology

Pedology deals with the study of soils, including their classification (pedogenesis) and soil changes according to the environment, such as weathering and biological processes. The gross weathering index measures the ratio of the Fe-weathering

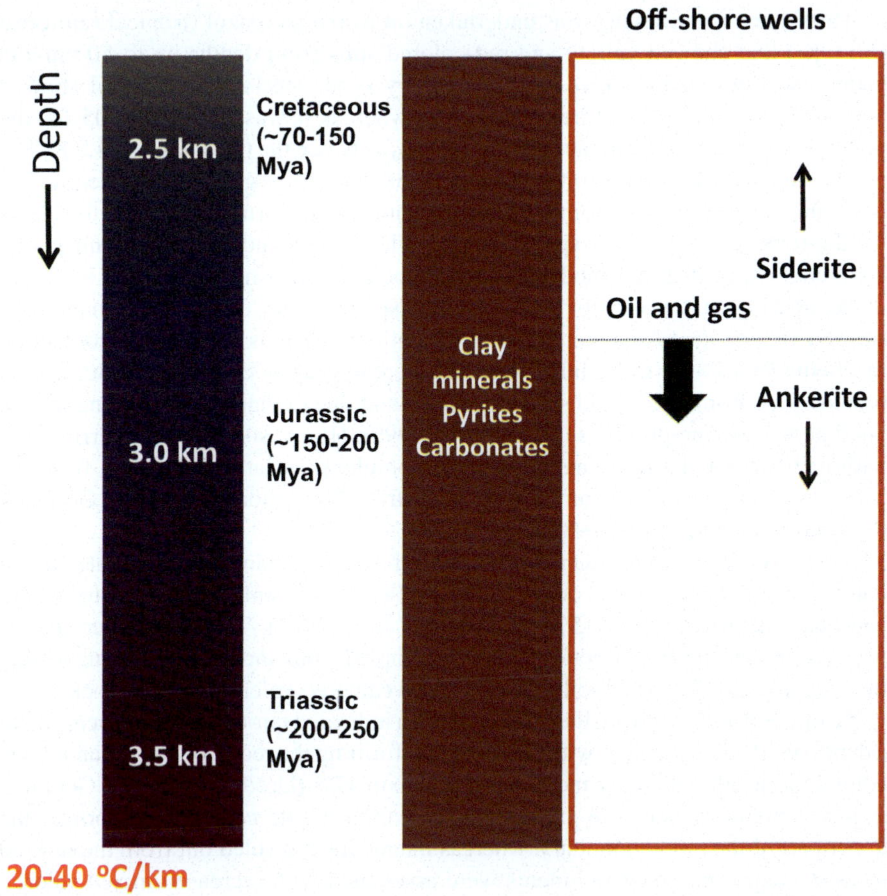

Fig. 10.15 Danish Mesozoic sediments that include Cretaceous, Jurassic and Triassic sediments at varying depths suggest potential oil and gas sites accompanied by the appearance of ankerite in off-shore oil wells (Mya = Mega years ago)

complex to the total Fe. Fe-bearing silicates, upon reacting with water and oxygen, produce goethite, α-FeO(OH) and further to Fe-organic ligands and other Fe-oxides such as hematite (α-Fe_2O_3) (Stucki et al. 1988). Hematite is widespread in warmer areas as reddish soil (southern hemisphere), whereas in cool, humid climates, the fraction becomes lower compared to goethite. The acidic and alkaline environment induces the transformation of ferrihydrite to goethite, while hematite results at an intermediate pH. Lepidocrocite, γ-FeO(OH), formed through oxidation of Fe^{2+}, transforms to goethite in Ca and Al-rich environments. Soil color is determined not only by the type of Fe-oxide but also according to the particle size, generally becoming darker with size reduction.

Mössbauer spectrometry was undertaken on Brazilian oxizol (tropical rainforest soil type) that showed yellow and red-colored soils from depths up to 50 cm and underneath up to 250 cm, respectively (Coey et al. 1991). The red soil showed hematite/(hematite + goethite) ratio > 0.2, whereas it was less than 0.05 for the yellow soil, estimated from the difference in B_{hf} between the two phases at 4.2 K. The weathering of the Fe-rich dolomite rock to Fe-oxides such as goethite, maghemite and hematite was established from the transformation of quadrupole doublet to sextets for the rock and its rinds, respectively. Particle size-dependent superparamagnetic doublet could be distinguished from low-temperature measurements.

Weathering studies employing Mössbauer spectrometry on Brazilian mafic rock showed magnetite to maghemite transformation, whereas stable soil containing magnetite was found to be inherited from steatite (Fabris et al. 1997). Studies on basaltic rock from Greenland showed reddish soil due to maghemite, magnetite and goethite weathering products (Mørup et al. 1990). Humic soil results in ferrihydrite with increasing C/Fe ratio and Fe-organic complexes at large C/Fe ratio from the Mössbauer studies on the samples obtained from Harz mountain range, Germany (Schwertmann and Murad 1988).

Loam from the Yokohama paddy field near Tokyo, Japan, showed a dark brown colored layer up to 15 cm depth with 8% Fe, brown soil beneath up to 40 cm and black further below (< 5% Fe) (Matsuo et al. 1994). The Mössbauer spectra revealed increasing Fe^{2+} fraction with soil sampled from various depths, suggesting the reducing conditions of soil derived from weathered volcanic silicate rocks.

Fe mineral analysis from the Peruvian basin indicated tan-colored sediments up to a depth of 20 cm associated with Fe^{2+} of 11%, forming the redox boundary and green color underneath, markedly increasing Fe^{2+} up to 37% (Drodt et al. 1997). Goethite, a trace amount of hematite and their superparamagnetic components formed the upper part of the redox boundary, whereas magnetite was ruled out from the in-field measurement. The green sediments were associated with increased mixed valence layered silicate minerals that were paramagnetic even at 300 mK. Red sea sediments include most of the Fe-oxides and hydroxides along with Fe-bearing clays, carbonates and pyrite (Goldman and Singer 2001). Permian to Triassic (PT) deep-sea pelagic sediments from Japan have revealed hematite, pyrite, goethite and silicates with more than 72% ferrous iron in the PT boundary, suggesting oxic to anoxic redox condition (Matsuo et al. 2002) as in most of the sea sediments. Sedimentary rock samples obtained from King George Island, Antarctica, from 30–70 m depths showed 0–20% hematite, 10–36% magnetite and the rest silicates from Mössbauer analysis (Kuzmann et al. 2002). Antarctica sea sediments at depths from 2–200 cm showed Mössbauer doublets assigned to Fe^{3+} glauconite along with Fe^{3+} clay and Fe^{2+} silicate even for the 5000 year-old sediments (Shozugawa et al. 2009).

10.12 Summary

This chapter has presented the role of Mössbauer spectrometry in the identification of crustal and mantle minerals, their phase stability under pressure, clay minerals, geochronology, archaeology, oil exploration and pedology. The Mössbauer parameters of the minerals under specific conditions are helpful references for researchers in Earth sciences.

References

G. Amthauer, H. Annersten, S.S. Hafner, Z. Kristallogr. **143**, 14 (1976)
D.L. Anderson, *New Theory of the Earth*, 2nd edn. (Cambridge University Press, Cambridge, 2007)
V.I. Bagin, T.S. Gendler, L.G. Dainyak, R.N. Kuz'min, Clay. Clay Miner. **28**, 188 (1980)
G.M. Bancroft, A.G. Maddock, R.G. Burns, Geochim. Cosmochim. Acta **31**, 2219 (1967)
F. Baron, S. Petit, M. Petrak, A. Decarreau, J.W. Stucki, Am. Mineral. **102**, 1501 (2017)
P. Blanc, H. Gailhanou, J. Rogez, G. Mikaelian, H. Kawaji, F. Warmont, S. Gaboreau, S. Grangeon, J.-M. Greneche, P. Viellard, C.I. Fialips, E. Giffaut, E.C. Gaucher, F. Claret, Phys. Chem. Miner. **41**, 603 (2014)
S. Bocquet, R.J. Pollard, J.D. Cashion, Phys. Rev. B **46**, 11657 (1992)
F. Bødker, M.F. Hansen, C.B. Koch, K. Lefmann, S. Mørup, Phys. Rev. B **61**, 6826 (2000)
O.K. Borggaard, H.B. Lindgreen, S. Mørup, Clays Clay Miner. **30**, 353 (1982)
J.M.D. Coey, J.D. Fabris, M. Resende, Hyperfine Interact. **66**, 51 (1991)
D.R. Cousins, K.G. Dharmawardena, Nature **223**, 732 (1969)
E. de Grave, R.E. Vandenberghe, Hyperfine Interact. **28**, 643 (1986)
J.C.P. de Oliveira, M.I. de Costa, Jr., W.H. Schreiner, A. Vasquez, J. Magn. Magn. Mater. **79**, 1 (1989)
M. Drodt, A.X. Trautwein, L. Konig, E. Suess, C.B. Koch, Phys. Chem. Miner. **24**, 281 (1997)
S.G. Eeckhout, E. de Grave, A. Lougear, M. Gerdan, C.A. McCammon, A.X. Trautwein, R. Vochten, Am. Mineral. **86**, 957 (2001)
J.D.Fabris, M.F. de J. Filho, J.M.D. Coey, W. da N. Mussel, A.T. Goulart, Hyperfine Interact. **110**, 23 (1997).
N. Fiet, X. Quidlleur, O. Parize, L.G. Bulot, P.Y. Gillot, Earth Planet. Sci. Lett. **246**, 407 (2006)
J.B. Forsyth, J.G. Hedley, C.E. Johnson, J. Phys. C: Solid State Phys. **1**, 179 (1968)
D.J. Frost, C.A. McCammon, Am. Mineral. **94**, 872 (2009)
D.J. Frost, C. Liebske, F. Langenhorst, C.A. McCammon, R.G. Tronnes, D.C. Rubie, Nature **428**, 409 (2004)
S.A. Fysh, P.E. Clark, Phys. Chem. Miner. **8**, 257 (1982)
N.H.J. Gangas, A. Kostikas, A. Simopoulos, Nature **229**, 485 (1971)
N.H.J. Gangas, I. Sigalas, A. Moukarika, J. Phys. Colloques **37**, C6-867 (1976)
N.T. Goldman, A. Singer, Clay. Clay Miner. **49**, 174 (2001)
C.V. Guidotti, M.D. Dyar, Am. Mineral. **76**, 161 (1991)
L. Häggström, R. Wappling, H. Annersten, Chem. Phys. Lett. **4**, 107 (1969)
J.M. Hazen, D. Papineau, W. Bleeker, R.T. Downs, J.M. Ferry, T.J. McCoy, D.A. Sverjensky, H. Yang, Am. Mineral. **93**, 1693 (2008)
G.R. Helffrich, B.J. Wood, Nature **412**, 501 (2001)
E.R.C. Hornibrook, F.J. Longstaffee, Clays Clay Miner. **44**, 1 (1996)
Y. Hsia, H. Huang, Hyperfine Interact. **150**, 33 (2003)
L.H. Kallai, I. Rozenson, Phys. Chem. Miner. **7**, 223 (1981)
J.F. Kasting, Science **259**, 920 (1993)

E.S. Kisseva, D.M. Vasiukov, B.J. Wood, C. McCammon, T. Stachel, M. Bykova, E. Bykova, A. Chumakov, V. Cerantola, J.W. Harris, L. Dubrovinsky, Nat. Geoscience **11**, 144 (2018)

O.C. Kistner, A.W. Sunyar, Phys. Rev. Lett. **4**, 412 (1960)

A. Kostikas, A. Simopoulos, N.H. Gangas, J. Phys. Colloques **35**, C1-107 (1974)

R. Kretz, Geochim. Cosmochim. Acta **46**, 411 (1982)

W. Kündig, H. Bömmel, G. Constabaris, R.H. Lindquist, Phys. Rev. **142**, 327 (1966)

E. Kuzmann, P.A. De Souza Jr., L.A. Schuch, A.C. de Olivera, R. Garg, V.K. Garg, Hyperfine Interact. **141/142**, 327 (2002)

I. Leonov, A.V. Ponomareva, R. Nazarov, I.A. Abrikosov, Phys. Rev. B **96**, 075136 (2017)

J.F. Lin, Z. Mao, J. Yang, J. Liu, Y. Xiao, P. Chow, T. Okuchi, Geophy. Res. Lett. **43**, 6952 (2016)

S. Longworth, N.H. Tennent, M.J. Tricker, P.P. Vaishnava, J. Archaeol. Sci. **9**, 261 (1982)

L. Lu, (ed.) *Iron Ore Mineralogy, Processing and Environmental Sustainability* (Elsevier, 2021).

K.J.D. Mackenzie, R.M. Berezowski, Thermochim. Acta **74**, 291 (1984)

M.V. Mansilla, R.D. Zysler, C. Arciprete, M.I. Dimitrijewits, C. Saragovi, J.-M. Greneche, J. Magn. Magn. Mater. **204**, 29 (1999)

Z. Mao, F. Wang, J.F. Lin, S. Fu, J. Yang, X. Wu, T. Okuchi, N. Tomioka, V.B. Prakapenka, Y. Xiao, P. Chow, Am. Mineral. **102**, 357 (2017)

M. Matsuo, K. Kubo, Y. Isozaki, Hyperfine Interact. (C), M.F. Thomas, J.M. Williams, T.C. Gibb (ed.) (Springer, 2002).

M. Matsuo, T. Kobayashi, M. Tsurumi, Hyperfine Interact. **84**, 533 (1994)

C.A. McCammon, Hyperfine Interact. **90**, 89 (1994)

C.A. McCammon, Science **308**, 807 (2005)

C. McCammon, K. Glazyrin, A. Kantor, I. Kantor, I. Kupenko, O. Narygina, V. Potapkin, C. Prescher, R. Sinmyo, A. Chumakov, R. Ruffer, I. Sergueev, G. Smirnov, L. Dubrovinsky, High Press. Res. **33**, 663 (2013)

S. Mørup, J. Franck, J. van Wonterghem, H.R. Poulsen, L. Larsen, Fuel **64**, 528 (1985)

S. Mørup, O. Saksager, M.B. Madsen, M.D. Benton, C.J.W. Koch, Hyperfine Interact. **57**, 2269 (1990)

E. Murad, Am. Mineral. **67**, 1007 (1982)

E. Murad, F.E. Wagner, Phys. Chem. Miner. **14**, 264 (1987)

E. Murad, U. Schwertmann, Clay Miner. **18**, 301 (1983)

E. Murad, U. Wagner, Clay Miner. **29**, 1 (1994)

H.S.C. O'Beukk, C.A. McCammon, D. Canil, D.C. Rubie, C.R. Ross II., F. Seifert, Am. Mineral. **78**, 456 (1993)

K. Ono, A. Ito, J. Phys. Soc. Jpn. **17**, 1012 (1962)

B. Orberger, C. Wagner, A. Tudryn, R. Wirth, R Morgan, J.D. Fabris, J.-M. Greneche, C. Rosire, Phys. Chem. Miner. **41**, 651 (2014)

D.P. Pevear, Proc. Natl. Acad. Sci. U.S.A. **96**, 3440 (1999)

U. Schwertmann, E. Murad, Clay Miner. **23**, 291 (1988)

I. Shinno, Phys. Chem. Miner. **7**, 91 (1981)

G. Shirane, D.E. Cox, W.J. Takei, S.L. Ruby, J. Phys. Soc. Jpn. **17**, 1598 (1962)

K. Shozugawa, A. Kuno, H. Miura, M. Matsuo, J. Nucl. Radiochem. Sci. **10**, 13 (2009)

J.R. Smyth, M.D. Dyar, H.M. May, O.P. Bricker, J.G. Acker, Clays Clay Miner. **45**, 544 (1997)

V.L. Sobolev, C.A. McCammon, L.A. Taylor, G.A. Snyder, N.V. Sobolev, Am. Mineral. **84**, 78 (1999)

S. Speziale, A. Milner, V.E. Lee, S.M. Clark, M.P. Pasternak, R. Jeanloz, Proc. Natl. Acad. Sci. U.S.A. **102**, 17918 (2005)

J.K. Srivastava, R.P. Sharma, Phys. Status Solidi B **49**, 135 (1972)

J. Srodon, Annu. Rev. Earth Planet. Sci. **27**, 19 (1999)

J. W. Stucki, B.A. Goodman, U. Schwertmann (ed.) *Occurrence and Formation of Iron Oxides in Various Pedoenvironments*. Iron in Soils and Clay Minerals. NATO ASI Series (Series C: Mathematical and Physical Sciences), vol. 217. (Springer, Dordrech, 1988).

T.A. Toth, S.J. Fritz, Clays Clay Miner. **45**, 564 (1997)

F. van der Woude, Phys. Status Solidi **17**, 417 (1966)
R.E. Vandenberghe, C.A. Barrero, G.M. da Costa, E.V. San, E. de Grave, Hyperfine Interact. **126**, 247 (2000)
G. Vanko, F.M.F. de Groot, Phys. Rev. B **75**, 177101 (2007)
G.Y.V. Victor, D. Ghosh, S. Ghose, Phys. Rev. B **64**, 144413 (2001)
U. Wagner, R. Gebhard, G. Grosse, T. Hutzelmann, E. Murad, J. Riederer, I. Shimada, F.E. Wagner, Hyperfine Interact. **117**, 323 (1998)
B.N. Warner, P.N. Shive, J.L. Allen, C. Terry, J. Geomag. Geoelec. **24**, 353 (1972)
C.E. Weaver, J.M. Wampler, T.E. Pecuil, Science **156**, 504 (1967)
C. Xu, J. Kynicky, R. Tao, X. Liu, L. Zhang, M. Pohanka, W. Song, Y. Fei, Sci. Adv. **3**, e1601589 (2017)
I.K. Yu, L.S. Dubrovinsky, C.A. McCammon, Phys. Rev. B **73**, 100101R (2006)
T.S. Zaitseva, A.B. Kuznetsov, V.M. Gorozhanin, I.M. Gorokhov, T.A. Ivanovskaya, G.V. Konstantinova, Stratigr. Geol. Correl. **27**, 573 (2019)
T.S. Zaitseva, I.M. Gorokhov, M.A. Semikhatov, A.B. Kuznetsov, T.A. Ivanovskaya, G.V. Konstantinova, O.V. Dorzhieva, Stratrigr. Geol. Correl. **26**, 611 (2018)

Chapter 11
Extraterrestrial Mössbauer Spectrometry

11.1 Solar System

Hydrogen, the remnant product of the big bang, is found throughout the universe. The molecular cloud is a giant interstellar gas cloud mostly made of hydrogen. The gravitational collapse of a giant molecular cloud, which is thousands of times more massive than the Sun, through a shock wave event such as a supernova explosion leads to the birth of a star. Figure 11.1 shows the schematic diagram illustrating the formation of our Solar System due to the gravitation collapse of a giant molecular cloud 4.5 billion years ago. The slowly spinning giant molecular cloud becomes a fast-spinning gravitationally contracted disk of the solar nebula, initiating nuclear fusion at the center. The Sun at the center has most of the hydrogen that undergoes nuclear fusion, and the remaining matter is bound to form planets, asteroids and dust particles through accretion.

Extraterrestrial Mössbauer studies are based on the mineral analysis of the following three sources: (i) meteorites, (ii) Moon samples and (iii) Mars exploration. Meteorites can originate from a comet, asteroid or planet fragments of our Solar System. As the meteoroid travels at very high velocity, it undergoes heating, shock and disintegrates into pieces as meteors upon entering the Earth's atmosphere. However, the preservation of the internal structure of a meteorite allows researchers to analyze the state of its early formation.

11.2 Chondrites

Chondrites are meteorites made of primitive materials formed during the genesis of the Solar System, while achondrites are those that have undergone transformations like planetary rocks. Chondrites are made of chondrules that are spherules of typical diameter 0.1–1 mm, calcium aluminum-rich inclusions (CAI), amoeboid olivine

R. J. Joseyphus and J.-M. Greneche, *Fundamentals of ^{57}Fe Mössbauer Spectrometry*,
https://doi.org/10.1007/978-981-99-8653-8_11

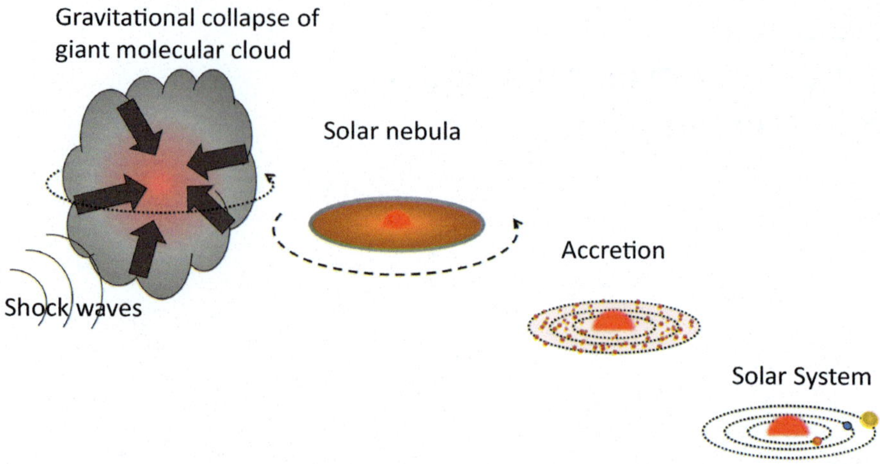

Fig. 11.1 The Solar System formed after the gravitational collapse of the giant molecular cloud almost 4.5 billion years ago, followed by nuclear fusion at the center of the solar nebula and accretion

aggregates (AOA), Fe–Ni, troilite, silicates, other minerals and volatile elements embedded in the matrix (Abreu 2018), as represented in Fig. 11.2. These components may vary according to the type of chondrite classifications like the CI type, where chondrules are absent. Chondrules are olivine-pyroxene components with varying Fe formed at the early stages of planetary accretion. The chondrites are classified according to the impact shock level from petrographic evidence in increasing order from S1 (< 5 GPa) to S6 (75–90 GPa).

The various classifications of meteorites and their sub-classes, based on their constituents and similarities among the collected meteorites, are shown in Fig. 11.3. Chondrites are classified into carbonaceous (C), ordinary (O) and enstatite (E) groups. The C-chondrites have the sub-groups that are CI, CM, CO, CV, CK, CR and CH among which CI is similar to the solar photosphere, excluding the volatile solar elements (Sears 2004). The CI, CM and CR are grouped together as they are rich in water and carbon content. CI type has the most water (~ 18 wt.%), carbon (~ 3 wt.%) and matrix among the various carbonaceous chondrites. CM is poor in Fe metal like CI but has lower carbon (~ 2 wt.%) and water content (~ 12 wt.%). Ordinary chondrites are classified according to the metal iron content among other Fe in silicates as, high (H) metal-rich group followed by low (L) and very low (LL). The metal Fe to total Fe atomic ratios range from 0.6 to 0.1. Ordinary chondrites constitute more than 80% of the falls and are less anhydrous than most other chondrites. EH and EL are Fe metal-rich pyroxene-silicate enstatite chondrites with Fe/Si of 0.95 and 0.62, respectively. Based on petrology, the chondrites are classified as types from 1 (aqueous altered without chondrules) to 6 (metamorphic).

Radiocarbon dating indicates the age of chondrites close to 4.57 Ga, coinciding with the age of formation of our Solar System from the collapse of the interstellar dust.

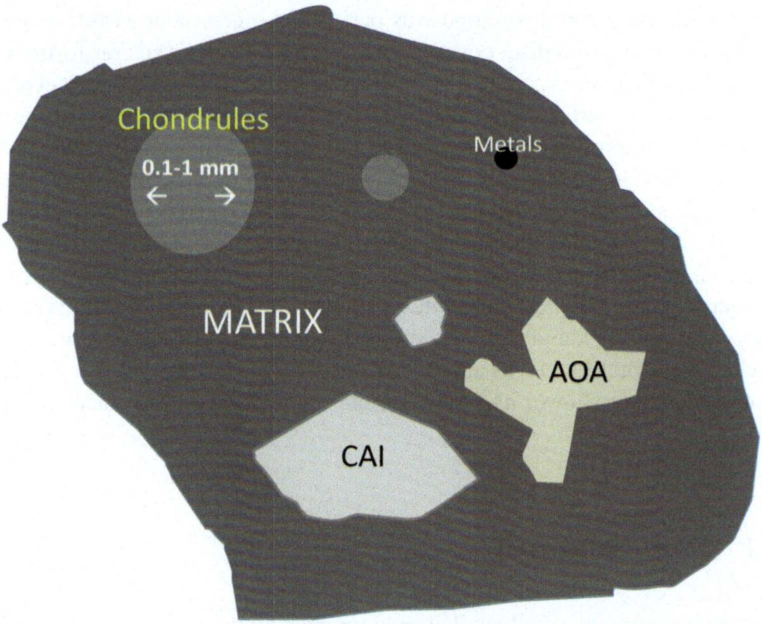

Fig. 11.2 Chondrites comprise chondrules, calcium aluminum-rich inclusions (CAI), amoeboid olivine aggregates (AOA) and metal inclusions in a silicate matrix

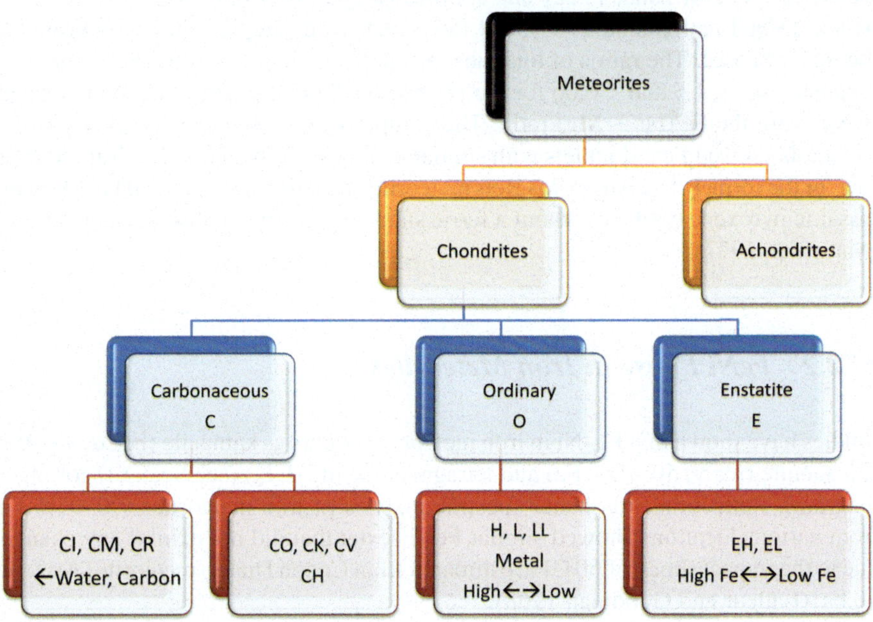

Fig. 11.3 The classification of meteorites based on their similarity and constituents

However, the interstellar dust cloud was part of the much older Milky Way galaxy. Recent studies from chondrite constituents suggest an enhanced star formation rate nearly 7 Ga ago that later became components of our Solar System (Heck et al. 2020).

Meteorites classified as stony and iron are rich in silicates and FeNi alloy, respectively. They are fragments that have undergone melting. The stony meteorites can be either chondrites or achondrites. Iron meteorites help us to understand the core of planets, while stony meteorites are that of the mantle (Knudsen 1989). Fortunately, they are available for study since they had shattered into space as asteroids and comets. The iron meteorites are of interest to Mössbauer researchers for the unusual FeNi ordered phase. The iron meteorites may have the Widmanstätten pattern in octahedrites consisting of kamacite and taenite lamellae, or it may be absent in high and low nickel chondrites known as ataxites and hexahedrites, respectively. The country/location of a meteor shower and the year/age are mentioned as {place, year} in the text.

11.2.1 Stony Meteorites

The meteorites are either collected at random sites by field specialists and amateurs that can be classified as *found* or collected from known impact events designated *fell*. The Mössbauer spectrum of three meteorites Plainview {*found*, Texas, 1917}, Holbrook {*fell*, Arizona, 1912} and Johnstown {*fell*, Colorado, 1924} were investigated (Segel and Hanna 1964). All three meteorite samples showed silicates as the major phases. The ratios of total absorption from olivine to orthopyroxene were obtained as 2.1, 1.5 and < 0.06 for the Holbrook, Plainview and Johnstown, respectively, from the Fe/(Fe + Mg) ratio. The comparison of olivine and orthopyroxene standards showed Fe^{2+} doublets with similar isomer shifts but large quadrupole splitting for the former. The Angra dos Reis meteorite {Brazil, 1869} showed Fe^{2+} bearing fassaitic pyroxene doublet without a ferric state and an optical absorption at 480 nm (Mao et al. 1977).

11.2.2 FeNi Phase in Iron Meteorites

Studies have shown that Fe–Ni in iron meteorites occurs as kamacite (bcc α, 90–95% Fe), taenite (fcc γ, 50–70% Fe) and tetragonal taenite ($L1_0$) (see Sect. 6.7 for phase diagram). Reflection Mössbauer spectral analysis of iron meteorites (90–95% Fe) from various locations showed similar FeNi sextet that did not exhibit any changes due to the impact shock of 60 GPa, estimated from Canon Diablo meteorite{Arizona, 50 ka}(Lafleur and Goodman 1968).

The Mössbauer studies on Cape York meteorite {Greenland, few ka} revealed a tetragonal $L1_0$ phase that could not be obtained under normal conditions (Petersen et al. 1977). The spectrum showed two sextets and a central paramagnetic line, which was interpreted due to the disordered taenite phase. The B_{hf} of the outer symmetric sextet was 34.2 T, and the inner asymmetric broadened sextet was 28.8 T with a quadrupole shift of 0.20 mm/s. The symmetric sextet was assigned to the martensite α_2 phase. The asymmetric inner sextet originated from the $L1_o$ phase with hyperfine parameters similar to neutron-irradiated samples. The ordered $L1_o$ structure was found to be from the 10–100 μm taenite lamellae in the Widmanstätten pattern (Peterson et al. 1977). A detailed analysis of the lamellae revealed the disordering to occur at 737 K for 10 h, resulting in an average B_{hf} of 30.6 T, suggesting the importance of cooling rate on the formation of $L1_o$ ordered alloy (Albertsen et al. 1978a, b). The bcc phase has 7–8% Ni, whereas the taenite phase has 30–50% Ni.

The Santa Catharina {Brazil, 123 Ma} meteorite with the equiatomic FeNi composition exhibited a singlet and a sextet of B_{hf} 29.1 T corresponding to $L1_o$ FeNi that disordered on heating at 732 K for 10 h, similar to Cape York lamellae (Danon et al. 1979). An appreciable change in the spectrum was not observed up to 4.2 K. The ordered phase was from the equiatomic FeNi that were visible as dark patches and the disordered phase had a lighter background of 27% Ni from the backscattered electron microscopy analysis.

While the sextet-singlet pattern in the iron meteorites arises due to the FeNi alloys in the ordered and disordered phases (Albertsen et al. 453, 1978), the composition-dependent ordering was studied further. Experiments with laboratory-prepared disordered FeNi of various compositions showed that the paramagnetic singlet begins close to 30% Ni and transform to broad sextets with increasing Ni, whereas the Fe-rich alloy results in narrow sextet (Albertsen et al. 1980). It was argued that the Cape York meteorite does not have an ordered Fe_3Ni phase as the ordering temperature for the phase, in that case, has to be less than 573 K, and $L1_2$ Fe_3Ni was ruled out as it would not result in quadrupole splitting. It was concluded that the taenite phase segregates into ordered and disordered phases below 593 K, with the ordered $L1_o$ corresponding to equiatomic FeNi composition, whereas the fcc phase with lower Ni gives a disordered component.

The Santa Catharina ataxite with the invar composition of 35% Ni showed a sextet and one central paramagnetic narrow peak that was investigated for the order–disorder characteristics (Danon et al. 1980). The B_{hf} of 28.8 T with the quadrupole splitting of + 0.23 mm/s was assigned to originate from FeNi alloy in the ordered $L1_o$ structure. A similar spectrum was obtained with the electron-irradiated FeNi invar alloy with a lower ordering parameter. It was suggested that different degrees of order would result in a spectrum with a sextet and paramagnetic peak in the phase-segregated invar composition.

The order–disorder studies were conducted by subjecting equiatomic taenite FeNi from Cape York meteorite to thermal treatment (Larsen et al. 1982). The ordered taenite was transformed to the disordered phase after repeated sudden cooling from an annealing temperature of 758–778 K/10–15 min in reducing atmosphere to RT. This could be observed from Fig. 11.4, where the RT Mössbauer spectra of the meteorites (a) Cape York and (b) Santa Catharina in its virgin state showed similar features, indicating ordered sextet and disordered paramagnetic peaks. The increase in the B_{hf} of the sextet due to disordering could be noticed with a higher heating duration from the RT Mössbauer spectrum of the Cape York meteorite after a heating duration of (c) 10 and (d) 110 min at 778 K. The B_{hf} of the FeNi with complete ordering was 28.5 T, which increased to 31.1 T for the fully disordered ($S = 0$) alloy. The complete transformation from order to the disorder required a total duration of 100 min at 778 K and a longer duration at lower annealing temperatures. The decrease in quadrupole shift with increasing disorder and a linear correlation with B_{hf} was also established. The ordered phase composition designated as tetrataenite was inferred to arise from the extreme cooling rate of around 1–10 K/Ma of the meteorite. However, tetrataenite was found even in terrestrial samples (Scorzelli and dos Santos 2019).

11.2.3 Carbonaceous Chondrites

Mössbauer spectra of various carbonaceous chondrites such as Orgueil {France, 1864}, Mighei {Ukraine, 1889}, Murchison {Victoria, 1969} and Allende {Mexico, 1969} are shown in Fig. 11.5 (Poulsen et al. 1981). These chondrites are an important class of carbonaceous chondrites that formed first, preceding the ordinary chondrites, each with its own characteristics, as listed in Table 11.1. The chondrite CI supports organic building blocks, CV is dry with refractory inclusions, whereas CM type is intermediate between the two.

Mössbauer analysis of Orgueil showed doublets from phyllosilicate chlorite in a ferric state and pyroxene in a ferrous state along with Fe_3O_4 sextet. Mighei and Murchison, both CM type, showed clear ferrous and ferric doublets. The Allende exhibited Fe^{2+} with $\Delta = 2.9$ mm/s and $\delta = 1.2$ mm/s, indicative of $Mg_{2-x}Fe_xSiO_4$ olivine compound. It also had weak ferric components from other minerals. Thus, the CI chondrites showed the largest ferric ion content, while the CV type had the largest ferrous content. The presence of water content and carbon–water content can be related to the phyllosilicate in the CI-type chondrite. The carbon content in chondrites of the order CI > CM > CV implies the oxidizing environment in the highly ferric chlorite in CI and dry ferrous olivine in CV.

The Orgueil was extensively studied for its superparamagnetic phase, first suggested to be due to magnetite (Wdowiak and Agresti 1984) and later inferred to be a ferric compound (Madsen et al. 1986). Petrographic and transmission electron microscopic examination revealed that the Orgueil CI matrix has Fe-bearing serpentine, saponite and ferrihydrite fine particles (Tomeoka and Buseck 1988). The superparamagnetism may be from the ferrihydrite fine particles that had undergone aqueous modification. Further studies showed Orgueil has phyllosilicate of saponite-serpentine as the major fraction and magnetite along with ferrihydrite (Bland et al. 2004). Allende was reported to have 75–80% of the total Fe as olivine, 15% pentlandite $(Fe, Ni)_9S_8$ with other minor constituents. Olivine was a minor constituent in random Orgueil samples whose fraction diminished with phyllosilicate. While the terrestrial or pre-terrestrial oxidization to Fe-oxides is debatable, it becomes clear that olivine transforms into phyllosilicates in the presence of water. Hence, the amino

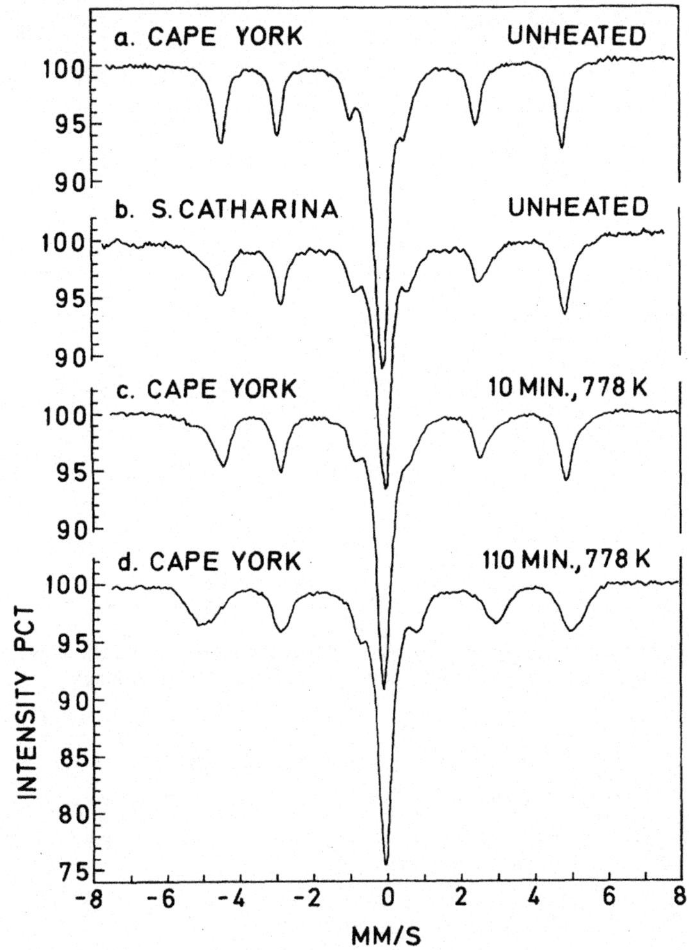

Fig. 11.4 The RT Mössbauer spectra of the meteorites **a** Cape York and **b** Santa Catharina in its virgin state with similar features indicating ordered sextet and disordered paramagnetic singlet. The RT Mössbauer spectra of Cape York meteorite after a heating duration of **c** 10 and **d** 110 min at 778 K. The increase in the B_{hf} of the sextet due to disordering could be noticed with higher heating duration. Reprinted with permission (Larsen et al. 1982), ©The American Physical Society

acids of extraterrestrial origin from a different parent body on Orgueil (Ehrenfreund et al. 2001), supported by its moist environment reaching Earth gain traction.

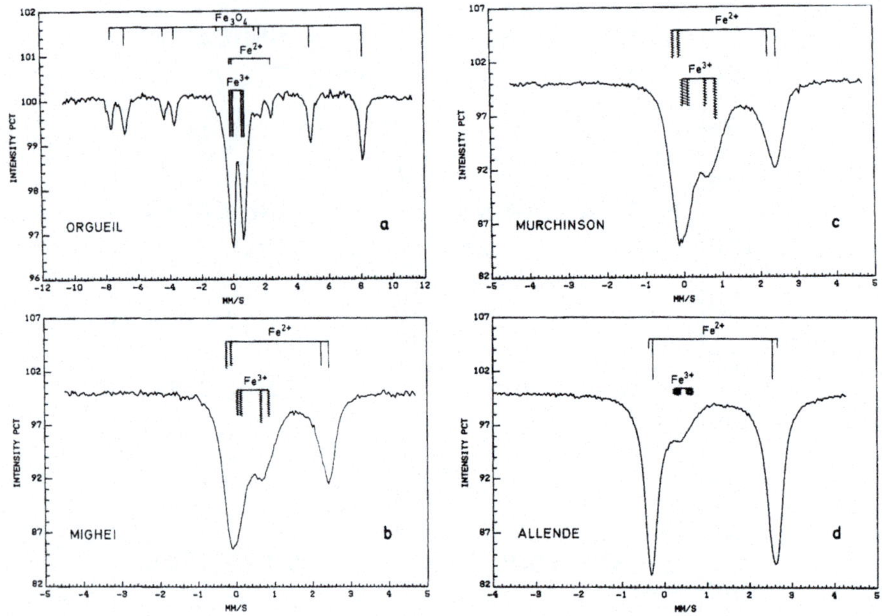

Fig. 11.5 RT Mössbauer spectra of meteorite samples **a** Orgueil, **b** Mighei, **c** Murchinson and **d** Allende. Reprinted with permission (Poulsen et al. 1981), © IOP Publishing

Table 11.1 Few carbonaceous chondrites and their characteristics

Chondrite	Type	Characteristics	References
Orgueil	CI	Amino acids from organic-water interaction in comets	Ehrenfreund et al. (2001)
Mighei	CM	Rich in organic and volatile elements{S, C (2.6%), H, N and O}, water content (12%), 4.54 Ga	Vdovyskin (1973)
Murchison	CM	Organic, Pre-solar grains that are part of old stars	Heck et al. (2020)
Allende	CV	4.57 Ga old refractory inclusions, precise age	Allegre et al. (1995)

11.2.4 Enstatite Chondrites

Mössbauer investigation of Abee {Alberta, 1952} enstatite chondrite meteorite (Fig. 11.6) showed kamacite as the principal phase along with taenite and sulfides (troilite and niningerite) (Dunlap 1997). The majority of the samples are Fe metal-rich and Fe-free enstatite, among a total of 32 wt.% Fe, and it could be classified as EH type chondrite that is rare. The RT B_{hf} of kamacite, taenite and troilite are 33.4, 31.1 and 30.6 T, respectively. The three sextets gave a total relative area of 82%,

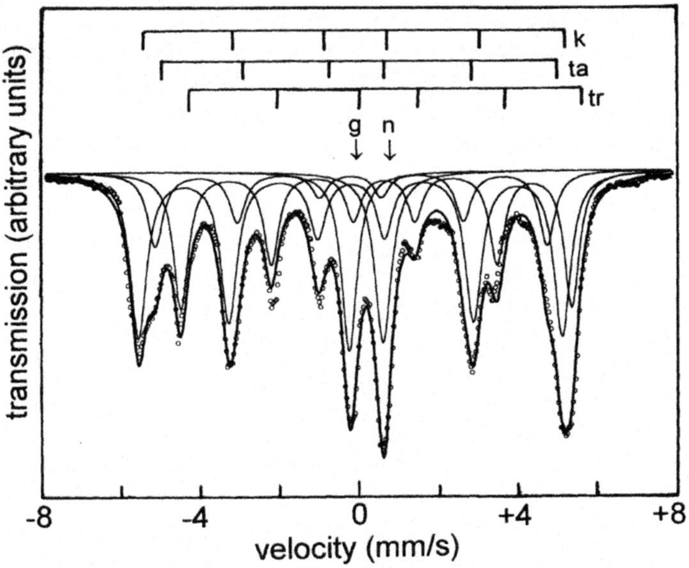

Fig. 11.6 Mössbauer spectrum of the Abee enstatite chondrite. The sextets corresponding to (k) kamacite, (ta) taenite (tr) troilite and singlets from (g) γ-FeNi and (n) niningerite are shown. Reprinted with permission (Dunlap 1997), ©Springer Nature

with kamacite giving 32%. Fe rich γ-FeNi and off-stoichiometric niningerite (XFeS phase, X = Mg, Mn, Ca, Cr, Zn) were assigned with singlets having isomer shifts of − 0.04 and + 0.81 mm/s, respectively, while troilite showed $\delta = +\ 0.75$ mm/s.

The EL enstatite meteorite fragments from Neuschwanstein {Bavaria, 2002} showed 55–63 wt. % kamacite and 37–48 wt.% troilite from the spectrum at RT that could be fitted only with sextets (Hochleitner et al. 2004). Enstatite and plagioclase are the silicate minerals present in the chondrite. The taenite sextet was absent in comparison with the Abee chondrite. The RT B_{hf} of kamacite was nearly 33.3 T, and that of troilite was 30.6–31.0 T. The variation in quadrupole splitting of troilite from − 0.08 to − 0.85 mm/s was explained due to other constituent elements such as Ni, Co, Mn, Cr or Ti.

11.2.5 Ordinary Chondrites

The LL3 type unequilibrated chondrite ALHA 77,278 {Antarctica, 1977–78} was sliced, and Mössbauer spectra were recorded, selecting the chondrules (Poulsen et al. 1982). The chondrules were made up of either olivine or pyroxene without troilite. The composition of the chondrules varies even within a short distance in a thin slice, and one of them showed the presence of kamacite. The matrix comprises silicates, troilite, kamacite and taenite, whereas tetrataenite is absent. The study suggested the formation of chondrules first, followed by their incorporation into the matrix.

The ordinary L chondrite, Putinga {Brazil, 1937} presented distinct Mössbauer spectra at RT and 80 K with doublets from paramagnetic taenite, orthopyroxene, olivine (fayalite), a ferric component in addition to sextets from troilite and kamacite (de Oliveira et al. 1988). The Δ of olivine (2.97 mm/s) showed higher values than orthopyroxene (2.12 mm/s), whereas the ferrous δ was nearly the same (1.14–1.15 mm/s). The hyperfine parameters of kamacite, taenite and troilite were also in accordance with previous reports. Olivine (44%) and orthopyroxene (21%) are the major constituents, along with troilite (15%), FeNi as paramagnetic taenite (6%) and ferric Fe (1%). After magnetic separation, the area fraction of the sextet components increased, and kamacite was identified. The chemical analysis showed 19 wt.% FeNi. Ferric components are usually not found in ordinary chondrites, whereas the presence of about 1% Fe^{3+} was suggested due to terrestrial oxidation.

The Mössbauer analysis of Jilin stony meteorite {China, 1976}, an ordinary chondrite classified as type H5, showed kamacite, troilite, olivine and pyroxene in Fe^{2+} and Fe^{3+} states (Grandjean et al. 1998). The ratio of olivine to pyroxene was 1.7–2.1. The kamacite area fraction ranged from 37–12% without any taenite fraction for different samples of the Jilin meteorite. Previous studies on the Jilin meteorite reported 11% kamacite and 3% taenite (Zhang et al. 1994). The ferric component was absent in many other ordinary meteorites, whereas many older samples showed Fe-oxides, suggesting the role of atmospheric weathering (Maksimova and Oshtrakh 2019).

The Didwana-Rajod meteorite {India, 1991}, which was earlier classified as an enstatite chondrite based on chemical studies alone, was identified as an H-type ordinary chondrite from the Mössbauer analysis (Paliwal et al. 2000). The olivine-pyroxene relative area was nearly 57%, whereas troilite and FeNi were 23 and 20%, respectively. The H-type was assigned since the FeNi fraction is lower than enstatite chondrites. The estimated phases of the Itawa-Bhopji meteorite {India, 2000} from the relative area of the RT Mössbauer spectrum are olivine (58%), pyroxene (20%), troilite (16%) and Fe–Ni (6%) (Verma et al. 2002). The Fe–Ni sextet was mostly kamacite, and other sextets corresponding to Fe-oxides were absent. This meteorite was classified as L/LL-type ordinary chondrite due to its low metallic content. While H and L/LL could be identified from the metallic Fe fraction, further distinctions between L and LL are difficult.

High-velocity resolution Mössbauer spectrometry revealed phases such as Fe(Ni, Co) and M1 and M2 sites in olivine and pyroxene (Oshtrakh et al. 2008). The quadrupole splitting of the sites and the phases are given as Δ(M1) > Δ(M2) in both olivine and pyroxene, besides Δ(olivine) > Δ(pyroxene) from the high-velocity resolution Mössbauer spectrometry with six different ordinary chondrites (Zhinganova et al. 2007). Chelyabinsk meteorite {Russia, 2013} was classified as LL type ordinary chondrite that showed Mg ferrite in few fragments apart from other minerals using high-velocity resolution (Oshtrakh et al. 2019).

A good linear correlation between the relative part of Fe^{2+} in olivine vs. pyroxene indicates that H and (L, LL) ordinary chondrites could be classified from the silicate sub-spectra using Mössbauer spectrometry (Maksimova and Oshtrakh 2019). H ordinary chondrites show a higher relative area of ferrous ions in pyroxene, whereas L and LL chondrites show higher ferrous olivine components at the expense of pyroxene. The understanding of the origin of Fe-oxides is important to identify the transformation of the chondrites from extraterrestrial aqueous weathering. However, according to the constituents, ordinary chondrites are more vulnerable to terrestrial weathering (Bland et al. 1996).

11.2.6 Achondrites

The achondrites that have undergone melting and differentiation can be primitive, asteroidal, Martian or planetary. The major classifications and sub-groups of achondrites are shown in Fig. 11.7. Achondrites are stony meteorites that belong to the class of meteorites formed after the chondrites. Diogenites, eucrites and howardites are subgroups of HED achondrites, and they are fragments of the asteroid Vesta. Angrites are named after Angra dos Reis stony meteorite, while aubrites are enstatite achondrites. SNC meteorites, named after the found locations Shergotty (India), Nakhla (Egypt) and Chassigny (France), that are of Martian origin can be grouped into shergottites (S), nakhlites (N) and chassignites (C). Shock-induced diamond formation is typical in some primitive achondrites, such as ureilites.

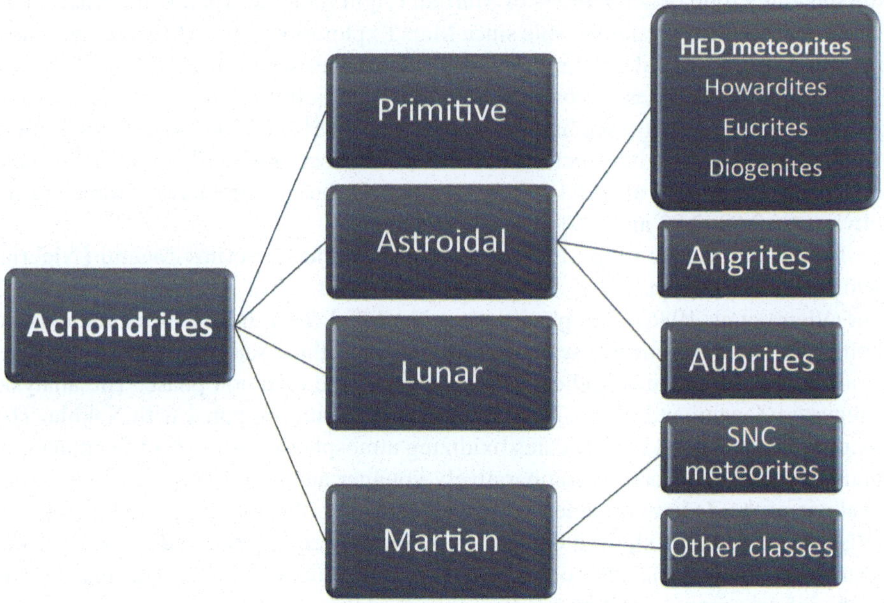

Fig. 11.7 The four major classifications of achondrites and sub-classifications based on their origin

The RT Mössbauer spectrum of the Ibitira meteorite {Brazil, 1957}, a eucrite achondrite, showed overlapping doublets from Fe^{2+} in clinopyroxene (pigeonite) (Costa et al. 1989). The hyperfine parameters from one inner and another outer doublet in M1 and M2 sites, respectively, correspond to the pigeonite $(Mg,Fe,Ca)_2Si_2O_6$. A third less intense doublet was assigned to ilmenite. The pigeonite doublets could be resolved at 55 K with Δ values 3.04 and 2.11 mm/s.

Diogenites are Mg rich orthopyroxene ($Mg_xFe_{1-x}SiO_3$) meteorites. The orthopyroxene in the diogenite ALHA 77,256 {Antarctica, 1977–78} could be identified from the Fe^{2+} doublets corresponding to the M1 and M2 sites, giving quadrupole splitting of ~ 2 mm/s (Vieira et al. 1983). Variation in the intensities of the first and second lines in the unresolved doublet arises due to differences in the preferential orientation of the crystallites. The eucrite meteorite, Stannern {Czech, 1808}, was fitted with four doublets corresponding to pyroxene, inferred as orthopyroxene with augite-type exsolution lamellae and the innermost doublet to a small fraction of ilmenite. In contrast to terrestrial pyroxene rocks, the achondrites did not exhibit a ferric state, implying its origin under reducing conditions.

11.2.7 Martian Achondrites

The study of SNC meteorites helps us to directly correlate with Martian mineralogy as Mars is within our reach for future human landing. The Allan Hills 84,001 belongs to the separate group of achondrites of Martian origin found in Allan Hills, Antarctica. The meteorites are distinguishable since Mars Exploration Rover (MER) extensively investigated Fe mineralogy through in situ Mössbauer spectrometry.

A summary of phases and oxidization states of various SNC meteorites studied by Mössbauer spectrometry are listed in Table 11.2. ALHA 77,005 {*found*, Antarctica, 1977} achondrite was interpreted as Fe^{2+} to Fe^{3+} shock (> 45 GPa) oxidized olivine meteorite (Ostertag et al. 1984). The Δ for two ferrous and the ferric doublets from olivine are 3.03, 2.83 and 0.54 mm/s, respectively.

The oxidation state from Mössbauer analysis of the meteorites Zagami {Nigeria, 1962}, Nakhla {Egypt, 1911} and EETA 79,001 {Antarctica, 1979–80} were carried out (Vieira et al. 1986). The phases were identified from the RT measurements on both bulk and magnetically separated samples. All three samples showed doublets from the RT Mössbauer studies and pyroxene as the common phase. The analysis suggested Zagami and EETA 79,001 in a reduced state compared with Nakhla. The oxidation state of Nakhla indicates oxidizing atmospheric exposure of the planetary body, presumably Mars, of comparatively younger age (~ 1.4 Ga). The Mössbauer analysis of the Zagami obtained from two different collectors from the UK and US (Vistisen et al. 1992) showed that one is pyroxene rich, as observed for the Zagami shergottite whereas the other one has considerable olivine (~ 23%). This may be due to the inhomogeneous welding of the igneous rock on Mars.

Mössbauer analysis of a series of SNC meteorite samples (Dyar et al. 2003) was undertaken to understand the correlation of Fe^{3+} to oxygen fugacity. Estimating

Table 11.2 The oxidization states of the phases in the SNC meteorites analyzed through Mössbauer spectrometry (1 Ma = 1 million year, 1 Ga = 1 billion year)

Meteorite	Type/Age	Phase/oxidization state	Findings	References
ALHA 77,005	Unique Age- 175 Ma	Olivine (Fe^{2+} in M1 and M2, Fe^{3+} in M2)	Shock oxidized	Ostertag et al. (1984)
Zagami	Shergottite Age ~ 170–230 Ma	Pyroxene (Fe^{2+} in M1 and M2) Pyrrhotite	Reducing condition	Vieira et al. (1986)
EETA79001	Shergottite Age ~ 170 Ma	Pyroxene (Fe^{2+}) Olivine (Fe^{2+})	Reducing condition	
Nakhla	Nakhlite Age ~ 1.4 Ga	Olivine (Fe^{2+}) Pyroxene (Fe^{2+} in M1 and M2, Fe^{3+}) Fe_3O_4 Ulvospinel Ilmenite	Oxidizing condition in the planet	
Zagami	Shergottite Age ~ 170–230 Ma	Pyroxene (Fe^{2+} in M1 and M2) Pyroxene (Fe^{2+} in M1 and M2) Olivine (Fe^{2+} 23%)	Absence of ferric component. Samples from different collectors suggest inhomogeneous igneous welded meteorite	Vistisen et al. (1992)
ALHA 77,005	Unique Age-175 Ma	Olivine (Fe^{2+} 81%) Pyroxene (Fe^{2+} 17%) Fe -oxide (Fe^{3+} 2%)	No relationship between Fe^{3+} and oxygen fugacity. Dehydrogenation rather than oxidation suggested for the presence of Fe^{3+}	Dyar (2003)
ALHA 84,001	Unique Age ~ 4.1 Ga	Pyroxene (Fe^{2+} 93%) Fe -oxide (Fe^{2+} 3%) Fe -oxide (Fe^{3+} 4%)		
Chassigny	Chassignite Age ~ 1.4 Ga Dunite, shocked	Olivine (Fe^{2+} 92%) Pyroxene (Fe^{2+} 8%)		
EETA79001	Shergottite Age ~ 170 Ma	Olivine (Fe^{2+} 64%) Pyroxene (Fe^{2+} 34%) Fe -oxide (Fe^{3+} 2%)		
LEW 88,516	Shergottite Age ~ 180 Ma	Olivine (Fe^{2+} 57%) Pyroxene (Fe^{2+} 40%) Fe -oxide (Fe^{3+} 3%)		
Los Angeles	Shergottite Basaltic Age ~ 3 Ma	Pyroxene (Fe^{2+} 90%) Fe -oxide (Fe^{2+} 7%) Fe -oxide (Fe^{3+} 3%)		
Nakhla	Nakhlite Age ~ 1.4 Ga	Olivine (Fe^{2+} 29%) Pyroxene (Fe^{2+} 67%) Fe-oxide (Fe^{3+} 4%)		
QUE 94,201	Shergottite Age ~ 330 Ma	Pyroxene (Fe^{2+} 93%) Fe-oxide (Fe^{3+} 4%) Fe -oxide (Fe^{2+} 3%)		

(continued)

Table 11.2 (continued)

Meteorite	Type/Age	Phase/oxidization state	Findings	References
Zagami	Shergottite Age ~ 170–230 Ma	Pyroxene (Fe^{2+} 98%) Fe -oxide (Fe^{3+} 2%)		
MIL 03,346	Nakhlite Age ~ 1.4 Ga	Clinopyroxene (Fe^{2+} and 24% Fe^{3+})	Most oxidized from magma	Dyar et al. (2005)
Y-984028	Shergottite Age ~ 190 Ma	Pyroxene (Fe^{2+}, Fe^{3+}) Olivine (Fe^{2+}, Fe^{3+}) Fe^{3+} (10%)	Interior and exterior studied separately. Olivine and pyroxene exterior oxidized. Most oxidized shergottite	Dyar et al. (2011)

olivine and pyroxenes is more straightforward than the identification of clino- and ortho-pyroxenes. However, the studies suggested no direct correlation between ferric ions and oxygen fugacity. Dehydrogenation rather than oxidation mechanism may also produce Fe^{3+} in olivine and pyroxene-rich samples. Studies on MIL 03,346 {*found*, Antarctica, 2003} proposed that the most oxidized augite, with almost 24% Fe^{3+} in the pyroxene, was due to the oxidizing environment of magma (Dyar et al. 2005). The Yamato 984,028 is the most oxidized shergottite-type meteorite (Dyar et al. 2011). The whole rock Fe^{3+} is nearly 10%, and the exterior of the olivine separate is more oxidized than the pyroxene separate. Most SNC meteorites show Fe^{3+} below 5%, except those oxidized under magmatic environments.

11.3 Lunar Mineralogy

The giant impact hypothesis (Hartmann and Davis 1975) suggests the formation of the Moon due to an oblique collision of a Mars-sized body, 'Theia', with the early Earth within ~ 30–200 million years of the Solar System formation. The lunar surface has dark-colored flat plains known as mare (sea) formed by volcanic lava. Apollo 11, the first manned Moon mission, landed in Mare Tranquillitatis, and the landing site is named Tranquility Base. Apollo 17, the last manned mission as of now, landed in the Taurus–Littrow valley southeastern of Mare Serenitatis. The lunar surface also has a lot of impact craters, boulders and highlands. The prominent features, such as the mare and the Apollo landing sites, are visible in the image of the Moon (Fig. 11.8). The light to dark grey color of the Moon, in contrast to the red-colored Mars, is due to the characteristics of the minerals present on the surface.

The Apollo missions carried back a total of 382 kg of samples from the Moon surface as well as from a depth up to 292 cm (Apollo 15–17) (Heiken et al. 1991). The major compounds on the lunar surface, in decreasing order, are SiO_2, Al_2O_3, CaO, FeO, MgO and TiO_2. Oxygen is the most abundant (> 60%) but chemically

Fig. 11.8 The lunar surface has impact craters, highlands and dark areas known as mare (sea). Apollo landing sites are marked. Image credit: NASA/Goddard Space Flight Center Scientific Visualization Studio

bound to the compounds. The atmosphere is inert with pressure as low as 1×10^{-10} Pa, equivalent to ultra-high vacuum conditions in research laboratories.

Figure 11.9 shows the mean mineral proportions on the soils obtained from the landing sites of the Apollo and Luna missions. The significant standard deviation in the mineral proportion is due to the varying soil conditions on the sites. The major mineral is found to be pyroxene with the chemical formula $(Ca,Fe,Mg)SiO_3$, followed by $(Ca,Na)(Al,Si)_4O_8$ from the mineral group plagioclase that is also abundant in the Earth crust, glass, olivine ($(Mg,Fe)_2SiO_4$), ilmenite and silica. The lunar impact and volcanic glasses with varying TiO_2 content exhibit colors from green to brown due to the varying degrees of crystallinity from olivine melt.

The first minerals to crystallize, Mg-rich olivine and pyroxenes, sank in the lunar magma ocean, and buoyant plagioclase ended as lunar highlands. Mössbauer analysis

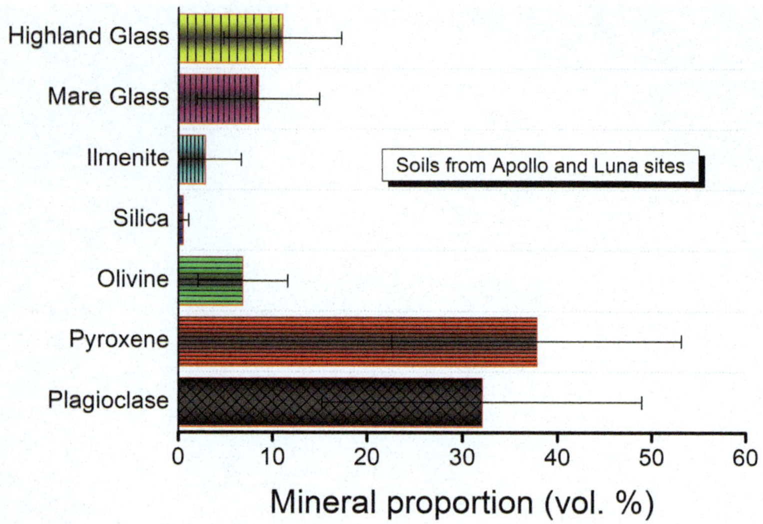

Fig. 11.9 Mean mineral proportion (vol.%) in soils from Apollo and Luna missions. (Heiken et al. 1991)

of lunar rocks and soils from Apollo 11 mission {landing site: Mare Tranquillitatis} for minerals plagioclase, pyroxene, olivine, ilmenite and metallic iron showed 60–68% area fraction for pyroxene followed by 20–27% for ilmenite (Gay et al. 1970). Minor fractions of magnetite in the lunar samples are of unclear origin, such as terrestrial contamination, meteorites or weathering. About 73% of the total iron was in the silicate phases, 17% in ilmenite and about 5% in metallic iron (Herzenberg and Riley 1970). Metallic iron in fine lunar samples is also alloyed with nickel and found as troilite. The Fe metal may be from the reduction of silicate during the formation of glass or of meteoritic origin (Muhir et al. 1970). Exposure of lunar fine soils to air for two months did not undergo significant changes in the Mössbauer spectrum.

Lunar rocks and soils from the Apollo missions are composed of three doublets attributed to (i) olivine and M1 site in pyroxene, (ii) glass and M2 site in pyroxene and (iii) ilmenite (Gibb et al. 1974). The RT quadrupole splitting was of the following order; $\Delta\{(\text{M1/Olivine}) = 2.81\text{–}3.00\ \text{mm/s}\} > \Delta\{(\text{M2/glass}) = 2.00\text{–}2.24\ \text{mm/s}\} > \Delta\{(\text{Ilmenite}) = 0.70\ \text{mm/s}\}$. The δ values of all the components were 1.1–1.2 mm/s due to the Fe^{2+} state. Ferric state minerals were absent in the lunar soil, suggesting the lack of oxidizing conditions. A combined analysis of the lunar samples revealed varying ilmenite from < 0.2% to 8% and less than 10% of metallic Fe among total Fe (Gibb et al. 1977). The pyroxene minerals present in the lunar samples are identified as the clinopyroxenes pigeonite and augite, apart from orthopyroxene. Another study pointed to larger quadrupole splitting for the pyroxene doublet for the brown glasses and the lowest for green glasses due to the varying degrees of amorphousness (Zemcik and Cimbalnikova 1986).

Figure 11.10 shows the RT transmission and backscatter Mössbauer spectra of fine soil obtained close to a mare basalt boulder from the Apollo 17 mission {landing site: Taurus-Littrow valley} (Morris et al. 1998). A part of the 71131 numbered < 1 mm soil (~ 86 g) was subjected to Mössbauer analysis. The Mössbauer spectrum of the fine soil is composed of quadrupole split doublets from silicates, ilmenite and weak sextet from Fe. The Apollo 17 mare basalt soil has high FeO (16–18 wt.%) and ilmenite contents. The Ti-rich basalts brought by Apollo 17 are 3.6 Ga old. The transmission and backscatter Mössbauer spectra showed similar patterns, suggesting that the measurement geometries would not significantly influence the analysis of lunar soil. The quality of the data from backscatter measurements is affected by weak radioactive sources.

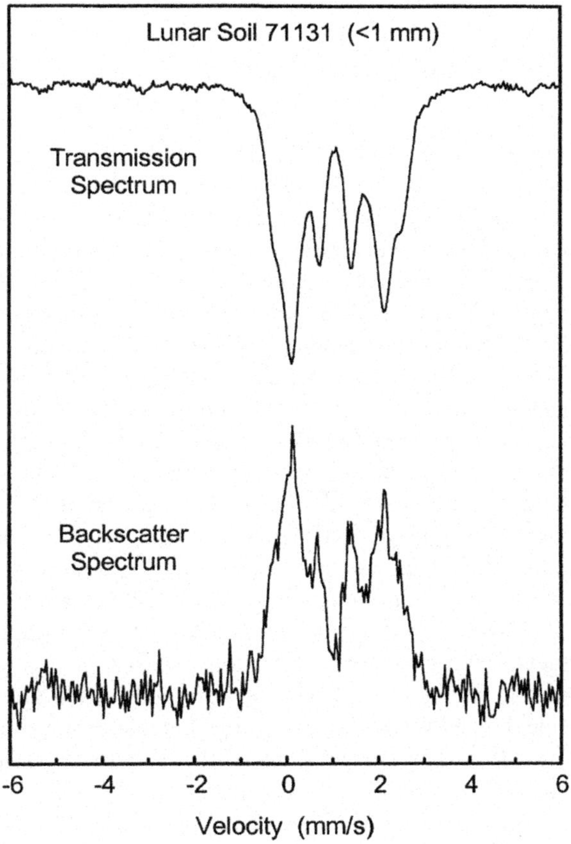

Fig. 11.10 The RT transmission and backscatter Mössbauer spectra of the lunar fine soil obtained close to a basalt boulder. The 71131 numbered < 1 mm soil (~ 86 g) was from the Apollo 17 mission. Reprinted with permission (Morris et al. 1998), ©Springer Nature

11.4 Mars Exploration

11.4.1 Mars Overview

Mars is a differentiated terrestrial planet with a mean density of 3.94 g/cc which is 1.4 times lower than Earth. It has a surface pressure of 610 Pa which is in the medium vacuum level. The atmosphere comprises 95% CO_2, N_2, Ar and trace amounts of oxygen and water vapor. Dust storms are common on the surface of Mars, which influences the performance of rovers. The only planet investigated through Mössbauer spectrometry is Mars, using the onboard instrument, miniaturized Mössbauer spectrometer (MIMOS) in Mars Exploration Rover (MER).

11.4.2 MIMOS

The Mössbauer spectrometry of Moon soils was based on the measurements from the samples brought by lunar missions. The Mars exploration included an in situ lightweight backscatter spectrometer, known as a miniaturized Mössbauer spectrometer, that was an outcome of many years of development (Klingelhöfer et al. 1992, 1996). The design of the MIMOS had a few features such as (i) source with an activity about 1 Ci/cm^2 with optimum shielding for narrow linewidth, (ii) small (40 mm length) and lightweight drive (50 g) that could withstand shocks up to 100 *g*, (iii) Si-PIN diode detector, (iv) reference sample for calibration, (v) low power consumption (30 mW) and (vi) total mass less than 500 g.

The Mars Express mission had the MIMOS-II payload in Beagle II that was expected to begin its operation at the end of December 2003. Although MIMOS landed on Mars, it could not be put to use due to Beagle's failed solar panel deployment. The features of MIMOS-II on Mars exploration rovers are (Klingelhöfer et al. 2003): it had narrow source linewidths, optimum shielding and a collimator within acceptable cosine smearing that could distinguish 1:1 hematite-magnetite mixtures. Simultaneous backscatter measurements of the sample and transmission measurements of ^{57}Fe enriched calibration standards, α-Fe_2O_3 and α-Fe, were possible using double source arrangement on both ends of the drive system. The aluminum tube shielding between the drive and pickup coil in the double-loudspeaker assembly was provided to prevent cross-talk. Velocity linearity was adjustable in the range 0.1 ~ 3% to operate in the Martian temperature range of 80–283 K. Four Si-PIN diodes with an energy resolution of 1.0–1.5 keV for backscatter measurements and another Si-PIN diode for the reference absorber together consumed less than 5 mW power. There were three temperature sensors, each measuring the sample temperature (Martian temperature), reference standard and the electronics assembly. The Mössbauer spectrum was recorded in a 512 velocity channel per spectrum that could store the data in EEPROM.

11.4.3 Mars Exploration Rover

Spirit and Opportunity are the rovers of the MER missions launched in June and July 2003, respectively, to study the surface of Mars by the National Aeronautics and Space Administration (NASA). A Martian solar day, referred to as sol, has a mean period of 24 h 39 min 35.2 s. The rover solar arrays generate about 35 W/h/sol under full illumination. The rovers are designed to withstand temperatures in the range 233–313 K. The rover body is protected by heaters as night time temperature on Mars may go down to 174 K. Spirit landed on 4th January 2004 in Gusev Crater and continued operation till 22nd March 2010 (sol 2208), ending its journey in Home Plate plateau. Opportunity landed 21 days after Spirit in Meridiani Planum and traveled 45 km from Eagle Crater to Perseverance Valley, lasting for 5352 sols (8 Martian years) until 2018.

Mössbauer spectrometer is one among the Athena science payloads with an objective to evaluate the conditions sustainable for life on Mars (Squyres et al. 2003). The artistic picture of the science payloads in the robotic arm of the Mars exploration rover, along with the MIMOS II Mössbauer spectrometer, is shown in Fig. 11.11. The typical duration for one Mössbauer measurement using MIMOS II is approximately 12 h within a temperature variation of $\pm$ 10 K, depending on the source strength and ^{57}Fe content, requiring the rover to be immobile during measurement. Before launch, the ^{57}Co source strength was ~ 300 mCi on the sample side and 5–15 mCi on the calibration side.

11.4.4 Gusev Crater Mineralogy

The Spirit landing site, the 160 km diameter Gusev Crater, has a plain terrain with basalt rocks strewn across and red soil overlooking a few hills. Figure 11.12 shows the Spirit panoramic camera image of Columbia Hills from Gusev Crater taken on sol 149 (3rd June, 2004). The Spirit was 300 m from the base of the hill and nearly 3 km from its landing site. The landing site was chosen since it may be a lake with water-related sediments.

MIMOS-II on board the Spirit rover was successful in determining the ^{57}Fe-mineralogy of Mars. Olivine, pyroxene, nano-ferric oxide, non-stoichiometric magnetite and hematite were detected from the measurements at Gusev Crater rock and soil using backscatter Mössbauer spectrometry (Morris et al. 2004). The counting statistics were improved by summing the spectra of a target at either overall temperature intervals or at the same temperature over similar targets. The hyperfine parameters were obtained from the data summed over the temperature range of 200–280 K. The differences in peak areas arising out of varying f-factors were corrected using the $f(Fe^{3+})/f(Fe^{2+})$ ratio of 1.21. The $\Delta\{(\text{Olivine}) = 3.5$ mm/s$\} > \Delta\{(\text{pyroxene}) = 2.11–2.14$ mm/s$\}$ correspond to Fe^{2+} octahedral doublets. The nano-ferric oxide was identified from the doublet that showed an isomer shift of 0.38 mm/s and quadrupole

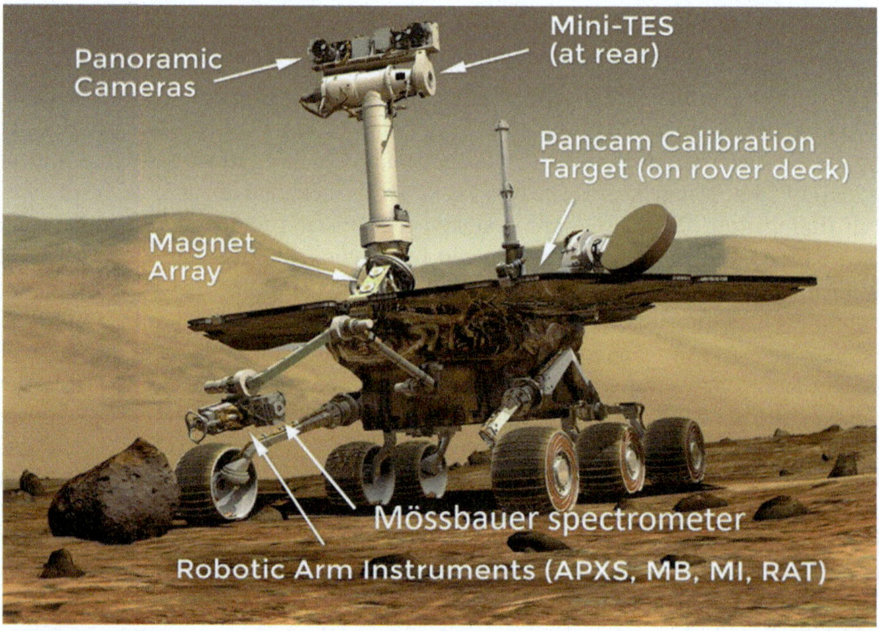

Fig. 11.11 Artistic view of the Mars exploration rover with the Mössbauer spectrometer. Image credit NASA/JPL-Caltech

Fig. 11.12 Mars exploration rover Spirit panoramic camera view of Columbia Hills from Gusev Crater. Image credit: NASA/JPL-Caltech

splitting of 0.81–0.88 mm/s. The sextet components showed B_{hf} of 52.2 T from hematite; 51.4 and 47.8 T from A- and B- sites of magnetite, respectively.

The relative fraction of nanophase ferric oxide was found at least 5 to 39% in rocks and soils, whereas hematite was found only in a few rock target locations ranging from 0 to 13%. The mean values of the f-factor corrected area fraction from the published data are 42%, 29%, 15%, 11% and 3% for the olivine, pyroxene, nano-ferric oxide, non-stoichiometric magnetite and hematite, respectively, in the rock target areas.

The mean value of the area fraction of olivine in soil targets reduced to an area fraction of 36%, while all other mineral area fractions except hematite increased. This suggested that the soil is primarily derived from olivine-bearing basalt from physical weathering. In contrast to the Moon soil, Mars soil has a substantial mean Fe^{3+}/total Fe ratio of 0.3. Preliminary Mössbauer analysis suggested no evidence of clay minerals or water-related sedimentation, although the Gusev Crater was once believed to be a lake.

Starting from the analysis of the Mössbauer spectra of soils and rocks during its first 90 sols, Spirit's journey continued crossing Columbia Hills (sol 156) and Husband Hill ascending Cumberland Ridge, providing a consolidated mineral analysis up to sol 510 (Morris et al. 2006a, b). Apart from the olivine, pyroxene, nanoferric oxide, magnetite and hematite, other minerals such as goethite and ilmenite were identified as the Spirit continued its journey to Husband Hill. Ferric sulfate was obtained in only two of the sampling areas at Paso Robles. The journey path of the Spirit rover with a few of the named landmarks is shown in Fig. 11.13. These areas and minerals are marked from the Spirit paths (Arvidson et al. 2006), and the results of the Mössbauer analysis from the corresponding locations. The area fraction obtained from the Mössbauer spectrum corresponds to only Fe minerals, which has to be born in mind. The increase in Fe-oxides area fraction from the Gusev plane to the hills was at the expense of olivine and pyroxene. The mean Fe^{3+}/total Fe increased to 0.6 from the sampling at West Spur compared with the Gusev plane. Kaolinite-type Al-rich phyllosilicates are proposed in West Spur, while it requires further analysis (Wang et al. 2006).

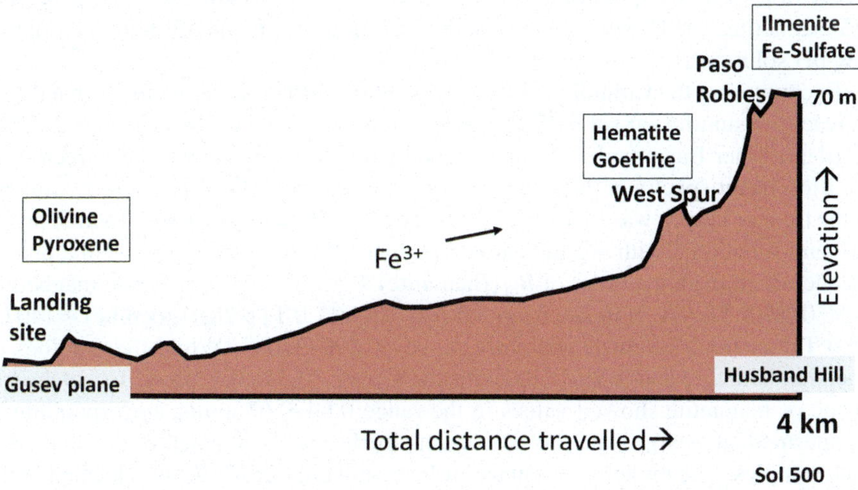

Fig. 11.13 The Spirit rover's travel summary from its landing site to the Cumberland Ridge up to 500 sol, adapted from (Arvidson et al. 2006). The identification of various Fe minerals on the way up to 4 km and elevation 70 m (ordinate) is shown

The rocks of the Gusev Crater are classified into five major types according to their mineral content. In the increasing order of discovery, they are (1) Adirondack Class, (2) Clovis Class, (3) Wishstone Class, (4) Peace Class, (5) Watchtower Class and (6) Backstay Class (Squyres et al. 2006). The Mössbauer spectra of various rock classes are shown in Fig. 11.14. Adirondack Class rocks are un-weathered olivine-rich basalts with the least ferric content (Fe^{3+}/total Fe ≈ 0.12). Clovis class is similar to Adirondack but an altered rock with higher C, Cl, Br and Ni, even up to a depth of ~ 9 mm from samples obtained with a rock abrasion tool. Wishstone Class, one of the less altered rocks, has lower iron but Al as high as 16 wt.%, TiO_2 (typical ~ 3 wt. %) and P_2O_5 up to 5.2 wt. %. Peace Class rocks are low in feldspar and have a substantial S content (12.9 wt.% SO_3) than any other rock. They also show bound water features and are the weakest rock. Watchtower Class is a heavily altered rock possibly from impact debris, with Fe-oxides as high as 82% in ferric state (Fe^{3+}/total Fe > 0.8). Backstay is a basalt/trachybasalt rock that has undergone little alteration. It has ferric ion state next to Adirondack Class (Fe^{3+}/total Fe ≈ 0.23). The rock studies suggested that the Columbia Hills was an impact-related uplift. It was also possible that the Peace Class and Clovis Class rocks had undergone rock-water interaction and that Mars had water.

The soils of Gusev Crater are classified into (Morris et al. 2006a, b) (1) Laguna Class, (2) Paso Robles Class and (3) Other Soils Class. The first class has a composition similar to basaltic bulk, and the second one has a high concentration of S (~3.2 mol/24(O + Cl)). The Other Soils Class has a composition similar to Wishstone and Watchtower Class rocks. There are a few sub-classes associated with the main classification with slightly varying compositions. The Mössbauer area fraction from Paso Robles Class soil showed 65% Fe^{3+}-sulfate, 14% hematite, 9% pyroxene, and 5% magnetite. The Fe-sulfate-rich soil deposit suggests precipitation from oxidized, low-pH solutions.

Encountering the Columbia Hills at West Spur, further traveling to Husband Hill to reach its summit on sol 618, Sprit descended via. Haskin Ridge (sol 662–702), Northern Inner Basin (sol 737) and reached Home Plate (sol 749–1500), recording the Mössbauer spectra at these landmarks (Morris et al. 2008). The source strength at the time of launch was ~ 300 mCi, reduced to ~ 150 mCi on sol 0, ~ 15 mCi on sol 510 and functional with ~ 2 mCi at Home Plate. The hyperfine fields of the phases are identified from the order of $B_{hf}\{(\text{hematite}) \approx 52.0\text{ T}\} > B_{hf}\{(\text{A- site magnetite}) \approx 50.0\text{–}50.5\text{ T}\} > B_{hf}\{(\text{B- site magnetite}) \approx 46.0\text{–}47.0\text{ T}\} > B_{hf}\{(\text{goethite}) \approx (40.0\text{–}35.0\text{ T})\}$ at the Martian temperature range of 190–270 K. While the δ values of goethite, hematite and A-site magnetite are between 0.3 to 0.4 mm/s, the octahedral B-site of magnetite showed values in the range 0.64–0.67 mm/s, corresponding to an intermediate charge state ($Fe^{2.5+}$) of non-stoichiometric magnetite encountered in rock samples. The highest percentage of Fe from hematite (78%) was located at the Low Ridge outcrop rock Montalva on sol 1073. The aqueous alteration of rocks was supported by the presence of goethite on the sides of Husband Hill summit. However, extensive water-induced sedimentation could not be confirmed.

While descending from the summit of Husband Hill, before Home Plate, on 690 sol, the erosion remnant outcrop named Comanche was identified as Mg-Fe carbonate

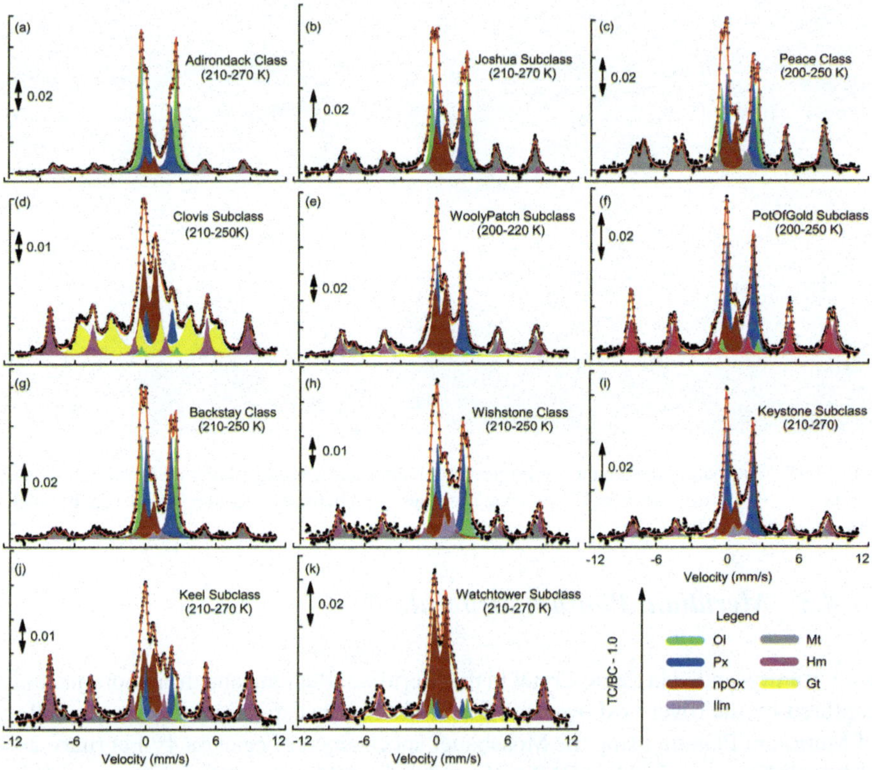

Fig. 11.14 Mössbauer spectra and subspectra of various classes of outcrop rocks encountered in Gusev Crater. **a** Adirondack Class, the least altered rock with the lowest ferric content **b** Joshu sub-class belongs to other unclassified rocks **c** Peace Class is the water-bound rock, **d**, **e** sub-classes of Clovis Class aqueous altered rock that has the highest goethite **f** belongs to other class of rocks, **g** the altered class of rock next to Adirondack, **h** Wishstone Class has ilmenite and low ferric state, **i**, **j** and **k** are sub-classes of Watchtower Class that has the highest hematite and ferric content without goethite and lower magnetite. Reprinted with permission (Morris et al. 2006a), ©Wiley

with 62% magnesite, 25% siderite, calcite and 2% rhodochrosite (Morris et al. 2010). The Mössbauer spectrum gave an absorption area corresponding to 49% ferrous olivine, 24% Mg-Fe ferrous carbonate, 19% nano-ferric oxide and 8% hematite. The Comache outcrop with up to 34 wt. % carbonate is associated with a neutral pH solution, suggesting water in early Mars. The presence of water on Mars nearly 3.5 Ga ago has been postulated with various scenarios, although there is no single model (Wordsworth 2016).

Fig. 11.15 The panorama image of Eagle Crater and surrounding plains of Meridiani Planum captured by Opportunity rover. The outcrop rocks are seen behind the lander. Image credit: NASA/JPL-Caltech/Cornell

11.4.5 Meridiani Planum Mineralogy

Opportunity landed in Eagle Crater in the Meridiani Planum, and the panorama image captured by the rover is shown in Fig. 11.15. The investigation of ^{57}Fe mineralogy of Meridiani Planum using the Mössbauer spectrometer began on 4th February 2004 (Klingelhöfer et al. 2004). MIMOSII spectra from Meridiani Planum were recorded with a source strength of ~ 150 mCi at landing and in the temperature interval of 10 K stored at separate memory locations between the temperature range 180–290 K. The instrumentation and analysis methodology were similar to the data obtained from Spirit.

The spectra of the outcrop at Eagle Crater were composed of doublet and octahedral ferric sextet, as shown in Fig. 11.16. Octahedral ferric state doublet with $\Delta = 1.22$ mm/s and $\delta = 0.33$ mm/s was obtained at 240 K. Jarosite, (K, Na)(Fe, Al)$_3$(SO$_4$)$_2$(OH)$_6$, was identified from the large Δ that is typical for K, Na and Al^{3+} substitutions in the ferric site. The other phases identified are hematite, pyroxene and olivine. Another basalt rock, named Bounce Rock, is unique as it had only pyroxene octahedral ferrous doublets characterized by $\Delta = 2.08$–2.79 mm/s and $\delta = 1.17$–1.20 mm/s in the temperature range 200–280 K, with a linear increase in Δ on decreasing temperature. The igneous type Bounce Rock is a pigeonite (medium to low Ca pyroxene) predominant impact ejecta from a crater located 75 km from the Eagle Crater (Zipfel et al. 2011).

The Meridiani Planum rocks are classified into two types (Morris et al. 2006a, b) (1) Burns Outcrop Class, a S-rich jarosite-hematite outcrop rock and (2) Other Rocks Class. The soils are classified into Laguna Class, which is similar to that found in Gusev Crater and Berry Class. Berry Class soil is made of hematite-rich (~40%) spherules and fragments of typical diameter 1–3 mm. The hematite content in

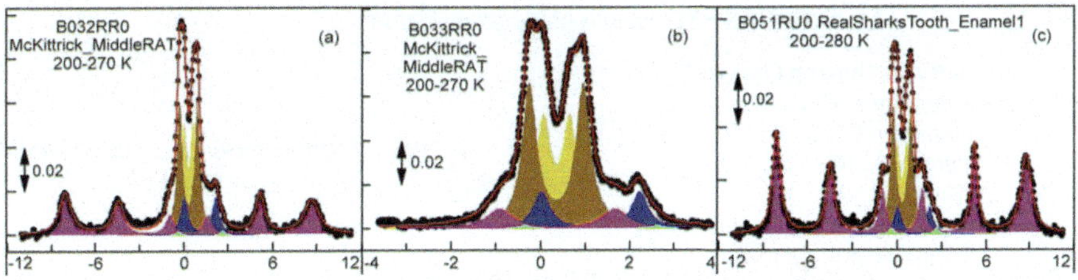

Fig. 11.16 Mössbauer spectra of rock outcrops from Meridiani Planum that are rich in **a** jarosite (full velocity scale), **b** jarosite in the low-velocity region and **c** hematite. (The fitted sub-spectrum is shown as: brown-jarosite, pink sextet–hematite, low-intensity blue-pyroxene, green-olivine; is very low in intensity, and the innermost doublet - unassigned ferric component). Reprinted with permission (Morris et al. 2006a), ©Wiley

spherules ranges from 24–100%. The Burns Outcrop Class showed similar features at all locations with a high Fe^{3+}/total Fe ratio of 0.8–0.9. On the other hand, isolated rocks such as Bounce Rock are the least oxidized without any ferric content. Rocks of different origins, such as iron meteorite rich in kamacite found near the rim of Endurance Crater and the stony meteorite Barberton, come under the Other Rocks Class on Mars. The iron meteorite and the hematite spherules are shown in Fig. 11.17 as left and right images, respectively.

The iron meteorite (~ 7 wt. % Ni) found on sol 339 by Opportunity was named *Meridiani Planum.* It exhibited kamacite sextet without any taenite and a low intense ferric doublet (van Cromphaut et al. 2007). The B_{hf} was 34.4 T for the sextet, and the Fe^{3+} doublet showed Δ of 0.73–0.78 mm/s in the temperature range 210–260 K. The ferric doublet, consistent with nano-ferric oxide found on Mars, may be an outcome of the environmental conditions.

Fig. 11.17 (Left image) The Meridiani Planum iron meteorite (~ 30 cm size), captured using the panoramic camera by Opportunity on sol 339 and (Right image) the hematite spherules pictured on Berry Bowl outcrop rock on sol 48. The dust-free circular impression was made by RAT brushing for 10 min. The Berry Class soil is seen surrounding the outcrop. Image credit: NASA/JPL-Caltech/Cornell

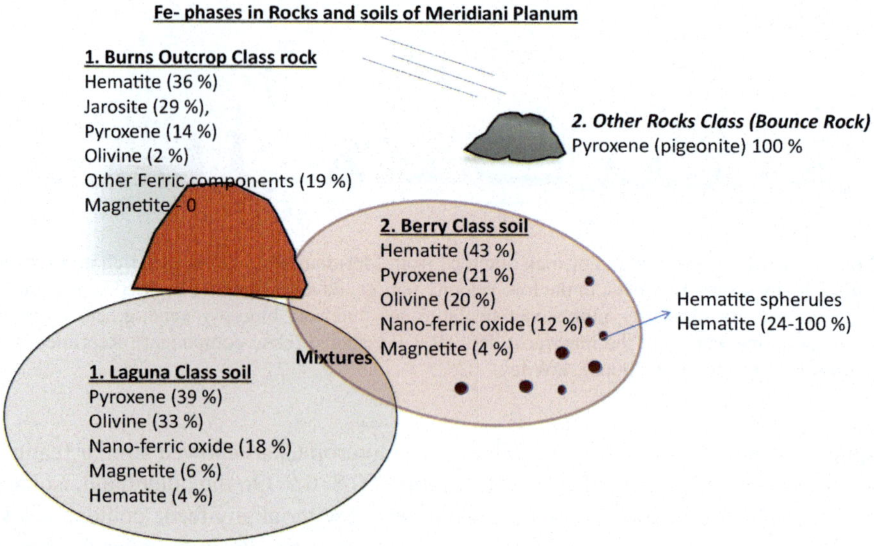

Fig. 11.18 Meridiani Planum has rocks and soils with different Fe-mineralogy, such as Bounce Rock made of only pyroxene, outcrop that is rich in jarosite and hematite, Laguna Class basaltic soil rich in pyroxene, and Berry Class soil that has more than 40% hematite

The summary of rock and soil analysis of Meridiani Planum by Opportunity based on ^{57}Fe Mössbauer spectrometry is illustrated in Fig. 11.18. It is to be noted here that Meridiani Planum has more pyroxene than olivine in its rocks and soils compared to the Gusev Crater. The Bounce Rock and rock fragments of different origins such as meteorites come under the Other Rocks Class. The ferric to total Fe ratio was the lowest for the Laguna Class soil and highest for the Burns Outcrop Class rock. The ferric ion contents in a few classes of rocks and soils are surprisingly large, suggesting their modification in a highly oxidative environment.

Laboratory studies indicated that the red color of Mars had been associated with the nano-hematite ever before the MER mission (Morris et al. 1989). The bright and dark soil found on Mars is facilitated by wind transfer as the soil nearby the rocks is not of the same composition (Yen et al. 2005). Analysis of the Martian soil indicates that the dark soil has low sulfur except for the hematitic soil. The bright soil having the nanophase ferric oxide with distinct properties is a global soil unit, unrelated to the rock composition, also present on the rock surface. The nanophase iron oxide component and Fe-sulfate found in the soils are closely associated with sulfur-related weathering. It was suggested that the bright and dark soil found in the Meridiani Planum (see Fig. 11.15) and Gusev Crater are transported by wind-related processes. The dark soil is from less altered basaltic material consisting of olivine and pyroxene from the mean percentage obtained through Mössbauer analysis. The low concentration of highly soluble Br ($< \sim 50$ ppm) in the soil, but enhanced in low-lying areas indicates the diurnal or seasonal cycling of a thin layer of water.

The MER studies, supported by the Mössbauer spectrometry, suggest liquid water activity, although Mars soil has not undergone extensive oxidative weathering by water.

11.5 Summary

This chapter has presented the Mössbauer spectrometry of extraterrestrial nature such as meteorites, Moon and Mars. The phase analysis from samples of extraterrestrial origin reveal similarities with Earth-based minerals that undergo changes under extreme environment or weathering. The results of the study of the Mars soil using a specially designed Mössbauer spectrometer and an experimental strategy for taking account of the planet's surface temperatures when analyzing the spectra are also discussed.

References

N. Abreu (ed.), *Primitive Meteorites and Asteroids* (Elsevier, MA, 2018)
J.F. Albertsen, G.B. Jensen, J.M. Knudsen, Nature **273**, 453 (1978a)
J.F. Albertsen, J.M. Knudsen, N.O.R. Poulsen, L. Visisen, Phys. Scr. **22**, 171 (1980)
J.F. Albertsen, M. Aydin, J.M. Knudsen, Phys. Scr. **17**, 467 (1978b)
C.J. Allegre, G. Manhes, C. Gopel, Geochim. Cosmochim. Acta **59**, 1445 (1995)
R.E. Arvidson, F. Poulet, R.V. Morris, J.P. Bibring, J.F. Bell III, S.W. Squyres, P.R. Christensen, G. Bellucci, B. Gondet, B.L. Ehlmann, W.H. Farrand, R.L. Fergason, M. Golombek, J.L. Griffes, J. Grotzinger, E.A. Guinness, K.E. Herkenhoff, J.R. Johnson, G. Klingelhöfer, Y. Langevin, D. Ming, K. Seelos, R.J. Sullivan, J.G. Ward, S.M. Wiseman, M. Wolff , J. Geophys. Res. **111**, E02S01 (2006)
P.A. Bland, F.J. Berry, T.B. Smith, S.J. Skinner, C.T. Pillinger, Geochim. Cosmochim. Acta **60**, 2053 (1996)
P.L. Bland, G. Cressey, O.N. Menzies, Meteor. Planet. Sci. **39**, 3 (2004)
T.V.V. Costa, V.W. Vieira, M.A.B. de Araujo, Phys. Scr. **40**, 702 (1989)
J. Danon, R.B. Scorzelli, I.S. Azevedo, J. Laugier, A. Chamberod, Nature **284**, 537 (1980)
J. Danon, R. Scorzelli, I.S. Azevedo, W. Curvello, J.F. Albertsen, J.M. Knudsen, Nature **277**, 283 (1979)
J.C.P. de Oliveira, M.I. de Costa Jr., A. Vasquez, A. Roisenberg, N. Vieira Jr., J.O. Chies, Phys. Scr. **37**, 185 (1988)
R.A. Dunlap, Hyperfine Interact. **110**, 209 (1997)
M.D. Dyar, A.H. Treiman, C.M. Pieters, T. Hiroi, M.D. Lane, V. O'Connor, J. Geophys. Res. **110**, E09005 (2005)
M.D. Dyar, Meteorit. Planet. Sci. **38**, 1733 (2003)
M.D. Dyar, T.D. Glotch, M.D. Lane, B. Wopenka, J.M. Tucker, S.J. Seaman, G.J. Marchand, R. Klima, T. Hiroi, J.L. Bishop, C. Pieters, J. Sunshine, Polar Sci. **4**, 530 (2011)
P. Ehrenfreund, D.P. Glavin, O. Botta, G. Cooper, J.L. Bada, Proc. Natl. Acad. Sci. U.S.A. **98**, 2138 (2001)
P. Gay, G.M. Bancroft, M.G. Brown, Science **167**, 626 (1970)
T.C. Gibb, R. Greatrex, N.N. Greenwood, J. Chem. Soc., Dalton Trans. 1148 (1974)
T.C. Gibb, R. Greatrex, N.N. Greenwood, Phil. Trans. Roy. Soc. Lond. A **284**, 157 (1977)

F. Grandjean, G.J. Long, D. Hautot, D.L. Whitney, Hyperfine Interact. **116**, 105 (1998)
W.K. Hartmann, D.R. Davis, Icarus **24**, 504 (1975)
P.R. Heck, J. Greer, L. Koop, R. Trappitsch, F. Gyngard, H. Busemann, C. Maden, J.N. Avila, A.M. Davis, R. Wieler, Proc. Natl. Acad. Sci. U.S.A. **117**, 1884 (2020)
G.H. Heiken, D.T. Vaniman, B.M. French (eds.), *Lunar Sourcebook* (Cambridge University Press, NY, 1991)
C.L. Herzenberg, D.L. Riley, Science **167**, 683 (1970)
R. Hochleitner, K.T. Fehr, G. Simon, J. Pohl, E. Schmidbauer, Meteorit. Planet. Sci. **39**, 1643 (2004)
G. Klingelhöfer, B. Fegley Jr., R.V. Morris, E. Kankeleit, P. Held, E. Evlanov, O. Priloutskii, Planet. Space Sci. **44**, 1277 (1996)
G. Klingelhöfer, J. Foh, P. Held, H. Jager, E. Kankeleit, R. Teucher, Hyperfine Interact. **71**, 1449 (1992)
G. Klingelhöfer, R.V. Morris, B. Bernhardt, C. Schroder, D.S. Rodionov, P.A. de Souza Jr., A. Yen, R. Gellert, E.N. Evlanov, B. Zubkov, J. Foh, U. Bonnes, E. Kankeleit, P. Gutlich, D.W. Ming, F. Renz, T. Wdowiak, S.W. Squyres, R.E. Arvidson, Science **306**, 1740 (2004)
G. Klingelhöfer, R.V. Morris, B. Bernhardt, D. Rodionov, P.A. de Souza Jr., S.W. Squyres, J. Foh, E. Kankeleit, U. Bonnes, R. Gellert, C. Schroder, S. Linkin, E. Evlanov, B. Zubkov, O. Prilutski, J. Geophy. Res. **108**, 8067 (2003)
J.M. Knudsen, Hyperfine Interact. **47**, 3 (1989)
L.D. Lafleur, C.D. Goodman, Science **162**, 1268 (1968)
L. Larsen, H.R. Poulsen, N.O.R. Poulsen, L. Vistisen, J.M. Knudsen, Phys. Rev. Lett. **48**, 1054 (1982)
M.B. Madsen, S. Mørup, T.V.V. Costa, J.M. Knudsen, M. Olsen, Nature **321**, 501 (1986)
A.A. Maksimova, M.I. Oshtrakh, J. Mol. Struct. **1186**, 104 (2019)
H.K. Mao, P.M. Bell, D. Virgo, Earth Planet. Sci. Lett. **35**, 352 (1977)
R.V. Morris, D.G. Agresti, H.V. Lauer Jr., J.A. Newcomb, T.D. Shelfer, A.V. Murali, J. Geophys. Res. **94**, 2760 (1989)
R.V. Morris, G. Klingelhöfer, B. Bernhardt, C. Schroder, D.S. Rodionov, P.A. de Souza Jr, A. Yen, R. Gellert, E.N. Evlanov, J. Foh, E. Kankeleit, P. Gutlich, D.W. Ming, F. Renz, T. Wdowiak, S.W. Squyres, R.E. Arvidson, Science **305**, 833 (2004)
R.V. Morris, G. Klingelhöfer, C. Schroder, D.S. Rodionov, A. Yen, D.W. Ming, P.A. de Souza Jr., I. Fleischer, T. Wdowiak, R. Gellert, B. Bernhardt, E.N. Evlanov, B. Zubkov, J. Foh, U. Bonnes, E. Kankeleit, P. Gutlich, F. Renz, S.W. Squyres, R.E. Arvidson, J. Geophy. Res. **111**, E02S13, (2006a)
R.V. Morris, G. Klingelhöfer, C. Schroder, D.S. Rodionov, A. Yen, D.W. Ming, P.A. de Souza Jr., T. Wdowiak, I. Fleischer, R. Gellert, B. Bernhardt, U. Bonnes, B.A. Cohen, E.N. Evlanov, J. Foh, P. Gutlich, E. Kankeleit, T. McCoy, D.W. Mittlefehldt, F. Renz, M.E. Schmidt, B. Zubkov, S.W. Squyres, R.E. Arvidson, J. Geophys. Res. **111**, E12S15 (2006b)
R.V. Morris, G. Klingelhöfer, C. Schroder, I. Fleischer, D.W. Ming, A.S. Yen, R. Gellert, R.E. Arvidson, D.S. Rodionov, L.S. Crumpler, B.C. Clark, B.A. Cohen, T.J. McCoy, D.W. Mittlefehldt, M.E. Schmidt, P.A. de Souza Jr., S.W. Squyres, J. Geophy. Res. **113**, E12S42 (2008).
R.V. Morris, G. Klingelhöfer, R.L. Korotev, T.D. Shelfer, Hyperfine Interact. **117**, 405 (1998)
R.V. Morris, S.W. Ruff, R. Gellert, D.W. Ming, R.E. Arvidson, B.C. Clark, D.C. Golden, K. Siebach, G. Klingelhöfer, C. Schorder, I. Fleischer, A.S. Yen, S.W. Squyres, Science **329**, 421 (2010)
A.H. Muhir Jr., R.M. Housley, R.W. Grant, M.A. Gawad, M. Blander, Science **167**, 688 (1970)
M.I. Oshtrakh, A.A. Maksimova, A.V. Chukin, E.V. Petrova, P. Jenniskens, E. Kuzmann, V.I. Grokhovsky, Z. Homonnay, V.A. Semionkin, Spectrochim. Acta A: Mol. Biomol. Spectrosc. **219**, 206 (2019)
M.I. Oshtrakh, E.V. Petrova, V.I. Grokhovsky, V.A. Semionkin, Meteorit. Planet. Sci. **43**, 941 (2008)
R. Ostertag, G. Amthauer, H. Rager, H.Y. McSween Jr., Earth Planet. Sci. Lett. **67**, 162 (1984)
B.S. Paliwal, R.P. Tripathi, H.C. Verma, S.K. Sharma, Meteorit. Planet. Sci. **35**, 639 (2000)
J.F. Petersen, M. Aydin, J.M. Knudsen, Phys. Lett. A **62**, 192 (1977)

H.R. Poulsen, L. Larsen, N.O.R. Poulsen, L. Vistisen, J.M. Knudsen, Phys. Scr. **23**, 1113 (1981)
H.R. Poulsen, M.L. Anderson, L. Larsen, N.O.R. Poulsen, L. Vistisen, J.M. Knudsen, Phys. Scr. **26**, 248 (1982)
R.B. Scorzelli, E. dos Santos, Geochem. **79**, 125547 (2019)
D.W.G. Sears, *The Origin of Chondrules and Chondrites* (Cambridge University Press, 2004), pp. 3–20
E.L.S. Segel, S.S. Hanna, Geochim. Cosmochim. Acta **28**, 1913 (1964)
S.W. Squyres, R.E. Arvidson, E.T. Baumgartner, J.F. Bell III., P.R. Christensen, S. Gorevan, K.E. Herkenhoff, G. Klingelhöfer, M.B. Madsen, R.V. Morris, R. Rieder, R.A. Romero, J. Geophy. Res. **108**, 8062 (2003)
S.W. Squyres, R.E. Arvidson, D.L. Blaney, B.C. Clark, L. Crumpler, W.H. Farrand, S. Gorevan, K.E. Herkenhoff, J. Hurowitz, A. Kusack, H.Y. McSween, D.W. Ming, R.V. Morris, S.W. Ruff, A. Wang, A. Yen, J. Geophys. Res. **111**, E02S11 (2006).
K. Tomeoka, P.R. Buseck, Geochim. Cosmochim. Acta **52**, 1627 (1988)
C. van Cromphaut, V.G. de Resende, E. de Grave, R.E. Vandenberghe, Meterorit. Planet. Sci. **42**, 2119 (2007)
G.P. Vdovyskin, Space Sci. Rev. **14**, 832 (1973)
H.C. Verma, A. Rawat, B.S. Paliwal, R.P. Tripathi, Hyperfine Interact. **142**, 643 (2002)
V.W.A. Vieira, J.M. Knudsen, N.O.R. Poulsen, J. Campsie, Phys. Scr. **27**, 437 (1983)
V.W.A. Vieira, T.V.V. Costa, H.G. Jensen, J.M. Knudsen, M. Olsen, L. Vistisen, Phys. Scr. **33**, 180 (1986)
L. Vistisen, D. Peterson, M.B. Madsen, Phys. Scr. **46**, 94 (1992)
A. Wang, R.L. Koortev, B.L. Jolliff, L.A. Haskin, L. Crumpler, W.H. Farrand, K.E. Herkenhoff, P. de Souza Jr., A.G. Kusack, J.A. Hurowitz, N.J. Tosca, J. Geophys. Res. **111**, E02S16 (2006)
T.J. Wdowiak, D.G. Agresti, Nature **311**, 140 (1984)
R.D. Wordsworth, Annu. Rev. Earth Planet. Sci. **44**, 381 (2016)
A.S. Yen, R. Gellert, C. Schröder, R.V. Morris, J.F. Bell III, A.T. Knudson, B.C. Clark, D.W. Ming, J.A. Crisp, R.E. Arvidson, D. Blaney, J. Brückner, P.R. Christensen, D.J. DesMarais, P.A. de Souza Jr., T.E. Economou, A. Ghosh, B.C. Hahn, K.E. Herkenhoff, L.A. Haskin, J.A. Hurowitz, B.L. Joliff, J.R. Johnson, G. Klingelhöfer, M. Bo Madsen, S.M. McLennan, H.Y. McSween, L. Richter, R. Rieder, D. Rodionov, L. Soderblom, S.W. Squyres, N.J. Tosca, A. Wang, M. Wyatt, J. Zipfel, Nature **436**, 49 (2005)
T. Zemcik, A. Cimbalnikova, Hyperfine Interact. **29**, 1105 (1986)
Y. Zhang, J.G. Stevens, Y. Li, Z. Li, Hyperfine Interact. **91**, 547 (1994)
E.V. Zhinganova, V.I. Grokhovsky, M.I. Oshtrakh, Phys. Status Solidi A **204**, 1185 (2007)
J. Zipfel, C. Schröder, B.L. Jolliff, R. Gellert, K.E. Herkenhoff, E. Rieder, R. Anderson, J.F. Bell III, J. Brückner, J.A. Crisp, P.R. Christensen, B.C. Clark, P.A. de Souza Jr., G. Dreibus, C. D'uston, T. Economou, S.P. Gorevan, B.C. Hahn, G. Klingelhöfer, T.J. McCoy, H.Y. McSween Jr., D.W. Ming, R.V. Morris, D.S. Rodionov, S.W. Squyres, H. Wänke, S.P. Wright, M.B. Wyatt, A.S. Yen, Meteorit. Planet. Sci. **46**, 1 (2011)

Chapter 12
Mössbauer Spectrometry in Life Sciences

12.1 Role of Iron in Human Body

^{57}Fe Mössbauer spectrometry has been extensively used in biology to study the role of iron in our body by analyzing Fe-bound proteins and enzymes. Iron is present in our body in the form of ions bound to hemoglobin (~ 66%), myoglobin (~ 10%), ferritin or hemosiderin (~ 23%) and in other enzymes and proteins (1%) (Lindley, 1996). The iron is available either as a ferrous or ferric ion bound to the proteins. Most ferrous iron is present in the oxygen-transporting hemoglobin and muscle-storage protein myoglobin. Ferric ion is present in the transporter, transferrin and stored in ferritin/hemosiderin. Ferritin is the ferric storage reservoir, which controls iron toxicity and is available for transport to other parts of the body. The iron is crystallized as biocompatible ferrihydrite in ferritin and subsequently transforms into hemosiderin.

The incorporation of ferric ions from ferrous ions into ferritin is facilitated by the ferroxidase enzyme ceruloplasmin. Serum transferrin helps in the control of iron ions transport and storage of ferric ions. The re-conversion to the ferrous state from the ferric state is through redox enzymes. The blood serum receives its ferrous content through dietary sources and hemoglobin. Ferroportin exports the ferrous ion found in the absorbed cells from dietary sources to the blood serum. Another ferrous iron source, hemoglobin, present in the erythrocytes (red blood cells), is produced in the bone marrow and decomposed in the liver. The whole process, as described above, is summarized in Fig. 12.1. The red blood cells (RBC) are removed and recycled in the spleen through the liver. The liver plays a dual role in removing toxins as well as the generation of vital proteins and RBC to some extent. Free iron is toxic, whereas the bound iron ions in the body proteins are either paramagnetic or diamagnetic. Most biological molecules are in the low spin state at their lowest energy.

R. J. Joseyphus and J.-M. Greneche, *Fundamentals of ^{57}Fe Mössbauer Spectrometry*,
https://doi.org/10.1007/978-981-99-8653-8_12

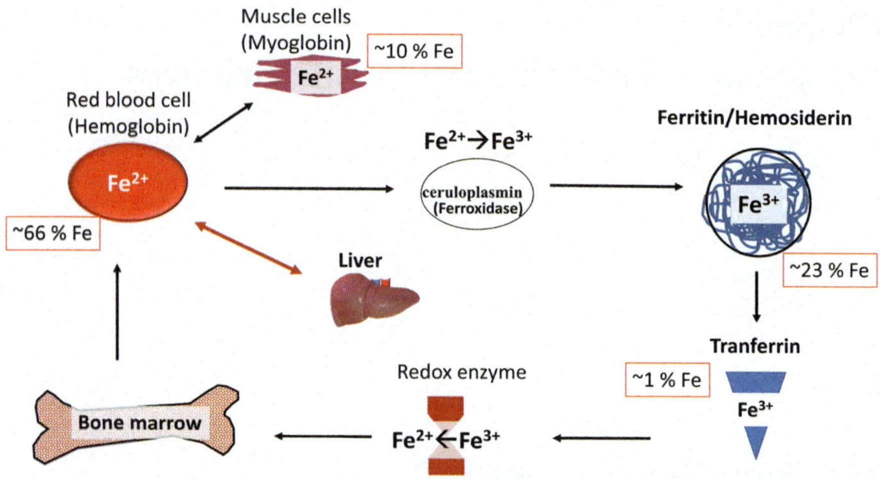

Fig. 12.1 Ferrous and ferric irons are present in the bound state in various protein compounds hemoglobin, myoglobin, ferritin/hemosiderin and transferrin. The Fe-bound proteins have multiple functions: storage, transport and redox reactions. The ferrous ions from cells are transported to the blood serum through ferroportin

12.2 Heme Group

12.2.1 Hemoglobin Overview

The lungs oxygenate our body with almost 9000 L of air per day. The oxygen molecule binds with hemoglobin (Hb) in the blood erythrocytes and is transported from the lungs to other body parts. In return, four carbon dioxide molecules can bind with the hemoglobin molecule as carbaminohemoglobin. The hemoglobin is made of two α and two β polypeptide amino acid chains, with each chain having a heme that can bind with four oxygen molecules. The Fe^{2+} ion in the heme is coordinated with porphyrin in the hemoglobin subunit, as shown in Fig. 12.2 (left) and is present in the α and β subunit (right) amino acid chains. The transfer of oxygen from the lungs to hemoglobin and the mitochondria of cells is mediated by myoglobin. Carbon monoxide (CO) has more than 100 times higher affinity to heme iron than oxygen. The oxygen-bound and oxygen-free hemoglobins are called oxyhemoglobin and deoxyhemoglobin, respectively. Hemoglobin and myoglobin contain heme B. Hemin is the ferric heme with a coordinating chlorine ligand. Oxidized ferric heme is called hematin, which has an $(OH)^{-}$ ligand. The spin state of Fe coordinated with the ligands in the heme group, and therefore its magnetic properties are explained by ligand field theory as the crystal field splitting energy predominates the spin–orbit coupling energy, with certain exceptions. The ferrous high spin state could be identified from its large temperature-dependent isomer shift and quadrupole splitting.

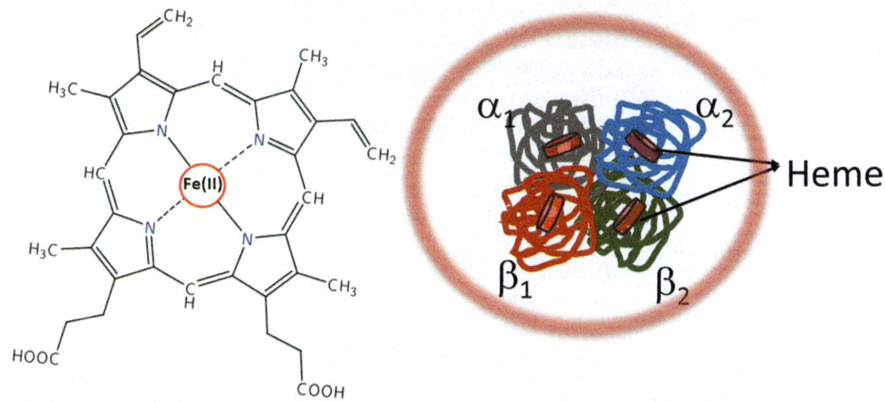

Fig. 12.2 The heme structure in the hemoglobin (left) and the representative image showing the α and β globins surrounding the heme in the hemoglobin (right) of red blood cells

12.2.2 Hemoglobin

Initial Mössbauer studies at 4 K on the hemoglobin sedimented in saline drawn from human venous blood and $\approx 3.5\%$ ^{57}Fe enriched cells of rats showed a strong doublet of low isomer shift and a weaker one with higher isomer shift for both the samples (Gonser et al. 1963). The strong doublet lines were identical in both specimens and suggested to be from the oxyhemoglobin. The weak doublet, almost one-sixth in intensity compared with oxyhemoglobin, had similar quadrupole splitting but a higher isomer shift attributed to CO_2-hemoglobin.

The Mössbauer spectra of hemoglobin (Hb) exposed to various atmospheres such as CO_2, N_2 and CO were recorded at low temperatures. Usually, heme compounds are preserved at low temperatures to prevent degradation, and experiments were conducted below 200 K. The isomer shift values were obtained relative to ^{57}Co/Pt at RT. The strong doublet at 4 K for the hemoglobin exhibiting $\Delta = 2.3$ mm/s and $\delta = -0.05$ mm/s was assigned the O_2-hemoglobin, and the weaker one with similar Δ except with $\delta = 0.55$ mm/s was suspected to be from H_2O attached to the hemoglobin (Gonser et al. 1964). Similar hyperfine parameters were exhibited by the hemoglobin exposed to CO_2 and N_2 at 77 K. On the other hand, the quadrupole splitting decreased drastically, showing a line resembling a singlet, due to a lower electric field gradient, for the hemoglobin exposed to CO. The isomer shift value with CO ligand was similar to O_2-hemoglobin. Further studies at 5 K revealed the quadrupole splitting of the Hb, hematic and hemin ligands (Gonser and Grant 1965) of the order,

Hb-H_2O ($\Delta = 2.36$ mm/s) > Hb-CO_2 > Hb-O_2 > hematin(OH)$^-$ > hemin-Cl$^-$ > Hb-CO ($\Delta = 0.34$ mm/s).

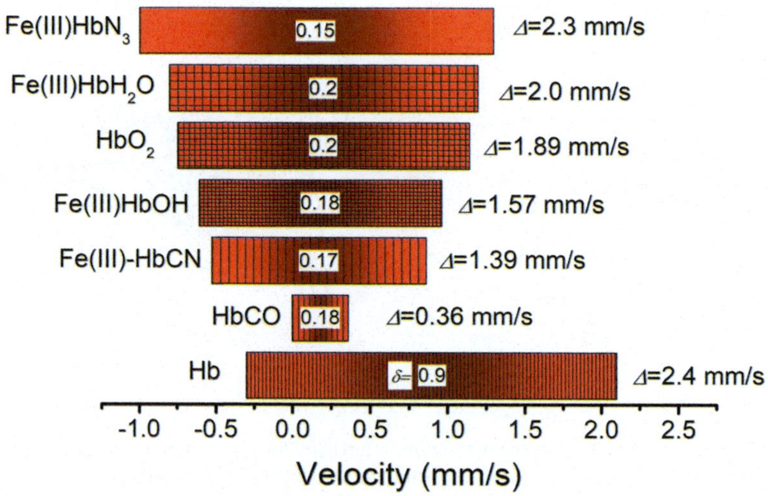

Fig. 12.3 The hyperfine parameters from the Mössbauer spectrum at 195 K for the hemoglobin (Hb) exposed to various compounds. Fe(III)Hb represents ferric hemoglobin. δ wrt Fe. Data converted to velocity scale from (Lang and Marshall 1966)

All the compounds showed $\delta < 0.1$ mm/s except the H_2O-hemoglobin ($\delta = 0.6$ mm/s). The sign of EFG is positive for the ferrous Hb, ferric ($S = 5/2$) hemin and hematin high spin compounds.

Figure 12.3 shows the hyperfine parameters at 195 K obtained from the Mössbauer spectra of ^{57}Fe enriched hemoglobin from albino rats and subsequently prepared ferrous and ferric hemoglobin compounds of Hb(Fe^{2+})CO, Hb(Fe^{3+})CN, Hb(Fe^{3+})OH, Hb(Fe^{2+})O_2, Hb(Fe^{3+})H_2O and Hb(Fe^{3+})N_3 (Lang and Marshall 1966). The samples were recorded using a 10–50 mCi ^{57}Co/Cr source. Covalent mixing-induced distortion in the spatial distribution of d electrons results in higher quadrupole splitting for Fe(III)HbH_2O. Fe(III)HbN_3 and Fe(III)HbCN have low-spin $S = 1/2$ state, and the EFG is influenced by their covalency.

The lower isomer shift of low spin $S = 0$ diamagnetic ferrous compounds HbCO and HbO_2 were explained due to their higher covalency. The quadrupole splitting of 0.36 mm/s in HbCO is based on the EFG contribution from enhanced Sternheimer antishielding factor as well as the difference in covalent bonding of d orbital Fe to the ligands. While the low quadrupole splitting in HbCO is expected due to the low spin ferrous charge distribution, the high quadrupole splitting in HbO_2 is puzzling.

HbO_2 exhibits temperature-dependent quadrupole splitting, suggesting thermal excitation as one of the reasons. The large quadrupole splitting of HbO_2 was based on the configuration and geometry of the oxygen ligand. The sign of EFG for oxyhemoglobin HbO_2 is negative, determined from applied field measurements that were explained based on the loss of an electron from d_{yz}–π_z bonding, resulting in a hole. This would produce a large quadrupole splitting similar to the high spin ferrous state. However, electronic excitation was doubtful as the compound remains diamagnetic even up to RT, and alternatively, conformational excitation was proposed from the studies of HbO_2 along with various Hb model compounds (Spartalian et al. 1975). The Δ value of HbO_2 was reported from 2.24 mm/s at 1.2 K to 1.89 mm/s at 195 K. The ferrous-oxygen coupling with different spin states resulting in the $S = 0$ state has been continuously studied by various researchers. The proposition includes an intermediate spin configuration with two unpaired d spins similar to the ozone model (Schuth et al. 2017) and more than 50% ferric character and a partially polarized Fe-O_2 π-bond in oxy hemes (Yan et al. 2019).

The reduced hemoglobin, Hb, shows a higher isomer shift ($\delta = 0.9$ mm/s) than the ferric state but an intermediate value corresponding to its covalent high spin ferrous state. The magnetic nature of Hb was manifested from the large magnetic moment and broadening of the spectrum in an applied magnetic field. It was shown that high spin-reduced Hb exhibits temperature-dependent quadrupole splitting from 2.26 to 1.94 mm/s as the temperature increases from 4 to 173 K (Eicher and Trautwein 1969). The ferrous Hb, in its high spin state, turns to a low spin ($S = 0$) diamagnetic state on binding with ligands such as CO or O_2. Although the reduced Hb is a ($S = 2$) high spin compound from its large values of Δ, δ, and its temperature dependence, a secondary doublet with lower values of hyperfine parameters is present in a few samples, especially in anhydrous hemoglobin resembling the $S = 0$ state.

12.2.3 Spin State of Hemoglobin Subunits

Anhydrous hemoglobin (AHb) was prepared by vacuum distillation of deoxyhemoglobin (Hb) flushed with nitrogen and adding sodium hydrosulfite. The Mössbauer spectra at 16 K for the AHb prepared from rat hemoglobin suggested high and low spin states (Grant et al. 1967). The isomer shift values were obtained relative to ^{57}Co/Pt at RT. In the absence of a ferric state, the high spin doublet showed $\Delta = 2.2$ mm/s and $\delta = 0.61$ mm/s (~1 mm/s wrt Fe) while the low spin values were $\Delta = 1.1$ mm/s and $\delta = 0.08$ mm/s (0.43 mm/s wrt Fe). The low spin ferrous state could be distinguished from the lower δ. The coexistence of the two spin states was postulated from the distribution of the spin states in α and β globin chains, as another possibility due to the smaller energy difference between the two states resulting in thermal equilibrium was ruled out from the temperature-dependent studies. Further studies were undertaken to clarify whether the inner doublet was from Hb, anhydrous Hb or from α, β subchain.

The ^{57}Co labeled Co (II) protoporphyrin IX dimethyl ester, (CoP-$(OMe_2)_2$, a heme compound, mentioned as CoP-$(OMe)_2(C_5H_5N)_2$ and the oxygenated compound oxy-CoP-$(OMe)_2(C_5H_5N)$ were studied through emission Mössbauer spectrometry (Marchant et al. 1972). Both the emission spectrum from the labeled compound and the absorption spectrum of a similar Fe heme compound, heme-$(OMe)_2(C_5H_5N)_2$, showed close quadrupole splitting of 1.12–1.25 mm/s and isomer shift of 0.45–0.47 mm/s at 4.2 K, indicating their low spin ferrous state. On the other hand, the oxygenated oxy-CoP-$(OMe)_2(C_5H_5N)$ resulted in a higher quadrupole splitting of

1.94 mm/s at 172 K that increased to 2.28 mm/s at 4.2 K. The δ values were in the range 0.22–0.30 mm/s. These values are similar to their oxygenated ferrous hemoglobin counterparts.

The Mössbauer spectra were obtained in the temperature range of 80–210 K for the deoxygenated hemoglobin (Hb) and their isolated α and β subunits (Huynh et al. 1974a, b). The deoxygenated Hb was prepared by adding an excess of $Na_2S_2O_4$ under N_2 atmosphere, and the deoxygenated subunits were obtained from the oxygenated hemoglobin using a series of procedures followed by separation and confirmation from the electrophoretic pattern. The quadrupole splitting decreased from 2.34 to 1.94 mm/s with increasing temperature in the 80–210 K range with similar values between the samples. Similarly, the isomer shift was invariant with a value of 0.8 mm/s relative to metallic Fe for all the samples. The quadrupole splitting and isomer shift values indicated that the deoxygenated Hb and the Hb-α and Hb-β subunits are all in the same high spin state. The results also confirmed that the oxygen affinity for all the subunits is similar. A less intense low-spin doublet was also inferred from the central dip between the high-spin doublet.

In one of the reports, it was presumed that the α chain shall be in its high spin state while the β chain favors a low spin state when it transforms to anhydrous hemoglobin (Trautwein et al. 1970). Deoxyhemoglobin and its isolated α and β subunits were found to show similar temperature-dependent quadrupole splitting (2.38–2 mm/s) and isomer shift values (0.85–0.78 mm/s) from the Mössbauer analysis in the range 4.2–210 K (Huynh et al. 1974a, b). Although this was due to the high spin ferrous nature, another temperature-independent doublet of lower intensity (~ 20%) was suspected from the low spin state. In order to study the preferential occupation of the high spin and low spin states in the α and β subunits of anhydrous Hb (AHb), Mössbauer spectra close to 80 K were obtained for the isolated AHb subunits (Papaefthymiou et al. 1975). Less than 10% ferric impurity was present in the samples. The hyperfine parameters were similar to the previously reported values for the high and low spin states. The low spin fraction for the AHb, AHb-α and AHb-β were found to be 35%, 55% and 51%, respectively. This study suggested no preference for the high or low spin components on the α and β subunits as both had almost equal fractions of $S = 2$ and $S = 0$ states. However, it was proposed that the dehydration step may convert the high spin to low spin state in the range 20–40%.

The absence of low spin components from the α and β chains were presented in the studies using anhydrous -Hb as well as its -α and -β separates (Chow and Mukherji 1975). The outer doublet at 85 K with $\Delta = 2.19$ mm/s and $\delta = 0.89$ mm/s was interpreted due to a ferrous high spin state, and the inner doublet with $\Delta =$ 1.10 mm/s and $\delta = 0.40$ mm/s was suggested to be from hemochromogen impurity. Further experiments with deoxygenated Hb and its subunits α, β -SH group chains, and (α_2-enriched)β_2 mass hybrids exhibited only high spin states without any low spin doublet (Trautwein et al. 1976). The symmetric doublets without any central line presented values corresponding to Δ of 2.39, 2.42, 2.49 and 2.38 mm/s at 4.2 K for the AHb, α_{SH}, β_{SH}, $\alpha_2(^{57}Fe)\beta_2$ samples, respectively. All the samples showed temperature-dependent quadrupole splitting within the accuracy limit of $\pm$ 0.01 mm/

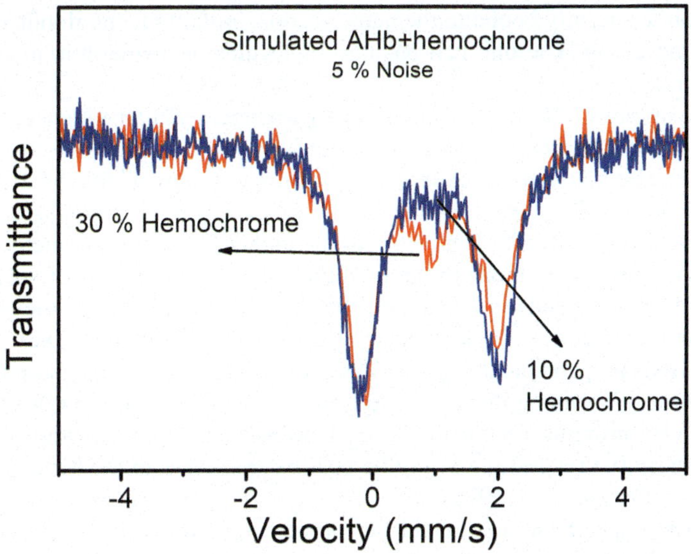

Fig. 12.4 The simulated line broadened Mössbauer spectra of high-spin anhydrous hemoglobin with 30 and 10% low-spin hemochrome

s. The difference in quadrupole splitting between α_{SH} and β_{SH} in the temperature range of 100–4.2 K was nearly 0.1 mm/s.

The comparison of Mössbauer spectra of the pure hemoglobin sample at 20 K and the samples at higher temperatures revealed that the central dip between the high intense doublet, arising from hemochrome, increased with temperature (Eicher et al. 1976). The simulated spectra with 30 and 10% hemochrome in AHb are presented in Fig. 12.4, where the central dip with a higher hemochrome fraction is evident. At temperatures beyond 180 K, the hemochrome fraction was significant, and therefore, preservation of samples at liquid nitrogen temperature is preferred. These studies clarified the similar high-spin nature of the subunits of Hb and AHb.

The interpretation of low spin state component in the Mössbauer spectrum, although appropriate, is not from hemoglobin, but from the low spin hemochrome. The molecular diffusion of the hemoglobin and hemochrome could be neglected up to 250 K, allowing Mössbauer studies on frozen solutions (Mayo et al. 1983).

12.2.4 Myoglobin

Myoglobin, an oxygen carrier found in muscle cells, belongs to the heme group protein comprising eight α helices connected loops. Myoglobin is present in red meat, well studied with muscle specimens from horse hearts, and available in a high proportion of nearly 5 wt.% in black meat of sperm whales. The ferrous ion in myoglobin is tightly bound to the heme group, similar to hemoglobin, except for a lone heme group. The unit cell has two molecules of myoglobin, in contrast to hemoglobin.

The myoglobin (Mb) presents hyperfine parameters similar to hemoglobin, and the results obtained from various studies for Mb, MbCO and MbO_2 are shown in Table 12.1. The myoglobin and lyophilized anhydrous myoglobin (AMb) showed similar Δ and δ(SS) at 83 K with the values 2.02 and 1.01 mm/s, respectively (Trautwein et al. 1970). The doublet with the higher hyperfine parameters was due to the high spin state similar to hemoglobin. Mössbauer spectra of sperm whale myoglobin single crystals enriched with 80% ^{57}Fe were used to study the quadrupole interaction parameters at 77 K (Gonser et al. 1974). Measurement was conducted with the *b*-axis perpendicular to γ ray. The sample showed a small amount of MbO_2 as an impurity. Assuming the asymmetry parameter $\eta = 0$, the V_{zz} was obtained as positive and at an angle of 40° to the *b*-axis. However, a modified theoretical approach using η nearly 0.4 revised the V_{zz} to be positive and close to the heme plane at an angle of 14° (Maeda et al. 1976).

Mössbauer spectra in an applied field (6 T, $B \perp \gamma$) for the ^{57}Fe enriched horse myoglobin frozen samples were undertaken in the temperature range 4.2–195 K (Kent et al. 1977). The applied magnetic field broadened the symmetric doublet lines considerably due to more than seven independent hyperfine interaction parameters. From the fit of the temperature-dependent internal field coefficient, V_{zz} was arrived as negative with temperature independent η value of 0.7. The EFG principal axis does not lie near the plane normal to the heme. These results are also found to be independent of the myoglobin sources such as mammals, fish or birds. The deoxy Mb exhibits a temperature-dependent quadrupole doublet corresponding to its high spin state (Kent et al. 1979). The Mb spectra in an applied field, compared with its synthetic ferrous 1,2 dimethyl analog, could be fitted with $\eta = 0.7$ at 77 K that decreased to 0.4 at 195 K.

The Mössbauer spectra with an applied field of 4.7 T at 4.2 K were obtained for the ^{57}Fe enriched sperm whale MbCO (Trautwein et al. 1974). The broad doublet was split in the presence of an applied field, and resulted in Δ and δ values of 0.36 and 0.27 mm/s, respectively, assuming η in the range 0–0.4. It was concluded that the EFG is positive, and the ferrous iron is situated in the heme plane at an angle of about 135° with the CO.

Nuclear resonant scattering (NRS) on enriched frozen MbCO single crystal was demonstrated by analyzing the crystal along the $60\overline{3}$ orientation (Parak et al. 1976). The asymmetry in the spectrum due to orientation difference could be obtained, and the study demonstrated the capability for protein crystallography. In addition to the largest EFG component perpendicular to the heme plane, $\eta = 0.75$ associated with strong rhombic distortion of the Fe environment, from the axially bound CO ligand was obtained from NRS (Parak et al. 1977).

The quadrupole interaction parameter (eqV_{zz}) of MbCO is positive in contrast to the MbO_2 from measurements on single crystals prepared through CO and O_2 diffusion (Maeda et al. 1981). The η was limited to 0.2–0.4 in MbCO and 0.3–0.4 in MbO_2. The Mössbauer spectra recorded in the temperature range 4.2–200 K in

Table 12.1 The hyperfine parameters for the myoglobin (Mb), MbCO and MbO_2

Sample	T (K)	Δ (mm/s)	δ (mm/s)	Characteristics	References
Mb (horse)	83	2.02	1.01 (SS)	High spin quadrupole splitting	Trautwein et al. (1970)
AMb (horse)	83	2.02	1.01 (SS)	Similar to Mb. Δ decreases with T	
Mb single crystals	77	2.29	0.90	Assymmetric doublet; small MbO_2 present with $\Delta \approx$ 2.27 mm/s, $\delta \approx$ 0.2 mm/s	Gonser et al. (1974)
Mb (horse)	4.2 16 35 60 77 140 195	2.22 2.18 2.23 2.18 2.17 1.95 1.78	0.92 0.92 0.92 0.90 0.91 0.88 0.84	All measurements with $B = 6$ T. $B \perp \gamma$. $\eta = 0.7$ (T independent); V_{zz} negative	Kent et al. (1977)
Mb	4.2 16 40 77 195	2.22 2.18 2.17 2.17 1.78	0.92 0.93 0.93 0.91 0.86	Data for $B = 0$. $\eta =$ 0.7 at low T, decreases to 0.4 at 195 K from applied field measurements. Similar with different Mb analogues. (see the above column)	Kent et al. (1979)
MbCO	4.2	0.37	0.27	$B \perp \gamma$. EFG positive; Fe in heme plane; Fe-CO angle 135°; Small MbO_2 present	Trautwein et al. (1974)
MbCO Single crystal	80	0.36	0.19	Nuclear resonant scattering. Oriented along $60\overline{3}$. (δ wrt ^{57}Co/Rh)	Parak et al. (1976)
MbO_2 Single crystal b-axis$\perp\gamma$	77	2.13	≈ 0.3	T dependent. Negative quadrupole interaction. $\lvert V_{zz}\rvert$ parallel to heme plane; $\eta = 0.3$–0.4	Maeda et al. (1981)
MbCO Single crystal b-axis$\perp\gamma$	77	0.35	≈ 0.3	Positive quadrupole interaction. $\lvert V_{zz}\rvert$ perpendicular to heme plane; $\eta =$ 0.2–0.4	
MbO_2 (Sperm whale)	4.2 50 100 150 200	2.28 2.27 2.11 2.02 1.93	0.27 0.28 0.28 0.27 0.25	All measurements with $B = 4$ T. Negative quadrupole interaction. $\eta =$ 0–0.18. Diamagnetic line shape with B	Boso et al. (1984)

Mb—Myoglobin, AMb—Anhydromyoglobin, B—applied external field, η—asymmetry parameter an applied field of 4 T proved the low spin diamagnetic nature of MbO_2 and HbO_2 (Boso et al. 1984). The analysis also presented the negative quadrupole interaction with η values between 0 and 0.18 for the MbO_2. The divergent nature of V_{zz} was also evident with its component perpendicular and parallel to the heme plane in MbCO and MbO_2, respectively, in the temperature range 77–195 K.

12.2.5 *Ferric Myoglobin*

Most ferrous heme ligands are in the low spin state, unlike pure ferrous hemoglobin or myoglobin, which are of high spin. The ferric states of Hb and Mb are known to be methemoglobin and metmyoglobin, respectively. The ferric high spin and low spin states are $S = 5/2$ and $S = 1/2$, respectively.

The lyophilized ^{57}Fe enriched dehydrated metmyoglobin exhibited a broad spectrum with two quadrupole doublets, one with higher quadrupole splitting (1.96–2.04 mm/s) and another with lower quadrupole splitting but comparatively smaller width in the temperature range 4.6–298 K (Caughey et al. 1966). The large linewidths of the outer doublet were more prominent below 77 K, and therefore, the hyperfine parameters could not be determined. The temperature-dependent quadrupole and isomer shift values are influenced by the linewidth. Although the lyophilized metmyoglobin had a high spin impurity, the low spin ferric component was identified from its hyperfine parameters. The low spin ferric component had a large quadrupole splitting, smaller isomer shift (Cu matrix) (− 0.12 to − 0.18 mm/s) and significant line broadening (0.5–1.4 mm/s). The line broadening is proportional to the unpaired s electron spin density changes due to the d-shell, which influences the internal field. Therefore, low-spin ferric lines are expected to be broader than their low or ferrous-spin counterparts. Most of the low-spin ferric hemes thus show widely split hyperfine field lines at 4.2 K due to strong dipolar and orbital coupling (Lang 1970).

The low-spin ferrous and high-spin ferric states have spherically symmetric charge distribution and, therefore, cannot give large quadrupole splitting. As seen from our previous section on high spin Hb and Mb, the δ and Δ values for high spin ferrous ions are larger. The quadrupole splitting of high-spin ferric metmyoglobin is reported to show a lower quadrupole splitting of 1.33 mm/s at 195–243 K (Maeda et al. 1971). Conclusions could be drawn from the hyperfine patterns as illustrated in Fig. 12.5. The low-spin ferric myoglobin exhibits large quadrupole splitting at high temperature and broadening at low temperature, whereas the high-spin compound exhibits Δ lower than its low-spin counterpart but a magnetic hyperfine splitting at low temperature. The saturation hyperfine fields of metmyoglobin are 49.3 and 49.8 T in their meso and proto prosthetic groups, and they indicate strong axial fields (Lang et al. 1970). The quadrupole splitting in the presence of an applied field is only 0.65 mm/s, and V_{zz} is positive.

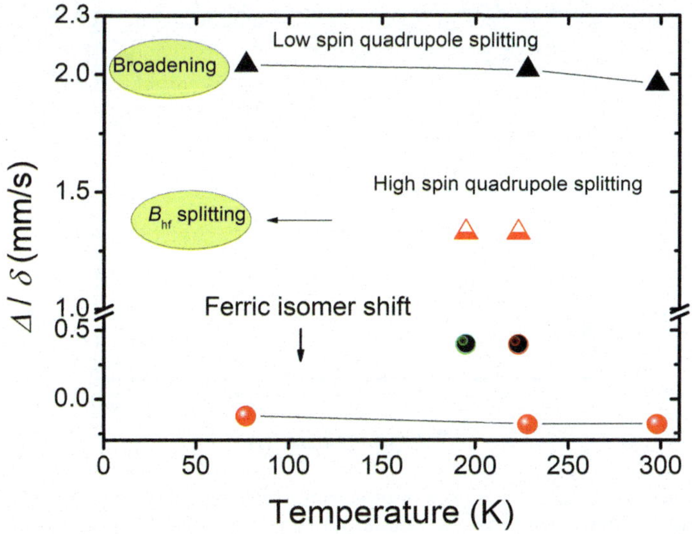

Fig. 12.5 The temperature-dependent hyperfine parameters of high spin and low spin components of metmyoglobin. Data from (Caughey et al. 1966; Lang et al. 1970; Maeda et al. 1971)

12.2.6 Protein Dynamics

The Mössbauer *f*-factor is based on the mean square displacement $\langle x^2 \rangle$ of atomic positions connected by the relation

$$f = \exp\left(-4\pi^2 \frac{\langle x^2 \rangle}{\lambda^2}\right) \tag{12.1}$$

where $\lambda = 0.86$ Å for Fe. The *f*-factor could be obtained from the absorption area in the Mössbauer time scale of 140 ns, coinciding with the molecular motion of proteins. The plot of the *f*-factor and mean square displacement of myoglobin crystal is shown in Fig. 12.6. (Parak and Formanek 1971). The value of $\langle x^2 \rangle$ increases from 0.005 Å^2 at 4.2 K to 0.09 Å^2 at RT for the myoglobin crystals. The *f*-factor decreases drastically from 0.8 at 4.2 K to 0.01 at RT.

The mean square displacement of protein crystals could be determined from the X-ray diffraction with better refinement for most of the non-hydrogen atoms in a protein. For example, the mean square displacement,

$$\langle x^2 \rangle_X = \langle x^2 \rangle_{conf} + \langle x^2 \rangle_{diff} + \langle x^2 \rangle_{latt.diso} + \langle x^2 \rangle_{vib} \tag{12.2}$$

in the order given as conformation-substate, diffusion, lattice-disorder and vibration state, was used to obtain $\langle x^2 \rangle_X$=0.091 and 0.108 Å^2 at 250 and 300 K, respectively, from X-ray diffraction in metmyoglobin (Frauenfelder et al. 1979). The $\langle x^2 \rangle_{vib}$ of

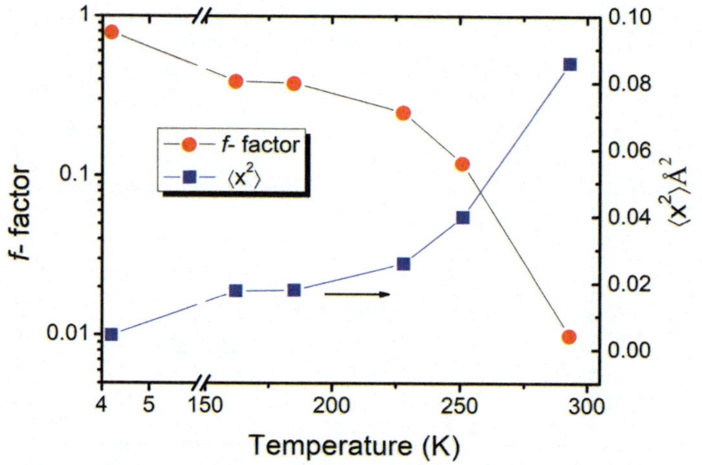

Fig. 12.6 The *f*-factor and mean square displacement of myoglobin crystal, obtained from Mössbauer spectra, at various temperatures. Drastic variations are noticed beyond 200 K. Data from (Parak and Formanek 1971)

the iron atom from Mössbauer data gives a value of 0.02 Å^2 at 300 K. Assuming conformational motion perpendicular to the heme plane in the myoglobin crystal,

$$\langle x^2 \rangle_{cv} = \langle x^2 \rangle_{vib} + \langle x^2 \rangle_{conf} \tag{12.3}$$

that gives 0.04 Å^2 leading to $x_{rms} = 0.35$ Å for the heme iron. This is in the region of an average tunneling distance of 0.3 Å in ferrous myoglobin. The $\langle x^2 \rangle_{cv}$ was as large as 0.25 Å^2 and increased continuously from the inside to the surface, indicating substantial interaction between the protein and the solvent.

The water adsorbed inside the metmyoglobin crystals fluctuates between the conformational substates with a higher relaxation rate as the temperature increases up to room temperature (Singh et al. 1981). The corresponding conformational transition time of Fe and dielectric relaxation time for the adsorbed water in metmyoglobin are $\approx$ 240 ns and 80 ps, respectively, at 300 K. The reorientation of water below 250 K was observed through an increase in dielectric relaxation time with reducing temperature from the studies. Therefore, these fluctuations with fast time scales could be observed only above $\approx$ 200 K through Mössbauer spectroscopy in most of the heme proteins.

Compared with X-ray diffraction, Mössbauer analysis gives information about much smaller displacements of Fe atoms from the relaxation broadening. The mean square displacement $\langle x^2 \rangle$ drastically increases beyond 200–250 K, resulting in the reduction of absorption and appearance of quasi-elastic line broadened Mössbauer spectra of myoglobin crystals (Parak et al. 1981; Mössbauer 1987). For most heme proteins, the f factor is 0.8 at the lowest temperature and extremely small at high temperatures, although it could be obtained from the absorption area.

The $\langle x^2 \rangle$ values are obtained from (Parak et al. 1982)

$$k^2\langle x^2 \rangle = \ln(\sigma_o n) - \ln t_o \tag{12.4}$$

and $t_o = n\sigma_o f_o$

where $k = 2\pi/\lambda$, $\sigma_o = 2.56 \times 10^{-18}$ cm^2 is the cross section area of ^{57}Fe, n is the number of ^{57}Fe nuclei per cm^2, and f_o is the f-factor of the non-expanded spectrum. For the myoglobin, from the broadened spectra in the temperature range 210–293 K, the $\langle x^2 \rangle$ increases from 0.02 to 0.06 Å^2 with temperature (Parak et al. 1981). Rayleigh scattering of Mössbauer radiation is another method used in protein dynamics (Krupyanskii et al. 1980; Goldanskii et al. 1980; Mössbauer 1987; Parak et al. 2007).

The coupling of sucrose/water to myoglobin was demonstrated using Mössbauer spectrometry from the quasi-elastic line broadening (Lichtenegger et al. 1999). The experiments were conducted using ^{57}Fe enriched horse myoglobin-CO (MbCO) and adding 80% sucrose/water in the temperature range 100–265 K. Potassium ferrocyanide sucrose/water was used for comparison that showed only a single line in the entire temperature range due to the octahedral symmetry of the low spin ferrous iron environment. Figure 12.7 shows the Mössbauer spectra of the MbCO-80% sucrose/water mixture in the temperature range from 100 to 265 K at close velocity range of $\pm$ 3 mm/s (left) and a larger velocity of $\pm$ 12 mm/s (right). The spectra taken in the larger velocity ranges are fitted with two doublets, one with narrow linewidth and another one broadened. The sharp and distinct doublet lines (100 K) broadened and almost merged at 265 K. The linewidth, less than 0.3 mm/s, increased drastically beyond 250 K and reached above 0.6 mm/s at 265 K, accompanied by an increase in quadrupole splitting. The $\langle x^2 \rangle$ of MbCo/sucrose/water mixture resulted in values lower than 0.03 Å^2 at the highest temperature, whereas the potassium ferrocyanide/sucrose/water mixture and deoxy-Mb crystal showed values in the range 0.04–0.06 Å^2 due to shift in their onset transition temperatures. The temperature dependence of the spectra was similar for the compared Mb-solvent mixtures. The Fe displacement of the myoglobin in the range 10–20% in comparison with dehydrated Mb was suggested by the study.

Various methods were introduced to interpret the f-factor relaxation spectrum, such as fluctuations in the protein's hydration shell using the Havriliak-Negami function (Young et al. 2011) and quantum mechanical spatial motion model (Frauenfelder et al. 2014).

12.3 Ferritin/Hemosiderin

Ferritin is a protein shell of a typical diameter 10 nm capable of storing up to 4500 ferric ions as ferric hydroxyphosphate (~ 8 nm diameter), a water-soluble, non-toxic compound (Crichton 2001). The ferric ion is incorporated inside the apoferritin

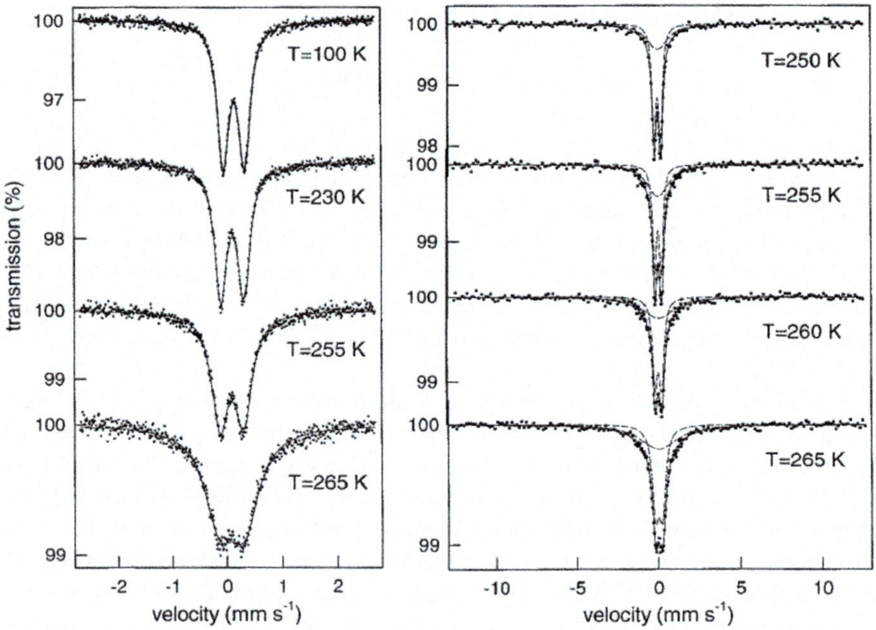

Fig. 12.7 The increase in the broadening of the Mössbauer spectra with temperature for the MbCO-sucrose/water mixture. The right panel shows the spectra recorded at higher velocities and fitted with two doublets. Reprinted with permission (Lichtenegger et al. 1999), © Elsevier

protein shell through ferroxidase. The steps involved are the oxidation of ferrous iron by oxygen-producing H_2O_2, migration of ferric ions from ferroxidase into the protein shell, ferrihydrite crystallization and growth. The chemical formula for ferritin corresponds to $(FeOOH)_8(FeO.OPO_3H_2)$, resembling the soil ferrihydrides.

Mössbauer spectra of the ferritin, with an isomer shift of 0.5 mm/s corresponding to the ferric state, showed a doublet at 77 K (Blaise et al. 1965). The Mössbauer and susceptibility studies indicated superparamagnetic transformation of the antiferromagnet. The ferritin made into a thin disk showed a quadrupole splitting of 0.37 mm/s, ferric isomer shift of 0.47 mm/s (SS) at 77 K and a hyperfine field of 49.3 T at 4.2 K (Boas and Window 1966). The hyperfine parameters are identical to FeOOH, exhibiting superparamagnetic behavior with the transition temperature depending on the size and shape.

Ferritin and hemosiderin isolated from horse spleen showed ferric sextet with B_{hf} of 47.0 and 49.0 T, respectively, at 11 K and doublets above 86 K as shown in Fig. 12.8. The ferritin and hemosiderin micelles were distributed in the size range 5–7 nm. Complete sextet transformation at low temperatures and a doublet at higher temperatures for both samples could be noticed. Similar temperature-dependent characteristics were observed from *pseudomonas aeruginosa* bacteria, hemolymph of *patella vulgate* (sea snail) and human spleen corresponding to the storage ferritin protein, although of varying crystallinity and particle size (Pierre et al. 1986). The major fraction of ferritin is found in the high spin state along with the low spin component. In addition to the bacteria, ferritin is also found in fungi and plants, omnipresent as a storage protein in both prokaryotes and eukaryotes.

Cultured Aquaspirillum magnetotacticum bacterial strain in bog water or ferric culture medium produced Fe_3O_4 of nearly 50 nm as a major fraction along with ferrihydrite, unlike freshwater non-magnetic bacteria (Frankel et al. 1979, 1983). The ferrimagnetic Fe_3O_4 was produced by the bacteria reducing antiferromagnetic ferrihydride. The spectrum of magnetoactic cells at 80 K showed B_{hf} of 50.0 and 47.0 T with δ of 0.44 and 0.74 mm/s, respectively, corresponding to A- and B- sites of Fe_3O_4.

The fraction of ferritin/hemosiderin in cells shall be affected by disease conditions such as thalassemia in humans. The Mössbauer spectra of both ferritin and hemosiderin, isolated from the iron-overloaded spleen of a thalassemia patient, in an applied field ($B \perp \gamma$) of 10 T at 4.2 K showed a considerably broadened spectra with a slight increase in hyperfine splitting that are characteristics of antiferromagnetic materials (Bell et al. 1984). The zero-field ferric sextets showed an almost similar hyperfine field of 50.0 T at 1.3 K. Both the samples were completely superparamagnetic above 160 K. However, the hemosiderin showed a smaller volume (40–172 nm^3) than the ferritin (84–182 nm^3) and a lower blocking temperature for the former.

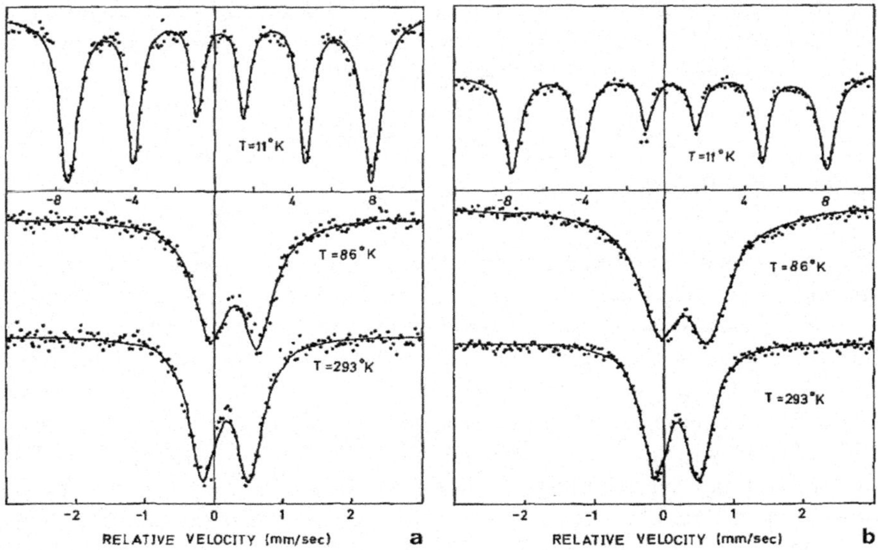

Fig. 12.8 The Mössbauer spectra of 5–7 nm ferritin (left) and hemosiderin (right) micelles at 11, 86 and 293 K show similar characteristics. Reprinted with permission (Fischbach et al. 1971), © Elsevier

12.4 The Transporter

The iron transport protein, transferrin, binds ferric ions after the oxidation of ferrous ions and unloads according to the pH conditions (Crichton 2008). The transferrin has two ferric binding sites in octahedral geometry with four protein ligands that have N- and C- lobes and two carbonate anions. The ferric ion binding is assisted by the carbonate, and the release of the ion is through protonation. The apotransferrin fetches the iron ions that bind at slightly acidic pH and forms the transferrin, which has a higher affinity to ferric ions for transport throughout the body. Ferrireductace cytochrome enzymes undertake the conversion to the ferrous state. The ferric to ferrous transformed ions are transported to the cytoplasm by the divalent metal-ion transporter 1 (DMT1) which is a ferrous transporter (Sherman et al. 2018). Another family of non-transporter transferrin is lactoferrin, found in milk.

The Mössbauer spectra of ^{57}Fe enriched transferrin were recorded at 77 and 4.2 K in applied magnetic fields (Tsang et al. 1973). The hyperfine split sextet could be observed at 77 and 4.2 K and exhibited broadening in applied longitudinal and transverse magnetic field of 0.01 T at 4.2 K. The low field spectrum had contributions from ferric high spin Kramers doublets. The zero-field spectrum of fully saturated transferrin showed doublet at RT due to an increase in spin relaxation (Spartalian and Oosterhuis 1973). The Δ and δ values were 0.75 and 0.38 mm/s with η close to 1. The hyperfine field at 4.2 K was determined as 22.0 T. Further Mössbauer experiments with mono, di and labeled C and N transferrins in applied magnetic fields suggested similar ferric environments (Kretchmar et al. 1988).

12.5 Cytochrome

Fe-S clusters and cytochromes are the two redox pathways of Fe ions in biological systems. Cytochrome c is an electron-transfer heme protein, while cytochrome p450 is a monooxygenase heme whose structures are shown in Fig. 12.9.

The CO-bound ferrous cytochrome exhibits a low spin state, giving an absorption maximum at 450 nm and, hence, its name P450.

The ferricytochrome c and ferrocytochrome c exhibit large quadrupole splitting of 1.93 and 1.3 mm/s, respectively, at RT, indicating an asymmetric ligand field (McDermott et al. 1967). Their respective δ(Pd) values are 0.01 and 0.25 mm/s which signified the low spin nature. The oxidized cytochrome c exhibits $\Delta > 2.0$ mm/s that shows temperature dependence, while the reduced sample shows a value less than 1.3 mm/s (Cooke and Debrunner 1968). The Mössbauer analysis of ferrous cytochrome in an applied magnetic field concluded that V_{zz} is positive and $\eta \approx 0.5$ (Lang et al. 1968). The large quadrupole splitting of the low-spin ferrous cytochrome is expected due to the covalent charge transfer from the ligands. The oxidized ferric cytochrome P450 from gram-negative bacterium exhibited a 77% high spin component in the presence

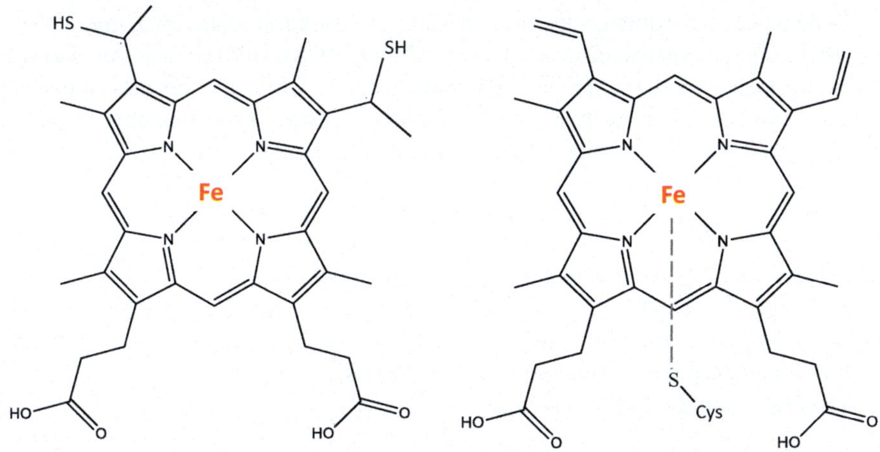

Fig. 12.9 The structure of cytochrome c (left) and cytochrome P450 (right) with different Fe coordination

Table 12.2 The hyperfine parameters of cytochrome c and P450

Sample	Δ, δ/ (mm/s)	T	References
Ferricytochome c (solution)	2.14, 0.21	195 K	Lang et al. (1968)
Ferrocytochome c (solution)	1.20, 0. 42 1.17, 0.45	195 K 4.2 K	
Ferricytochrome-P450 Oxidized + camphor	2.66, 0.31 High spin 0.79, 0.35 Low spin	~ 200 K ~ 200 K	Sharrock et al. (1973)

of camphor, while it was less than 35% with camphor depletion (Sharrock et al. 1973) and the values of the hyperfine parameters at 200 K are given in Table 12.2.

12.6 Bacterial and Plant Proteins

12.6.1 Iron-Sulfur Cluster

Iron-sulfur clusters are biological protein components bound to the amino acids. Rubredoxin [1Fe-0S] and ferredoxin [2Fe-2S] are the bacterial and plant-type iron-sulfur clusters, respectively. The other Fe-S clusters, such as [3Fe-4S] and [4Fe-4S] show mixed valence state and complex spectra with increasing Fe species requiring spin-coupling and antisymmetric exchange models to extract well-founded hyperfine parameters (Kent et al. 1980, 1985; Munck and Kent 1986; Sanakis et al. 2000; Hans et al. 2008; Pandelia et al. 2015).

Proteins are coordinated by iron in various oxidation states that give special hyperfine characteristics discussed below (Munck 1978). Odd valence iron and spin coupled clusters with ½ spin are Kramer systems, while the integral spin even valence and zero spin clusters are non-Kramer systems. The electronic hyperfine characteristics can be described by the spin Hamiltonian given as

$$\hat{H} = \hat{H}_{\mathrm{E}} + \hat{H}_{\mathrm{HF}} = (\hat{H}_{\mathrm{FS}} + \hat{H}_{\mathrm{ZS}}) + \hat{H}_{\mathrm{HF}} \tag{12.5}$$

where H_{FS} and H_{ZS} are the electronic fine-structure splitting and Zeeman splitting terms, respectively, while H_{HF} is the nuclear hyperfine interaction parameter. For weak applied field of nearly 1 mT, the H_{ZS} is comparable to H_{HF} and at large fields, both H_{FS} and H_{ZS} dominate hyperfine interactions.

The H_{HF} could be further given as

$$\hat{H}_{\mathrm{HF}} = \hat{H}_{\mathrm{M}} + \hat{H}_{\mathrm{Q}} + \hat{H}_{\mathrm{Z}} \tag{12.6}$$

where H_{M}, H_{Q} and H_{Z} are the nuclear interaction terms describing magnetic interaction with electronic cloud, quadrupole interaction and Zeeman interaction in an applied field, respectively. For $\hat{H}_{\mathrm{E}} > \hat{H}_{\mathrm{HF}}$, the H_{HF} expression becomes

$$\hat{H}_{\mathrm{HF}} = -g_{\mathrm{N}}\mu_{\mathrm{N}}(H_{\mathrm{int}} + H)I + I \cdot \hat{Q}.I \tag{12.7}$$

Including the quadrupole coupling constant and isomer shift, (12.7) can be written as

$$\hat{H}_{\mathrm{HF}} = -g_{\mathrm{N}}\mu_{\mathrm{N}}(H_{\mathrm{int}} + H)I + \frac{e^2qQ}{2}\sqrt{1 + \frac{\eta^2}{3}} + \delta \tag{12.8}$$

$$H_{\mathrm{int}} = -\frac{\langle S\rangle \cdot \tilde{A}}{g_{\mathrm{N}}\mu_{\mathrm{N}}} \tag{12.9}$$

where $\tilde{A}$ is the magnetic hyperfine coupling tensor, $\langle S\rangle$ is the spin expectation value, g_{N} is the g-factor and μ_{N} is the nuclear magneton. I, eQ, eq and η are the nuclear spin, nuclear quadrupole moment, z-component of the EFG (V_{zz}) and asymmetry parameter, respectively. The parameters in H_{HF} can be obtained from Mössbauer spectra from the temperature-dependent behavior as the effective field is the sum of internal field and the external applied field. Using electron paramagnetic resonance (EPR), we could obtain the fine structure parameters from $\hat{H}_{\mathrm{E}} = \hat{H}_{\mathrm{FS}} + \hat{H}_{\mathrm{ZS}}$ given as

$$\hat{H}_{\mathrm{E}} = \hat{H}_{\mathrm{FS}} + \mu_{\mathrm{B}}S \cdot \tilde{g} \cdot H \tag{12.10}$$

$$\hat{H}_{\mathrm{E}} = D\left(S_z^2 - \frac{1}{3}S(S+1) + \frac{E}{D}(S_x^2 - S_y^2) + \mu_{\mathrm{B}}S \cdot \tilde{g} \cdot H\right) \tag{12.11}$$

where D is the zero-field splitting, E/D is the rhombicity and g is the g-factor.

The characteristics of rubredoxin and [2Fe-2S] clusters and their Mössbauer parameters are listed in Table 12.3. The simple one, iron rubredoxin is a non-heme redox protein that takes part in electron transfer reactions found in bacteria and archaea. The iron atom is tetrahedrally coordinated to four cysteinyl sulfur atoms, as shown in Fig. 12.10. The [2Fe-2S] cluster has two iron sites attached to cysteine residues as in ferredoxin or as cysteine-histidine residue pairs in Rieske protein as shown in Fig. 12.11. It can also exist as three cysteine residues and one histidine or one arginine ligation as in biotin synthase (Berkovitch et al. 2004).

The Mössbauer spectra of the rubredoxin obtained in the oxidized ($S = 5/2$) and reduced form ($S = 2$) from *clostridium pasteurianum* were recorded in the temperature range 4.2 and 300 K (Phillips et al. 1970). The 300–4.2 K doublet spectra for the reduced rubredoxin showed Δ and δ in the range 3.24–3.38 mm/s and 0.36–0.38 mm/s (wrt ^{57}Co/Cu), respectively, originating from the tetrahedrally coordinated high spin ferrous state. The oxidized rubredoxin showed doublet at RT with Δ of 0.74 mm/s and δ of 0.15 mm/s (wrt ^{57}Co/Cu) that transformed to a sextet of B_{hf} 37.5 T at 4.2 K, consistent with high spin ferric state. The oxidized rubredoxin with a B_{hf} of 37.0 T was not significantly affected by an applied magnetic field of 3.0 T below 4.2 K, whereas spin-relaxation rates are rapid beyond 77 K (Rao et al. 1972). On the other hand, considerable changes in splitting could be observed for the reduced ferrous rubredoxin. The negative internal field of – 21.0 T arising from the sum of other hyperfine field contributions (spin, B_s; dipole, B_D; orbital, B_L) could be evaluated for the reduced rubredoxin while other hyperfine parameters are listed in Table 12.3.

The oxidized spinach ferredoxin showed a temperature-independent component with a low isomer shift, and the reduced spinach ferredoxin had two doublet components, one with a lower isomer shift and another corresponding to a ferrous high spin state (Johnson and Hall 1968). Later studies showed single doublet corresponding to a ferric state in oxidized form, whereas it was ferrous and ferric in the reduced state in green algae ferredoxin (Rao et al. 1971) as well as spinach ferredoxin (Dunham et al. 1971) with hyperfine parameters as shown in Table 12.3. The oxidized spinach ferredoxin has two ferric sites of $S = 5/2$ antiferromagnetically coupled to each other giving a net spin of zero. The reduced spinach ferredoxin has both ferrous and ferric high spin states antiferromagnetically coupled, giving a net spin of ½. The reduced state was temperature dependent with hyperfine interactions present at low temperatures and pure doublet pair at high temperatures. The studies also showed tetrahedral symmetry in all iron sites and $d_z{}^2$ as the ferrous ground state.

The Mössbauer analysis of [2Fe-2S] purified Rieske protein from *Thermus thermophilus* membranes showed two distinct ferric sites in the oxidized form with temperature independent Δ in the range 4.2 K < T < 195 K (Fee et al. 1984). For the reduced form, the fast spin relaxation at a temperature above 170 K could average hyperfine interaction and the doublet parameters from one ferrous and one ferric site could be evaluated. The $\Delta = 2.81$ mm/s and $\delta = 0.65$ mm/s at 230 K corresponding to Fe^{2+} showed temperature-dependence increasing the Δ to 2.90 mm/s at 195 K. The most significant component of the ferric EFG tensor is positive.

Table 12.3 The hyperfine parameters of redox proteins and derivatives from bacteria and plants

Compound/Fe-S cluster/spin/ valence	Hyperfine parameters	References
Rubredoxin (reduced) $S = 2$ Fe^{2+}	$\Delta = 3.24$ mm/s, $\delta = 0.36$ mm/s (300 K) $\Delta = 3.36$ mm/s, $\delta = 0.37$ mm/s (4.2 K) δ wrt ^{57}Co/Cu	Phillips et al. (1970)
Rubredoxin (oxidized) $S = 5/2$ Fe^{3+}	$\Delta = 0.74$ mm/s, $\delta = 0.15$ mm/s (300 K) $\delta = 0.13$ mm/s, $B_{hf} = 37.5$ T (4.2 K) δ wrt ^{57}Co/Cu	
Rubredoxin (reduced) $S = 2$ Fe^{2+}	$\Delta = 3.16$ mm/s, $\delta = 0.65$ mm/s (77 K) $B_s = -30.0$, $B_D = 5.4$, $B_L = 3.6$, $B_{int} = -21.0$ in T EFG - negative	Rao et al. (1972)
$[2Fe\text{-}2S]^{2+}$ $S = 0$ Fe^{3+} Oxidized spinach ferredoxin	$\Delta = 0.60$ mm/s, $\delta = 0.20$ mm/s (77 K) T independent. Non-magnetic state not affected by external field	Johnson and Hall (1968)
$[2Fe\text{-}2S]^{1+}$ $S = 5/2 - 2 = 1/2$ Fe^{2+}, Fe^{3+} Reduced spinach ferredoxin	$\Delta = 0.6$ mm/s, $\delta = 0.25$ mm/s (4.2 K) $\Delta = 3.0$ mm/s, $\delta = 1.5$ mm/s (4.2 K) Fe^{2+} T dependent and affected by B	
$[2Fe\text{-}2S]^{2+}$ $S = 0$ Fe^{3+} Oxidized green algae ferredoxin	$\Delta = 0.60$ mm/s, $\delta = 0.20$ mm/s (195 K) Fe^{3+} Similar to spinach ferredoxin	Rao et al. (1971)
$[2Fe\text{-}2S]^{1+}$ $S = 5/2 - 2 = 1/2$ Fe^{2+}, Fe^{3+} Reduced green algae ferredoxin	$\Delta = 0.59$ mm/s, $\delta = 0.22$ mm/s (195 K) Fe^{3+} $\Delta = 2.75$ mm/s, $\delta = 0.56$ mm/s (195 K) Fe^{2+} Similar to spinach ferredoxin Hyperfine splitting below 77 K	
$[2Fe\text{-}2S]^{2+}$ $S = 0$ Fe^{3+} Oxidized spinach ferredoxin	$\Delta = 0.65$ mm/s, $\delta = -0.08$ mm/s (4.2 K) δ wrt ^{57}Co/Pt T independent	Dunham et al. (1971)

(continued)

Table 12.3 (continued)

Compound/Fe-S cluster/spin/ valence	Hyperfine parameters	References
$[2Fe\text{-}2S]^{1+}$ $S = 5/2 - 2 = 1/2$ Fe^{2+},Fe^{3+} Reduced spinach ferredoxin	$\Delta = 0.64$ mm/s, $\delta = -0.06$ mm/s (250 K) Fe^{3+} $\Delta = 2.63$ mm/s, $\delta = 0.21$ mm/s (250 K)Fe^{2+} $\Delta = 0.64$ mm/s, $\delta = -0.10$ mm/s (4.2 K) Fe^{3+} $\Delta = -3.0$ mm/s, $\delta = 0.19$ mm/s (4.2 K)Fe^{2+} δ wrt ^{57}Co/Pt	
$[2Fe\text{-}2S]^{2+}$ Rieske protein $S = 0$ Fe^{3+} Oxidized (*Thermus thermophilus*)	$\Delta = 0.91$ mm/s, $\delta = 0.32$ mm/s (4.2 K) Fe^{3+} $\Delta = 0.52$ mm/s, $\delta = 0.24$ mm/s (4.2 K) Fe^{3+} Δ independent of T in the range 4.2 K < T < 195 K $V_{zz}(Fe^{3+})$ positive, $\eta(Fe^{3+}) < 0.5$ Isotropic ferric $\tilde{A}$	Fee et al. (1984)
$[2Fe\text{-}2S]^{1+}$ Rieske protein $S = 1/2$ Fe^{2+},Fe^{3+} Reduced (*Thermus thermophilus*)	$\Delta = 0.61$ mm/s, $\delta = 0.22$ mm/s (230 K) Fe^{3+} $\Delta = 2.81$ mm/s, $\delta = 0.65$ mm/s (230 K) Fe^{2+} $\Delta = 2.90$ mm/s (195 K), T dependent Fe^{2+} Fast spin relaxation for $T > 170$ K Ferrous EFG axial symmetry $V_{zz}(Fe^{2+})$ positive	

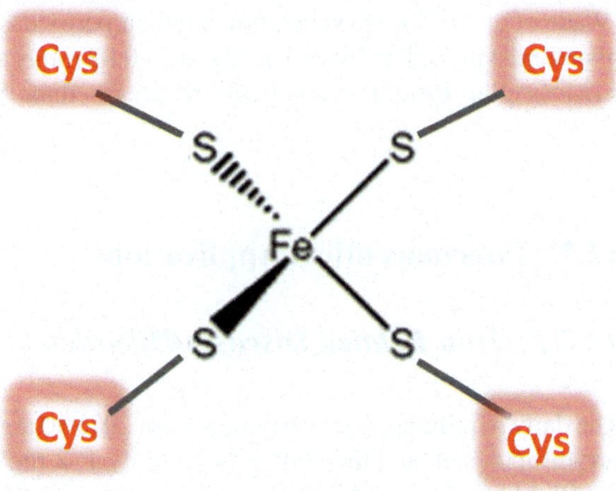

Fig. 12.10 The structure of rubredoxin showing tetrahedrally coordinated Fe with cysteine residues

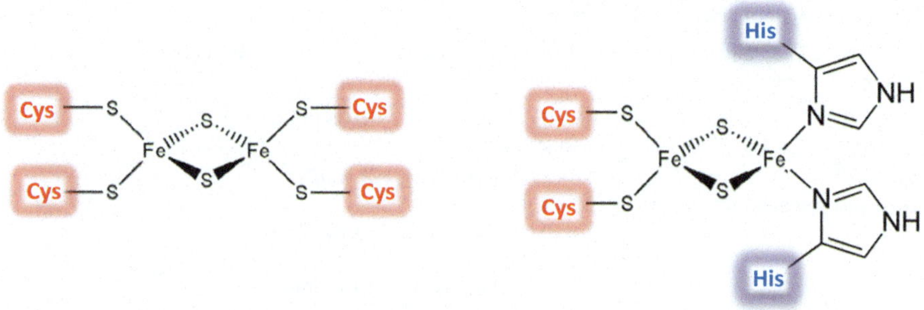

Fig. 12.11 The [2Fe-2S] ferredoxin cluster has cysteine residues (left), and the Rieske protein iron-sulfur cluster has two cysteine and two histidine residues (right)

12.6.2 Plant Parts

Apart from the ferritin and ferredoxin extracted and analyzed from the plants, various other parts such as roots, stems, seeds and leaves were studied. The Mössbauer spectra of duckweed stem, leaf and root at 77 K showed Δ and δ close to 0.7 and 0.5 mm/s, respectively, analogous to the ferric values of ferritin (Goodman and DeKock 1982). Similarly, the tomato plant leaf, stem, root and seed were of ferric state (Ambe 1989). The dry leaves and roots of soybean plants showed a ferric state Δ of 0.7 mm/s and δ of 0.5 mm/s at 77 K, whereas the cotyledon and seeds showed high spin ferrous fraction ranging from 4 to 17% (Ambe et al. 1987). The ferrous fraction decreased to 4% for the ripe yellow seed, whereas it was the highest for the immature green seed. The pea-seed ferritin is mostly amorphous with broader magnetic sextets at 4.1 K that has an average hyperfine field of 44.5 T and wide field distribution of 6.0 T (Wade et al. 1993). However, horse radish peroxidase derivatives present in higher plants exist in the Fe^{4+} with δ as low as $-$ 0.02 mm/s and Δ in the range 1.2–1.4 mm/s (Moss et al. 1969) that can be in the intermediate $S = 1$ ferryl state (Schulz et al. 1979).

12.7 Pharmaceutical Applications

12.7.1 Iron Related Diseases/Disorder

The daily requirement of iron ranges from 20 to 30 mg/d for red blood cell production (erythropoiesis), and the body absorbs only 1–2 mg/d of dietary iron (Muñoz et al. 2009). Most of our body's iron requirement is met through recycled internal sources as there is no effective means to excrete iron. Iron deficiency may arise from blood loss, insufficient dietary intake, genetic, endocrinal disorders, and underlying chronic inflammatory diseases such as cancer, inflammatory bowel diseases and rheumatoid arthritis. The diseases and disorders that result in acute/chronic anemia, iron overload conditions and the related drug ingredients are given in Table 12.4. The normal ranges of hemoglobin, serum iron, transferrin and ferritin are 12–17 g/dL, 50–180 μg/dL 200–360 mg/dL and 30–300 ng/mL, respectively. Iron deficiency anemia is characterized by values below the normal range for all the blood diagnostic iron indicators (Vieth et al. 2017). The hormone hepcidin regulates the intake of iron.

Fe as free radicals produce enormous damage, such as DNA damage resulting in cancer, Fenton reaction adding double bonds to DNA bases, atherosclerotic accumulation, and oxidative stress-induced neurodegenerative disorders (Jomova and Valko 2011). Although these disease progressions are not limited to Fe alone, the detrimental effect of the essential element under certain circumstances has to be understood. Neurodegeneration is the symptom of Parkinson's and Alzheimer's diseases. Oxidative stress due to the overproduction of divalent iron and the associated Fenton reaction is a consequence (Friedman and Friedman 2017). The Fenton reaction is given by

$$Fe^{2+} + H_2O_2 \rightarrow Fe^{3+} + OH^* + OH^- \tag{12.12}$$

The ferritin defect and L-ferritin iron leakage from the protein shell leads to neurodegeneration (Friedman and Friedman 2012). As a result, these disease conditions require iron-chelating drugs to remove iron overload. Deferoxamine, a hexadentate chelator and deferiprone, a bidentate chelator, deactivate the free iron ions. Certain disease conditions, such as rheumatoid arthritis, result in anemia but iron accumulates in the synovial fluid of arthritic joints. Mössbauer spectrometry is being used to understand the origin of these diseases and characterize the drugs.

12.7.2 Iron Supplements

Oral hematinic drugs are usually ferrous since they are soluble, pH independent and available in the intestine for absorption (Umbreit 2005). Common low-cost, well-tolerated oral medication to treat iron deficiency anemia is 300–325 mg ferrous sulfate given 3–4 times a day, which could elevate the Hb concentration to 1 g/dL/week (Weiss and Gordeuk 2005). A few other drugs that replaced ferrous sulfate are ferrous fumarate, ferrous gluconate and ferrous succinate. The availability of various drugs and their combinations are useful alternatives if a particular drug is not tolerant. Iron dextran and sodium ferric gluconate complex are a few parenteral drugs to treat anemia.

The Mössbauer investigation of the ferrous supplements resulted in RT quadrupole splitting of the order ferrous sulfate (hyd) ($\Delta \approx$ 3.3 mm/s) > ferrous gluconate > ferrous sulfate > ferrous succinate > ferrous fumarate (hyd) ($\Delta \approx$ 2.4 mm/s) as shown in Fig. 12.12. All the samples exhibited an isomer shift of 1.2–1.3 mm/s corresponding to its ferrous state. The RT quadrupole splitting of the dehydrated ferrous succinate doublet increases to 2.9 mm/s compared with 2.4 mm/s for the

Table 12.4 The disease/disorder that requires either iron supplements or iron chelating compounds

Disease/disorder	Iron related effect	Drug ingredient
Iron deficiency anemia Inflammatory bowel diseases (ulcerative colitis, Crohn's disease) Paroxysmal nocturnal hemoglobinuria Congestive heart failure Chronic kidney disease Hypochromic anemia	Chronic anemia	***Hematinic*** Ferric citrate Ferumoxytol Iron dextran Ferric carboxymaltose Ferumoxytol Iron isomaltoside Iron sucrose Iron polymaltose
Intestinal nematode infections	Anemia in children	***Oral drugs*** Ferrous fumarate Ferrous gluconate Ferrous succinate Ferrous sulfate Ferrous bisglycinate chelate Ferrous fructose Ferrous lactate Iron choline Ferrous carbonate Ferrous ammonium citrate Ferrous glycine citrate
Hypermenorrhea	Anemia in women	
Hereditary spherocytosis	Hemolytic anemia	
		Fe chelating
Iron overload Hemochromatosis	Fe as free radical	Deferoxamine (Desferrioxamine B, desferal) Ferriprox (deferiprone) 2,3-Dihydroxybenzoic acid Hydroxypyridone L1 Clioquinol (5-chloro-7-iodoquinolin-8-ol)
β-thalassemia	Anemia Secondary iron overload	
Parkinson's disease	Fe oxidative stress	
Alzheimer's disease	Fe oxidative stress	
Progressive supranuclear palsy	Fe oxidative stress	
Friedreich's ataxia	Excess Fe in mitochondria	
Huntington's disease	Fe accumulation in the central nervous system	
Neuroferritinopathy	Fe accumulation in the brain	
Alcoholic liver disease	Hemochromatosis risk	
Cancer	Free radical DNA damage	Bleomycin
Malaria	Anemia, Hemazoin synthesis	Chloroquine Quinine Deferrioxamine Deferiprone Spermidine hydrobromide Ferroquine

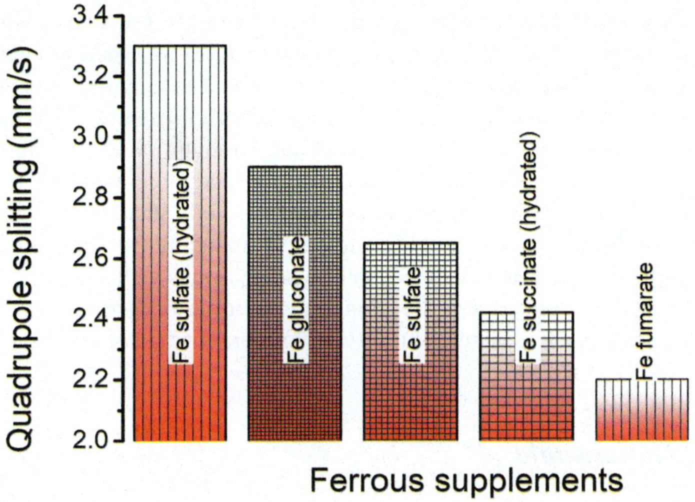

Fig. 12.12 The room temperature quadrupole splitting of a few ferrous supplements used to treat iron deficiency anemia. Data from (Coe et al. 1995; Nikumbh et al. 1994)

hydrated ferrous succinate (Nikumbh et al. 1994). The dehydrated form was obtained by heating the compound at 380–490 K, which further decomposes successively to Fe_3O_4, γ-Fe_2O_3 and α-Fe_2O_3 as a function of temperature that showed their corresponding sextets. The RT Mössbauer spectra of various ferrous hematinic drugs sold by various manufacturers were analyzed (Coe et al. 1995). The hydrated $FeSO_4.7H_2O$ showed a quadrupole splitting of 2.37 mm/s, and it was 2.69 mm/s for the dehydrated one ($FeSO_4$). The ferrous fumarate and ferrous gluconate had nearly 10% ferric content. Most of the ferrous drugs are less than 10% ferric but above 2%, whereas ferrous bisglycinate chelate formulation showed more than 50% ferric high spin fraction ($\Delta = 0.66$ and $\delta = 0.39$ mm/s at RT (Oshtrakh et al. 2006).

The Mössbauer studies from two different injectable drugs showed that the iron-dextran complex is similar to the ferritin while the iron-chondroitin-sulfate complex drug had another component (15–20%) close to hematite along with the ferritin-like major component (Yang et al. 1986). The ferric nature was confirmed from the lower isomer shift of 0.5 mm/s at 20 K for all the samples, while the hyperfine field of 52.7–53.1 T was assigned to the hematite for the iron-chondroitin-sulfate complex. The Mössbauer analysis of iron sucrose, iron polymaltose, iron dextran and iron dextrin, obtained between 4.2 and 295 K, resulted in blocking temperatures of 55, 65, 115 and 215 K, respectively (Funk et al. 2001). These drug formulations have 1–4 nm particles of ferric ions in akagneite or ferrihydrite-like structures exhibiting different pharmacokinetic behavior due to their shell.

The Mössbauer analyses of more stable recent drugs, such as ferric carboxymaltose (FCM), ferumoxytol (FMX) and iron isomaltoside (IIM) introduced for intravenous (IV) administration, were undertaken at 77 and 4.2 K (Neiser et al. 2015). All the samples showed ferric isomer shift, and the core material in IIM and FCM is akaganeite (β-FeOOH) while it is maghemite (γ-Fe_2O_3) in FMX. All three showed relaxation spectra at 77 K, and the blocking temperatures were determined as 56, 73 and 114 K for the IIM, FMX and FCM, respectively. The maximum recommended doses and serum terminal half-lives were significantly different due to the difference in their particle sizes and core–shell bond strength. New drug formulations are focused on better absorption and tolerance. The body tolerance to intravenously administered drugs is enhanced by biocompatible shells such as carbohydrates or proteins. Food fortification is an efficient way of reducing iron deficiency anemia in targeted populations (Hurrell 1997). While heme iron is well absorbed (10–35%) compared with non-heme, the absorption from the drug or fortification forms also depends on factors such as water solubility, ferrous or ferric states and bioavailability.

12.7.3 Thalassemia

The normal hemoglobin levels in β-thalassemia patients decrease due to low β-globin chain synthesis. Iron is not bound to transferrin or ferritin in thalassemia patients. As a result, iron overload in most of the body parts is commonly observed. The comparison of the Mössbauer spectra of RBC at 4.1 K for the normal person and from β-thalassemia intermedia patient revealed that an additional ferritin-like component was present in the RBC of the patient with the disease (Bauminger et al. 1979). In contrast to the normal RBC exhibiting only two doublets (oxy-Hb and deoxy-Hb) with less than 1% ferritin, the β-thalassemia intermedia RBC showed nearly 60% ferritin-like sextet component. The ferritin accumulation may be due to intracellular Hb denaturation. Ferritin deposits in cardiac walls were earlier reported in β-thalassemia subjects on comparison of the normal samples from autopsy analysis (Papaefthymiou et al. 1978; Kaufman et al. 1980). The fraction of ferritin/hemosiderin in heart tissues may be even 75% (Anusorn et al. 2000).

12.7.4 Neurodegenerative Diseases

The Mössbauer studies could not conclude the elevated brain iron levels due to Parkinson's disease, whereas there were contradictory reports from other studies. Mössbauer analysis of substantia nigra from Parkinsonian and normal control samples indicated 90% of ferritin-like ferric sextet (Bauminger et al. 1994). The Mössbauer spectrum at 4.2 K for the neuromelanin polymer pigment from the normal brain showed a 92% major sextet component with B_{hf} of 49.5 T and ferric isomer shift of 0.51 mm/s along with superparamagnetic relaxation component (Double et al. 2003).

Neurodegeneration symptoms are substantially from substantia nigra in Parkinson's, hippocampal cortex in Alzheimer's and globus pallidus in progressive supranuclear palsy and therefore, Mössbauer spectra of these tissue specimens were examined (Friedman et al. 2012). All the brain samples showed ferric isomer shift (0.46–0.49 mm/s), and quadrupole split doublet at 90 K with Δ of 0.67–0.72 mm/s similar to ferritin. Similarly, the sextet with hyperfine field of 49.2–49.8 T is similar to ferritin (49.4 T) at 4.1 K. The superparamagnetic nature of the samples was identified with blocking temperature varying according to the size of the particles that could influence the estimation of the sextet relative area fraction from the Mössbauer studies. Ferrous iron could not be detected in any of the samples. Although the increase in the concentration of brain iron is not conclusive, iron-mediated oxidative stress may be triggered even by low concentrations of free iron, a common factor in these neurodegenerative diseases. The free ion sources may be from the L-ferritin or neuromelanin (Friedman et al. 2004).

12.7.5 Deferoxamine

Deferoxamine is a chelating agent that removes excess iron from intra-lysosomal ferritin and siderin by forming ferrioxamine. The structure of deferoxamine, as shown in Fig. 12.13 (top) can bind with Fe ions under different coordination environments (bottom). The absorption capacity of deferoxamine (desferrioxamine B) in methanesulfonate form was tested with Fe^{3+} and Fe^{2+} solutions using Mössbauer spectroscopy (Bock and Lang 1972). The spectrum at 4.2 K indicated Fe^{3+} sextet in similar environments with deferoxamine, whereas additional sextet and zero velocity central line were observed for the ferric control solution. On the other hand, the ferrous-deferoxamine and ferrous control samples had almost similar doublets with distribution in Fe environments. Although the complex formation with ferric-deferoxamine was clear, the ferrous sample's spectrum did not exhibit any features to distinguish complex formation.

12.7.6 Bleomycin

Bleomycin is an anticancer drug whose structure has a metal binding domain, carbohydrate domain and DNA binding domain connected with a linker region, as shown in Fig. 12.14. The metal binding domain with ferrous ions forms complexes of O_2, CO, C_2H_5NC or NO. The Mössbauer spectra of Fe enriched Fe(II)-bleomycin, Fe(II)-bleomycin-NO and Fe(II)-bleomycin-CN complexes recorded at 110 K are shown in Fig. 12.15 (Sugiura et al. 1982). The bleomycin-Fe(II) show Mössbauer parameters corresponding to a high spin ferrous state with $\Delta = 2.99$ mm/s and $\delta = 1.10$ mm/s. The quadrupole splitting of the -CN complex (0.89 mm/s) is higher than the -NO complex (0.78 mm/s). Although the hyperfine parameters of the CN complex vary from the hemoglobin-CN proteins, other complexes of bleomycin are similar to the hemoprotein. The Mössbauer analyses of ferrous, ferric and liganded ^{57}Fe enriched bleomycin were undertaken at low temperatures and with high magnetic

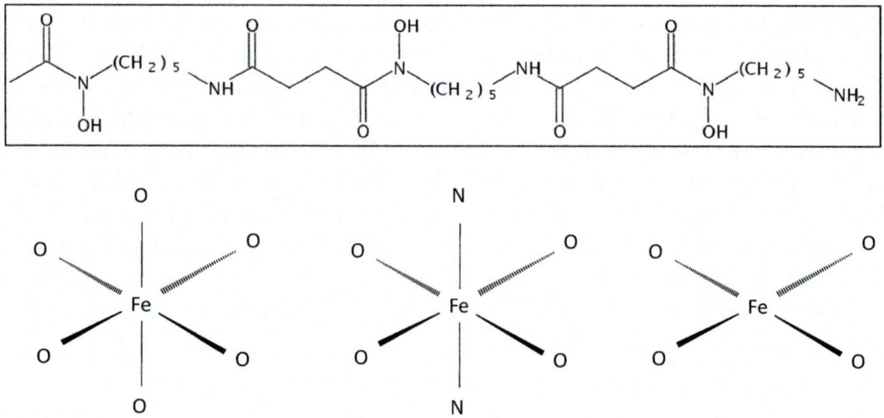

Fig. 12.13 The structure of deferoxamine (top) and a few of its Fe coordination on forming a complex with Fe

fields (Burger et al. 1983). The ferrous bleomycin buffered at pH 7 exhibited high spin values with $\Delta = 2.83$ mm/s and $\delta = 1.22$ mm/s at 4.2 K. Low spin ferric bleomycin exhibited sharp spectra without significant relaxation broadening at 200 K. The spectrum at 4.2 K showed $\Delta = 3$ mm/s and $\delta = 0.2$ mm/s with negative quadrupole splitting and $\eta = 0.3$. High-field Mössbauer studies showed that CO and oxygenated ferrous bleomycin are diamagnetic $S = 0$ complexes. The activated bleomycin was prepared by either adding air to ferrous bleomycin or H_2O_2 to ferric bleomycin, which is a low-spin ferric compound.

12.7.7 Malaria

Malaria is a disease caused by Plasmodium parasites that consume most of our hemoglobin, converting the heme to hemozoin (β-hematin) inside the parasite's food vacuole (Corondado et al. 2014). In malaria-infested RBC, a third ferric high spin doublet was observed in addition to the oxy-Hb and deoxy-Hb doublets at 90 K (Bauminger et al. 1983). The ferric high spin doublet transformed to a high spin sextet with a B_{hf} of 49.4 T at 4.1 K. The ferric ions in the hematin-like pigment are strongly bound, as observed from the temperature-dependent *f*-factor, not resembling hematin and unaffected by chloroquine. The conversion rate of the hematin-like pigment from Hb increased with progressive disease stages. Later studies showed that the malaria pigment is similar to β-hematin with Δ of 0.37 mm/s while that of hematin is 0.85 mm/s from measurements at 78 K (Adams et al. 1996).

The comparison of the β-hematin malaria pigment, the normal one termed R-hematin and another B-hematin were undertaken through Mössbauer analysis. The β-hematin showed the lowest quadrupole splitting of 0.59 mm/s at 4.1 K among the

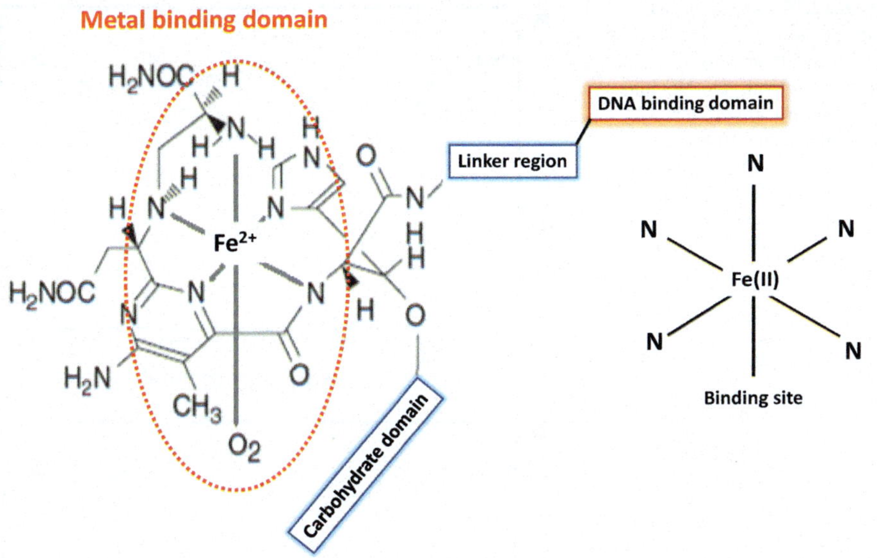

Fig. 12.14 The structure of oxygenated bleomycin in the metal binding domain. The ferrous binding site has five nitrogen coordination (right)

various hematins, indicating symmetric charge distribution (Bauminger et al. 1999). The β- and B- hematins showed a relaxation spectrum at 90 K, whereas the R- hematin exhibited distinct doublets, as shown in Fig. 12.16. The hemozoin from P. falciparum exhibits $\Delta = 0.5$ mm/s, $\delta = 0.39$ mm/s and narrow linewidth at 4.1 K similar to the ferric high spin β-hematin. Further studies showed that the Mössbauer spectra of freeze-dried trophozoites resembled that of β-hematin exhibiting distinct quadrupole doublets at low temperature and broad features at 90 K (Egan et al. 2002). The studies also showed that almost 95% of Hb was digested as hemozoin inside the food vacuole in trophozoite. Quinacrine, chloroquine and amodiaquine are potential β-hematin inhibitors that increase free hemin and promote peroxidative reactions (Herraiz et al. 2019). Ferrocene conjugate, especially ferroquine, is a potential antimalarial drug alternative due to increasing drug-resistant strains (Ludwig et al. 2019).

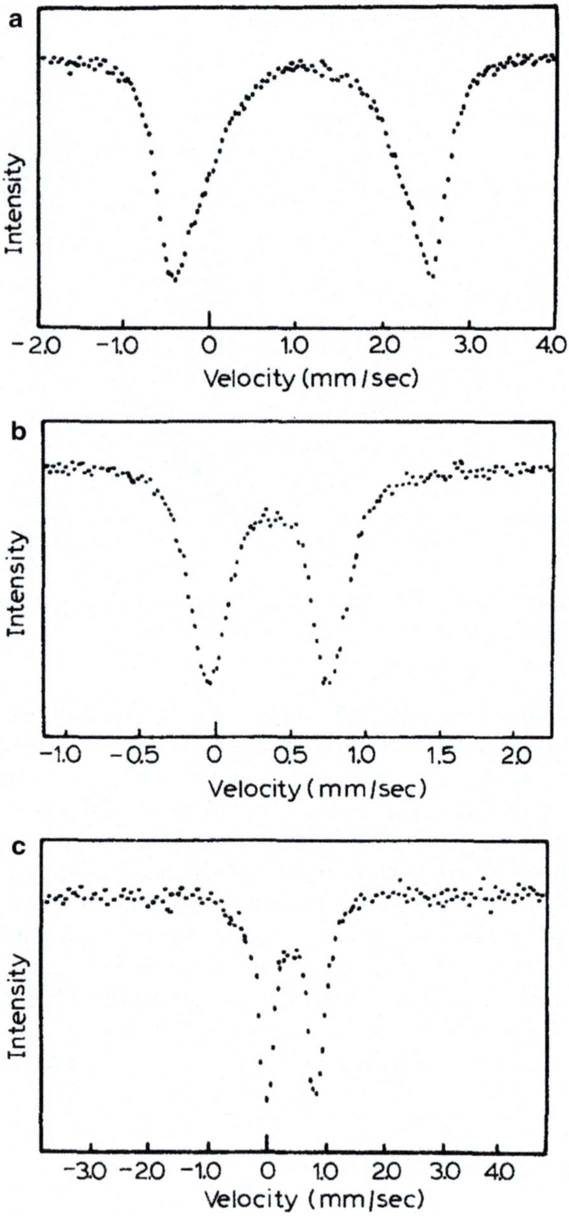

Fig. 12.15 The Mössbauer spectra of **a** Fe(II)-bleomycin **b** Fe(II)-bleomycin-NO and **c** Fe(II)-bleomycin-CN complexes at 110 K. Reprinted with permission (Sugiura et al. 1982), © Elsevier

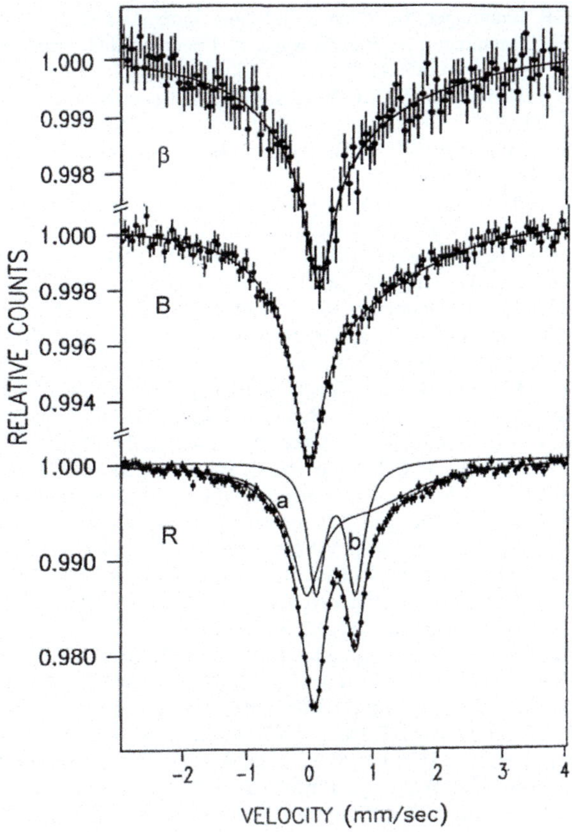

Fig. 12.16 The Mössbauer spectra at 90 K of β-, B- and regular (R) hematins (top to bottom). Reprinted with permission (Bauminger et al. 1999), © Elsevier

12.8 Summary

The role of Mossbauer spectrometry in identifying the Fe-based compounds of biological systems is discussed in this chapter. The Mossbauer parameters of Fe-based compounds in the human body, bacterial and plant proteins are presented. The identification of oxidation states in the diagnosis of diseases and pharmaceuticals has shown the extended application of Mössbauer spectrometry in life sciences.

References

P.A. Adams, T.J. Egan, D.C. Ross, J. Silver, P.J. Marsh, Biochem. J. **318**, 25 (1996)

S. Ambe, Appl. Radiat. Isot. **40**, 671 (1989)

S. Ambe, F. Ambe, T. Nozaki, J. Agric. Food Chem. **35**, 292 (1987)

W.C. Anusorn, K.C. Tran, J. Webb, D.J. Marcey, T.G. St Pierre, Inorganica Chim. Acta 300–302, 932 (2000)

E.R. Bauminger, H. Ginsburg, S. Ofer, A. Yayon, Hyperfine Interact. 15/16, 885 (1983)

E.R. Bauminger, M. Akkawi, G. Blauer, Inorganica Chim. Acta **286**, 229 (1999)
E.R. Bauminger, M. Barcikowska, A. Friedman, J.G. Friedman, D. Hechel, I. Nowik, Hyperfine Interact. **91**, 853 (1994)
E.R. Bauminger, S.G. Cohen, S. Ofer, A. Rachmilewitz, Proc. Natl. Acad. Sci. U.S.A. **76**, 939 (1979)
S.N. Bell, M.P. Weir, D.P.E. Dickson, J.F. Gibson, G.A. Sharp, T.J. Peters, Biochim. Biophys. Acta **787**, 227 (1984)
F. Berkovitch, Y. Nicolet, J.T. Wan, J.T. Jarrett, C.L. Drennan, Science **303**, 5654 (2004)
A. Blaise, J. Chappert, J. Giradet, C.R. Acad. Sci. Paris **261**, 2310 (1965)
J.F. Boas, B. Window, Aust. J. Phys. **19**, 573 (1966)
J.L. Bock, G. Lang, Biochim. Biophys. Acta **264**, 245 (1972)
B. Boso, P.G. Debrunner, G.C. Wagner, T. Inbushi, Biochim. Biophys. Acta **791**, 244 (1984)
R.M. Burger, T.A. Kent, S.B. Horwitz, E. Munck, J. Peisach, J. Biol. Chem. **258**, 1559 (1983)
W.S. Caughey, W.Y.F. Fujimoto, A.J. Bearden, T.H. Moss, Biochemistry **5**, 1255 (1966)
Y.W. Chow, A. Mukerji, Biochem. Biophys. Res. Commun. **62**, 989 (1975)
E.M. Coe, L.H. Bowen, R. Bereman, J. Inorg. Biochem. **58**, 291 (1995)
R. Cooke, P. Debrunner, J. Chem. Phys. **48**, 4532 (1968)
L.M. Corondado, C.T. Nadovich, C. Spadafora, Biochim. Biophys. Acta **1840**, 2032 (2014)
R. Crichton, *Inorganic Biochemistry of Iron Metabolism* (John Wiley&Sons Ltd., England, 2001)
R.R. Crichton, *Biological Inorganic Chemistry, An Introduction* (Elsevier, Oxford, 2008)
K.L. Double, M. Gerlach, V. Schunemann, A.X. Trautwein, L. Zecca, M. Gallorini, M.B.H. Youdim, P. Riederer, D.B. Shachar, Biochem. Pharma. **66**, 489 (2003)
W.R. Dunham, A.L. Bearden, I.T. Salmeen, G. Palmer, R.H. Sands, W.H.O. Johnson, H. Beinert, Biochim. Biophys. Acta **253**, 134 (1971)
T.J. Egan, J.M. Combrinck, J. Egan, G.R. Hearne, H.M. Marques, S. Ntenteni, B.T. Sewell, P.J. Smith, D. Taylor, D.A. van Schalkwyk, J.C. Walden, Biochem. J. **365**, 343 (2002)
H. Eicher, A. Trautwein, J. Chem. Phys. **50**, 2540 (1969)
H. Eicher, D. Bade, F. Parak, J. Chem. Phys. **64**, 1446 (1976)
J.A. Fee, K.L. Findling, T. Yoshida, R. Hille, G.E. Tarr, D.O. Hearshen, W.R. Dunham, E.P. Day, T.A. Kent, E. Munck, J. Biol. Chem. **259**, 124 (1984)
F.A. Fischbach, D.W. Gregory, P.M. Harrison, T.G. Hoy, J.M. Williams, J. Ultrastruct. Res. **37**, 495 (1971)
R.B. Frankel, G.C. Papaefthymiou, R.P. Blakemore, W. O'Brien, Biochim. Biophys. Acta **763**, 147 (1983)
R.B. Frankel, R.P. Blakemore, R.S. Wolfe, Science **203**, 1355 (1979)
H. Frauenfelder, G.A. Petsko, D. Tsernoglou, Nature **280**, 558 (1979)
H. Frauenfelder, P.W. Fenimore, R.D. Young, Proc. Natl. Acad. Sci. U.S.A. **111**, 12768 (2014)
A. Friedman, J.G. Friedman, J. Neural Transm. **119**, 1507 (2012)
A. Friedman, J.G. Friedman, Hyperfine Interact. **238**, 62 (2017)
J.G. Friedman, E.R. Bauminger, D. Koziorowski, A. Friedman, Biochim. Biophys. Acta **1688**, 130 (2004)
J.G. Friedman, E.R. Bauminger, K. Szlachta, A. Friedman, J. Phys. Condens. Matter **24**, 244106 (2012)
F. Funk, G.J. Long, D. Hautot, R. Buchi, I. Christl, P.G. Weidler, Hyperfine Interact. **136**, 73 (2001)
U. Gonser, R.W. Grant, Biophys. J. **5**, 823 (1965)
U. Gonser, R.W. Grant, J. Kregzde, Appl. Phys. Lett. **3**, 189 (1963)
U. Gonser, R.W. Grant, J. Kregzde, Science **143**, 680 (1964)
U. Gonser, Y. Maeda, A. Trautwein, F. Parak, H. Formanek, Z. Naturforsch. **296**, 241 (1974)
G.A. Goodman, P.C. DeKock, J. Plant Nutr. **5**, 345 (1982)
R.W. Grant, J.A. Cape, U. Gonser, L.E. Topol, P. Saltman, Biophys. J. **7**, 652 (1967)
M. Hans, W. Buckel, E. Bill, J. Biol. Inorg. Chem. **13**, 563 (2008)
T. Herraiz, H. Guillen, D.G. Pena, V.J. Aran, Sci. Rep. **9**, 15398 (2019)
R.F. Hurrell, Nutr. Rev. **55**, 210 (1997)

B.H. Huynh, G.C. Papaefthymiou, C.S. Yen, J.L. Groves, C.S. Wu, J. Chem. Phys. **61**, 3750 (1974a)
B.H. Huynh, G.C. Papaefthymiou, C.S. Yen, J.L. Groves, C.S. Wu, Biochem. Biophys. Res. Commun. **60**, 1295 (1974b)
C.E. Johnson, D.O. Hall, Nature **446**, 446 (1968)
K. Jomova, M. Valko, Toxicology **283**, 65 (2011)
K.S. Kaufman, G.C. Papaefthymiou, R.B. Frankel, A. Rosenthal, Biochim. Biophys. Acta **629**, 522 (1980)
T.A. Kent, B.H. Huynh, E. Muck, Proc. Natl. Acad. Sci. U.S.A. **77**, 6574 (1980)
T.A. Kent, K. Spartalian, G. Lang, T. Yonetani, C.A. Reed, J.P. Collman, Biochim. Biophys. Acta **580**, 245 (1979)
T.A. Kent, M.H. Emptage, H. Merkle, M.C. Kennedy, H. Beinert, E. Munck, J. Biol. Chem. **260**, 6871 (1985)
T. Kent, K. Spartalian, G. Lang, T. Yonetani, Biochim. Biophys. Acta **490**, 331 (1977)
S.A. Kretchmar, M. Teixeira, B.H. Huynh, K.N. Raymond, Biol. Met. **1**, 26 (1988)
Y.F. Krupyanskii, F. Parak, E.E. Gaubamn, F.M. Wagner, V.I. Goldanskii, R.L. Mössbauer, I.P. Suzdalev, F.J. Litterest, H. Vogel, J. Phys. Colloques **41**, C1-489 (1980)
G. Lang, D. Herbert, T. Yonetani, J. Chem. Phys. **49**, 944 (1968)
G. Lang, Q. Rev. Biophys. **3**, 1 (1970)
G. Lang, T. Asakura, T. Yonetani, Biochim. Biophys. Acta **214**, 381 (1970)
G. Lang, W. Marshall, Proc. Phys. Soc. **87**, 3 (1966)
H. Lichtenegger, W. Doster, T. Kleinert, A. Birk, B. Sepiol, G. Vogl, Biophys. J. **76**, 414 (1999)
P.F. Lindley, Rep. Prog. Phys. **59**, 867 (1996)
B.S. Ludwig, J.D.G. Correia, F.E. Kuhn, Coord. Chem. Rev. **396**, 22 (2019)
Y. Maeda, T. Harami, A. Trautwein, U. Gonser, Z. Naturforsch. B **31**, 487 (1976).
Y. Maeda, T. Harami, Y. Morita, A. Trautwein, U. Gonser, J. Chem. Phys. **75**, 36 (1981)
Y. Maeda, Y. Morita, C. Yoshida, J. Biochem. **70**, 509 (1971)
L. Marchant, M. Sharrock, B.M. Hoffman, E. Munck, Proc. Natl. Acad. Sci. U.S.A. **69**, 2396 (1972)
K.H. Mayo, D. Kucheida, F. Parak, J.C.W. Chien, Proc. Natl. Acad. Sci. U.S.A. **80**, 5294 (1983)
P. McDermott, L. May, J. Orlando, Biophys. J. **7**, 615 (1967)
T.H. Moss, A. Ehrenberg, A.J. Bearden, Biochemistry **8**, 4159 (1969)
R.L. Mössbauer, Hyperfine Interact. **33**, 199 (1987)
E. Munck, *Methods in Enzymolology*, vol. 54 (Elsevier, 1978) p. 346
E. Munck, T.A. Kent, Hyperfine Interact. **27**, 161 (1986)
M. Muñoz, I. Villar, J.A. García-Erce, World J. Gastroenterol. **15**, 4617 (2009)
S. Neiser, D. Rentsch, U. Dippon, A. Kappler, P.G. Weidler, J. Gottlicher, R. Steininger, M. Wilhelm, M. Braitsch, F. Funk, E. Phillip, S. Burckhardt, Biometals **28**, 615 (2015)
A.K. Nikumbh, M.M. Phadke, A.A. Latkar, J. Magn. Magn. Mater. **131**, 189 (1994)
M.I. Oshtrakh, O.B. Milder, V.A. Semionkin, J. Pharm. Biomed. Anal. **40**, 1281 (2006)
F. Parak, U.F. Thomanek, D. Bade, B. Wintergerst, Z. Naturforsch. C **32**, 507 (1977)
M.E. Pandelia, N.D. Lanz, S.J. Booker, C. Krebs, Biochim. Biophys. Acta **1853**, 1395 (2015)
G.C. Papaefthymiou, B.H. Huynh, C.S. Yen, J.L. Groves, C.S. Wu, J. Chem. Phys. **62**, 2995 (1975)
G.C. Papaefthymiou, K.S. Kaufman, R.B. Frankel, A. Rosenthal, J. Appl. Phys. **49**, 1815 (1978)
E. Parak, E.W. Knapp, D. Kucheida, J. Mol. Biol. **161**, 177 (1982)
F.G. Parak, K. Achterhold, S. Croci, M. Schmidt, J. Biol. Phys. **33**, 371 (2007)
F. Parak, E.N. Frolov, R.L. Mössbauer, V.I. Goldanskii, J. Mol. Biol. **145**, 825 (1981)
F. Parak, H. Formanek, Acta Cryst. A **27**, 573 (1971)
F. Parak, R.L. Mössbauer, W. Hoppe, U.F. Thomanek, D. Bade, J. Phys. Colloques **37**, C6-703 (1976)
W.D. Phillips, M. Poe, J.F. Weiher, C.C. McDonald, W. Lovenberg, Nature **227**, 574 (1970)
K.K. Rao, M.C.W. Evans, R. Cammack, D.O. Hall, C.L. Thompson, P.J. Jackson, C.E. Johnson, Biochem. J. **129**, 1063 (1972)
K.K. Rao, R. Cammack, D.O. Hall, C.E. Johnson, Biochem. J. **122**, 257 (1971)

Y. Sanakis, A.L. Macedo, I. Moura, J.J.G. Moura, V. Papaefthymiou, E. Munck, J. Am. Chem. Soc. **122**, 11855 (2000)
C.E. Schulz, P.W. Devaney, H. Winkler, P.G. Debrunner, N. Doan, R. Chiang, R. Rutter, L.P. Hager, FEBS Lett. **103**, 102 (1979)
N. Schuth, S. Mebs, D. Huwald, P. Wrzolek, M. Schwalbe, A. Hemschemeier, M. Haumann, Proc. Natl. Acad. Sci. U.S.A. **114**, 8556 (2017)
M. Sharrock, E. Munck, P.G. Debrunner, V. Marshall, J.D. Lipscomb, I.C. Gunsalus, Biochemistry **12**, 258 (1973)
H.G. Sherman, C. Jovanovic, S. Stolnik, K. Baronian, A.J. Downard, F.J. Rawson, Front. Mol. Biosci. **5**, 1 (2018)
G.P. Singh, F. Parak, S. Hunklinger, K. Dransfeld, Phys. Rev. Lett. **47**, 685 (1981)
K. Spartalian, G. Lang, J.P. Collman, R.R. Gagne, C.A. Reed, J. Chem. Phys. **63**, 5375 (1975)
K. Spartalian, W.T. Oosterhuis, J. Chem. Phys. **59**, 617 (1973)
T.G. St. Pierre, S.H. Bell, D.P.E. Dickson, S. Mann, J. Webb, G.R. Moore, R.J.P. Williams, Biochim. Biophys. Acta, **870**, 127 (1986)
Y. Sugiura, T. Suzuki, H. Kawabe, H. Tanaka, K. Watanabe, Biochim. Biophys. Acta **716**, 38 (1982)
A. Trautwein, H. Eicher, A. Mayer, J. Chem. Phys. **52**, 2473 (1970)
A. Trautwein, Y. Alpert, Y. Maeda, H.E. Marcolin, J. Phys. Colloques **37**, C6-191 (1976)
A. Trautwein, Y. Maeda, F.E. Harris, H. Formanek, Theor. Chim. Acta **36**, 67 (1974)
C.P. Tsang, A.J.F. Boyle, E.H. Morgan, Biochim. Biophys. Acta **328**, 84 (1973)
J. Umbreit, Am. J. Hematol. **78**, 225 (2005)
J.T. Vieth, D.R. Lane, Hematol./Oncol. Clin. N. Am. **31**, 1045 (2017)
V.J. Wade, A. Treffry, J.P. Laulhere, E.R. Bauminger, M.I. Cleton, S. Mann, J.F. Briat, P.M. Harrison, Biochim. Biophys. Acta **1161**, 91 (1993)
G. Weiss, V.R. Gordeuk, Eur. J. Clin. Invest. **35**, 36 (2005)
J.J. Yan, T. Kroll, M.L. Baker, S.A. Wilson, R. Decreau, M. Lundberg, D. Sokaras, P. Glatzel, B. Hedman, K.O. Hodgson, E.I. Solomon, Proc. Natl. Acad. Sci. U.S.A. **116**, 2851 (2019)
C. Yang, A.M. Bryan, E.C. Theil, D.E. Sayers, L.H. Bowen, J. Inorg. Biochem. **28**, 393 (1986)
R.D. Young, H. Frauenfelder, P.W. Fenimore, Phys. Rev. Lett. **107**, 158102 (2011)

Chapter 13
Mössbauer Spectrometry of Superconductors and Manganites

13.1 Superconductors

Conventional superconductors show pure diamagnetic nature where magnetic impurities destroy their superconductivity. The diamagnetic response depends on the applied magnetic field where superconductivity is destroyed with a higher applied field and with increasing temperature. The magnetic transition temperature is not encountered in such superconductors, and therefore, we have only the superconducting transition temperature. The superconducting transition temperature is generally denoted as T_c, whereas we have used T_{sc} here to avoid confusion with the T_c used for Curie temperature. Fe-based superconductors such as U_6Fe and rare-earth iron silicides exhibit T_{sc} below 6–7 K without magnetic ordering. With the advent of high T_{sc} YBCO superconductors in 1986, many studies were carried out with Fe doping to understand the structure. Fe-based superconductors became prominent after the discovery of Fe-pnictides with T_{sc} exceeding 25 K in 2008. Mössbauer spectrometry became an essential tool for characterizing Fe-based superconductors thereafter.

13.2 Conventional Superconductors

13.2.1 U_6Fe

U_6Fe is one of the first Fe-based superconductors with high T_{sc} (3.86 K) among its uranium-transition metal counterparts (Chandrasekhar and Hulm 1958). It is one among the strongly interacting Fermi liquid Type II superconductors exhibiting a large upper critical field, $B_{c2}(0\ \text{K}) \rightarrow 10$ T (DeLong et al. 1983). The moderately heavy Fermi liquid system (effective mass $m^* \approx 20m_e$) fall between heavy fermion and cuprate superconductors (White et al. 2015).

R. J. Joseyphus and J.-M. Greneche, *Fundamentals of ^{57}Fe Mössbauer Spectrometry*,
https://doi.org/10.1007/978-981-99-8653-8_13

U_6Fe belongs to the tetragonal *I4*/mcm space group that has Fe atoms as linear chains at the center and corners of the unit cell along the *c*-axis. Mössbauer spectrum at 26 K revealed the paramagnetic nature of the sample with a quadrupole splitting of 0.74 mm/s from noncubic symmetry and a small isomer shift, indicating the metallic nature (Blow 1969). The doublet often has asymmetry that may be due to either oxide with larger δ or ferromagnetic UFe_2 with smaller Δ (Lemon et al. 1987). Mössbauer analysis of U_6Fe single crystal showed $\delta = -0.36$ mm/s at RT, suggesting a higher s electron density than α-Fe due to s/d-f hybridization (Tsustsui et al. 2000). The application of an external magnetic field up to 7 T did not induce any hyperfine field at 4.2 K in accordance with their Pauli paramagnetic nature. Large negative quadrupole splitting of -0.74 mm/s with $\eta = 0$ originating from incompletely hybridized $3d_{z^2-r^2}$ and the principal axis of EFG along [001] has been obtained from the study.

13.2.2 Th_7Fe_3

The hexagonal (*P6₃mc*) Th_7Fe_3 is another Fe-based superconductor with a T_{sc} of 1.96 K (Matthias et al. 1961). The Mössbauer spectrum of the Pauli paramagnet has a Δ of 0.5 mm/s and isomer shift indicative of its metallic nature while the other thorium compounds $ThFe_3$, $ThFe_5$ and Th_2Fe_{17} exhibit magnetic ordering (Blow 1970). The RT Mössbauer spectra of ^{57}Fe substituted $Th_7Co_{2.5}Fe_{0.5}$ and $Th_7Ni_{2.5}Fe_{0.5}$ showed Δ of 0.44 and 0.42 mm/s, respectively, that are smaller than Th_7Fe_3 (Idczak and Tran 2019). This was suggested due to the increased EFG with asymmetric 3d and its corresponding spin–orbit coupling as substituents from Co to Fe.

13.2.3 $R_2Fe_3Si_5$

The rare-earth (R) ternary alloys of the composition $R_2Fe_3Si_5$ (tetragonal, *P4/mnc*) exhibit variable T_{sc} from 2 to 6 K with Y, Sc and Lu (Braun 1980). The $Sc_2Fe_3Si_5$ exhibited superconducting transition at 4.6 K, while Mössbauer analysis with a longitudinal applied field of 5.6 T at 4.2 K did not induce any hyperfine field, suggesting the absence of magnetic moment (Cashion et al. 1980). The spectrum exhibited two doublets of intensity ratio 2:1 with δ values 0.14 and 0.22 mm/s corresponding to the 8h and 4d sites, respectively. The magnitude of $\Delta = 0.37$ mm/s is equal for both sites, while 4d site showed a negative value. $Lu_2Fe_3Si_5$ has a relatively higher T_{sc} of 6 K that could be fitted with δ values of 0.13–0.22 and 0.28–0.22 mm/s at RT corresponding to the 4d and 8h sites, respectively (Ma et al. 2015). The almost temperature-independent quadrupole splitting suggested the ligand charge distribution to the EFG.

13.3 High T_C YBCO Superconductors

YBCO or Y123 unconventional superconductors show superconducting transition temperature, T_{sc} above 90 K. The YBCO superconductor has the orthorhombic perovskite structure with layers of CuO, BaO, CuO_2 and Y, as shown in Fig. 13.1. The Cu(I) and Cu(II) sites in the CuO and CuO_2 layers are usually substituted with transition metal ions to understand the structure. The Cu(I) chain site has square planar ligand coordination that is prone to variation with oxygen deficiency, while the Cu(II) planar site is of capped square pyramidal coordination. Superconductivity in YBCO is determined by the structure of $YBa_2Cu_3O_{7-y}$, with y = 0.31 and 0.68 exhibiting orthorhombic and tetragonal phases, respectively (Maeno et al. 1987). The oxygen-deficient structure, producing vacancies in the Cu–O chain, is antiferromagnetic, with the Néel temperature reaching up to 500 K (Tranquada et al. 1988). With increasing oxygen, holes are produced in the CuO_2 layer, decreasing the Néel temperature and superconductivity sets in. Unlike conventional superconductors, the superconductivity is not destroyed with Fe substitution in high temperature superconductors such as YBCO.

Fe substituted $YBa_2Cu_{2.94}Fe_{0.06}O_{7-\delta}$ powder enriched with ^{57}Fe resulted in a spectrum fitted with three doublets of δ values 0.04, 0.01 and 0.27 mm/s and Δ values 1.99, 1.22 and 0.52 mm/s, respectively (Nasu et al. 1987). These values were not affected much at the measured temperatures of 300, 77 and 4.2 K. From the hyperfine parameters, the Fe sites were identified in Cu(II) and Cu(I). Cu(II) is located between the Y and Ba layers, while Cu(I) is between the Ba layers with oxygen-deficient coordination. The δ values in these sites are identified to be of high spin to intermediate ferric spin octahedral configurations. The application of a longitudinal external magnetic field of 4.5 T at 4.2 K induced a broadening of 20.0 T, indicating the localized magnetic moment, although superconductivity was not destroyed.

Fe substituted $YBa_2(Cu_{1-x}Fe_x)_3O_{7-y}$ ceramic pellets showed decreasing T_{sc} from 89 to 39 K as x is increased from 0.01 to 0.067 (Tamaki et al. 1988). The orthorhombic to tetragonal phase transformation takes place with Fe substitution above x = 0.02. Consequently, the T_{sc} decreases while the magnetic broadening temperature, indicative of hyperfine field distribution, increases (6–20 K) with x. The RT spectra of all the compositions are composed of three quadrupole doublets A, B and C, with close-lying hyperfine parameters given as A(δ = 0.04 mm/s, Δ = 2.0 mm/s), B(δ = − 0.02 mm/s, Δ = 1.1 mm/s) and C(δ = 0.33 mm/s, Δ = 0.6 mm/s). The intensity ratio of the doublets I(B)/I(A) decreased with Fe content, whereas that of I(C) is independent of composition. The hyperfine parameters of component C indicate covalent high spin Fe^{3+} that also showed a B_{hf} of 43.0 T at 4.2 K for x = 0.067.

Studies with various substituents (M) such as Co, Ni, Al and Zn in addition to Fe in $YBa_2Cu_{3-x}M_xO_{7-y}$ have shown that the T_{sc} decreases irrespective of their magnetic nature, whereas the structure becomes tetragonal with Fe, Co and Al when x > 0.05 (Tarascon et al. 1988). The T_{sc} decreases from 92 K for the undoped $YBa_2Cu_3O_{6.85}$ to 31 K with Fe substitution when x = 0.4, resulting in the composition $YBa_2Cu_{2.6}Fe_{0.4}O_7$. The ^{57}Fe Mössbauer spectra were composed of three quadrupole

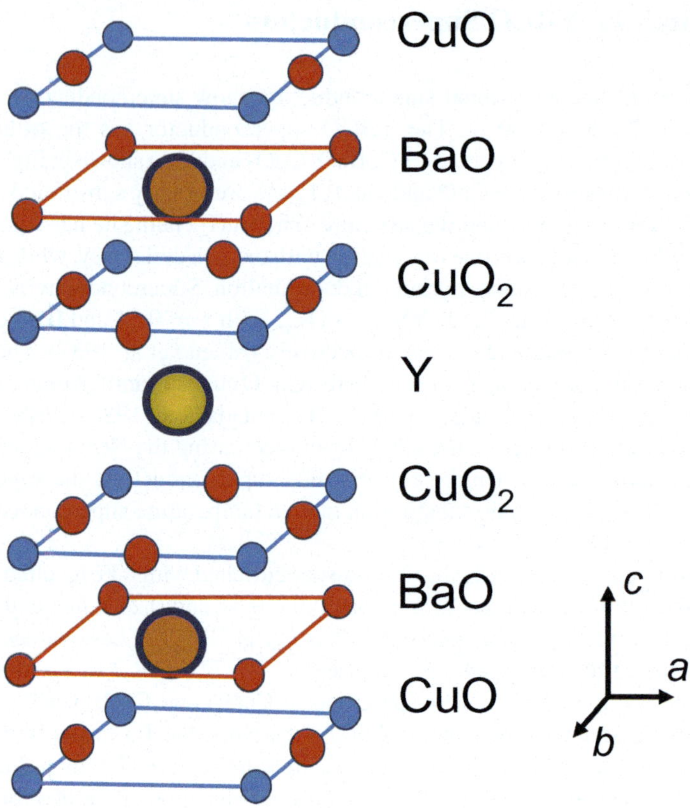

Fig. 13.1 The CuO and CuO_2 layers of the orthorhombic YBCO structure has different oxygen coordination

doublets with Δ and δ values, respectively, given as Δ(1.94–1.98, 1.10–0.98, 0.54–0.56) mm/s and δ(0.02–0.05, – 0.02, 0.30) mm/s for x values 0.1, 0.2 and 0.3. The isomer shift values indicated the Fe^{3+} state, while the lower values for the first two sets of doublets indicated either localized Fe^{3+} or itinerant nature. From the magnetic moment (3.4 μ_B/Fe^{3+}) and Mössbauer parameters, it was suggested that the ferric ions may have an intermediate spin structure.

According to the synthesis conditions, more than three doublets were also identified with features such as more negative δ values suggesting high spin Fe^{4+}, varying site occupancies, orthorhombic-tetragonal phase mixture, oxygen fugacity and aging effects (Saitovitch et al. 1988; Sedykh et al. 1988; Blue et al. 1988; Saitovitch 1991). It was shown that up to six different Fe environments could be identified in the literature, as listed in Table 13.1. (Lines and Eibschütz 1990; Boolchand 1991). The Cu(II) planar site could be easily distinguished from the large δ, smaller Δ and saturation hyperfine field of 50.0 T, all pointing to the high spin ferric state ($S = 5/2$). The fitting with two to four components for the Cu(II) and Cu(I) sites is influenced by

Table 13.1 Hyperfine parameters of Fe substituted YBCO superconductor (Lines and Eibschütz 1990; Boolchand et al. 1991)

Site	Spin state	δ (mm/s)	$\Delta/2\varepsilon$ (mm/s)	$B_{hf}(0)$ (T)
Cu(II)	Fe^{3+}, $S = 5/2$	$+0.3 \sim 0.4$	$0.2 \sim 0.5$	50.0
		0.2	0.8	
Cu(I)	Fe^{3+}, $S = 3/2$	0.00	$1.1 \sim 1.2$	23.0
Cu(I)	mixed spin and valency	$-0.15 \sim -0.10$	1.6	24.0
Cu(I)	Fe^{3+}, $S = 3/2$	$+0.05$	$1.9 \sim 2.0$	28.0
Cu(I)	Fe^{3+}, $S = 5/2$	< 0.2	0.28	–

the synthesis conditions of the sample that affect the oxygen coordination and spin state. The intensity ratios of these sites are, therefore, variable.

The chain transformation model suggests four chain sites (A, C, D, E) and two planar sites (B, B′) (Boolchand et al. 1991). According to this model, the octahedrally coordinated Fe^{3+} (site E) could transform to site C (penta-coordinated oxygen) and subsequently to site A (square planar) with oxygen desorption. The C-site formed at twin planes may transform into the D-site (quasi tetrahedral). If site E is identified as high spin Fe^{3+}, other sites could not have a higher valence state. However, the Fe favors the Cu(I) in three to five ligand coordination only, with the higher ones for highly oxygenated conditions, thus supporting Fe^{3+}, $S = 5/2$ (Lines and Eibschütz 1990). Other Cu(I) sites can have mixed valent intermediate spin states. The RT spectrum of the Fe substituted YBCO prepared fresh was fitted with A-E subspectra including B and B′ while the aged sample showed the absence of E subspectrum as shown in Fig. 13.2.

Later studies with lower ^{57}Fe (0.167 at.%) concentration replaced for Cu in fully oxygenated YBCO showed a T_{sc} of 92 K and local antiferromagnetic ordering of the Fe in Cu(II) plane (Linden et al. 2007). The spectrum was fitted with three components corresponding to Cu(I) site and a third one of 16% relative intensity to Cu(II). The spectra at 77 K and above showed hyperfine field components in the range 0.5–1.5 T from the Cu(II) site and constant $eQV_{zz} \approx -0.4$ mm/s, $\eta \approx 0.6$ and spin orientation angle, $\alpha \approx 60°$. The internal field was maximum just above the T_{sc} although canted at an angle of 45° at all T and vanishes on approaching the Néel temperature, T_N close to 410 K. Thus, the coexistence of superconductivity and antiferromagnetism was apparent in the YBCO superconductor.

13.4 Iron Based Superconductors

Although Fe-based heavy element superconductors have been known for long, the Fe based superconductors are usually represented by the FeX type superconductors that have relatively higher T_{sc}, such as Fe pnictides, Fe chalcogenides, FeSe and FeTe (11) superconductors (Stewart 2011). The unconventional superconductors having the 2D

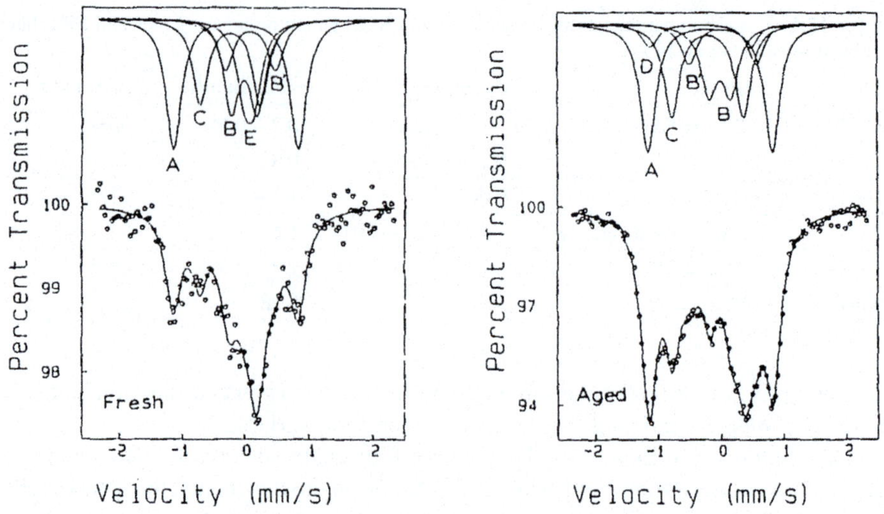

Fig. 13.2 RT Mössbauer spectra of $YBa_2(Cu_{0.97}Fe_{0.03})O_7$ fresh sample (left) and three-day aged sample (right). Reprinted with permission (Boolchand et al. 1991), ©Springer Nature

FeX (X = As or P) pnictide layers are known as iron pnictide superconductors. The RFeXO (R- rare-earth) and oxygen-free AFe_2X_2 (A = Ba, Ca, Sr, Eu) iron pnictides are known as 1111 and 122 superconductors, respectively. Superconductivity appears with Fermi surface distortion induced by electron, hole and isovalent doping in Fe pnictides.

First-order structural phase transformation involving tetragonal to orthorhombic phase and second-order magnetic transformation with reducing temperature is commonly observed in Fe pnictides. The superconductivity is believed to originate from the d orbitals in FeX layers, and the Fe-based superconductors have the following properties (Wang and Lee 2011) (i) Cooper pair in the *ab* plane ~ 3 nm, (ii) Type II superconductivity with London penetration depth ~ 300 nm, (iii) Fermi surface distortion and charge-density-wave order, and (iv) antiferromagnetic spin fluctuations, apart from various other characteristics that are being refined with continuing experiments. The static antiferromagnetic or spin density wave magnetic and superconducting phases vary according to the concentration of the dopant in the parent compound. In a few of the Fe-based superconductors, the magnetic and superconducting region coexists, as shown in Fig. 13.3. The antiferromagnetic to superconducting transition occurs with hole doping as in $Ba_{1-x}K_xFe_2As_2$ (Johrendt and Pöttgen, 2009) and from helimagnetic ordering to superconducting with pressure near 8 GPa in MnP (Cheng et al. 2015).

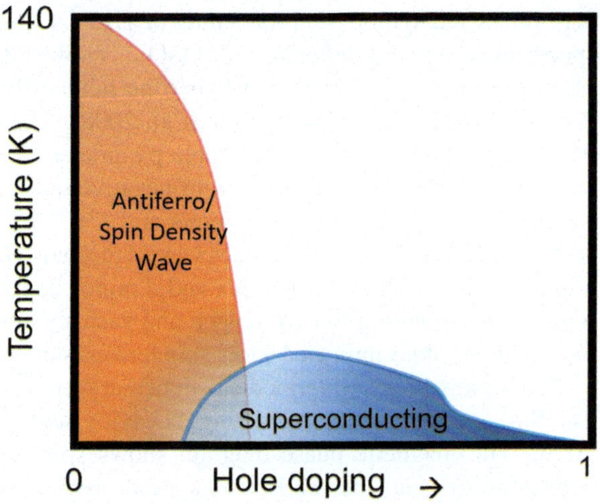

Fig. 13.3 The general representation of magnetic and superconducting phases with temperature and doping. Antiferromagnetic or incommensurate magnetic structure does not overlap with superconductivity, whereas dynamic antiferromagnetic spin density wave phase may overlap in a few Fe-based superconductors

13.4.1 RFeXO (1111) Superconductor

LaOFeP superconductor (tetragonal, *P4/nmm*) has alternate stacks of LaO and tetrahedrally coordinated Fe^{2+} by P^{3-} anions as pnictide layers and has T_{sc} below 5 K (Kamihara et al. 2006). F doping in LaOFeP increases the T_{sc} to ~ 7–10 K whereas 5–12 at.% F doped LaOFeAs increases the T_{sc} to ~ 26 K (Kamihara et al. 2008). A tetragonal to monoclinic (*P112/n*) phase change was observed below 160 K and long-range spin density wave antiferromagnetic order near 137 K in LaFeAsO (de la Cruz et al. 2008).

Mössbauer spectra of polycrystalline LaFeAsO showed a single line with δ = 0.52 mm/s, characteristics of low or intermediate Fe^{2+} whereas hyperfine field was observed below 140 K with saturation field $B_{hf}(0) = 4.86$ T (Klauss et al. 2008). Only slight broadening of the spectra suggests commensurate antiferromagnetic ordering below $T_N = 138$ K with a small magnetic moment of 0.25 μ_B. The spectra were fitted with two sextets with $\Delta = 0.3$ mm/s to account for the distorted $FeAs_4$ tetrahedron and variation in oxygen content. Comparison of LaFeAsO with F doped sample showed a singlet with similar δ of 0.45–0.55 mm/s in the temperature range 298–150 K (Kitao et al. 2008). The magnetic order was absent in $LaFeAsO_{0.89}F_{0.11}$ with persistent singlet up to 4.2 K, unlike LaFeAsO that showed hyperfine broadening below 140 K with B_{hf} of 5.3 T at 4.2 K. Mössbauer studies with $LaO_{1-x}F_xFeAs$ showed hyperfine fields below 10 K with a small magnetic moment of 0.22 μ_B for $x < 0.05$, associated with orthorhombic phase change whereas higher concentration showed paramagnetic singlet down to 3 K (Luetkens et al. 2009). The T_N decreases from 134 to 120 K with x up to 0.04, where the orthorhombic phase transition happens at 151 K and the superconducting tetragonal phase in $0.04 < x < 0.2$.

The 77 K and RT spectra of LaFePO showed relatively lower δ of 0.3 mm/s due to the high electron density as in P compounds, and at 4.2 K presented weak Δ of

0.19 mm/s and symmetric broadening of 1.15 T (Tegel et al. 2008). The Mössbauer spectrum of oxygen-deficient $LaFeAsO_{1-x}$ showed doublet and a singlet from two Fe environments at 200 K and hyperfine field splitting at 95 K, below its T_N as demonstrated in Fig. 13.4 (Nowik et al. 2008). The hyperfine field components at 95 K showed one without quadrupole parameter ($B_{hf} = 5.1$ T, $\delta = 0.57$ mm/s, QS $= 0$ and $I = 84\%$) and another with negative quadrupole parameter ($B_{hf} = 0.8$ T, $\delta = 0.46$ mm/s, QS $= -0.86$ mm/s and $I = 16\%$). A partially incommensurate spin density wave due to vacancies was suggested. The spectrum at 200 K exhibited a major singlet of $\delta = 0.51$ mm/s and a minor doublet ($\delta = 0.35$ mm/s, $\Delta = 0.89$ mm/s) corresponding to two oxygen and vacancy neighborhoods, respectively. The isomer shift values indicated a low spin ferrous state for all the sites from the spectra.

Co-doping in the intermediate concentration ranges in $RFe_{1-x}Co_xAsO$ (R = La, Sm) exhibits superconducting behavior with T_{sc} ~ 13–17 K (Wang et al. 2009). The magnetic phase diagram shows spin density wave behavior at lower concentrations in the antiferromagnetic region, metallic at higher temperatures and semiconducting-like behavior at intermediate temperatures above the superconducting region. It was shown that magnetic fluctuation coexists with superconductivity in $SmFe_{0.95}Co_{0.05}AsO$ ($T_{sc} = 5.1$ K) from the Mössbauer studies with an applied magnetic field (Long et al. 2011). The Mössbauer spectrum of the SmFeAsO at 4.2 K with a longitudinal applied field of 5 T showed a broadening of the spectrum along with a reduction of the intensity compared with the zero-field spectrum indicating the strong anisotropic spin arrangement of the randomly oriented sample. The SmFeAsO showed a T_N of 137 K, B_{hf} of 4.98 T, $\delta = 0.42$ mm/s

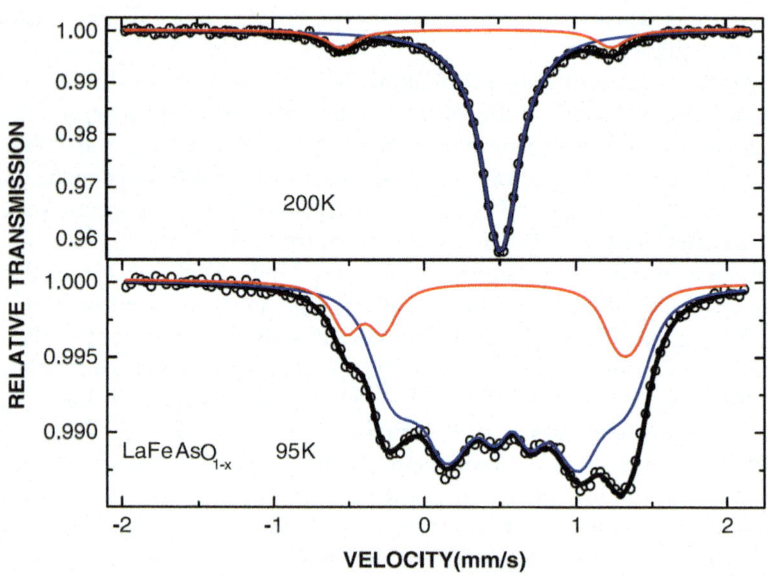

Fig. 13.4 The Mössbauer spectra of oxygen-deficient $LaFeAsO_{1-x}$ below and above its T_N. Reprinted with permission (Nowik et al. 2008), ©IOP Publishing

and a small quadrupole contribution of 0.1 mm/s. The $SmFe_{0.95}Co_{0.05}AsO$ displayed a singlet where the linewidth decreased with temperature. A longitudinal magnetic field of 5 T in the temperature range of 4.2–27.5 K showed a four-line pattern exhibiting decreasing linewidth with temperature. Such variations were absent in $SmFe_{0.9}Co_{0.1}AsO$ ($T_{sc} = 17$ K), suggesting the temperature-dependent hyperfine field fluctuation with the lower Co. The 5 T applied field is above the critical field, and therefore, the Mössbauer measurements indicate the normal state of the sample.

High-pressure Mössbauer analysis up to 24 GPa on polycrystalline ^{57}Fe enriched LaFeAsO was undertaken at various temperatures (Kawakami et al. 2009). The magnetic ordering temperature was determined from the additionally broadened component in the singlet at various temperatures with pressure. The T_N and B_{hf} decreased gradually with pressure reaching ~ 50 K and 2.2 T at 20 GPa, respectively, and with increasing pressure reached negligible values. The onset temperature for superconductivity reached a maximum of 21 K at a pressure of 12 GPa and vanished beyond 34 GPa. High-pressure synchrotron Mössbauer experiments with single crystal LaFeAsO at 13 K showed suppression of hyperfine field at a pressure of 7.5 GPa itself (Materne et al. 2018). The pressure-dependent hyperfine field values were 1.3–1.6 T lower than the polycrystalline samples. Similarly, the reduction in T_N was observed in the single crystal. The difference in magnitudes is suggested due to the inhomogeneities and vacancies in polycrystalline samples.

13.4.2 AFe_2As_2

$BaFe_2As_2$ belongs to the 122 type Fe pnictides that have the tetragonal structure (*I4/mmm*) and exhibit orthorhombic structural (*Fmmm*) and magnetic transition near 140 K, resulting in spin density wave anomaly (Rotter et al. 2008). The Mössbauer spectrum at RT is a singlet with $\delta = 0.31$ mm/s, whereas it exhibits a spin density wave influenced hyperfine splitting of 5.23 T and negligible Δ of -0.03 mm/s at 77 K. The B_{hf} and Δ at 4.2 K are 5.47 T and -0.04 mm/s, respectively. Previously it has been reported higher ordering temperature of 200 K and B_{hf} of 8.5 T at 4.2 K in $EuFe_2As_2$ (Raffius et al. 1993). The tetragonal to orthorhombic (*Fmmm*) structural transition in $EuFe_2As_2$ and $SrFe_2As_2$ are found to occur at 190 and 203 K, respectively, where a second-order phase transformation is suggested in 122 compounds from the structural and Mössbauer studies (Tegel et al. 2008).

The spin density wave anomaly below 138 K in $BaFe_2As_2$ is accompanied by a single hyperfine field component and a strong increase in the internal field with a reduction in the temperature, as shown in Fig. 13.5 (Rotter et al. 2009). The spectrum with a hyperfine field at 4.2 K reaches 5.47 T, markedly different from KFe_2As_2 or higher K doping, showing only a single line with very small quadrupole splitting. The transformation from hyperfine field split spectrum to single line spectrum with K doping in $(Ba_{1-x}K_x)Fe_2As_2$ could be observed from the spectra measured at 4.2 K as shown in Fig. 13.6 (Johrendt and Pottgen 2009). The T_{sc} is below 2 K in $BaFe_2As_2$ whereas it is 38 K in $(Ba_{0.6}K_{0.4})Fe_2As_2$ where spin density wave anomaly is absent.

The KFe_2As_2, on the other hand, shows a T_{sc} of only 3.8 K, suggesting the role of hole doping with K in $BaFe_2As_2$. The charge density variation with one electron transfer to the Fe-pnictide layer could be understood from the δ values at 4.2 K that are 0.44–0.43 mm/s for $BaFe_2As_2$ and lightly doped with K, whereas it decreases to 0.39 and 0.34 mm/s for $(Ba_{0.6}K_{0.4})Fe_2As_2$ and KFe_2As_2, respectively. The $(Ba_{1-x}K_x)Fe_2As_2$ at x = 0.1 and 0.2 falls in the overlapping region of superconductivity and spin density wave anomaly and hence exhibits a mixed behavior.

Co or Ni doping did not induce any s electron density changes as the δ value with the dopants was 0.52 mm/s at 78 K, comparable to that of $BaFe_2As_2$ (Khasanov et al. 2011a, b). The adsorbed oxygen in $BaFe_2As_2$ may result in the formation of relaxation spin clusters that grow with higher oxygen partial pressures (>1 mTorr), producing a hyperfine field of nearly 4.1–5.0 T at 90 K, which is otherwise a singlet (Khasanov et al. 2011a, b). High-pressure synchrotron Mössbauer experiments have shown that magnetic ordering precedes the tetragonal to orthorhombic phase transformation with pressure, and the hyperfine field consequently vanishes at all temperatures with 16.7 GPa in $BaFe_2As_2$ (Wu et al. 2013). The frustrated near-neighbor magnetic interaction ends up in orientation order, resulting in the so-called nematic phase, which is observed with pressure before the long-range ordering temperature.

Mössbauer spectrum of $Ba_{0.75}K_{0.25}Fe_2As_2$ single crystal revealed the coexistence of magnetism and superconductivity from the broad hyperfine field spectra down

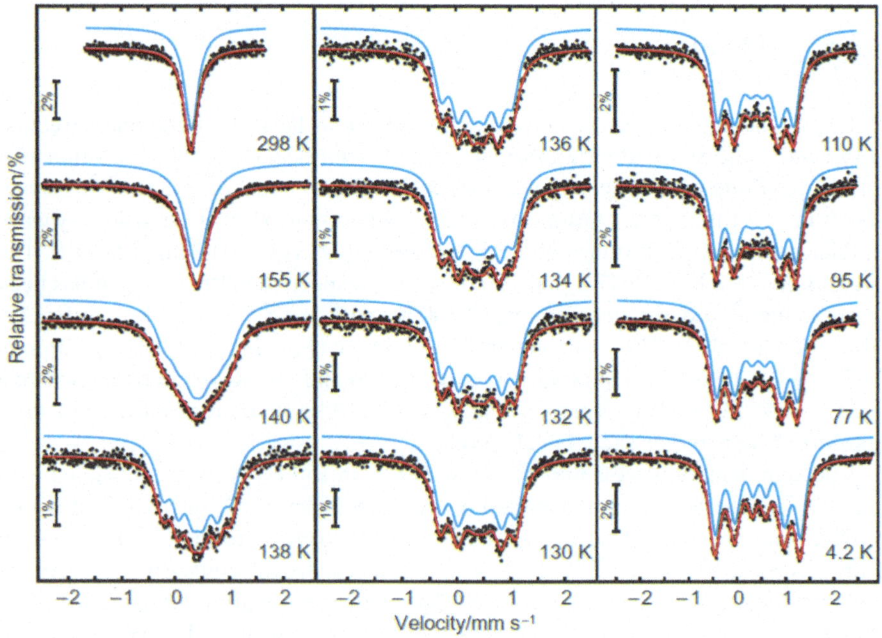

Fig. 13.5 The Mössbauer spectra of $BaFe_2As_2$ at various temperatures. Spin density wave anomaly is accompanied by the variation in the hyperfine field pattern at 138 K (Rotter et al. 2009)

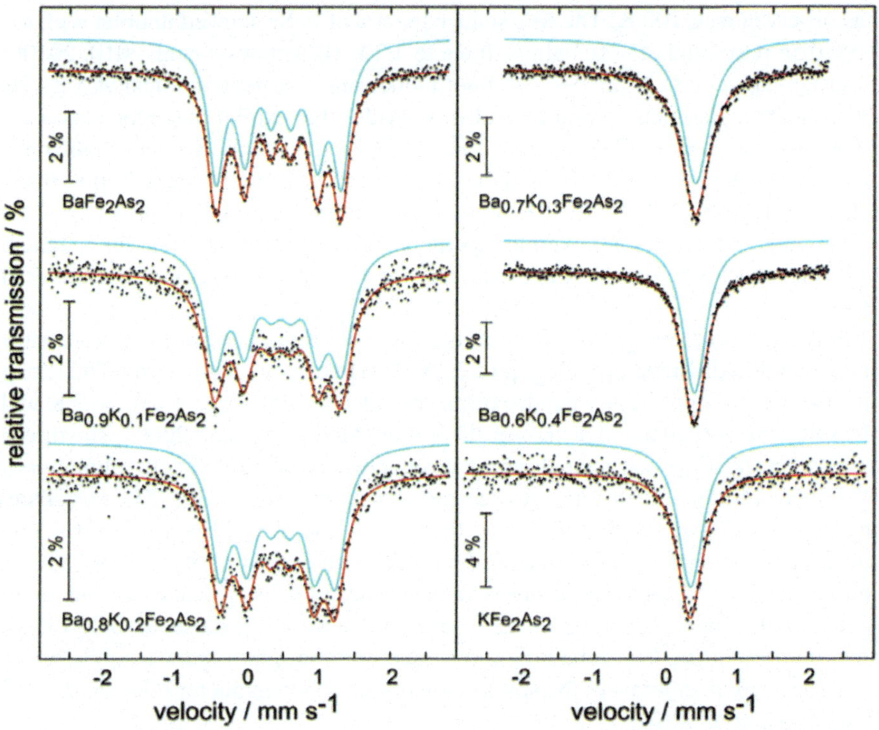

Fig. 13.6 The Mössbauer spectra at 4.2 K of the $Ba_{1-x}K_xFe_2As_2$ for various concentrations *x* (Johrendt and Pöttgen 2009)

to 4.35 K whereas T_N(90 K) > T_{sc}(~30 K) (Munevar et al. 2013). The V_{zz} was found to be parallel to the *c*-axis and hyperfine field component along the *ab* plane with isomer shift indicative of Fe^{2+}. The non-magnetic component in the form of a doublet with a small quadrupole splitting was nearly 20% below T_{sc}. Consequently, a reduction in the hyperfine field by almost 0.5–1.0 T was observed below T_{sc} due to the superconductivity coexistence, with a superconducting coherence length of nearly 2 nm. The coexistence of magnetism and superconductivity was previously shown in $Ba(Fe_{0.961}Rh_{0.039})_2As_2$ where $T_N = 55.5$ K, $T_{sc} = 14$ K and $B_{hf}(0) = 2.23$ T (Wang et al. 2011). The hyperfine field component was present as broadened single-line spectrum with a distribution below T_N.

13.4.3 FeSe (11) Superconductor

The slightly Se deficient, layered tetragonal FeSe was found to be superconducting below 8 K with an upper critical field of 16.3 T and coherence length of 4.5 nm (Hsu et al. 2008). The tetragonal (*P4/nmm*) to triclinic (*P-1*) phase transformation was observed near 100 K. The Mössbauer spectra of FeSe showed doublet with $\delta =$ 0.55–0.56 mm/s and $\Delta \approx 0.3$ mm/s down to 4.2 K (Blachowski et al. 2010, 2010b). External magnetic field up to 9 T did not induce any magnetic moment at 5 K. The FeSe samples are often accompanied by Fe-oxides that exhibit magnetic sextet as a minor fraction at 80 K. The tetragonal $FeSe_{0.5}Te_{0.5}$ superconductor shows the onset of superconductivity at 12.5 K, and the Mössbauer spectrum measured up to 6 K is a doublet (Linden et al. 2011). The non-stoichiometric FeTe with slightly excess Fe, on the other hand, does not exhibit superconductivity but shows complex magnetic hyperfine spectra below 73 K according to the Fe concentration (Mizuguchi et al. 2012).

Polycrystalline tetragonal FeS exhibits a T_{sc} of 4.3 K that further decreases continuously with hydrostatic pressure up to 2.2 GPa (Holenstein et al. 2016). The tetragonal β-$Fe_{1.01}$Se, on the other hand, exhibits broader and enhanced T_{sc} of 36.7 K with pressure up to 8.9 GPa and decreases thereafter (Medvedev et al. 2009). The drastic increase in the T_{sc} is associated with the reduction in tetragonal cell volume up to 1.5 GPa that further transforms into a hexagonal structure at 12 GPa. The Mössbauer spectrum at RT is a doublet with $\delta = 0.44$ mm/s and $\Delta = 0.28$ mm/s at 0.2 GPa while the hexagonal component was observed with $\delta = 0.84$ mm/s, $\Delta = 0.21$ mm/s and $I = 20\%$ at 7.2 GPa that has the characteristics of a less covalent ferrous state.

The 111 superconductors LiFeP and LiFeAs with T_{sc} close to 6 K exhibit quadrupole splitting lower than 0.2 mm/s, and therefore, they show a single broadened spectrum in contrast to the split doublet in 11 superconductor (Gao et al. 2010; Blachowski et al. 2010).

13.5 Manganites

Manganites having the general formula $RMnO_3$ (R = La, Sr, Ca, Ba), are mixed valent cubic perovskite compounds and exhibit a variety of magnetic and electrical characteristics (Salamon and Jaime 2001). The $LaCaMnO_3$ (LCMO) manganite, which has trivalent and tetravalent Mn, displays colossal magnetoresistance near its T_c (Jin et al. 1994; Schiffer et al. 1995). The YBCO/LCMO superlattice interfaces are known for its magnetic proximity effect.

The Mössbauer spectra of the $La_{0.75}Ca_{0.25}Mn_{0.98}{}^{57}Fe_{0.02}O_3$ recorded at various temperatures showed unresolved doublet at RT, which developed into a broad magnetic component below 210 K and a single sextet at 4.2 K (Pissas et al. 1997). The RT spectrum showed $\delta = 0.36$ mm/s and $\Delta = 0.18$ mm/s while the 4.2 K spectrum showed $B_{hf} = 53.3$ T, $\delta = 0.52$ mm/s and $2\varepsilon = -0.02$ m/s. The hyperfine parameters indicate high spin Fe^{3+} ($S = 5/2$) in octahedral coordination. The LCMO system has a ferromagnetic phase coexistence with the paramagnetic phase at intermediate temperatures due to the relaxation exhibited by the size distribution of ferromagnetic clusters. Emission Mössbauer spectroscopy of $La_{0.8}Ca_{0.2}Mn(^{57}Co)O_3$ showed both paramagnetic and ferromagnetic phases, with the paramagnetic phase increasing drastically beyond 120 K up to its T_c of 200 K (Chechersky et al. 1997).

The Mössbauer spectra of $La_{0.75}Ca_{0.25}Mn_{0.955}Fe_{0.045}O_3$ were fitted with two doublets at 300 K and three sextet components at 4.2 K (Hannoyer et al. 2000). The spectra at 300 and 4.2 K for the $La_{0.75}Ca_{0.25}Mn_{0.955}Fe_{0.045}O_3$ sample are shown in Fig. 13.7. The doublets at RT with $\Delta = 0.12$ and 1.42 mm/s are due to the orthorhombic distortion and the Fe in different Mn environments at the grain boundaries, respectively. The B_{hf} of two sextets at 4.2 K are 53.0 and 49.0 T, and the third one showed a distribution with an average B_{hf} of 44.0 T. Application of a longitudinal magnetic field of 6 T at 100 K showed magnetic moment alignment along the applied magnetic field. On the other hand, at 8 K in an applied field of 6 T, it showed three well-defined components, two due to antiparallel alignment and one with a lower effective field of 39.0 T due to parallel alignment. In addition to the mixed ferromagnetic and paramagnetic behavior due to the clusters, grain boundary ferromagnetic component in a strained environment was also inferred from the Mössbauer studies.

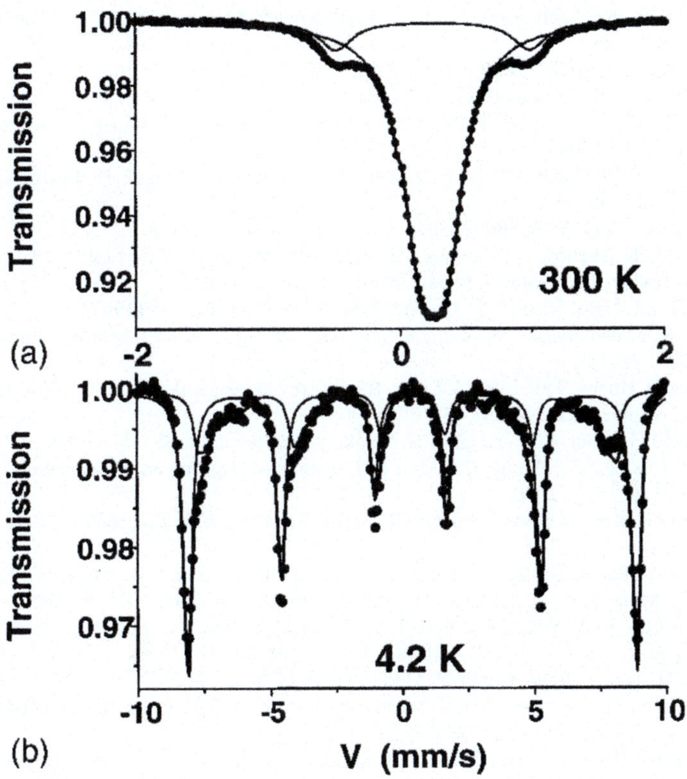

Fig. 13.7 Mössbauer spectra of $La_{0.75}Ca_{0.25}Mn_{0.955}Fe_{0.045}O_3$ at **a** 300 and **b** 4.2 K. Reprinted with permission (Hannoyer et al. 2000), ©The American Physical Society

13.6 Summary

This chapter has introduced Mössbauer spectrometry in the analysis of various superconducting compounds. The coexistence of superconductivity and antiferromagnetism in certain new-generation superconducting materials and the hyperfine parameters are presented.

References

A. Błachowski, K. Ruebenbauer, J. Zukrowski, J. Przewoznik, J. Marzec, J. Alloys Compd. **505**, L35 (2010a)

A. Błachowski, K. Ruebenbauer, J. Zukrowski, J. Przewoznik, K. Wojciechowski, Z.M. Stadnik, J. Alloys Compd. **494**, 1 (2010b)

A. Błachowski, K. Ruebenbauer, P. Zajdel, E.E. Rodriguez, M.A. Green, J. Phys. Condens. Matter **24**, 386006 (2012)

S. Blow, J. Phys. Chem. Solids **30**, 1549 (1969)

S. Blow, J. Phys. C: Solid State Phys. **3**, 158 (1970)

C. Blue, K. Elgaid, I. Zitkovsky, P. Boolchand, D. McDaniel, W.C.H. Joiner, J. Oostens, W. Huff, Phys. Rev. B **37**, 5905 (1988)

P. Boolchand, D. McDaniel, C. Blue, Y. Wu, R. Enzweiler, K. Elgaid, R. Burrows, Hyperfine Interact. **68**, 15 (1991)

H.F. Braun, Phys. Lett. A **75**, 386 (1980)

J.D. Cashion, G.K. Shenoy, D. Niarchos, P.J. Viccaro, C.M. Falco, Phys. Lett. A **79**, 454 (1980)

B.S. Chandrasekhar, J.K. Hulm, J. Phys. Chem. Solids **7**, 259 (1958)

V. Chechersky, A. Nath, H. Ju, R.L. Greene, Low Temp. Phys. **23**, 545 (1997)

J.G. Cheng, K. Matsubayashi, W. Wu, J.P. Sun, F.K. Lin, J.L. Luo, Y. Uwatoko, Phys. Rev. Lett. **114**, 117001 (2015)

C. de la Cruz, Q. Huang, J.W. Lynn, J. Li, W. Ratcliff II., J.L. Zarestky, H.A. Mook, G.F. Chen, J.L. Luo, N.L. Wang, P. Dai, Nature **453**, 899 (2008)

L.E. DeLong, J.G. Juber, K.N. Yang, M.B. Maple, Phys. Rev. Lett. **51**, 312 (1983)

W.B. Gao, J. Linden, X.C. Wang, C.Q. Jin, T. Tohyama, M. Karppinen, H. Yamauchi, Solid State Commun. **150**, 1525 (2010)

B. Hannoyer, G. Marest, J.-M. Greneche, R. Bathe, S.I. Patil, S.B. Ogale, Phys. Rev. B **61**, 9613 (2000)

S. Holenstein, U. Pachmayr, Z. Guguchia, S. Kamusella, R. Khasanov, A. Amato, C. Baines, H.H. Klauss, E. Morenzoni, D. Johrendt, H. Luetkens, Phys. Rev. B **93**, 1040506 (2016)

F.C. Hsu J.Y. Luo, K.W. Yeh, T.K. Chen, T.W. Huang, P.M. Wu, Y.C. Lee, Y.L. Huang, Y.Y. Chu, D.C Yan, M.K. Wu, Proc. Natl. Acad. Sci. U.S.A. **105**, 14262 (2008)

R. Idczak, V.H. Tran, J. Phys. Condens. Matter **31**, 495802 (2019)

S. Jin, T.H. Tiefel, M. McCormack, R.A. Fastnacht, R. Ramesh, L.H. Chen, Science **264**, 413 (1994)

D. Johrendt, R. Pöttgen, Physica C **469**, 332 (2009)

Y. Kamihara, H. Hiramatsu, M. Hirano, R. Kawamura, H. Yanagi, T. Kamiya, H. Hosono, J. Am. Chem. Soc. **128**, 10012 (2006)

Y. Kamihara, T. Watanabe, M. Hirano, H. Hosono, J. Am. Chem. Soc. **130**, 3296 (2008)

T. Kawakami, T. Kamatani, H. Okada, H. Takahashi, S. Nasu, Y. Kamihara, M. Hirano, H. Hosono, J. Phys. Soc. Jpn. **78**, 123703 (2009)

A. Khasanov, S.C. Bhargava, J.G. Stevens, J. Jiang, J.D. Weiss, E.E. Hellstrom, A. Nath, J. Phys. Condens. Matter **23**, 202201 (2011b)

A. Khasanov, J. Jiang, E.E. Hellstrom, A. Nath, J. Phys. Condens. Matter **23**, 342201 (2011)

S. Kitao, Y. Kobayashi, S. Higashitaniguchi, M. Saito, Y. Kamihara, M. Hirano, T. Mitsui, H. Hosono, M. Seto, J. Phys. Soc. Jpn. **77**, 103706 (2008)
H.H. Klauss, H. Luetkens, R. Klingeler, C. Hess, F.J. Litterst, M. Kraken, M.M. Korshunov, I. Eremin, S.L. Drechsler, R. Khasanov, A. Amato, J.H. Borrero, N. Leps, A. Kondrat, G. Behr, J. Werner, B. Buchner, Phys. Rev. Lett. **101**, 077005 (2008)
G. Lemon, P. Boolchand, M. Stevens, M. Marcuso, L.E. DeLong, J.G. Huber, J. Less Common Met. **127**, 329 (1987)
J. Linden, M. Karppinen, I. Grigoraviciute, H. Yamauchi, Phys. Rev. Lett. **98**, 067001 (2007)
J. Linden, J.P. Liback, M. Karppinen, E.L. Rautama, H. Yamauchi, Solid State Commun. **151**, 130 (2011)
M.E. Lines, M. Eibschütz, Physica C **166**, 235 (1990)
G. Long, M. DeMarco, M. Chudyk, J. Steiner, D. Coffey, H. Zeng, Y.K. Li, G.H. Cao, Z.A. Xu, Phys. Rev. B **84**, 064423 (2011)
H. Luetkens, H.H. Klauss, M. Kraken, F.J. Litterst, T. Dellmann, R. Klingeler, C. Hess, R. Khasanov, A. Amato, C. Baines, M. Kosmala, O.J. Schumann, M. Braden, J.H. Borrero, N. Leps, A. Kondrat, G. Behr, J. Werner, B. Büchner, Nature Mater. **8**, 305 (2009)
X. Ma, S. Ran, H. Pang, F. Li, P.C. Canfield, S.L. Budko, J. Phys. Chem. Solids **83**, 58 (2015)
Y. Maeno, T. Tomita, M. Kyogoku, S. Awaji, Y. Aoki, K. Hoshino, A. Minami, T. Fujita, Nature **328**, 512 (1987)
P. Materne, W. Bi, J. Zhao, M.Y. Hu, R. Kappenberger, S. Wurmehl, S. Aswartham, B. Buchner, E.E. Alp, Phys. Rev. B **98**, 174510 (2018)
B.T. Matthias, V.B. Compton, E. Corenzwit, J. Phys. Chem. Solids **19**, 130 (1961)
S. Medvedev, T.M. McQueen, I.A. Troyan, T. Palasyuk, M.I. Eremets, R.J. Cava, S. Naghavi, F. Casper, V. Ksenofontov, G. Wortmann, C. Felser, Nature Mater. **8**, 630 (2009)
Y. Mizuguchi, T. Furubayashi, K. Deguchi, S. Tsuda, T. Yamaguchi, Y. Takano, Physica C **470**, S338 (2010)
J. Munevar, H. Micklitz, J. Aguero, G. Tan, C. Zhang, P. Dai, E.B. Saitovitch, Phys. Rev. B **88**, 184514 (2013)
S. Nasu, H. Kitagawa, Y. Oda, T. Kohara, T. Shinjo, K. Asayama, F.E. Fujita, Physica B **148**, 484 (1987)
I. Nowik, I. Felner, V.P.S. Awana, A. Vajpayee, H. Kishan, J. Phys. Condens. Matter **20**, 292201 (2008)
M. Pissas, G. Kallias, E. Devlin, A. Simopoulos, D. Niarchos, J. Appl. Phys. **81**, 5770 (1997)
H. Raffius, E. Morsen, B.D. Mosel, W.M. Warmuth, W. Jeitschko, L. Terbuchte, T. Vomhof, J. Phys. Chem. Solids **54**, 135 (1993)
M. Rotter, M. Tegel, D. Johrendt, Phys. Rev. B **78**, 020503 (2008)
M. Rotter, M. Tegel, I. Schellenberg, F.M. Schappacher, R. Pottgen, J. Deisenhofer, A. Gunther, F. Schrettle, A. Loidl, D. Johrendt, New J. Phys. **11**, 025014 (2009)
E.B. Saitovitch, Hyperfine Interact. **66**, 231 (1991)
E.B. Saitovitch, I.S. Azevedo, R.B. Scorzelli, H. Saitovitch, S.F. da Cunha, A.P. Guimaraes, P.R. Silva, A.Y. Takeuchi, Phys. Rev. B **37**, 7967 (1988)
M.B. Salamon, M. Jaime, Rev. Mod. Phys. **73**, 583 (2001)
P. Schiffer, A.P. Ramirez, W. Bao, S.-W. Cheong, Phys. Rev. Lett. **75**, 3336 (1995)
V. Sedykh, S. Nasu, F.E. Fujita, Solid State Commun. **67**, 1063 (1988)
G.R. Stewart, Rev. Mod. Phys. **83**, 1589 (2011)
T. Tamaki, T. Komai, A. Ito, Y. Maeno, T. Fujita, Solid State Commun. **65**, 43 (1988)
J.M. Tarascon, P. Barboux, P.F. Miceli, L.H. Greene, G.W. Hull, M. Eibschütz, S.A. Sunshine, Phys. Rev. B **37**, 7458 (1988)
M. Tegel, M. Rotter, V. Weiss, F.M. Schappacher, R. Pottgen, D. Johrendt, J. Phys. Conden. Matter **20**, 452201 (2008)
M. Tegel, I. Schellenberg, R. Pottgen, D. Johrendt, Z. Naturforsch. B **63**, 1057 (2008)

J.M. Tranquada, A.H. Moudden, A.I. Goldman, P. Zolliker, D.E. Cox, G. Shirane, S.K. Sinha, D. Vaknin, D.C. Johnston, M.S. Alvarez, A.J. Jacobson, J.T. Lewandowski, J.M. Newsam, Phys. Rev. B **38**, 2477 (1988)
S. Tsutsui, Y. Kobayashi, M. Nakada, S. Nasu, E. Yamamoto, Y. Haga, Y. Onuki, J. Phys. Soc. Jpn. **69**, 1764 (2000)
F. Wang, D.H. Lee, Science **332**, 200 (2011)
C. Wang, Y.K. Li, Z.W. Zhu, S. Jiang, X. Lin, Y.K. Luo, S. Chi, L.J. Li, Z. Ren, M. He, H. Chen, Y.T. Wang, Q. Tao, G.H. Cao, Z.A. Xu, Phys. Rev. B **79**, 054521 (2009)
P. Wang, Z.M. Stadnik, J. Zukrowski, A. Thaler, S.L. Budko, P.C. Canfield, Phys. Rev. B **84**, 024509 (2011)
B.D. White, J.D. Thompson, M.B. Maple, Physica C **514**, 246 (2015)
J.J. Wu, J.F. Lin, X.C. Wang, Q.Q. Liu, J.L. Zhu, Y.M. Xiao, P. Chow, C. Jin, Proc. Natl. Acad. Sci. U.S.A. **110**, 17263 (2013)

Index

R. J. Joseyphus and J.-M. Greneche, *Fundamentals of ^{57}Fe Mössbauer Spectrometry*,
https://doi.org/10.1007/978-981-99-8653-8

D

E

R

S

SPRINGER NATURE

GPSR Compliance

The European Union's (EU) General Product Safety Regulation (GPSR) is a set of rules that requires consumer products to be safe and our obligations to ensure this.

If you have any concerns about our products, you can contact us on ProductSafety@springernature.com

In case Publisher is established outside the EU, the EU authorized representative is:

Springer Nature Customer Service Center GmbH
Europaplatz 3
69115 Heidelberg, Germany

The manufacturer's authorised representative in the EU is Springer Nature Customer Service Centre GmbH, Europaplatz 3, 69115 Heidelberg, Germany. If you have any concerns regarding our products, please contact ProductSafety@springernature.com

Printed and bound by CPI Group (UK) Ltd, Croydon, CR0 4YY
01/10/2026
02225902-0018